Handbook of Research on Holistic Optimization Techniques in the Hospitality, Tourism, and Travel Industry

Pandian Vasant
Universiti Teknologi PETRONAS, Malaysia

Kalaivanthan M.
Rasa Sayang Resort and Spa, Malaysia & Golden Sands Resort, Malaysia

A volume in the Advances in Hospitality, Tourism, and the Services Industry (AHTSI) Book Series

IGI GLOBAL
DISSEMINATOR OF KNOWLEDGE

www.igi-global.com

Published in the United States of America by
IGI Global
Business Science Reference (an imprint of IGI Global)
701 E. Chocolate Avenue
Hershey PA, USA 17033
Tel: 717-533-8845
Fax: 717-533-8661
E-mail: cust@igi-global.com
Web site: http://www.igi-global.com

Library of Congress Cataloging-in-Publication Data

Names: Vasant, Pandian, editor. | M., Kalaivanthan, 1974- editor.
Title: Handbook of research on holistic optimization techniques in the hospitality, tourism and travel industry / Pandian Vasant and Kalaivanthan M, editors.
Description: Hershey, PA : Business Science Reference, [2016] | Series: Advances in hospitality, tourism, and the services industry | Includes bibliographical references and index.
Identifiers: LCCN 2016037435| ISBN 9781522510543 (hardcover) | ISBN 9781522510550 (ebook)
Subjects: LCSH: Hospitality industry--Management--Mathematics--Handbooks, manuals, etc. | Mathematical optimization--Handbooks, manuals, etc.
Classification: LCC TX911.3.M27 H364 2016 | DDC 647.94--dc23 LC record available at https://lccn.loc.gov/2016037435

This book is published in the IGI Global book series Advances in Hospitality, Tourism, and the Services Industry (AHTSI) (ISSN: Pending; eISSN: Pending)

British Cataloguing in Publication Data
A Cataloguing in Publication record for this book is available from the British Library.

For electronic access to this publication, please contact: eresources@igi-global.com.

Advances in Hospitality, Tourism, and the Services Industry (AHTSI) Book Series

ISSN:Pending
EISSN:Pending

Mission

Globally, the hospitality, travel, tourism, and services industries generate a significant percentage of revenue and represent a large portion of the business world. Even in tough economic times, these industries thrive as individuals continue to spend on leisure and recreation activities as well as services.

The Advances in Hospitality, Tourism, and the Services Industry (AHTSI) book series offers diverse publications relating to the management, promotion, and profitability of the leisure, recreation, and services industries. Highlighting current research pertaining to various topics within the realm of hospitality, travel, tourism, and services management, the titles found within the AHTSI book series are pertinent to the research and professional needs of managers, business practitioners, researchers, and upper-level students studying in the field.

Coverage

- Travel Agency Management
- Leisure & Business Travel
- Hotel Management
- Sustainable Tourism
- Service Design
- Cruise Marketing and Sales
- Health and Wellness Tourism
- Service Management
- Food and Beverage Management
- Casino Management

IGI Global is currently accepting manuscripts for publication within this series. To submit a proposal for a volume in this series, please contact our Acquisition Editors at Acquisitions@igi-global.com or visit: http://www.igi-global.com/publish/.

The Advances in Hospitality, Tourism, and the Services Industry (AHTSI) Book Series (ISSN Pending) is published by IGI Global, 701 E. Chocolate Avenue, Hershey, PA 17033-1240, USA, www.igi-global.com. This series is composed of titles available for purchase individually; each title is edited to be contextually exclusive from any other title within the series. For pricing and ordering information please visit http://www.igi-global.com/book-series/advances-hospitality-tourism-services-industry/121014. Postmaster: Send all address changes to above address.

Titles in this Series

For a list of additional titles in this series, please visit: www.igi-global.com

Opportunities and Challenges for Tourism and Hospitality in the BRIC Nations
Mohinder Chand Dhiman (Kurukshetra University, India)
Business Science Reference • copyright 2017 • 426pp • H/C (ISBN: 9781522507086) • US $205.00 (our price)

Strategic Place Branding Methodologies and Theory for Tourist Attraction
Ahmet Bayraktar (Bozok University, Turkey) and Can Uslay (Rutgers University, USA)
Business Science Reference • copyright 2017 • 394pp • H/C (ISBN: 9781522505792) • US $205.00 (our price)

Global Place Branding Campaigns across Cities, Regions, and Nations
Ahmet Bayraktar (Bozok University, Turkey) and Can Uslay (Rutgers University, USA)
Business Science Reference • copyright 2017 • 368pp • H/C (ISBN: 9781522505761) • US $205.00 (our price)

Global Dynamics in Travel, Tourism, and Hospitality
Nikolaos Pappas (The University of West London, UK) and Ilenia Bregoli (University of Lincoln, UK)
Business Science Reference • copyright 2016 • 372pp • H/C (ISBN: 9781522502012) • US $205.00 (our price)

Corporate Social Responsibility in the Hospitality and Tourism Industry
Lipika Kaur Guliani (Panjab University, India) and Syed Ahmad Rizwan (Tourism Recreation Research, CTRD, India)
Business Science Reference • copyright 2016 • 356pp • H/C (ISBN: 9781466699021) • US $190.00 (our price)

Strategic Tools and Methods for Promoting Hospitality and Tourism Services
Alexandru-Mircea Nedelea (Stefan cel Mare University of Suceava, Romania) Maximiliano Korstanje (University of Palermo, Argentina) and Babu George (Fort Hays State University, USA)
Business Science Reference • copyright 2016 • 326pp • H/C (ISBN: 9781466697614) • US $195.00 (our price)

Handbook of Research on Global Hospitality and Tourism Management
Angelo A. Camillo (Woodbury University, USA)
Business Science Reference • copyright 2015 • 621pp • H/C (ISBN: 9781466686069) • US $345.00 (our price)

Emerging Innovative Marketing Strategies in the Tourism Industry
Nilanjan Ray (Royal Thimphu College, Bhutan)
Business Science Reference • copyright 2015 • 425pp • H/C (ISBN: 9781466686991) • US $215.00 (our price)

Editorial Advisory Board

List of Reviewers

List of Contributors

Table of Contents

Detailed Table of Contents

The aim of this chapter is to present current accessibility to the UNESCO World Heritage Sites on the example of Cracow and Warsaw, the most popular historical cities in Poland. The disabled represent a significant proportion of the cultural tourism group and might have considerable importance to the development of tourist industry. However, their needs are not always met. Especially in Warsaw specific problems for tourists with disabilities include: poor information, inaccessible streets and sidewalks, only partly accessible public transport, hotels or dining places, museums and architectural relics. In Poland tourism development in the heritage sites could be a positive force for generating revenue, enhancing cultural education and helping to preserve the historic values. However, it might be done under condition of re-examining current trends of planning and introducing a universal, holistic design philosophy, innovative technologies enhancing accessibility to all tourist infrastructure, facilities and services.

This chapter examines hotel yield management from a game perspective using a duopoly situation of two hotels. The hotel yield management determines strategies by considering the number of available rooms in the Bertrand situation. Each hotel does not know the strategies adopted by a competitive hotel. We derive the strategy that realizes a maximum profit under a given situation and constraints. Furthermore, we validate the game-based strategy developed for hotel yield management. In the real world, a business manager adopts an optimum strategy of yield management to gain profits in the current conditions; after the initial strategy is chosen, however, managers continuously weigh new strategies and investments. Therefore, we import the method of real option. Such maneuvers and investments are required to build new strategies amidst competition in the industry.

This chapter aims to analyze the different tourism search and metasearch engines for online booking based on supplier perspective (accommodation, flights, leisure and package deals) and taking into account the possible relations generated by the structure and content variables of web offers. A literature review is carried out to study user experience and provide greater in-depth knowledge which will, in turn, in addition to observation of online booking, trends, data, and profile and user preferences. With this information, the study establishes the main attributes of strucuture and offer content of webs focused on the user preferences. The empirical analysis is based on a representative sample of e-commerce of tourism websites, its main characteristics and possible correlations according to the structure and content of their offers. Based on this, the work identifies competitive advantages that will set the trends for the sector and future short-term strategies.

Large scale infrastructure expansions in hotels are exposed to uncertainty. Since the costs involved in these expansion projects are high and often irreversible, hotels would benefit from analyses that incorporate uncertainty along with traditional valuation techniques like the discounted cash flow (DCF) method. Decision tree analysis (DTA) and real options analysis (ROA) have been in use for the past couple of decades to handle uncertainties and optimize investment decisions. DTA provides a distinct approach to strategic investments that quantitatively takes into account the uncertainties involved in the investments. Under uncertainty, the decision about whether to expand is analogous to the decision about whether to exercise an American call option. By using ROA to the hotel expansion scenario, managers can incorporate and quantify, flexibility and timing in their analysis. The objective of this chapter is to detail the DCF, DTA and ROA methodologies and their applications specific to hotel expansion investments.

This chapter presents the utilization of optimization techniques in the hospitality and tourism industry; the perspectives on customer service and customer service teams; customer service and service leadership; customer service behavior in the travel industry; the overview of customer experience; customer experience, customer satisfaction, and service quality in the airline industry; customer experience and budget hotel in the hospitality industry; the issues with customer orientation; and the relationship between customer orientation and service innovation in the hotel industry. Optimization techniques are the essential approaches to many real-world problems, and can be applied to promote the performance of hospitality and tourism industry regarding customer service, customer experience, and customer orientation. The chapter argues that enhancing customer service, customer experience, and customer orientation through optimization techniques has the potential to increase organizational performance and reach strategic goals in the hospitality and tourism industry.

Booking cancellations in the hospitality industry not only generate revenue loss and affect pricing and inventory allocation decisions, but they also, in overbooking situations, have the potential to affect the hotel's online social reputation. By employing data sets from four resort hotels and addressing this issue as a classification problem in the scope of data science, the authors demonstrate that it is possible to build models for predicting booking cancellations with accuracy results in excess of 90%. This research also demonstrates that despite what was alleged by Morales and Wang (2010), it is possible to predict with high accuracy whether a booking will be canceled. Results allow hotel managers to act on bookings with high cancellation probability and contain the associated revenue losses, produce better net demand forecasts, improve overbooking/cancellation policies, and have more assertive pricing and inventory allocation strategies.

Around the world tourism industry represents economic benefits to the countries in where the tourist attractions are located. The purpose of this chapter is to search and review recent researches related to the area of tourism under supply chain management and optimization models perspectives. The main aim of this chapter is to identify and discuss how the tourism supply chain is studied when it is subject to different economic, market, and optimization strategies. Considering the period of 2005 to 2016, a systematic review was performed using research studies in the area of tourism supply chain management. The results show that game theory is used as a theoretical base in the majority of the cases, but several novel approaches are also incorporated to the analysis. This review can be used as a complement of the previous works and a valuable information source for the decision makers involved in the tourism area.

Organisations constantly have to address the problem of identifying tangible and intangible benefits that can be achieved in result of investments made in Information Systems/Information Technology (IS/IT). This paper has the objective of approaching the problem in the perspective and context of Enterprise Resource Planning (ERP) systems in the Hospitality Industry. It focuses on implementation project strategies, namely in the particular and relevant aspects of their return on investment as well as benefits expected and delivered. When an organisation is confronted with the question of knowing if a certain investment in IS/IT may obtain a considered positive financial revenue, and if that investment means valuable benefits for the business, the answer is often far from being conclusive. Projects are often implemented on-time, on-budget and are technically appropriate, nevertheless the expected benefits are not achieved. This paper seeks to answer these questions in the context of ERP systems, in particular in the sector of hospitality industry, through a case study developed in Portugal. The investigation was developed on a group of hotel units, held by a Portuguese company, for which ERP meant an opportunity to harmonize and improve business processes.

The present chapter discusses to what extent the rise of new risks and dangers, modernity has brought, has causing the end of tourism. The answer to the above formulated questions is not easy, and of course, it exhibits a fertile ground to be explored in other approaches. The ceaseless news about violence, cruelty, wars and deaths have serious negative impacts on audience worldwide. Though policy makers have devoted their time and efforts in looking for new alternative segments where death replaces the allegory of beautiness, giving as a result new products as dark-tourism, slum-tourism, disaster-tourism or doom-tourism, other problems arise. The curiosity for tourists in visiting spaces of mass death or extreme suffering contradicts not only its nature as a mechanism of revitalization and relaxes but may very well lead to narcissism which in fact is the end of tourism.

The hotel industry is capital-intensive with high operating leverage in providing accommodations. Examining final costs play very important role in determining of overcapacity and undercapacity. Turkey's experiencing growth in tourism and has remarkable increases in accommodation. This situation possesses great importance for the capital groups investing in this area. How these groups and local tradesman implement a strategy on capacity management is the main question. Single period inventory model is used at the analysis phase and several interests are examined on this topic. Future Room demand is estimated by ARIMA and in estimating cost of undersupply earnings before taxes is used per room night sold. Combining the derived cost ratio with the future room demand and probability distribution estimated from the ARIMA method, the optimal hotel room capacity can be determined. The purpose of the study is to examine hotel room supply and demand for optimal capacity in Turkey.

Defining the boundaries of tourism destinations has been long recognised as a problem in tourism research. The authors aim to define the spatial configuration of tourism areas including different destinations within a same region. Tourist mobility is employed as a methodological criterion to reveal the network relationships among destinations and explain how tourism areas are being shaped and reshaped. The study combines Network Analysis methods and multinomial logistic regression models, in an approach to processing the data of a sampling survey, carried out in Sicily. The results show that the network structures among destinations affect the shape and dimension of tourism areas. Useful evidence for the spatial planning of tourism regions and destination management strategies are derived.

Optimization algorithms are a powerful tool for solving many problems of engineering applications from different fields of real life. They are usually used where the solution of a given problem analytically

Foreword

Holistic Optimization Techniques in the Hospitality, Tourism, and Travel Industry proves that optimization techniques could be valuable ally for problem solving in domain of non-engineering area.

Professor Pandian Vasant and a Hotel Practitioner Kalaivanthan as editors selected chapters which shows how formal mathematically oriented methods opens new way of thinking in area of Hospitality, Tourism, and Travel Industry.

In modern business, cost cutting, better decision making, optimal usage of disposable resources are a base for survival in turbulent market conditions. *Holistic Optimization Techniques in the Hospitality, Tourism, and Travel Industry* gives contribution how to achieve those aims and how to be better as a company in area of Hospitality, Tourism and Travel Industry by using power of optimization techniques.

The contributing authors of this book showed fresh ideas and enthusiasm which connects academic and practical aspects in area of optimization techniques and specific industries. Almost each chapter opens window into unrevealed optimization world landscape, where reader can discover many interesting details colored by creativity.

Books like this, spreads horizons, not only in area of optimization methods, but also in area of economy, IT, machine learning and lead us on thinking and awaking creativity regarding interdisciplinary topics.

Showed synergy effect between IT, machine learning, optimization methods, economy, can be great inspiration for the book readers, not only for practical applying of presented concepts in everyday business, but also for some new great book like *Holistic Optimization Techniques in the Hospitality, Tourism, and Travel Industry* is.

Goran Klepac
University College for Applied Computer Engineering Algebra, Croatia & Raiffeisenbank Austria Inc., Croatia
June 2016

Goran Klepac, *PhD, University College Professor, works as Head of Chair for Information Systems at University College for Applied Computer Engineering Algebra, and as a head of Strategic unit in Sector of credit risk in Raiffeisenbank Austria Inc, Croatia. In several universities in Croatia, he lectures subjects in domain of data mining, predictive analytics, decision support system, data science, banking risk, risk evaluation models, expert system, database marketing and business intelligence. As a team leader, he successfully finished many data mining projects in different domains like retail, finance, insurance, hospitality, telecommunications, and productions. He is an author/coauthor of several books published in Croatian and English in domain of data mining, as well as scientific papers. Goran Klepac is Editorial Review Board member in International Journal of Swarm Intelligence Research (IJSIR), Global Journal of Technology and Optimization (GJTO), International Journal of Ambient Computing and Intelligence (IJACI). www.goranklepac.com.*

Foreword

Holistic optimization techniques became a powerful tool in engineering, science and technology. However applications of corresponding techniques in social sciences are rather limited. The book *Holistic Optimization Techniques in the Hospitality, Tourism and Travel Industry* aims to reduce the gap between the use of holistic techniques in exact and social sciences, focusing on globalization and interdisciplinary aspects of real world applications.

The book presents current trends and the state-of-the art in using holistic optimization techniques in the hospitality, tourism and travel industry. The book chapters may be divided into three large groups. The first covers traditional issues typical for general logistic and supply chains management: optimal hotel expansion, capacity management, using ERP systems to minimize investments. The second focuses on more specific hospitality issues such as yield management strategies, online booking and cancellation, analyzing and modeling tourists' opinions. The third group concentrates on risk, terrorism and security issues which became crucial issues in the last years. Altogether, the chapters demonstrate how holistic and general systems engineering approaches may lead to successful applications in various areas of tourism and travel industry.

In my opinion, the book *Holistic Optimization Techniques in the Hospitality, Tourism and Travel Industry*, edited by Dr. Pandian Vasant and a Hotel Practitioner Kalaivanthan, presents a carefully selected text collection in the area. This volume is highly recommended and can be useful not only for researches, but also for tourism industry practitioners, decision makers, executives, hoteliers and agents. The reader will find a wide variety of interesting and inspiring techniques and applications in this book.

Igor Litvinchev
Computing Center, Russian Academy of Sciences, Russia

Igor S. Litvinchev *received his MSc degree in Applied Mathematics from Moscow Institute of Physics and Technology (Fizteh), Moscow, Russia, PhD in Systems Theory and Operations Research, and DrSci (Habilitation) in Systems Modeling and Optimization, both from Computing Center, Russian Academy of Sciences, Moscow. He has held visiting positions at universities in Brazil, Mexico, and Norway, as well as positions at various universities and research centers in Russia. He is currently a Head of Department at Computing Center Russian Academy of Sciences and Professor at Nuevo Leon State University, Monterrey, Mexico. He is the author of 4 books and more than 60 refereed papers focused on large-scale system modeling, optimization, and control. His research was supported by more than 30 grants from NATO Scientific Affairs Division and European Community, ISF (USA) and RFBR (Russia), CNPq and FAPESP (Brasil), BRFBR (Belarus), CONACYT, PROMEP, and PAICYT (Mexico). Professor Litvinchev is a member of Russian Academy of Natural Sciences and Mexican Academy of Sciences.*

Foreword

The purpose of this book *Holistic Optimization Techniques in the Hospitality, Tourism and Travel Industry* is to deeply inquire, holistically reflect, and comprehensively expose this current and emerging cutting edge technology in important fields of investigation and facilities for original, innovative, and novel real-world applications of Optimization in the modern world related to the emerging area of the Hospitality, Tourism and Travel Industry. During the previous decades, the analytical tools and the methodological toolbox of computer science and applied mathematics, of informatics and statistics, of emerging analytics and information technologies has gained the attention of numerous researchers and practitioners from all over the world, providing a strong impact also in natural sciences, engineering, economics and finance, the IT sector and, since recent years, the industry of *hospitality, tourism and travel*. Here, Optimization is turning out to be a key technology from an integrated perspective, and it is closely connected with further areas of modern Operational Research such as Development, Societal Complexity and Ethics, in order to also appropriately address aspects of living conditions, culture and environment in relation with that new and fast growing industry.

Holistic optimization techniques are more and more well-known and used in research areas of engineering, science and technology. Current trends on application problems in engineering, science and technology are also expanded into the areas of economy, finance and, especially, into the hotel, tourism and travel industry across the globe. This book focuses on best and high-quality selected papers in a rich variety of research topics. Well-known and new methodologies of holistic optimization techniques have been used to resolve some of the complicated and hard problems in the areas of hotel, tourism and travel industry.

This book *Holistic Optimization Techniques in the Hospitality, Tourism, and Travel Industry* welcomed submissions by interested authors from hotel, hospitality and tourism management areas and from social science. It also invited for best chapters from the research areas of information technology and computer science which are related to hotel, tourism and travel industry. The book chapters focus on new development and contribution of current research to the body of the knowledge in the areas of hotel, tourism and travel industry, supported by internet and computer technology, Operational Research - and Optimization as one of its core fields. This book could become a helpful compendium for decision makers, managers, supervisors, executives, travelers, tourist, hoteliers, agents, researchers, directors, financiers, economics as well as industrialist.

The contributing authors of this book are experienced scientists and practitioners from all over the world; they use and further refine the deep model-based methods of mathematics and the less model-based, also so-called smart or intelligent algorithms with their roots in the engineering disciplines, in computer science, and informatics. The second ones are often called as heuristics and model-free; they are less rigorous mathematically, released from firm calculus, in order to integrate nature- and, especially, bio-inspired approaches to efficiently cope with hard problems. The rise of these algorithms from Artificial Intelligence happened in parallel to the powerful progress in mathematics that is model-based mainly. Today, labeled by names like Statistical Learning and Machine Learning, Metaheuristics and Matheuristics, and by Operational Research, model-free and model-based streamlines of traditions and approaches meet and exchange in various centers of research, at important congresses, and in leading projects and agendas in all over the world to overcome misunderstandings and misperceptions between those two academic avenues, but to benefit from synergy effects, to jointly advance scientific progress and to provide a common service to the solution of urgent real-life challenges. Those huge problems exist in every area of the modern world and academics, in engineering, the economy, social sciences, life and human sciences, and development and the improvement of living conditions and future perspectives. Among the variety of these subjects and, in fact, including all of them, the editors selected the area of *Holistic Optimization Techniques in the Hospitality, Tourism, and Travel Industry*.

To all the authors of these valuable chapters, I extend my appreciation and gratitude for having shared their devotion, knowledge and vision with the academic community and mankind. I am very thankful to the publishing house IGI Global, and to the editor. Dr. Pandian Vasant, for having provided the chance for these experts to make public best achievements and contributions. I express my thankfulness to them for having ensured a premium book of a high standard academic, applied, and social importance!

Now, I wish all of you a lot of joy in reading this interesting work, and I hope that a great benefit is gained from it both personally and societally.

Gerhard-Wilhelm Weber
Middle East Technical University, Turkey
May 2016

Gerhard-Wilhelm Weber *is a Professor at IAM, METU, Ankara, Turkey. His research is on optimization and optimal control (continuous and discrete), Operational Research, financial mathematics, on life, bio and human sciences, dynamics, data mining, statistical learning, inverse problems, environment and development. He is involved in the organization of scientific life internationally. He received both his Diploma and Doctorate in mathematics and economics / business administration at Aachen University of Technology (RWTH Aachen), and his Habilitation at Darmstadt University of Technology (TU Darmstadt). He held Professorships by proxy at the University of Cologne, Germany, and Chemnitz University of Technology, Germany, before he worked at Cologne Bioinformatics Center and then, in 2003, went to Ankara. At IAM, METU, he is in the Programs of Financial Mathematics and Scientific Computing. He is Assistant to the Director of IAM and a member of five further graduate schools, departments, and institutes of METU, and he has different affiliations to University of Siegen (Germany), University of Aveiro (Portugal), Federation University (Ballarat, Australia) and University of North Sumatra (Medan, Indonesia).*

Preface

Holistic optimization techniques are well known in the research areas of engineering, science and technology. The current trends on application problems in engineering, science and technology are expanded in the areas of hotel, tourism and travel industry across the globe. This book chapter focuses on the best and high quality selected papers on the following research topics. The well-known and new methodologies on the holistic optimization techniques have been used to solve some of the complicated and hard problems in the areas of hotel, tourism and travel industry.

This book chapter project will be open for interested authors in hotel management, social science, hospitality management areas to submit their manuscripts. It would also cover the best chapters from the research areas of information technology and computer science which are related to hotel, tourism and travel industry. The book chapter mainly focuses on new development and contribution of the current research to the body of the knowledge in the areas of hotel, tourism and travel industry which are related to internet and computer technology.

The objective of the book is to explore the emerging innovative and novel technology on Holistic Optimization Techniques in the Hospitality, Tourism, and Travel Industry.

The prospective audience of the book "Holistic Optimization Techniques in the Hospitality, Tourism, and Travel Industry," will be decision makers, managers, supervisors, executives, travelers, tourist, hoteliers, agents, researchers, directors, financiers, economics as well as industrialist.

The book chapters cover the following topics:

- Cultural Tourism,
- Tourist Mobility,
- Metaheuristic,
- Ethiopian Tourism,
- Game Theory,
- Hotel Management,
- E-Commerce,
- Business to Customer (B2C).

The book is organized into 20 chapters. A brief description of each of the chapters is as follows:

Chapter 1 is about the current accessibility to the UNESCO World Heritage Sites on the example of Cracow and Warsaw, the most popular historical cities in Poland. The disabled represent a significant proportion of the cultural tourism group and might have considerable importance to the development of tourist industry. However, their needs are not always met. Especially in Warsaw specific problems for tourists with disabilities include: poor information, inaccessible streets and sidewalks, only partly accessible public transport, hotels or dining places, museums and architectural relics. In Poland tourism development in the heritage sites could be a positive force for generating revenue, enhancing cultural education and helping to preserve the historic values. However, it might be done under condition of re-examining current trends of planning and introducing a universal, holistic design philosophy, innovative technologies enhancing accessibility to all tourist infrastructure, facilities and services.

Chapter 2 examines hotel yield management from a game perspective using a duopoly situation of two hotels. The hotel yield management determines strategies by considering the number of available rooms in the Bertrand situation. The authors have derived the strategy that realizes a maximum profit under a given situation and constraints. Furthermore, it will validate the game-based strategy developed for hotel yield management. In the real world, a business manager adopts an optimum strategy of yield management to gain profits in the current conditions; after the initial strategy is chosen, however, managers continuously weigh new strategies and investments. Therefore, it's necessary to import the method of real option. Such maneuvers and investments are required to build new strategies amidst competition in the industry.

Chapter 3 aims to analyze the different tourism search and metasearch engines for online booking based on supplier perspective (accommodation, flights, leisure and package deals) and taking into account the possible relations generated by the structure and content variables of web offers. A thorough literature review study establishes the main attributes of structure and offer content of webs focused on the user preferences. The empirical analysis is based on a representative sample of e-commerce of tourism websites, its main characteristics and possible correlations according to the structure and content of their offers. Based on this, the work identifies competitive advantages that will set the trends for the sector and future short-term strategies.

Chapter 4 deals with the decisions on the large scale infrastructure expansions in hotels are often taken while future developments are uncertain. As these projects take many years to be implemented, the hotels are exposed to multiple uncertainties. Since the costs involved in these expansion projects are quite high and most often irreversible, hotels would benefit from analyses that incorporate uncertainty in addition to traditional valuation techniques like the discounted cash flow (DCF) method. Decision tree analysis provides a distinct approach to strategic investments that quantitatively takes into account the uncertainties involved in the investments. By using real options analysis managers can incorporate and quantify flexibility and timing in their analysis. The objective of this chapter is to detail the discounted cash flow, decision tree and real options methodology and their applications specific to hotel expansion investments.

Chapter 5 focuses on the implications of branding on customer satisfaction and loyalty through two mediators:

- Brand image and
- Destination brand loyalty.

The study took place in North Cyprus as a case to investigate and explore the factors that determine the brand definition for the island and how such branding can become a marketing tool. The study revealed that the customer involvement which is the plan for tourism agencies to involve tourists in trip plan to make customer feel that they are part of the process themselves. The results revealed that customer satisfaction is a determinant factor at creating destination image.

Chapter 6 talks about booking cancellations in the hospitality industry not only generate revenue loss and affect pricing and inventory allocation decisions, but they also, in overbooking situations, have the potential to affect the hotel's online social reputation. By employing data sets from four resort hotels and addressing this issue as a classification problem in the scope of data science, the authors demonstrate that it is possible to build models for predicting booking cancellations with accuracy results in excess of 90%. Results allow hotel managers to act on bookings with high cancellation probability and contain the associated revenue losses, produce better net demand forecasts, improve overbooking/cancellation policies, and have more assertive pricing and inventory allocation strategies.

Chapter 7 aims to analyze referral and conversion traffic from user-generated content (UGC) to other key online/offline information sources and when consumers use UGC along with their information-search and the related cross-visitation. Manual coding of open-ended questions and descriptive statistics were applied. Findings revealed that UGC plays a relevant role, along with all the online searching and booking behaviour, particularly UGC within online travel agencies (OTAs), which appears to be significant in the early stage of the tourist information search process. Contributions to the body of knowledge and managerial implications are discussed and suggestions for further research are given.

Investing in a hotel involves a big-ticket transaction. Traditionally, a number of cash-surplus entities are selected to pool-in the capital necessary for the acquisition. A critical consideration is, then, to minimize the overall cost of capital resulting from the return expectations of the surplus units. In Chapter 8, the authors introduce the readers to typical sources of hotel finance and considerations about minimizing the weighted average cost of capital (WACC). First, they provide a commentary on various types of funding sources. Further, they provide rationale for why a particular surplus unit specifies certain constraints to the investment manager. Finally, they walk the readers through various steps of the optimization process using the SOLVER function in MS Excel.

Chapter 9 presents essay review explores the problem of terrorism and security in tourism fields. Certainly, plans and policies provided by guide-books are not being followed in disaster-contests simply chaos and disorder are the nature of emergencies. Beyond any protocol, crises and security are not properly defined by scholars. They have explored not only the roots of terrorism but security over 20 years. Despite the criticism, they deserve recognition for this legacy. Based on substantial point of divergence, these specialists are concerned by the financial dependency of societies respecting to mass-media and its coverage of terrorist attacks.

The purpose of Chapter 10 is to search and review recent researches related to the area of tourism under supply chain management and optimization models perspectives. The main aim of this chapter is to identify and discuss how the tourism supply chain is studied when it is subject to different economic, market, and optimization strategies. Considering the period of 2005 to 2016, a systematic review was performed using research studies in the area of tourism supply chain management. The results show that game theory is used as a theoretical base in the majority of the cases, but several novel approaches are also incorporated to the analysis. This review can be used as a complement of the previous works and a valuable information source for the decision makers involved in the tourism area.

Chapter 11 has the objective of approaching the problem in the perspective and context of Enterprise Resource Planning (ERP) systems in the Hospitality Industry. It focuses on implementation project strategies, namely in the particular and relevant aspects of their return on investment as well as benefits expected and delivered. Projects are often implemented on-time, on-budget and are technically appropriate, nevertheless the expected benefits are not achieved. This chapter seeks to answer these questions in the context of ERP systems in the sector of hospitality industry, through a case study developed in Portugal.

Organizations constantly have to address the problem of identifying tangible and intangible benefits that can be achieved in result of investments made in Information Systems/Information Technology (IS/IT). Chapter 12 has the objective of approaching the problem in the perspective and context of Enterprise Resource Planning (ERP) systems in the Hospitality Industry. It focuses on implementation project strategies, namely in the particular and relevant aspects of their return on investment as well as benefits expected and delivered. Projects are often implemented on-time, on-budget and are technically appropriate, nevertheless the expected benefits are not achieved. This chapter seeks to answer these questions in the context of ERP systems in the sector of hospitality industry, through a case study developed in Portugal.

The purpose of Chapter 13 is to examine hotel room supply and demand for optimal capacity in Turkey. The hotel industry is capital-intensive with high operating leverage in providing accommodations. Turkey's experiencing growth in tourism and has remarkable increases in accommodation. This situation possesses great importance for the capital groups investing in this area. Single period inventory model is used at the analysis phase and several interests are examined on this topic. Future Room demand is estimated by ARIMA and in estimating cost of undersupply earnings before taxes is used per room night sold. Combining the derived cost ratio with the future room demand and probability distribution estimated from the ARIMA method, the optimal hotel room capacity can be determined.

The authors of Chapter 14 aim to define the spatial configuration of tourism areas including different destinations within a same region. Tourist mobility is employed as a methodological criterion to reveal the network relationships among destinations and explain how tourism areas are being shaped and reshaped. The study combines Network Analysis methods and multinomial logistic regression models, in an approach to processing the data of a sampling survey, carried out in Sicily. The results show that the network structures among destinations affect the shape and dimension of tourism areas. Useful evidence for the spatial planning of tourism regions and destination management strategies are derived in this chapter.

The majority of the problems of real life applications can be defined as optimization problems, for example, finding the optimum trajectory of a robot, optimal data flows in various processes like city traffic optimization or modeling and optimization of the seasonal variances of supply, traffic and facilities occupation in tourism among the others. The structure of Chapter 15 is such that on the beginning are introduced bio-inspired algorithms, then parallelization of algorithms and parallel hardware and at the end, open research on Ho Chi Minh City traffic optimization real world example is introduced. In Conclusion are discussed possibilities of mutual combinations of introduced methods.

In Chapter 16 the study on a game theoretical model is applied to explain the decisional process of entrepreneurs about forming partnerships in formal tourism networks. The model shows that, in terms of Nash equilibrium, the cooperative optimization of a tourism network should be achieved when entrepreneurs have the same business goals and a common tourism vision. The chapter presents an empirical solution that arises from the case of Business Networks in Italy, which represents an innovation in Italian Law.

Chapter 17 researched on profit optimisation in the travel sector of hospitality industry has been dominated by the development of effective revenue management techniques to help managers in situations where demand is variable, variable costs are low, assets are fixed and perishable. These techniques have been extended to the restaurant sector, with recognition that variable costs are a larger proportion of total costs and not fixed. In fine dining restaurants substantive wine and spirit inventories present operators with the challenge of how to optimise the sales per dollar of inventory. Data from a successful fine dining restaurant are used to illustrate how the model can be used to improve the sales efficiency of the wine list and inventory. Opportunities to extend the model to spirit inventories are proposed in this chapter.

Generation Y has been considered to be a sizeable new market. Chapter 18 is based on a sample of 1131 Italian travellers from Gen Y, investigates their views for and against disintermediation, and analyses how their choices are influenced by user generated content (UGC), rather than by information provided by high street travel agencies. The factor analysis uncovers three dimensions:

- "Benefits of Travel Agency,"
- "Benefits of Online Reservation," and
- "Online Trust and Search Behaviour."

The findings suggest that hotel managers and travel agencies should monitor Gen Y perceptions of the benefits and constraints of using the Internet, UGC and travel agencies for hotel booking. The accommodation providers should use online channels to create affective commitment in their young customers. Limitations of the study are discussed and suggestions for further research are given in this chapter.

In Ethiopia, a country which has been named as best tourism destination for 2015 by the European Council on Tourism and Trade (ECTT), there are many tourist destinations. The problem of determining the optimum route to visit all the tourist sites with minimum traveling time can be formulated as a travel salesman problem. Chapter 19 studies 17 of the famous tourist destination will be selected and a travel salesman model will be formulated. Due to the NP hardness of the travel salesman problem, metaheuristic based algorithms are found to be more effective. Hence, a recently introduced swarm based metaheuristic algorithm, called prey predator algorithm has been used to deal with the formulated problem.

Chapter 20 is attempting to use "Relative to an Identified Distribution" (RIDIT) algorithms based modelling for analysing real-time empirical data relating to tourists' attitude and preference for a better understanding of the tourists' motivation and behaviour. RIDIT approach for evaluating the factors that influence tourist behaviour is not a very common approach in tourism sector. This chapter on modelling tourists' opinions and perceptions with RIDIT analysis would try to guide the empirical research in the domains of hospitality, tourism and travel research and analytics process in generating Optimized research outcomes.

The book editors are very thankful to the editorial team and entire staff of IGI Global, for their confidence, interest, continuous encouragement and support at all levels of preparation of this book's preparation.

We editors wish all the readers a pleasant and enjoyable, insightful and inspiring lecture of the contributions of this IGI Global book. In fact, we cordially hope that this special issue will present and value IGI Global as a premium publishing house in optimization, hotel, tourism, travel and hospitality industry which strongly fosters very much needed intellectual advances and their contributions to humanity and mankind across the planet.

Pandian Vasant
Universiti Teknologi PETRONAS, Malaysia

Kalaivanthan M.
Rasa Sayang Resort and Spa, Malaysia & Golden Sands Resort, Malaysia

Acknowledgment

The editors would like to sincerely thank Universiti Teknologi PETRONAS, Malaysia and Shangri-La's Rasa Sayang Resort and Spa and Golden Sands Resort Penang for their valuable encouragement and support on the publication of this book chapters. The valuable contribution of the reviewers is highly appreciated and acknowledged.

Chapter 1
Holistic Technical Solutions to Enhance Accessible Tourism in the UNESCO World Heritage Sites

Anna Pawlikowska Piechotka
AWF University, Poland

Anna Ostrowska Tryzno
AWF University, Poland

Natalia Lukasik
AWF University, Poland

Maciej Piechotka
ECO CLASSIC Architects, Poland

Karolina Sawicka
University Warsaw, Poland

ABSTRACT

The aim of this chapter is to present current accessibility to the UNESCO World Heritage Sites on the example of Cracow and Warsaw, the most popular historical cities in Poland. The disabled represent a significant proportion of the cultural tourism group and might have considerable importance to the development of tourist industry. However, their needs are not always met. Especially in Warsaw specific problems for tourists with disabilities include: poor information, inaccessible streets and sidewalks, only partly accessible public transport, hotels or dining places, museums and architectural relics. In Poland tourism development in the heritage sites could be a positive force for generating revenue, enhancing cultural education and helping to preserve the historic values. However, it might be done under condition of re-examining current trends of planning and introducing a universal, holistic design philosophy, innovative technologies enhancing accessibility to all tourist infrastructure, facilities and services.

INTRODUCTION

Responsible Planning of Tourist Space

Today tourist activities are becoming more competitive, extensive and demanding of infrastructure and services than ever before. In order to respond to these growing needs positively, it is necessary to practice

DOI: 10.4018/978-1-5225-1054-3.ch001

tourism planning in the host destination areas to answer these challenges, avoid damage or permanent alteration of natural features, social values or historical valuable resources (Buhalis et al., 2012; Dallen et al., 2003; Hall, 2000; Lovelock et al., 2013, Page et al., 2003, Shaw et al., 2004).

The UN WTO resolutions and declarations recognize good management and rational spatial planning as key factors to sustainable development of reception areas. Principles of integrated planning are considered as an effective approach, especially when protective areas are in question (of natural, social and/ or cultural values). An integrated planning, understood as holistic activity, is a planning approach that ensures participation of all stakeholders and affected departments. The objective of integrated planning is to examine all economic, social and environmental costs and benefits in order to select and implement the best possible option and to plan a suitable course of action. In the area of tourism planning an integrated planning approach is highly recommended, as being responsive not only to the various tourist groups' expectations, but also the local community desires and needs (Inskeep, 1991; Newman, 2011; Shelton 2005; Smith, 2006; Sigala, 2003). Moreover, the heritage sites need special attention and planning rules, tourism policy and spatial planning there should be developed in the innovative ways. Therefore, integrated planning approach seems to be ideal to respond the needs of sustainable planning at the UNESCO World Heritage Sites (Buhalis et al., 2012; Dallen et al., 2003). Accordingly to its theoretical basis, contemporary integrated tourism planning include: economic planning, human resources planning, social and community planning, environmental planning, business planning, public administration, infrastructure planning and design (Hall, 2000; Inskeep, 1991; Newman, 2011). As integral planning encompasses a very wide range of issues, in this chapter the authors decided to focus on the tourist infrastructure planning questions (urban and architectural design) with the aim to obtain accessibility for disabled tourists at destinations of high cultural values.

The future vision of tourism is a 'mobile tourism' imperative and the governments are obliged to provide the non – discriminatory access to the tourist destinations. Today about 15% of the world population is estimated to have some form of disability and one can observe the growing number of tourists with special requirements (ENAT, 2015; UN WHO, 2011; UN WTO, 2015). Studies carried in many countries have shown that tourists with disabilities are becoming an important part of tourism market (Australia – 11%, UK – 12%) and for example American adults with disabilities or reduced mobility spend on average 13.6 billion USD annually on travelling (UN WTO, 2013). Therefore, accessible tourism market presents a challenge to the global travel industry in terms of providing barrier – free reception areas, improvement of services, accommodation and transport (Darcy, 2010; Shaw et al., 2004).

Destinations Accessibility in Poland

Accessible tourist destination depends on a variety of determinants. All over the world, in Australia, Canada, US and most of the European countries,

- UK,
- Spain,
- Sweden,
- Switzerland,
- Malta,
- Norway,
- Holland,

- France,
- Italy,
- Germany,
- Denmark, and
- Austria.

many aspects of spatial settings and services have been improved to shape 'all tourist friendly' destinations. They include material conditions (technical barriers, physical environment) and intangible conditions as social responsibility, behavioral and personal determinants, services and other social factors. It has resulted in respect and social inclusion, which are reflected as the accessibility of the buildings (hotels, museums, terminals), public spaces around tourist attractions (as old towns streets and squares), pre – visit and in situ information and the wide range of opportunities that the tourist destination could offer to its visitors. In these countries social participation (in tourism and other sectors) is based on accessibility and safety.

However, in Poland, as in other post-communist countries, there are many signs that most of disabled tourists are keen to travel, but unnecessary barriers, including:

- Inaccessible transport,
- Barriers of outdoor space,
- Poor communication,
- Insufficient information, and
- Negative experiences of catering or accommodation services in the past travels

could effectively discourage potential travelers.

It is true, that Polish cities have traditionally been characterized by stronger exclusion of marginalized groups compared to Western Europe. Mostly as the consequence of weaker economy and welfare systems in the past.

Although in Poland over the last two decades average living standards have increased, including efforts to implement the 'National Accessible Scheme Program', there are visible social inequalities. It is seen as difficulties and inabilities to access public buildings and outdoor space, including some tourist destination services (transportation, information, communication, accommodation). These are results equally of economic troubles and poor social concern. Therefore, to allow stronger social inclusion, it is necessary to improve accessibility to information and attractions, local transport and accommodation together with changing social sensitivity towards responsibility and deeper concern.

Such an 'inclusion spatial policy' implemented at the tourist destinations, including the world heritage sites, might result in economic benefits, in better education and help disabled tourists to achieve social integration. As Buhalis underlined - business interests and social responsibility are not contradictive (Buhalis et al., 2012). It means that in Poland the wide range of necessary changes in the tourist services and destination area settings should be examined, recognized and re-evaluated.

The research aim, undertaken by the authors in 2012-2014, was to examine the host space, identify the main accessibility barriers (using the check-list) and provide a far – reaching overview of the concerns facing disabled tourists at the UNESCO heritage sites in Poland.

This chapter presents some generalized outcomes of the research and is focused both on theoretical holistic concepts of sustainable tourism planning and technical practical solutions, aimed to create an optimum level of accessibility at tourist destination, friendly core tourist features suitable 'for all'.

THEORETICAL BACKGROUND AND LAW REGULATIONS

International Documents and Regulations

Tourist accessible destinations, including infrastructure, facilities, information and services, might enable all people to participate and enjoy tourist experience (Darcy, 2010; Shaw et al., 2004; Shelton et al., 2005). Products and services, including information, connected with travelling, accommodation, dining, visiting tourist attractions must be prepared to ensure that are accessible to all people regardless of limitations, different form of disabilities, family situation or age. Therefore, responsible and holistic techniques in planning and management solutions should be applied to shape a sustainable development of barrier - free tourist destinations (Darcy, 2009, 2010, 2011; Dredge, 1996; Drost, 1998; ENAT, 2015; Harrison, 2005; Lovelock et al. 2013, Preiser, 2001; Rifai, 2013; Shelton, 2005; Small et al., 2012; Timothy, 2009; UN WTO, 2013, 2014, 2015; Vael, 2002).

Law regulations on universal access and technical standards evolved within the last few decades intensely. First were the United Nations declarations of general character and then the UN WTO more detailed regulations, which influenced the domestic legislation in Poland in order to ensure universal accessibility and to foster the removal of mental, architectural or urban barriers. In December 1975 the General Assembly adopted the "Declaration on the Rights of Disabled Persons" (UN, 1975). The recent document was the issue of the "International Convention on the Rights of Persons with Disabilities" (UN, 2007), contributing significantly to the dignity and equality of all human beings. In 1980 the UN WTO approved the 'Manila Declaration', pioneering statement about non-discriminatory access to all tourist services and facilities. In 2005 the UN WTO issued the document "Creating Tourism Opportunities for Handicapped People in the Nineties" (known under the title "Accessible Tourism for All"). This document provides a series of guidelines grouped into several segments:

1. Tourist information and publicity,
2. Staff training,
3. Common requirements and requirements concerning specific facilities (UN WTO, 2005).

The recent document issued by the UN WTO is titled "Manual on Accessible Tourism for All: Public – Private Partnership and Good Practices" (UN WTO, 2015). In 1981 the European Council approved a resolution "Social Integration of Handicapped People" with a view to foster actions in favour of the disabled. The recent European Council document was issued in 2010 under the title "The Charter of Fundamental Rights of the European Union". It recognizes and respects the right of persons with disabilities to benefit from measures, designed to ensure their independence, social integrity and right to participate in community life.

In 2011 the UNESCO ICOMOS issued the document "The Valetta Principles for the Safeguarding and Management of Historic Cities, Towns and Urban Areas" with a hope it will serve as a guideline for planners and help to avoid rising conflicts between the historic monuments conservatory, city council, investors, local community and tourists. It is to underline that this document considers needs and special requirements of disabled tourists. The main objective of the declaration is to propose principles and strategies enabling to safeguard the unique values of historic towns and ensuring harmonious development of tourism, including disabled individuals (UNESCO ICOMOS, 2011).

Law Regulations and Building Practices in Poland

Poland is a signatory of all these mentioned above international law acts, which have been bilaterally ratified by its government. In the last few decades, Polish legal regulations and building practices have sought to establish a universal access to public space, also accessibility in the tourism sector. The basic law regulation that Polish architects and urban planners have to obey, is the Ministry of Infrastructure recommendations on building requirements issued in 2002 with amendments (MI, 2002). Although progress in law regulations and technical standards is being made, still several questions and issues, especially in the technical sphere of building regulations, remain to be improved. A series of challenges must be met, as for example sustainable and harmonious development of the popular tourist destinations. There is no denial of the importance of shaping barrier – free reception space in more integrated way.

Therefore, the scientific researches, especially applied ones, designed on the base of case studies are crucial to make a contribution in supporting legislation efforts aimed to provide and secure barrier - free access, including information, to the tourist reception areas.

The aim of the research was to examine current accessibility of the UNESCO World Heritage Sites on the example of Cracow and Warsaw case studies. It was based on the checklist of the core tourist features (buildings, outdoor space and selected services). The Old Town in Warsaw and the Old Town in Cracow are the most popular tourist destinations in Poland, visited by million guests per year. Tourists with mobility, visual, hearing or cognitive impairments require special aids to visit historical cities and tourist attractions safely and relatively independently (Darcy 2009, 2011; UN WTO, 2015; Vael, 2002). The researchers intended to critically examine the access requirements to the historic sites and also to propose some concrete technical solutions, including accessible information, highlighting the best practices enabling people with disabilities to visit the UNESCO heritage sites with equality and dignity (ENAT, 2015; UNESCO, 2011).

This research study included consideration of all tourist groups, also families travelling with children in prams, people with disabilities and seniors.

MEASURING THE ACCESSIBILITY TO THE UNESCO HERITAGE SITE

Research Hypothesis Assumption

The basic ground for the research hypothesis assumption was the belief that it is possible to develop a universal access to the UNESCO heritage sites in Poland through integrated and holistic spatial planning approach to achieve optimum level of proposed technical solutions. These planning goals were understood as a part of 'tourism for all' ideology and policy, as it is expressed in the United Nation declarations and resolutions (UN WTO, 2013) and renown researchers findings (Darcy, 2009, 2010, 2011; Dredge, 1996; Drost, 1998; Harrison, 2005; Lovelock et al., 2013, Preiser, 2001; Rifai, 2013; Shaw et al., 2004; Shelton, 2005; Small et al., 2012;). According to these documents and statements persons with disabilities have a right, on equal basis with the others, to take part in cultural tourism experience, have an access to heritage sites, transportation, information and other facilities provided at destination.

For the research needs the definition of term "persons with disabilities" has been adapted in a very broad sense, on the base of the UN WHO document (UN WHO, 2007) and it means tourists suffering a limitation in their abilities and having special needs during travel (as accommodation, catering and transport). It includes people with:

- Physical disabilities,
- Sensory impairments (visual, hearing),
- Individuals with intellectual disabilities or other medical conditions requiring special care (seniors),
- As well as child carers (travelling with small children in prams or toddlers),
- The elderly,
- People carrying luggage and people with temporary disabilities (as tourists with crutches used during a temporary period).

Research Material

The research was conducted between 2012 and 2014 in the Warsaw Old Town and in the Cracow Old Town, both sites inscribed on the World Heritage List. The 13th century historic centre (the Old Town) of Cracow was included in the UNESCO World Heritage List in 1978 and it covers the area of 150 ha. The medieval Old Town of Warsaw was included in the UNESCO World Heritage List in 1980 and it covers the area of 26 ha.

'Accessibility of tourist space', for the research needs, was connected with accessible products, environment, programme and services, creation of destination area without architectural or urban barriers. This definition is consistent with the UN WTO official terminology (UN WTO, 2013). Information about tourist accessibility to these two UNESCO World Heritage sites was gathered on the base of the authors' frequent visits to the surveyed sites (the Cracow Old Town and the Warsaw Old Town), direct observations, structured and unstructured interviews with tourists and local community members (primary data), undertaken during field research. Secondary data analysis was undertaken on the base of law regulations, government statistic, published materials and planning strategies prepared by cities governments in aim to improve tourist accessibility to both old towns (quantitative data). The researchers tried to:

1. Identify and classify the factors to be taken into account when one considers the reception area accessibility (infrastructure, facilities, services, transport).
2. Consider planning principles and design (urban and architecture technical barriers).
3. Suggest some best practices and practical technical solutions to overcome barriers.

The researchers were able to recognize a wide range of factors, which should be considered under the process of integral and holistic planning imperative. A sustainable, universally planned and accessible reception area concept should be connected with provision tourist information about services and facilities available to persons with disabilities, staff training to prepare staff to deal with problems of disabled customers, infrastructure and facilities answering special needs. Therefore, it was necessary to narrow the survey study and focus the research on the specific needs connected with the following aspects, answering special requirements during travel:

- Barrier-free infrastructure and facilities (as hotels, catering, shops, art galleries, public toilets),
- Accessible local transport,
- Barrier-free tourist attractions (as museums, historic urban layout),
- Accessible information, including information about accessible tourist attractions and services,
- Public safety.

Research Techniques and Instruments

The interviews conducted in Warsaw (2012 – 2014) were undertaken during spring semester, designed as combined qualitative and quantitative research focusing on small – scale micro aspect of project:

- Motives on visiting the heritage site and opinions on accessibility,
- Informal comments about tourist infrastructure and services (see Figure 1 in Appendix).

The research survey was carried out with the use of structured and unstructured interviews, done with tourists, local community members by students of the Faculty of Tourism and Recreation from the Joseph Pilsudski Academy of Physical Education in Warsaw. The participants were informed about the main goals of the research and agreed to be interviewed. In three consecutive years as many as 600 questionnaire forms were completed and later analysed (each year 200 forms were collected). The interviews were conducted on in situ, face to face basis, each year carried out at the same locations (Main Old Market and Castle Square). Structured interviews, conducted with the usage of standardised questionnaires (designed as closed questions) were the source of quantitative data for analysis (see Figure 1 in Appendix). Unstructured interviews (exploratory interviews) were conducted in the form of informal talks and were the source of qualitative data for analysis, helping to understand the perspective of focus group - the tourists and local community members visiting the Old Town in Warsaw. Participant observations, field notes, photo documentations completed the research survey (see Tables 1 – 6 in Appendix).

The authors believed that this approach to applied research and its final result, achieved on the base of the collected material analysis (combined quantitative and qualitative data, primary and secondary data) would provide the possibility of deep and broad insight into the complexities of accessible tourism in the two surveyed historic sites, as recommended by literature (Darcy, 2009; Darcy, 2011; ENAT, 2015; Finn, 2000; Phillimore, 2009; UN WTO, 2013, 2014, 2015).

Tourists Mobility in the Cracow and Warsaw

Accessible tourism in the cultural sites means providing products and services accessible to all people, regardless of physical limitations, sensor disabilities or age. The research agenda for case studies in the Old Town historic districts of Cracow and Warsaw was focused on the four segments of tourist services and facilities:

- Barrier-free world heritage sites (infrastructure and facilities),
- Accessible local transport (including traffic – free zones of the old towns),

- Barrier-free accommodation,
- Catering services and tourist attractions (as for example museums) and accessible tourist information (see Tables 1-6 in Appendix).

In Poland, as in the other European Union countries, accordingly to the building and spatial planning law regulations, all newly erected public buildings and public places must be accessible for persons with hearing, visual or motor impairments (as for example for wheelchair users) (MI, 2002). But so far there are many historic places and historic buildings, including two surveyed cases of the UNESCO heritage sites, which are not entirely barrier – free. For example both in Warsaw and Cracow historic urban layouts specific problems for tourists with disabilities include:

- Inaccessible streets and sidewalks,
- Uneven curbs,
- Inaccessible main entrances to the shops, restaurants, art galleries,
- Only partly accessible public transport,
- Troubles with finding barrier-free hotel or dining place.

Furthermore, almost 50% of the museums located in Warsaw and 20% in Cracow are inaccessible to persons with motor impairments. On the base of the case studies authors found out, that visual information in the outdoor and indoor public space (museums, public transport) is not always accompanied by records in the Braille alphabet or with audio information. In Warsaw and Cracow there is lack of well-adopted hotel rooms, lack of accessible restaurants, bars and public toilets. Not all train and bus terminals are barrier – free and only the airports are fully accessible, which is due to the International Air Transport Association (IATA) separate regulations and requirements (see Tables 1 - 6 in Appendix).

On the other hand, one must admit a visible progress of improvements, as recently access to tourist information is becoming easier, especially in Cracow and Warsaw, due to various institutions and organisations publishing a significant amount of information. Moreover, dedicated to disabled tourists websites, present information in increasingly diverse forms (e.g., audio clips, large print, visualisations and helpful mobile applications), making the presentation more effective. Tourist Information (TI) centres in Cracow provide also CDs with binaural recordings (which make the listeners feel as if they were standing in the described area) and smart-phone applications. One of the priorities of the project was to increase the degree of accessibility of (TI) centres for persons with disabilities and special seats have been introduced to the selected TI units. The seats come with equipment that allows persons with disabilities to browse and obtain materials they are interested in without the need for assistance (in print or as mp3 recordings). In Cracow tourists with visual impairment can take advantage of large-print publications, publications in Braille and models of the most important historical monuments. As a further support information kiosks were introduced, operating in a convenient for tourists system '24/7'.

Despite all these technological aids, local tourist information centres and tourist information points (TI) are still crucial and still much has to be done. For example, in Cracow, in addition to freely available tourist information units, there is the Cultural Information Centre for Persons with Disabilities. The centre staff provides information on interesting cultural events taking place in Cracow and its vicinity and materials about the accessibility of cultural and historical attractions for persons with disabilities. Unfortunately, the centre opening hours are limited and it operates only during office hours from Tuesday to Friday.

Cracow participates in an interesting project that has been carried out as part of tourist infrastructure development in the Malopolska Province. The project called 'Building the Malopolskie Province Tourist Information System' was developed under the Malopolska Province Regional Operational EU Programme for 2007–2013. As a result of the project, a comprehensive website and materials in seven languages have been developed, and 34 new TI centres, 90 information boards, 150 tourist traffic signs were established. Another noteworthy project that contributes to create Cracow urban space and tourist attractions accessible to persons with disabilities is the Royal Road for the Disabled Tourist project, carried out by the Cracow City Council under the Malopolska Province Regional Operational Programme. The main aim of the enterprise is to create an attractive tourist route for persons with visual and motor impairment. The project encompasses 12 models;

- In form of bronze cast examples of historical monuments with descriptions in Latin alphabet (in Polish and English) and in Braille (also in Polish and English).

The models were designed to allow persons in wheelchairs to approach them.

Thanks to the effective cooperation between the Cracow City Council and the NGOs (as the Polish Association of the Blind, the Polish Special School and Educational Centre for Blind and Visually Impaired Children in Cracow), tourist maps have been developed that make travelling and discovering the city tourist attractions easier for blind and visually impaired persons. The maps come with books that explain abbreviations and describe historical monuments in large print and in Braille alphabet.

In Cracow an especially dedicated GPS system has recently been introduced and users can listen to information about a given monument. Information about available aids provided for persons with disabilities could be obtained via a mobile phone.

As a final remark, it is worth to underline that the aforementioned initiatives on the part of Cracow City Council authorities have been acknowledged by international institutions. Cracow was placed highly in the finals of the Access City Award 2012 Competition. The competition is organised by the European Commission and aims to reward cities that significantly contribute to the accessibility of the urban space for persons with disabilities.

As it was mentioned, so far Warsaw is below Cracow achievements in providing equal access to pre-visit and on-site information about tourist attractions and services. But accordingly to the document 'Historical Centre of Warsaw - Management Plan Strategy' (City Council, 2012), a series of improvements are being considered. As it says, the City Council will focus on modifying not only the public transport, dedicated parking system, improvement of accessibility to tourist attractions and services (museums, restaurants and banks), adapting pedestrian routes to persons in wheelchairs, but also will consider preparing integrated information systems for tourists with disabilities. For example, the future changes will include marking routes and buildings in Braille and in other manners in order to help persons with motor and visual impairment, 3D maps, models and information provided via audio description and ensuring good illumination of routes and buildings.

SOLUTIONS AND RECOMMENDATIONS

Tourism development in Warsaw and Cracow could be a positive force for generating revenue, for cultural education and preservation of these historic sites, however, under the condition of re-examining the current trends of planning (which is so far inadequate, especially in Warsaw). The research material analysis has shown the poor results of city councils management, weak signs in current tourism planning process and, as a consequence, inadequate effects of reception area accessibility. The situation in Polish cities is varied and so far clearly Cracow has much better spatial policy than others. On the base of the research material analysis, the authors would like to suggest some technical solutions to enhance the accessibility to the historical sites in Warsaw, which should be undertaken on the base of holistic, integral planning approach (see Figures 1-8 and Tables 1-6 in Appendix).

The disabled represent a significant proportion of Polish population (10% of total population) and might have considerable importance to the tourist industry. Moreover, we will soon face the challenge of adapting the public space to the needs, demands and requirements of an ageing population, elderly residents that live longer, are accustomed to active, social and cultural lives and wish to travel. However, as the research outcomes showed, these needs are not always met. As it is against the United Nations declarations and resolutions, for example "Accessible Tourism for All" (2005) or preamble to the UN "Convention on the Rights of Persons with Disabilities" (2007), saying that "disability issues are an integral part of sustainable development strategies", it had to be improved and carefully considered in the local spatial plans as soon as possible. The reception areas management policy of tomorrow will have to ensure that all tourist groups are well integrated and marginalized groups are supported (elderly, disabled, families with small children).

Accordingly to the integral and holistic planning theory, providing the accessible UNESCO World Heritage Sites, involve a collaborative process between all stakeholders - government, local governments, international agencies, tour operators and NGOs. The active participation of all groups of interests is necessary:

- Local community and persons with disabilities,
- Government and various NGOs organizations.

Modern society and decision-making stakeholders (as local community, city council, architects, urban planners, operators in tourism sector) are aware of the need of integration of impaired persons. However, removing barriers for disabled tourists in the historical places and buildings is often difficult not only due to the high costs of such projects, but also due to strict requirements, which are part of building conservation plans, often in consequence preventing extensive intervention in historical buildings (Dredge, 1996; Harrison, 2005; UNESCO, 2011). Moreover, local communities might be not so keen to welcome deep changes in the public space and would rather try to preserve the status quo image and atmosphere of the old town streets and squares.

Therefore, the future projects for Cracow and Warsaw need a new approach and better understanding of the nature of necessary integration of heritage conservation, impaired tourists needs, local community demands and investors expectations to develop planning concepts. The city councils not only should identify stakeholders and induce them to collaborate, but also be responsive to their voices and be trusted. The strategic problem of successful implementation of plan is not only a deep knowledge about the site and interrelationships between stakeholders, but also a skill to negotiate and persuade.

Naturally, there are many issues which should be discussed. For example, technically there are two possible approaches to improve the accessibility of tourist destination. First involves the specially designed and provided facilities and services that specifically target disabled. The second approach is based on the idea of modifications applied to general facilities and services so that they can accommodate people with special requirements alongside other tourists (accordingly to the holistic and integrated tourism philosophy). It encompassed the wide range of issues:

- Accessible transport,
- Pre-visit and in situ information,
- comfortable and safe pavements,
- Safe pedestrian crossings,
- Adequate public toilets,
- Accessible buildings (hotels, museums and terminals),
- Public safety in all open spaces and buildings.

And that is why, in authors' opinion, this research agenda in future should be focused on the several segments, to identify options for improvement accessible tourist services and shape policy for sustainable, barrier - free development of tourist destinations space – created as harmonious and fully inclusive.

FUTURE RESEARCH DIRECTIONS

The researchers are sure that the further research project should be dedicated to broader goal, supporting accessible tourism not only in the UNESCO World Heritage Sites, but also to be focused on universal design questions of host destination. It may help to alleviate social segregation process and combat spatial exclusion of marginalised groups. All tourist reception areas must be designed with regard 'to all' respect and be able to secure accessible facilities and services. For example, with regard to the innovative tourist information to answer what issues are important to create a suitable "for all" host space, accordingly to the holistic and integrated tourism idea, the following questions need to be answered:

- What information is important for disabled tourists (including pre-visit information)?
- What are the constrains to use standard information services (pre-visit, on site)?
- What technologies and innovations could help to overcame the problem?
- What solutions and design character could be the most suitable to be provided at the smart destination area?

On the base of these feedbacks, the concrete technical design of innovative tourist information facilities to be located on the site and accessible 'for all', could be developed.

It is to be underlined that the future research theme is not a luxury and managerial approach – it is a growing need, driven also by senior tourism development, due to the increasing life expectancy in the future. It is also due to the global increase in chronic health conditions linked to disability that the worldwide growth rate of persons with disabilities or with special needs is higher than population growth (UN WHO, 2007, 2011).

As the authors hope, the outcomes might provide a solid base for planning smart tourist destinations theory, together with recommendations focused on universal design philosophy, as the idea of accessibility to all tourist infrastructure, facilities and services which should be a central part of responsible tourist policy at all kinds of tourist destinations.

CONCLUSION

On the base of research material authors concluded that in Poland further sought is necessary to outline a better base for legal regulations, town planning approach, built on the 'consensus platform', enabling to search compromising technical solutions, ready to meet a challenge of mobile tourism, understanding the meaning of social inclusion and the crucial role of barrier – free destination space (infrastructure and services). It requires a new governance mode as both horizontal and vertical coordination is necessary to reinforce cooperation and networking within the relevant stakeholders.

Furthermore, to develop in Poland a barrier – free space quality with high architectural quality, it is important that planning a smart and accessible tourist destination should not be undertaken as an isolated activity. All stakeholders need to understand the value of collaboration and help to establish the responsible planning process in aim to create smart host space:

- Decent,
- Suitable,
- Secure,
- Efficient, and
- Affordable for low-income visitors.

It is to be underlined that the bottom-up participatory approach, involving tourist with disabilities (together with elder tourists and child caregivers) in analysing and expressing their situation should be emphasised. Certainly they need to be empowered to contribute and participate in the decision – making process as consultants. These people make observations based on everyday experiences and could provide information not reported by even the most meticulous research studies focused on social participation and wellbeing.

To shape effectively the barrier – free UNESCO World Heritage Sites, tourism planning policy at the Old Towns in Warsaw and in Cracow (and other heritage sites) must be coordinated within the urban master plans and directions of spatial development in the whole city structure. The role of both the City Council and local governments (gminas) is undeniable. The authors are sure that efficient and creative planning of accessible urban space should be followed by smart and innovative management decisions – responding to the constant and inevitable changes. It includes the wide range of elements:

- Tourism destination management,
- Tourism information,
- Urban environment,
- Buildings and their facilities,
- Public open space,
- Transport,
- Accommodation and catering services,
- Cultural activities, and
- Events.

The researchers believe that a holistic approach is necessary, aiming to accommodate the needs of people with different disabilities, considered as integrated part of policies and planning sustainable tourism.

One of the key factors is providing clear and reliable information for disabled tourists, as well pre-visit and on-site. Disability has many levels and individual limitations, the reception area might be both inaccessible and accessible for tourists with different impairments. Good information on current accessibility allows disabled tourists to judge individually whether the site and its facilities are accessible to them. Therefore, reliable pre-visit information may increase the market potential for the tourism sector.

The complexity of approaches, law regulations, resource demands, different groups of interests, local community aspirations, social interrelationships, disabled tourists requirements, conservatory recommendations - all make integrated and holistic tourism planning a project which could have a base in form of set of principles. On the base of collected and analysed research material (see Table 1 – 6 in Appendix), the authors have tried to formulate some recommendations of general and universal character (See Figures 1 – 3 in Appendix). However, it is necessary to underline that the implementations and applications should be designed and tailored individually, depending on uniqueness and individual characteristic of each heritage site, stage of tourism activity in the given UNESCO World Heritage Site.

Although this issue is beyond the chapter main theme, authors considered it as important and feel this last comment is worth to be mentioned. Accessible destination depends on many determinants: material and technical along with social, what includes affordability. The cost is viewed as a significant factor affecting disabled tourist decision to travel and was frequently reported in the case study interviews. For example, in Poland most of the well adapted for disabled hotels are 4 or 5 star offers and hostels dedicated to low-income tourists are inaccessible. As it seems, without government support (subsidizing policy, discounted travel), tourist mobility might be very difficult to improve.

ACKNOWLEDGMENT

The presented material is the result of research conducted on the base of the grant offered by the Ministry of Higher Education and Sciences (ds-144 AWF), carried out by the authors at the Joseph Pilsudski Academy of Physical Education in Warsaw, Poland (2012 – 2015).

REFERENCES

Buhalis, D., & Darcy, S. (2010). *Accessible Tourism. Concept and Issues*. Bristol: Channel View Publications.

Buhalis, D., Darcy, S., & Ambrose, I. (2012). *Best Practice in Accessible Tourism: Inclusion, Disability, Ageing*. Bristol: Channel View Publications.

Dallen, J. T., & Stephen, W. B. (2003). *Heritage Tourism*. Harlow, UK: Prentice Hall.

Darcy, S. (2010). Inherent complexity: Disability, accessible tourism and accommodation information preferences. *Tourism Management Journal, Elsevier.*, *31*(6), 816–826. doi:10.1016/j.tourman.2009.08.010

Darcy, S., & Buhalis, D. (2011). *Accessible tourism: concepts and issues. Aspects of Tourism*. Bristol: Channel View Publications Ltd.

Darcy, S., & Dickson, T. (2009). A Whole – of – Life Approach to Tourism: The Case of Accessible Tourism Experiences. *Journal of Hospitality and Tourism Management, 16*(1), 32–44. doi:10.1375/jhtm.16.1.32

Dredge, D., & Moore, S. (1992). A methodology for the integration of tourism in town planning. *Journal of Tourism Studies, 3*(3-1), 8-27.

Drost, A. (1996). Developing sustainable tourism for world heritage sites. *Annals of Tourism Research, 23*(2), 479–484. doi:10.1016/0160-7383(96)83345-7

European Commission. Regional Policy. (2011). *Cities of tomorrow. Challenges, visions, ways forward.* Retrieved February 29, 2016 from: http://www.europa.eu

European Commission. Enterprise and Industry. (2015). *Accessible tourism.* Retrieved December 15, 2015 from: http://www.europa.eu

European Network for Accessible Tourism (ENAT). (n.d.). Retrieved November 10, 2015 from: http://www.accessibletourism.org

Finn, M., Elliott-White, M., & Walton, M. (2000). *Tourism and Leisure Research Methods. Data Collection, Analysis and Interpretation.* Harlow, UK: Pearson – Longman.

Flavigny, C., & Pascal, M. (1995). Tourism among the disabled: Dissatisfied clients. Espaces (Paris), (134), 36-42.

Hall, M. C. (2000). *Tourism Planning*. Harlow, UK: Prentice Hall.

Harrison, D., & Hitchcock, M. (2005). *The politics of world heritage. Negotiating tourism and conservation*. Paris: UNESCO World Heritage Centre.

Inskeep, E. (1991). *Tourism planning: an integrated and sustainable development approach.* New York: Van Nostrand Reinhold.

Korstanje, M. E. (2011). Influence of history in the encounter of guests and hosts. *Anatolia, 22*(2), 282–285. doi:10.1080/13032917.2011.597991

Lovelock, B., & Lovelock, K. M. (2013). *The Ethics of Tourism. Critical and Applied Perspectives.* New York: Routledge.

Ministry of Infrastructure. (2002). *Ministry of Infrastructure Recommendations on Building Design and Practices.* Warsaw, Poland: Ministry of Infrastructure.

Murrey, M., & Sproats, J. (1990). The disabled traveller: Tourism and disability in Australia. *Journal of Tourism Studies, 19*(1), 9–14.

Newman, P., & Thornley, A. (2011). *Planning World Cities. Globalisation and Urban Politics.* New York: Palgrave - Macmillan.

Packer, T.L., Small, J., & Darcy, S. (2008). *Tourist Experiences of Individuals with Vision Impairment.* Gold Coast Australia, Sustainable Tourism CRC Tech Report.

Page, S. J., & Hall, M. C. (2003). *Managing Urban Tourism.* Harlow, UK: Prentice Hall.

Pawlikowska-Piechotka, A. (2013). *Planowanie przestrzeni turystycznej* [Planning Tourism]. Gdynia: Novae Res.

Phillimore, J., & Goodson, L. (2009). *Qualitative Research in Tourism*. London: Routledge.

Preiser, W. F. E., & Ostroff, E. (2001). *Universal Design Handbook*. New York: McGraw – Hill.

Reichhart, F. (2009). Means of access to tourism activities for disabled people. *Teoros Journal. Revue de Recherche en Tourisme*, (2), 73-81.

Rifai, T. (2013, August). *Accessible Tourism for All.* UN Secretary General speech, UN General Assembly Resolution of August 13, Zimbabwe (A/RES/637(XX).

Shaw, G., & Coles, T. (2004, June). Disability, holiday making and the tourism industry in the UK: A preliminary survey. *Tourism Management*, *25*(3), 397–403. doi:10.1016/S0261-5177(03)00139-0

Shelton, E., & Tucker, H. (2005). Tourism and disability: Issues beyond access. *Tourism Review International*, *8*(3), 211–219. doi:10.3727/154427205774791528

Sigala, M., & Leslie, D. (2006). *International Cultural Tourism. Management, Implication and Cases*. Oxford: Elsevier.

Small, J., Darcy, S., & Packer, T. (2012, August). The embodied tourist experiences of people with vision impairment: Management implications beyond the visual gaze. *Tourism Management*, *33*(4), 941–950. doi:10.1016/j.tourman.2011.09.015

Smith, K. M. (2003). *Issues in Cultural Tourism Studies*. London: Routledge. doi:10.4324/9780203402825

Tantawy, A., Kim, W. G., & Pyo, S. S. (2004). Evaluation of hotels to accommodate disabled visitors. *Journal of Quality Assurance in Hospitality & Tourism*, *5*(1), 91–101. doi:10.1300/J162v05n01_07

Timothy, D. J., & Nyaupane, G. P. (2009). *Cultural heritage and tourism: a regional perspective*. New York: Routledge.

UN. (2006). *The Convention on the Rights of Persons with Disabilities (CRPD)*. New York: UN.

UN ENABLE. (2013). *Promoting Accessible Tourism for All. Development and human rights for all.* New York: UN ENABLE.

UN WHO. (2007). *UN WHO Global Age-friendly Cities: A Guide*. Retrieved February 29, 2016 from: http://www.who.int

UN WHO. (2011). *UN WHO World report on disability*. New York: Routledge.

UN WTO. (2013). *UN WTO Recommendations on Accessible Tourism for All*. New York: UN WTO.

UN WTO. (2014). *San Marino Declaration on Accessible Tourism*. San Marino: UN WTO.

UN WTO. (2015). *Manual on Accessible Tourism for All: Public – Private Partnership and Good Practice*. New York: UN WTO.

UNESCO ICOMOS. (2011). *The Valetta Principles for the Safeguarding and Management of Historic Cities, Towns and Urban Areas*. Paris: UNESCO.

Vael, A. J. (2002). *Leisure and Tourism. Policy and Planning*. London: Routledge; doi:10.1079/9780851995465.0000

KEY TERMS AND DEFINITIONS

Accessible Tourist Destination: The possibility to travel and to take part in different activities connected with tourist experience - tourist reception area and public space without functional barriers, products and services accessible to all tourist groups, regardless the limitations due to the age, impairment form or disability level.

Disabled Person: physical, cognitive, mental, sensory, emotional or developmental impairments.

Mobile Tourism: Tourist could be mobile and visit any destination, unless the infrastructure, facilities, information and services are accessible to all people, regardless limitations, different form of disabilities, family situation or age.

Tourist Destination: A place for tourists to visit and stay, could be a country, state, region or city – usually due to its cultural or natural values.

Tourist Information (TI): Tourist information centers or points have several functions, they provide pre-visit information about destination (for example a 'book a bed ahead' scheme) as well as information to be obtained in situ about reception area, facilities and services, tourist attractions.

Tourist Infrastructure: Facilities and services necessary to develop tourist reception area (accommodation, catering, transport, information, museums and tourist attractions).

UNESCO World Heritage Sites: It might be a building, city, urban layout, landscape or architectural monument, that is listed by the United Nations Educational, Scientific and Cultural Organization (UNESCO).

APPENDIX

Figure 1. Visitors interest in the UNESCO World Heritage Site (the Old Town, Warsaw): motives for visiting and opinion on accessibility: 1- came to learn about and understand the history of Warsaw, 2 – various motives, with some interest in the town history, 3 – other reasons, as leisure and passing time; 4 – site is not fully accessible for disabled, 5 – site is accessible for impaired

Source: The presented material is the result of research conducted on the base of the grant issued by the Ministry of Higher Education and Sciences (ds-144 AWF), carried out by the authors (2012 – 2015).

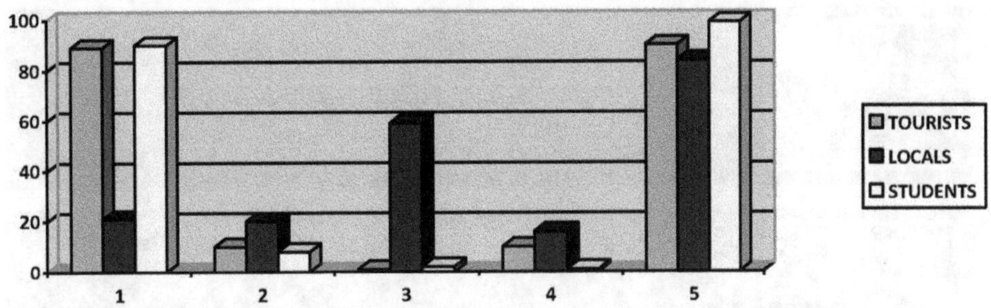

Figure 2. Holistic planning approach to create accessibility to the UNESCO World Heritage Site

Source: The presented material is the result of research conducted on the base of the grant issued by the Ministry of Higher Education and Sciences (ds-144 AWF), carried out by the authors (2012 – 2015).

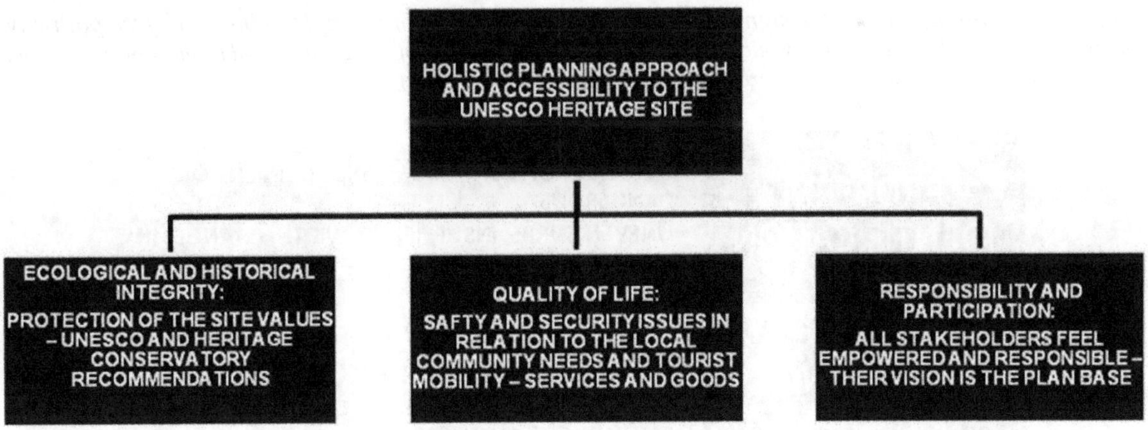

Figure 3. General principles and universal conditions of barrier-free access to UNESCO World Heritage Sites to be considered in the holistic planning approach
Source: The presented material is the result of research conducted on the base of the grant issued by the Ministry of Higher Education and Sciences (ds-144 AWF), carried out by the authors (2012 – 2015).

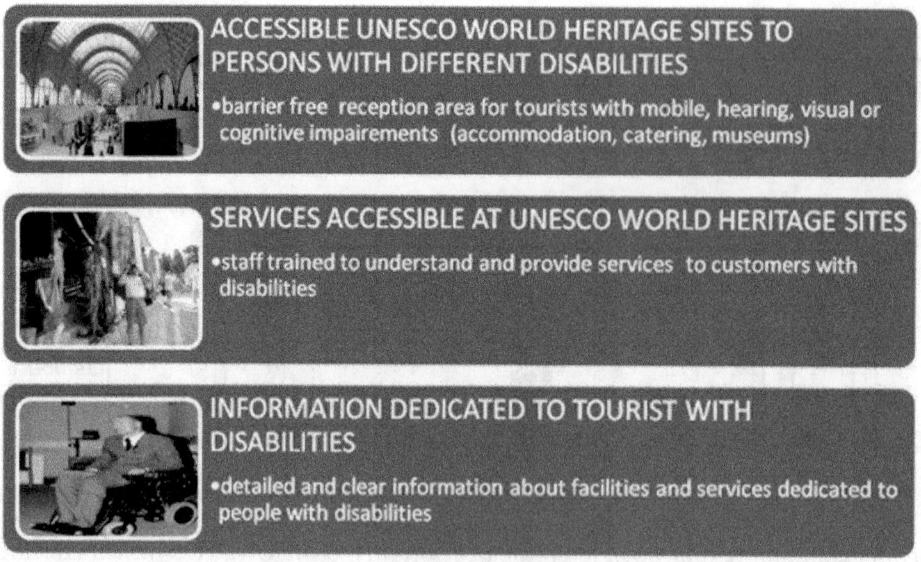

Figure 4. More detailed considerations of barrier-free access to heritage site: a) problem identification, b) planning principles and c) design consideration (here there is an example of barrier - free pathways)
Source: The presented material is the result of research conducted on the base of two grants issued by the Ministry of Higher Education and Sciences (ds-144 AWF), carried out by the authors (2012 – 2015).

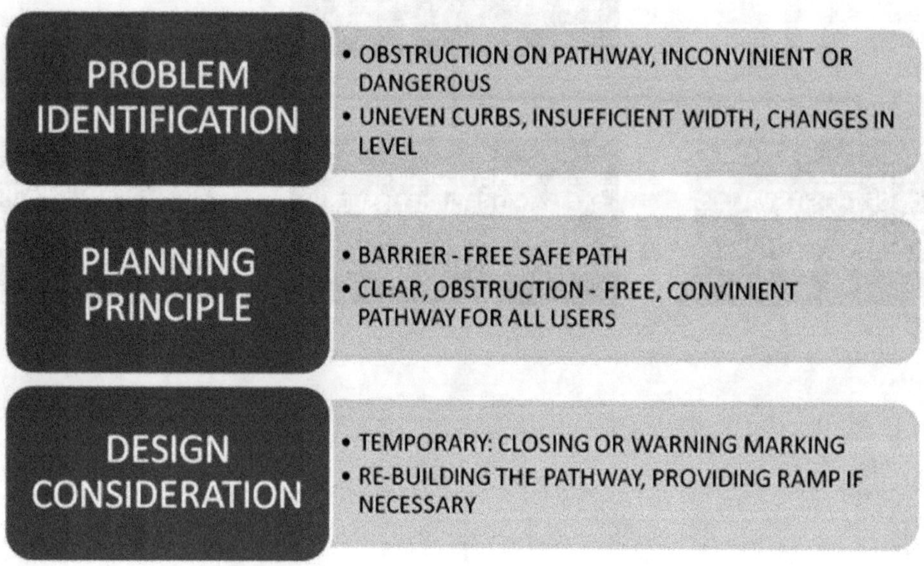

Figure 5. Accessible tourist reception area - consideration at heritage site: Practical solutions for barrier – free access improvement (example: pathways), the minimum width of an unobstructed sidewalk should be of 150 cm for two- way wheelchair traffic passage (recommended width is 180 cm), surface must not be interrupted by steps or sudden changes, without loose gravel, cobbles, smooth but not slippery.
Source: The presented material is the result of research conducted on the base of the grant issued by the Ministry of Higher Education and Sciences (ds-114 AWF and ds-144 AWF), carried out by the authors (2012 – 2015).

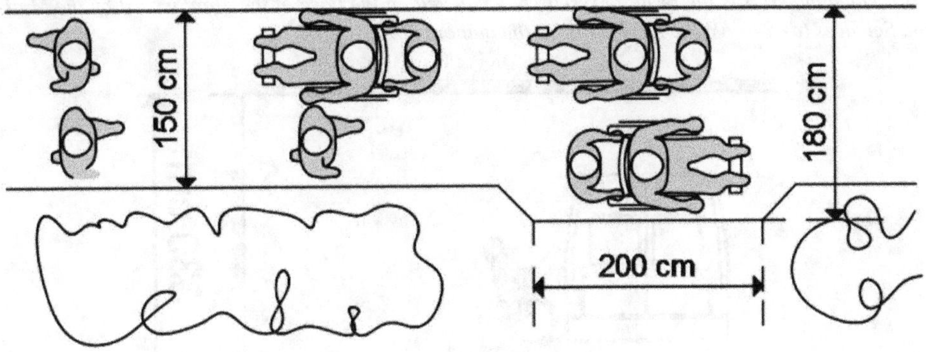

Figure 6. Accessible tourist reception area - consideration at heritage site: problem is identified as inaccessible building entrance due to the difference between indoor and outdoor level. Practical solutions for barrier – free access improvement (example: stairways vs. ramps), ramps should be provided wherever stairs obstruct the free passage of pedestrians, mainly wheelchair users and people with mobility problems (seniors, individuals with crutches, temporary disabled, toddlers, child carers with prams, people with luggage): 1- ramp, 2 - landing, 3 – mid landing, 4 – railings, 5 – curbs, 6 - stairs. Ramp rails should have smooth surface, extending min 30 cm beyond beginning and end of ramp ridge, handrails required 95 cm high, walking surface should be smooth but not slippery, one-way ramp should be minimum 90 cm wide, 2-way ramp minimum 180 cm wide. Recommended maximum gradient is 8% for outdoor ramp. No changes of gradient along the ramp are allowed. Plant should be located enough far that shadows do not prevent sun light from melting snow or ice on the ramp surface and will provide natural light. The above standards may vary accordingly to the local building law regulations.
Source: The presented material is the result of research conducted on the base of the grant issued by the Ministry of Higher Education and Sciences (ds-114 AWF and ds-144 AWF), carried out by the authors (2012 – 2015).

Figure 7. Accessible tourist reception area - consideration at heritage site: problem is identified as parking space requirements. Practical solutions for barrier – free access improvement: parking space should be adequate to enable a wheelchair user safely get in and get out into the van. Wheelchair users need wider parking space (minimum 360 cm x 500 cm, recommended 500 cm x 500 cm). Dedicated for disabled parking lot should be nearest the building (for example museum) and clearly marked. Sheltered parking for wheelchair users is very desirable.
Source: The presented material is the result of research conducted on the base of the grant issued by the Ministry of Higher Education and Sciences (ds-144 AWF), carried out by the authors (2012 – 2015).

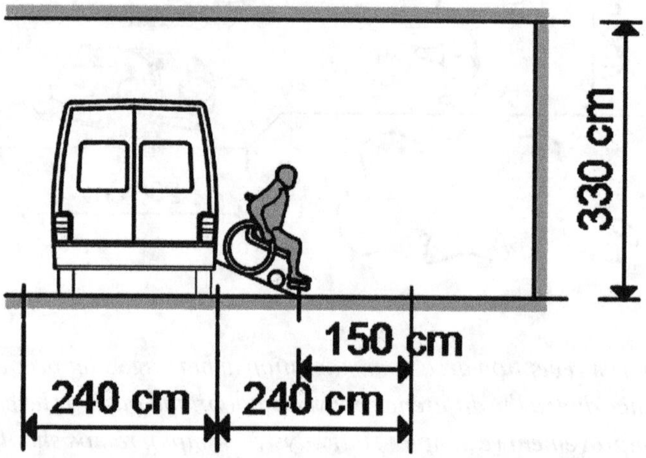

Figure 8. Accessible tourist reception area - consideration at heritage site: problem is identified as parking space special requirements. Practical solutions for barrier – free access improvement: parking space should be enough spacious to enable a wheelchair user safely get in and get out the car. Recommended size of parking lot is 500 cm x 500 cm, minimum is 360 cm x 500 cm. Sheltered parking spot dedicated for disabled is very desirable
Source: The presented material is the result of research conducted on the base of the grant issued by the Ministry of Higher Education and Sciences (ds-144 AWF), carried out by the authors (2012 – 2015).

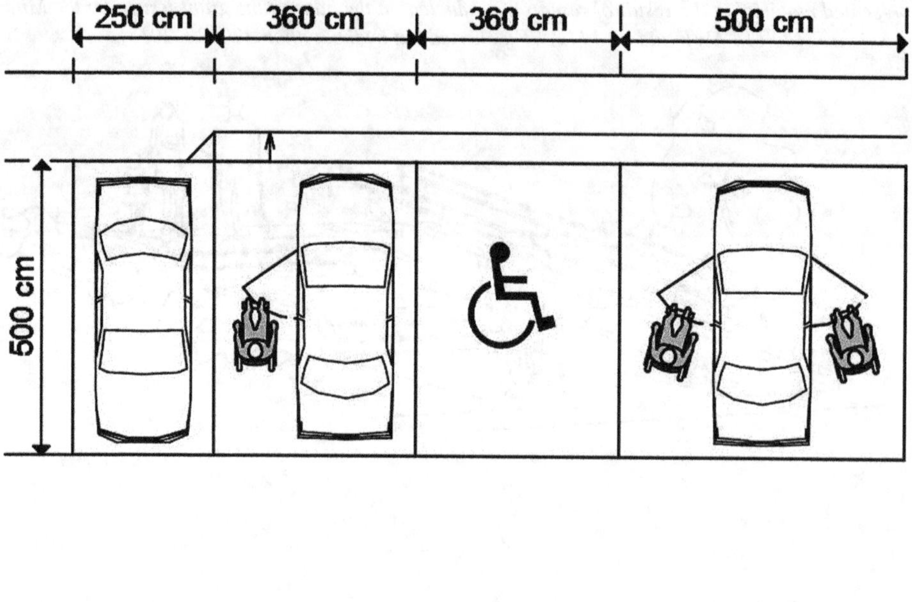

Figure 9. Accessible tourist reception area - consideration at the UNESCO heritage site: problem is identified as public transport stop special requirements. Practical solutions for barrier – free access improvement: 1 – bus stop platform, 2 – sidewalk with distinctive colour texture to mark the walking direction, 3 – dedicated lowered curbs with distinctive colour or / and texture, 4 – colour design to mark the obstruction and provide the safe cordon from the busy street, 5 – bus stop shelter, 6- sitting, 7 – space for wheelchair user next to the sitting area, 8 – platform surface should be clear, smooth, but not slippery, without sudden changes of level, distinctive texture might be very helpful for blind or sight impaired.
Source: The presented material is the result of research conducted on the base of the grant issued by the Ministry of Higher Education and Sciences (ds-114 AWF and ds-144 AWF), carried out by the authors (2012 – 2015).

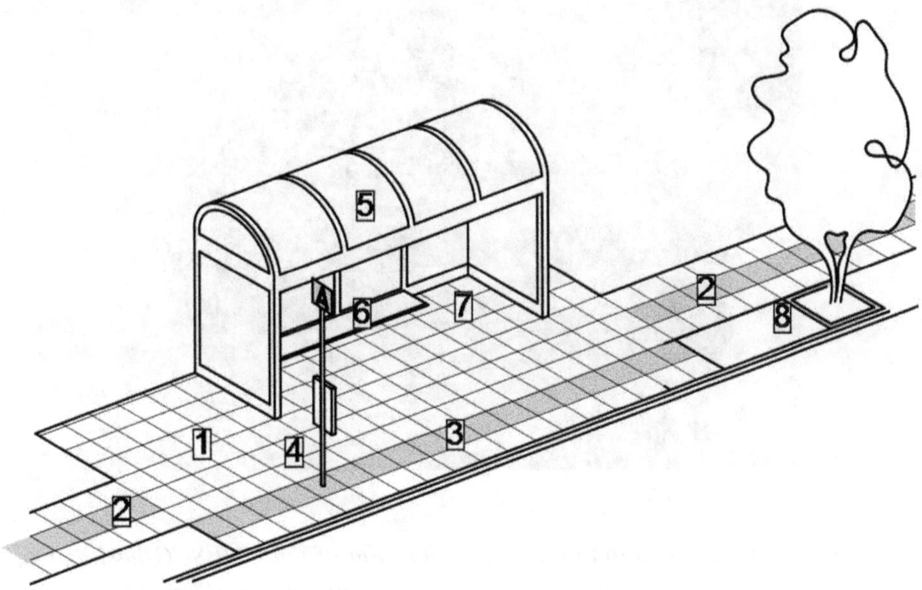

Figure 10. Accessible tourist reception area - consideration at the UNESCO heritage site: tourist information imperative. Solution: clear and graphically consistent visual information system. a – direction to the nearest accessible public telephone, b – direction to the nearest accessible public toilet, c – direction to the nearest accessible parking, d – direction to the accessible lift
Source: The presented material is the result of research conducted on the base of the grant issued by the Ministry of Higher Education and Sciences (ds-144 AWF), carried out by the authors (2012 – 2015).

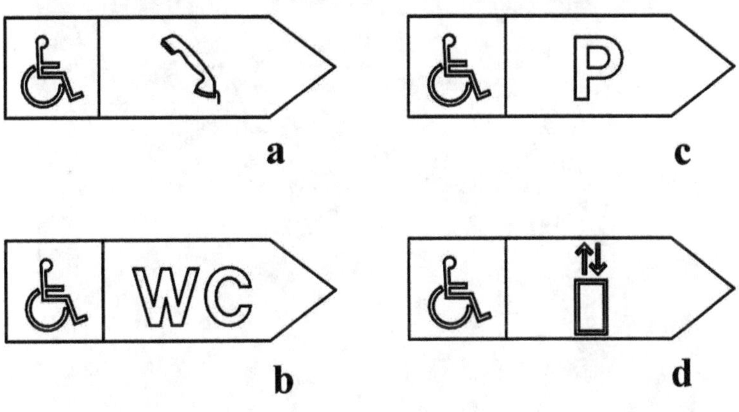

Figure 11. Accessible tourist reception area - consideration at the UNESCO heritage site (Old Town in Warsaw, Poland): tourist information imperative. Solution: clear and graphically consistent visual information system in form of the bronze model, showing historic urban layout, description in Polish and English (unfortunately there is no description in Braille alphabet). Project carried out by the UNESCO and Warsaw City Council under the Regional Operational Programme

Source: The presented material is the result of research conducted on the base of the grant issued by the Ministry of Higher Education and Sciences (ds-144 AWF), carried out by the authors (2012 – 2015) .

Figure 12. Accessible tourist reception area - consideration at the UNESCO heritage site (Old Town in Warsaw, Poland): tourist information imperative. Solution: clear and graphically consistent visual information system in form of the bronze model, showing historic urban layout, description in Polish and English (unfortunately there is no description in Braille alphabet). Project carried out by the UNESCO and Warsaw City Council under the Regional Operational Programme

Source: The presented material is the result of research conducted on the base of the grant issued by the Ministry of Higher Education and Sciences (ds-144 AWF), carried out by the authors (2012 – 2015).

Figure 13. Accessible tourist reception area - consideration at the UNESCO heritage site (Old Town in Warsaw, Poland): sidewalks. Solution: even and lowered curbs, accessible main entrances to the shops, restaurants, art galleries. Project carried out by the UNESCO and Warsaw City Council under the Regional Operational Programme

Source: The presented material is the result of research conducted on the base of the grant issued by the Ministry of Higher Education and Sciences (ds-144 AWF), carried out by the authors (2012 – 2015).

Figure 14. Accessible tourist reception area - consideration at the UNESCO heritage site (Old Town in Warsaw, Poland): sidewalks. Solution: even and lowered curbs, accessible main entrances to the shops, restaurants, art galleries. Project carried out by the UNESCO and Warsaw City Council under the Regional Operational Programme. Notice: the flower pots located by the front door of restaurant, are not only aesthetically questionable but also creating a danger for disabled (too narrow sidewalk and unexpected obstacle)

Source: The presented material is the result of research conducted on the base of the grant issued by the Ministry of Higher Education and Sciences (ds-144 AWF), carried out by the authors (2012 – 2015).

Table 1. Solutions and recommendations on the base of the comparative study of the UNESCO World Heritage Sites and basic tourist services accessibility (case studies: the Old Town in Warsaw vs. the Old Town in Warsaw). Imperative: accommodation services without barriers. Requirements: access to the main entrance, elevators, sanitary facilities, hotel room facilities, horizontal communications, signing_ and hotel services accessible (as catering, shops)

Accommodation Services	The Old Town in Cracow	The Old Town in Warsaw
Hotels without barriers	There are 50 hotels in the vicinity of the Old Town: Accessible: 8 hotels Partially accessible: 1 hotel Inaccessible: 41 hotels	There are 11 hotels in the vicinity of the Old Town: Accessible: 2 hotels Partially accessible: 1 hotel Inaccessible: 8 hotels
Conclusion	Accessible for disabled tourists is about 15% of total accommodation capacity in the surveyed district area. It shows the proportion of disabled tourists in population (UN WTO 2013). All accessible hotels are of high standard (4 and 5 star). There are no affordable accommodation offers for low-income tourists with mobility constrains.	Accessible for disabled tourist is about 18% of total accommodation capacity in the surveyed district area. It shows the proportion of disabled tourists in population (UN WTO 2013) The accessible hotels are: • The newly erected 3 star Hotel Ibis Old Town Warsaw • The newly erected 5- star Hotel La Regina Partially accessible is 2- star Dom Literatury Guesthouse
Solutions and recommendations	More hotels without barriers of lower standard, 2- and 3- star (including youth hostels) should be founded to answer the needs of ageing society	More hotels without barriers of different standard (including youth hostels) should be founded to answer the needs of ageing society

Source: The presented material is the result of research conducted on the base of the grant issued by the Ministry of Higher Education and Sciences (ds-144 AWF), carried out by the authors (2012 – 2014).

Table 2. Solutions and recommendations on the base of the comparative study of the UNESCO World Heritage Sites and basic tourist services accessibility (case studies: the Old Town in Warsaw vs. the Old Town in Warsaw). Imperative: museums and museums services without barriers. Requirements: elevators, access to main entrance, horizontal communication accessible, accessible sanitary units, information, signing, audio and Braille alphabet information, accessible museum services (as catering, shop, cinema)

Museums	The Old Town in Cracow	The Old Town in Warsaw
Museum and museum services without barriers	There are 12 museums in the vicinity of the Old Town: Accessible: 7 museums Partially accessible: 4 museums, Inaccessible: 1 museum	There are 17 museums in the vicinity of the Old Town: Accessible: 7 museums; Partially accessible: 2 museums Inaccessible: 8 museums.
Conclusion	Accessible for disabled tourists is about 85% of total number museum institutions, which is highly above average town in Poland	Accessible for disabled tourists is 50% of total number of museum institutions, which is better than in an average Polish town.
Solutions and recommendations	More museums should be adapted to answer the impaired tourists needs	More museums should be adapted to answer the impaired tourists needs

Source: The presented material is the result of research conducted on the base of the grant issued by the Ministry of Higher Education and Sciences (ds-144 AWF), carried out by the authors (2012 – 2014).

Table 3. Solutions and recommendations on the base of the comparative study of the UNESCO World Heritage Sites and basic tourist services accessibility (case studies: the Old Town in Warsaw vs. the Old Town in Warsaw). Imperative: innovative tourist information services without barriers. Requirements: the information in public space, signing, audio information, information in Braille alphabet, information centres accessible for people with disabilities, accessible standards in reservation system (accommodation, transport)_ and promotional materials dedicated to tourist with different forms of impairments

Tourist Information	The Old Town in Cracow	The Old Town in Warsaw
Innovative tourist information services	There are offered: websites, guidebook for tourists with disabilities, tourist information centre adapted to persons with disabilities; publications in Braille, CDs with 'voice postcards', binaural recordings that give the illusion of being inside the described space, information and catalogues in large print, smartphone applications with tourist information, info-kiosks and bronze cast models of historical buildings with Braille alphabet description	There are offered: websites, a guidebook for tourists with disabilities, descriptions on walls of historical buildings and streets, a tourist information centre, bronze cast model of urban historic layout (without Braille alphabet description)
Conclusion	Rich offer, responding fully to the needs of disabled tourists	Poor and insufficient offer, not responding to the needs of disabled tourists
Solutions and recommendations	More tourist information centres to be opened 7/24	Richer offer is necessary as for example: websites and guidebooks in large prints, bronze model casts of monuments of architecture and architecture details (as capitals), information in Braille alphabet, binaural recordings, smartphone applications, info-kiosks to be opened 7/24, tourist information centres with 'smart seats'

Source: The presented material is the result of research conducted on the base of the grant issued by the Ministry of Higher Education and Sciences (ds-144 AWF), carried out by the authors (2012 – 2014).

Table 4. Solutions and recommendations on the base of the comparative study of the UNESCO World Heritage Sites and basic tourist services accessibility (case studies: the Old Town in Warsaw vs. the Old Town in Warsaw). Imperative: local transport services without barriers. Requirements: accessible terminals, accessible for wheel – chair users' tramway and bus stops, metro stations, special services and transport offers at the areas of traffic – free zones (pedestrian only)

Transport	The Old Town in Cracow	The Old Town in Warsaw
Local transport without barriers	There are provided: city tramway service (1 line), bus service (2 lines), night bus and tramway service (2 lines), taxis dedicated to disabled (with discounts), renting a motorised scooters; renting wheelchairs for travelling over large distances	There are provided: city metro service (2 lines), city tramway service (4 lines), bus service (15 lines), night bus service (2 lines), taxis dedicated to disabled (with discounts), horse carriages, bicycle rickshaws, 'tourist train' 'tourist bus' and 'tourist tramway' ('hop on-hop off' service)
Conclusion	Rather good offer, responding to the disabled tourists and local community needs	Good offer, responding to the disabled tourists and local community needs
Solutions and recommendations	More tramway and bus stops to be located near tourist attractions, improvement of service lines operation might be helpful, season alternative transport: rickshaws, horse carriages, tourist train, tourist tramway ('hop on – hop off' service)	Provision of renting a motorised scooters service might be helpful for disabled and elder tourists to travel over a large distance along traffic free zone

Source: The presented material is the result of research conducted on the base of the grant issued by the Ministry of Higher Education and Sciences (ds-114 AWF and ds-144 AWF), carried out by the authors (2012 – 2014).

Table 5. Solutions and recommendations on the base of the comparative study of the UNESCO World Heritage Sites and basic tourist services accessibility (case studies: the Old Town in Warsaw vs. the Old Town in Warsaw). Imperative: parking services without barriers. Requirement: dedicated for disabled parking space, location close to the sites of tourist interest

Parking Services	The Old Town in Cracow	The Old Town in Warsaw
Parking services dedicated to disabled	There are parking spots in the vicinity of the Old Town dedicated to tourists with disabilities (18 spots in 8 car parks located in the Old Town), good information	There are parking spots in the vicinity of the Old Town dedicated to tourists with disabilities (12 spots in 6 car parks located in the Old Town)
Conclusion	Rather satisfactory offer, responding to the needs of disabled tourists and local community members	Rather satisfactory offer, responding to the needs of disabled tourists and local community members, insufficient information
Solutions and recommendations	Further improvements, responding to the changing needs of impaired tourists and local community members	Better information about dedicated to disabled parking spots, further improvements, responding to the changing needs of impaired tourists and local community members

Source: The presented material is the result of research conducted on the base of the grant issued by the Ministry of Higher Education and Sciences (ds-144 AWF), carried out by the authors (2012 – 2014).

Table 6. Solutions and recommendations on the base of the comparative study of the UNESCO World Heritage Sites and basic tourist services accessibility (case studies: the Old Town in Warsaw vs. the Old Town in Warsaw). Imperative: sidewalks without barriers. Requirement: dedicated for disabled parking space, location close to the sites of tourist interest

Sidewalks	The Old Town in Cracow	The Old Town in Warsaw
Pedestrian routs without barriers	Urban tourist routes for persons with disabilities (Royal Road for the Disabled Tourist), renovated surface of the Main Market and the Old Town main streets	Not all streets and squares are accessible for impaired tourists and local community members
Conclusions	Rather satisfactory level of public space accessibility, responding to the needs of disabled tourists, enhancing the quality of life of the local community members	Several tourists attractions and historic sites are not accessible for impaired tourists, life quality of the local community members is not satisfactory;
Solutions and recommendations	Further improvements, responding to the changing needs of impaired tourists	Streets renovation to provide improvements in aim to respond impaired tourists and local community needs: The minimum _ of an unobstructed sidewalk should be of 150 cm for two- way wheelchair traffic passage (recommended width is 180 cm), Surface must not be interrupted by steps or sudden changes, without loose gravel, cobbles, smooth but not slippery, The historic Royal Road for the Disabled Tourist project to be developed (similar to successful Cracow project)

Source: The presented material is the result of research conducted on the base of the grant issued by the Ministry of Higher Education and Sciences (ds-144 AWF), carried out by the authors (2012 – 2014).

Chapter 2
A Bertrand Game-Based Approach to Hotel Yield Management Strategies

Junzo Watada
Universiti Teknologi PETRONAS, Malaysia

Koki Yoshimura
Waseda University, Japan

Pandian Vasant
Universiti Teknologi PETRONAS, Malaysia

ABSTRACT

This chapter examines hotel yield management from a game perspective using a duopoly situation of two hotels. The hotel yield management determines strategies by considering the number of available rooms in the Bertrand situation. Each hotel does not know the strategies adopted by a competitive hotel. We derive the strategy that realizes a maximum profit under a given situation and constraints. Furthermore, we validate the game-based strategy developed for hotel yield management. In the real world, a business manager adopts an optimum strategy of yield management to gain profits in the current conditions; after the initial strategy is chosen, however, managers continuously weigh new strategies and investments. Therefore, we import the method of real option. Such maneuvers and investments are required to build new strategies amidst competition in the industry.

1. INTRODUCTION

Business circumstances today are becoming increasingly competitive, particularly in the hotel industry, where demand has been shrinking during the recession. Furthermore, as Figures 1 and 2 show, the supply of hotel rooms in Japan, most notably in Tokyo, has been increasing because of the development of numerous foreign-affiliated hotels. Under these circumstances, each hotel must adopt useful measures

DOI: 10.4018/978-1-5225-1054-3.ch002

Figure 1. Volume of hotel industries

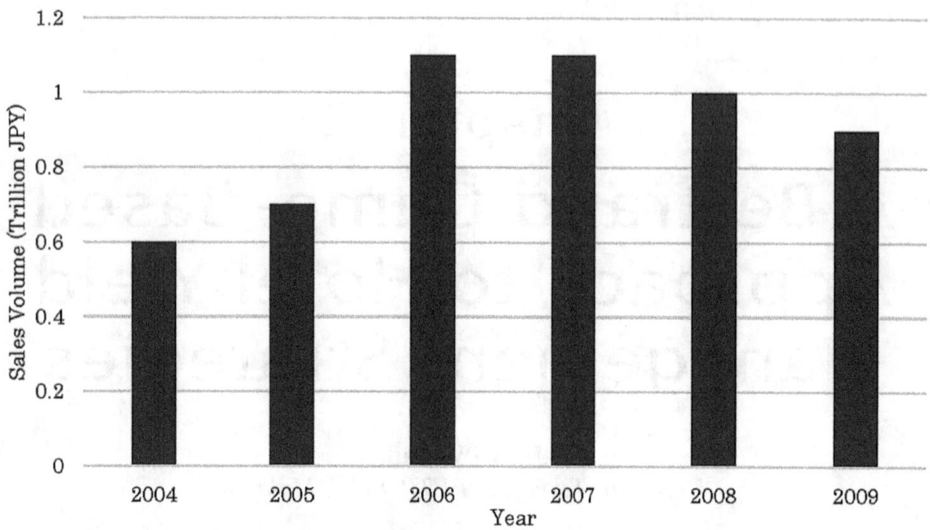

Figure 2. Number of rooms

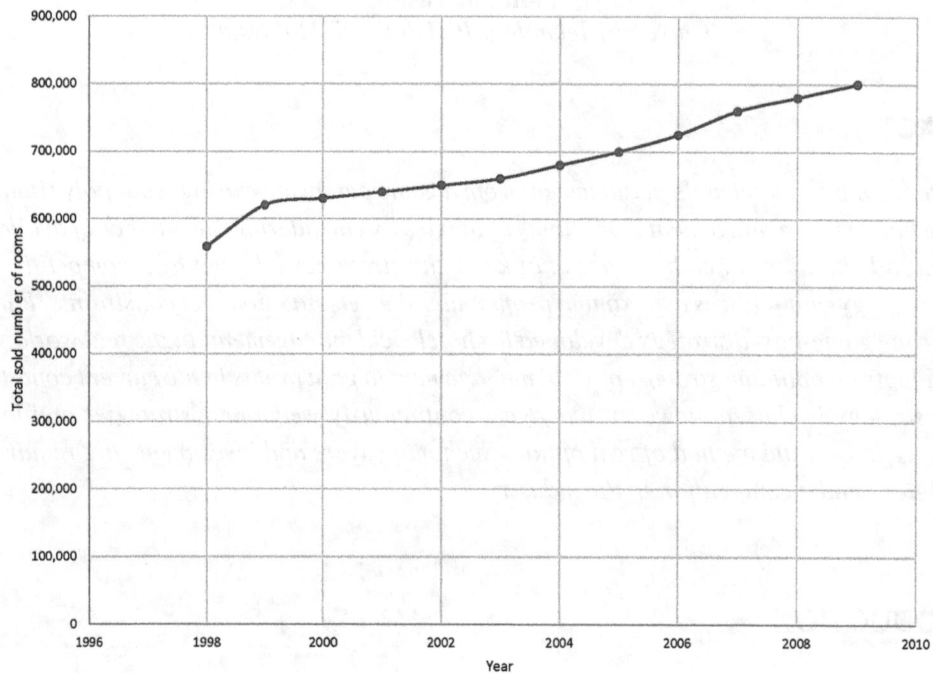

to remain competitive. The majority of hotels select yield management as one of these strategies, aiming to develop effective strategies to obtain optimal profits in today's competitive environment.

In this chapter, we propose the game-based development of strategies in yield management, that is, we model a competitive business circumstance to develop an optimal strategy using game theory under a duopolistic state. First, the duopoly state of two hotels is analyzed in a single year using game theory.

The Bertrand model is employed in this case. Bertrand model provides a simple equation in this work. The model uses the given number of rooms to obtain the maximum profit within the price. As we can see, each hotel in a certain area competes for how many guests choose to stay at the hotel each night, for the price of the room, and for overall demand from visitors in the area.

This research on hotel management focuses primarily on yield management. To obtain optimum solutions, we must consider not only room rates but also yield management strategies. Hotel yield management strategies decide that an early hotel booking generally warrants a higher discount on the room rate.

The theoretical result is compared with the one obtained by Monte Carlo simulation. To achieve these phases, we created and tested simple models using reference books, including the model described in Section 6. The hotel business environment is challenging, requiring exact strategies and uncertain future risks. If neither the demand nor the outside environment changes, hotels do not need to alter their initial strategies. However, by examining game theory, we suggest that when one hotel tends to improve, its competitors also must continuously change their strategies in response. After further testing, we introduce the "real option technique" such that the optimal strategy can be adopted. The novelty of this model is found in the game perspective treatment of yield management (Suzuki, K., 1999;Shigeo M., 2001;Watanabe, T., 2008).

The remaining of the chapter consists of the following structure. Section 2 provides literature review of game theory and real option theory. Section 3 explains the basic concept of yield management. Section 4 illustrates a game theoretical approach. Sections 5 and 6 provide the research result. Finally, in Section 7, we consider the strategy of continuous time length instead of a single term by employing real option model to decide strategies in hotel yield management. Then Section 8 summarizes this chapter.

2. LITERATURE REVIEW

Game Theory Literature

The first known discussion of game theory occurred in a letter written by James Waldegrave in 1713 (Walker, 2005, October). In this letter, Waldegrave provided a minimax mixed strategy solution to a two-person version of the card game known as le Her. James Madison described what we now recognize as a game-theoretic analysis of the ways states can be expected to behave under different systems of taxation. (Kobayashi, Matsushima, & Hishida, 2008; Kulatilaka & Perotti, 1998, August) In his 1838 work, 'Recherches sur les principles mathmatiques de la thorie des richesses (Research into the Mathematical Principles of the Theory of Wealth),' Antoine Augustin Cournot considered a duopoly and presented a solution that is a restricted version of Nash equilibrium. However, game theory did not really exist as a unique field until John von Neumann (1928a,1928b) published the paper. Von Neumann and Morgenstern (1944) published the Theory of Games and Economic Behavior. Since then, the theory has experienced incessant growth in both the number of theoretical results and the scope and variety of applications. In 1950, the first mathematical discussion of the prisoner's dilemma appeared, and an experiment was undertaken by notable mathematicians Merrill M. Flood and Melvin Dresher as part of the Rand Corporation's investigations on game theory. Rand pursued the studies because of possible applications to global nuclear strategy. Nash (1950, January) introduced a Nash equilibrium point, demonstrating that finite games have always had an equilibrium point at which all players choose actions that are best for them given their opponents' choices. Game theory examines competitors' behavior in strategic situations, in which individuals make decisions when they are aware of the choices of others. Cooperative and noncooperative games are the two main branches of game theory (Cachon, 2001; James, 2006; Watada, Waripan, & Wu, 2014). Following the original work of von Stackelberg (1934), the player who holds the powerful position in a decision problem is called the leader, and the other players who react (rationally) to the leader's decision (strategy) are called the followers. Yang and Zhou (2006) consider the pricing and quantity decisions of a manufacturer who supplies a single product to two competitive retailers. Their research analyzes the effects of the duopolistic retailers' different competitive behaviors—Cournot, Collusion and Stackelberg—on the optimal decisions of the manufacturer and the duopolistic retailers themselves. Their studies focus on a discussion of Cournot, Collusion and Stackelberg given various demand curves. These researchers found that in a downward-sloping, price-versus-demand relationship, the manufacturer's profit is double that of the retailer. Another gaming structure converse to the manufacturer-Stg process is the "retailer-Stg" process: the retailer acts as the Stackelberg leader and the manufacturer is the Stackelberg follower. Although this process is encountered less frequently in the literature, there is practical motivation for studying the process. For example, Choi (1991) solved a retailer-Stg system with two manufacturers and a linear demand curve. Messinger and Narasimhan (1995) provided an interesting discussion regarding how the Stackelberg power leader had shifted to the retailer (buyer) in the grocery channel. Ertek and Griffin (2002) developed a two-echelon supply chain model, in which they considered both the manufacturer-Stg process and the retailer-Stg process. For the retailer-Stg process, they considered a pricing scheme for the buyer that involves both a multiplier and a constant markup. They showed that it was optimal for the buyer to set the markup to zero and use only a multiplier. Lau and Lau (2003) also considered the retailer-Stg system with a single manufacturer and a single retailer and investigated the effect of the demand curve's shape on the optimal solution. Other related two-echelon price-sensitive-demand Stackelberg models focus primarily on incorporating stochastic demand, information asymmetry, and other factors. For instance, Cachon (2001) developed game-theoretic models for the competitive cases of continuous review and periodic review of the models. Parlar & Weng (2006) study the effects of coordinating pricing and production decisions on the improvement of a firm's position in a price-competitive environment. As mentioned above, they propose to combine yield management with the game theory to select optimum strategies in hotel yield management. However, in this research, because of the advantages of both yield management and the game theory, this model is able to adopt deference under the considerations for the "timing of reservation by a customer as well as feature of a customer'" as a novelty.

Real Option Literature

Steward Myers (1977, November) was the first who introduced "real options". The real options approach applies the theory of financial option pricing to real assets while financial options deals with buying and selling contracts and obligations in the future. It referred the application of option pricing theory to the valuation of non-financial or "real" investments, such as multi-stage R&D, manufacturing plant expansion. In strategic investments, a decision maker must understand the market evolves uncertainty. The real options method takes into consideration the managerial and strategic resilience in decision making and change the whole decision making process to a contingent one, which does not require the future investment decisions to be fixed at the outset. The managers can choose to expand if the market potential turns out to be large, or to contract if the past phases do not seem to be profitable. These options on strategy raise the value of the project, which have often been underestimated by the NPV approach.

Many researchers have put great effort in the application of the real options approach to various industries and most of them proved helpful for managers to make contingent strategic decisions.

According to the limitations of such conventional methods as NPV, a project discount rate is the opportunity cost of capital. All risks are completely accounted for by the discount rate. Therefore, the traditional NPV method has serious shortcomings in analyzing projects when information concerning future investment decisions is not yet known. This leads most of decision makers to minimize the discount rate as much as possible in order to improve the payoff of the project. However, the uncertainty also means the opportunity to get more payoffs because the value of options is real, the greater the future uncertainty, the greater the project value should be. Dixit and Pindyck (1994) provide a systematic treatment of an approach to capital investment decision and the basic theory of irreversible investment under uncertainty. The study of the optimal timing of investment in an irreversible project where the benefits from the project and the investment cost follow continuous time stochastic processes can be found in (McDonald & Siegel, 1986). The project value and the investment are assumed follow geometric Brownian motion (Dixit et al., 1994; McDonald et al., 1986).

For practitioners, the binomial tree model is the most widely used technique in the field of decision analysis. The details of such a technique can be found in (Trigeorgis, 1991). Trigeorgis develops a log-transformed variation of the Cox-Ross-Rubinstein binomial method. The major advantage of the procedure is that it can be flexibly applied to various real options problems encountered in corporate finance practices. This method is suitable for the situations that there are more than two options to be evaluated. In addition, there are multiple interacting options with the risk of rare event in the investment process (Trigeorgis, 1991; Cox et al., 1979, September).

Amram & Kulatilaka's describe an advantage of real options analysis as following: "Options are contingent decisions: an option is the opportunity to make a decision after seeing events unfolds. This means the payoff to an option is changes with your decision. Option valuations are aligned with financial market valuations:

- The real options approach uses financial market inputs and concepts to value complex payoffs across all types of real assets.

Options thinking can be used to design and manage strategic investments proactively: The nonlinear payoffs can also be a design tool. The first step is to identify and value the options in a strategic investment. The second step is to redesign the investment to better use the options. The third step is to manage the investment proactively through the options created (Amram & Kulatilaka, 1999)."

Many researchers had provided a strategic rationale for growth options under uncertainty and imperfect competition. They pointed out that higher uncertainty means more opportunity rather than simply larger risk and their results contradict the view that volatility was a strong disincentive for investment. In certain situations, an increase in uncertainty could actually increase the probability of investing, and thereby has a positive impact on investment (Kulatilaka & Perotti, 1998, August; Sarkar, 2000).

Reinganum (1981) was the first to analyze this kind of model. His literature considered a duopoly with identical firms. There is no uncertainty in the innovation process and one new technology is considered. If a company adopts the new technology before other companies, it can gain profits at the expense of the other companies. On the other hand, the investment cost being decreasing over time provides an incentive to a company for deferring investment decision until receiving clearer information. Fudenberg & Tirole (1985, July) presented the first strategic investment model that analyzed the effects of preemption in games. In their model, market uncertainty is assumed to be zero. However, their model forms the concrete analysis of strategic investment model. Two equilibriums which are a preemption equilibrium, one of the firms invests early, and a simultaneous one, where both firms delay their investment. Dixit (1989) analyzes stop-restart decision as "entry-exit". The decision of an entry and exit is examined. Smit & Ankum (1993) provide the real options approach by combining with game theory for project timing in a microeconomic framework. Their work focuses on the concept of economic rests or the excess profit above the opportunity cost of investment. A simplified model was discussed in (Dixit et al., 1994) where the firms are not active before the investment is undertaken. Stenbacka and Tombak (1994, September) develop a duopoly game model of timing adoption to analyze the effect of uncertainty in the time lag between adoptions of new technology. In this model, continuous and downward-sloping reaction functions were used in the model. They point out that a higher level of uncertainty increases the extent of dispersion between the equilibrium timings of adoption. Furthermore, they also found that the equilibrium timings are more dispersed when the leader takes the follower's reaction into account and it is possible for collusive adoption to improve payoff. Grenadier (1996) develops an equilibrium framework for strategic option games. The timing of real estate development is discussed. The model provides insights into the forces that shape a market behavior. It also explains why some markets may experience building booms in the face of declining demand and property values while such a behavior is often regarded. Baldursson (1998) studies the price of a perishable commodity produced in an oligopolistic industry where firms use capacity as strategic variable. The model uses the theory of irreversible investment under uncertainty combining with the technique of a fictitious social planner.

Huisman and Kort (1999) extends the previous model of Dixit & Pindyck in modeling symmetrical option game by considering demand uncertainty. It discusses an investment timing problem in a duopoly framework. Their model allows mixed strategies and analyze based on the possibility of non-binding collusion equilibrium. In this setting, two symmetrical companies have the option to make an investment compete with each other. Three scenarios are given.

1. Preemption equilibrium with dispersed investment timing is obtained.
2. equilibrium with joint investment prevails.
3. Preemption holds in case when uncertainty is low, and joint investment is the Pareto dominating equilibrium if uncertainty is large.

According to real option analysis, the company will invest when the NPV exceeds the option value of deferring investment. In the model, the real options investment rule is modified by adding strategic interactions. Comparing the option value of waiting in the monopoly case, the strategic option value of waiting is the same in the joint investment case and lower in the preemptions equilibrium. In the latter case it can even occur that investing is optimal, while the net present value is negative. Moreover, they present many propositions that establish the conditions for the occurrence of different equilibriums in a concrete approach.

Murto analyze an investment in an oligopoly market where there are a few major players in the market. Their model is formulated as a discrete time state-space game where the firms use feedback strategies. The example of the model in a duopoly market is provided to illustrate the tradeoff between the value of flexibility and economies of scale under competitive interaction. He also examines a declining duopoly in which the firms must choose when to exit from the market. The uncertainty is modeled by assumed that the revenue follows a geometric Brownian motion and considers the Markov-perfect equilibrium. It extends the standard real options model by adding the strategic interaction associated with abandonment options in oligopoly (Murto, Nasakkala, & Keppo, 2004; Murto, 2004).

Pawlina and Kort examine asymmetry in the investment. The literature analyzes the impact of investment cost asymmetry on the optimal real option exercise strategies and the value of companies in duopoly. If the asymmetry gap between companies is small, the company tends to invest simultaneously. In case when first-mover advantage is significant, the lower cost firm preempts the higher cost firm. If the asymmetry gap is large, their mutual decision tends to have less direct effect to each other (Pawlina & Kort, 2006).

Kong and Kwok (2007) analyze strategic investment games between two asymmetric companies compete each other in an investment project under uncertainty. The option value leads to delay in project investment if uncertainty increases. However, the greater uncertainty can rush the investment rather than delay.

The application for combining real options and game theory can be applied in many investment projects. The model applying in ICT industries can be found in (Angelou & Economides, 2009; Tsai & Hung, 2009, April). Smith (2003), and Smith and Trigeorgis (2006, January) illustrate the example of infrastructure investment decision making.

Kester (1984) and Trigeorgis (1986) propose a real options classification scheme that is motivated by similarities and differences with financial options. It illustrates a version of this real options classification that distinguishes between simple and compound options, and proprietary and shared options. To appreciate this, let us investigate basic problems managers may face. The strategic question concerns the value characteristics of the project:

- Does this business alternative realize its value primarily through direct measurable cash inflows, or does it have a strategic value?

Commercial one-stage projects that derive their value from expected cash flows would be classified as simple options. Other projects do not derive their value primarily from cash inflows, but from strategic value. For instance, a pilot project that might create a new market, R & D, or exploration investments in natural resources may derive their value from future multistage commercial opportunities and are classified as compound options (Herath & Barth, 2001).

3. YIELD MANAGEMENT

In this section, the basic concepts of yield management is provided. First, we explain the origin of yield management and concepts, after then their effects and necessity conditions are clarified.

The Origin of Yield Management

The deregulation of the American aviation industry in the 1970s led to derived yield management. By the mid-1970s, several charter flight airlines had entered the aviation market in the United States, led by a surge in liberalization. These companies provided several types of low-cost charter flights. Furthermore, they intended to enter the regular service market which had been ruled out for mega-carriers by public charter flights (periodical charter flights).

During this movement, American Airlines noticed that the majority of flights were sold off at the half price of a vacant seat, and the seat cost was much cheaper than the charter flight company. Therefore, the company chose to offer a special reduced fare only when passengers made a reservation three weeks or more before a flight. This fare system proved to be a success. Then, other rival mega-carriers immediately adopted this pricing scheme, and the menace of the public charter flight ended in imaginary fears. The company can be termed "yield management," which is a micro-market strategy to distinguish passengers with the urge to pay. Delta Air Lines called this situation "revenue management."

In addition to its use in the aviation industry, yield management is also used today in various fields including the hotel and transportation industry.

Basic Concept of Yield Management

Yield management is a management technique to maximize profits by changing the price and quota of products and services based on the customers' urge to pay. Moreover, yield management can be a sales technique used in sales that have an upper limit of supply capacity such as aviation (periodical flight), hotels, restaurants, rental cars, sports and concerts, and for products that are not to be carried over as stock. The aviation and hotel businesses generally depend on equipment and facilities. This type of business has a high fixed-cost ratio and a lower variable costs rate despite the quantity of users. Furthermore, if the goods (e.g., seats in aviation and guest rooms in hotel) become leftover (e.g., vacant seats in aviation and empty rooms in hotels), they cannot be sold after the next day. Instead, the businesses suffer a total loss. Therefore, these industries usually offer discounts to increase usage rates. Although wealthy visitors still use the product even if the price is normal and high, the income from wealthy visitors is decreased. In short, the profit may not always rise. Initially, therefore, the type of company ought to secure the sales quantity of wealthy visitors such as people on business trips visitor who are willingly pay the regular rate. Next, the companies should sequentially lower the price to sell the remainder. By adopting such a strategy, companies can maximize their profits. This scenario describes the basic theory of yield management.

Effect of Yield Management

The effect of yield management is explained using a theoretical demand curve. In perfect competition, the law of indifference is satisfied. In this case, the profit loss or sale loss becomes large. However, companies can increase their profits significantly if customers are segmented by their urge of payment and the appropriate pricing policy is implemented.

When only a single price policy is prepared for one product (Figure 3), it results in producing two types of loss:

- [A] and [B].

[A] means that it is possible to sell even if a more expensive price is offered to customers, whereas [B] means that it is possible to sell if a cheaper price is offered to customers. [A] is profit loss and known as consumer-related surplus. [B] is sale loss. Therefore, companies can reduce profit loss and sale loss if multiple price ranges are prepared (Figure 4).

Figure 3. Demand curve

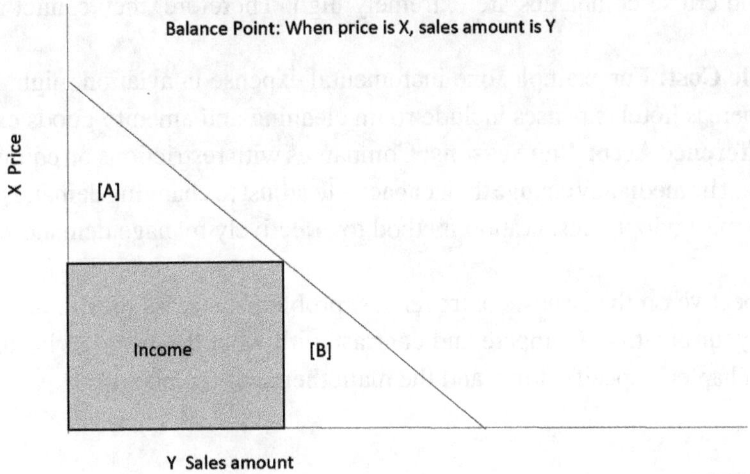

Figure 4. Demand curve of yield management

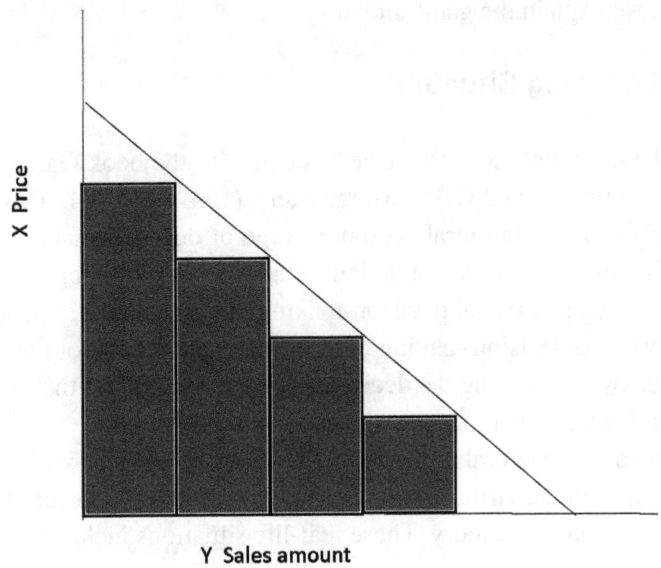

Necessary Conditions of Yield Management

1. **Fixed Equipment and Facilities Capacity:** Yield management is designed for the companies which have constraints in equipment and facilities capacity. In the case when equipment and facilities capacity are not restricted, companies can hold some stock as a buffer to handle a change of demand. Conversely, the companies with certain constraints must cover the change with the limitation until they are ready.

2. **Highly Fixed Cost:** The additional expenses to boost equipment and facilities capacity in aviation, hotel and cruise companies are extremely high. Therefore, they cannot adjust the capacity immediately.
3. **Low Variable Cost:** For example, one incremental expense in aviation might be additional meal expenses, whereas hotel expenses include room cleaning and amenity goods expenses.
4. **Demand Difference According Seasons:** Companies with restrictions on equipment and facilities capacity cannot immediately change their capacity to adjust to changing demand patterns. Therefore, the company must adopt a reservation method to effectively manage demand.

Present your perspective on the issues, controversies, problems, etc., as they relate to theme and arguments supporting your position. Compare and contrast with what has been, or is currently being done as it relates to the chapter's specific topic and the main theme of the book.

4. GAME THEORY

The game theory is illustrated in this section before discussing the game based yield management in Sections 5 and 6. First, we explain the game situation.

Game Theory and Gaming Situation

Game theory, one of the decision-making theories, began with the book Game Theory and Economic Behavior by John Von Neumann and Oskar Morgenstern (1947) in 1944. This theory is treated as a method of making decisions among plural persons, a type of decision-making that differs essentially from other decisions. In our study, we adopt an individual decision-making scenario for our purposes. The result depends not only on personal decision-making but also on other people's decisions. Therefore, in game theory, even if a decision-making process is intended only for the purpose of one person, it always has to be made by considering the decisions of other people for their own purpose. This type of circumstances is considered a "gaming situation."

Game theory began as a behavioral analysis of two decision makers with opposing interests. For instance, the theory is represented virtually by games such as chess. However, there are also numerous daily situations that involve gaming theory. These real-life situations include:

- Economic news such as the mergers or bankruptcies of companies,
- Salary negotiations between labor and management,
- The release of new products, and
- Changes in pricing, which are seen and heard on tv and in newspapers nearly every day.

Political examples include the division or junction of political parties, election results, and the formation of coalition governments. International examples include cross-national negotiations, territory negotiations and trade negotiations. Other applications include the construction or re-operation of nuclear power plants, the construction of dams, and the acceptance and processing of garbage. All of these situations involve competition and cooperation among various organizations in which the interests

Figure 5. Demand curve of yield management

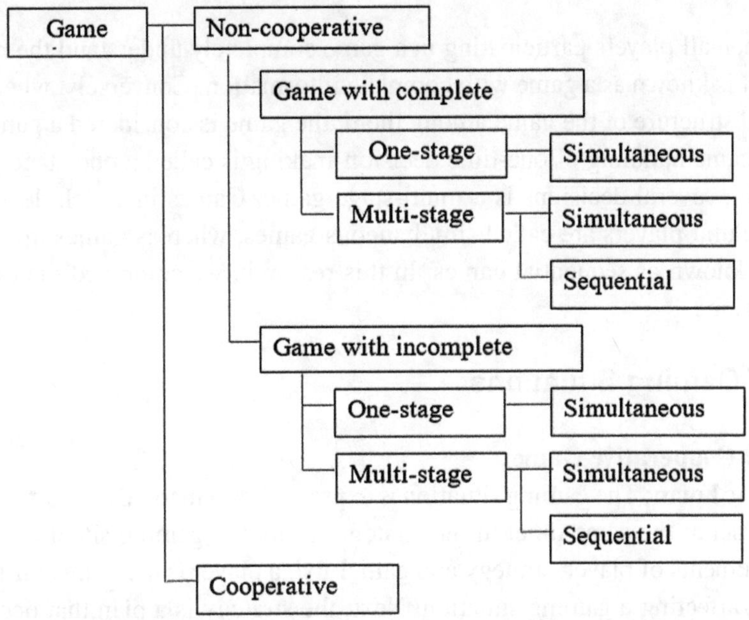

of companies, political parties, nations, groups, and local inhabitants are not always in accordance. In other words, they are generated from gaming situations among various decision makers.

Gaming situations include various cases. For example, in the competition between companies, each company makes decisions individually without discussions among companies. In price setting, for example, the Fair Trade Commission would have started a collusion inquiry if it suspected that companies had talked with each other.

However, when making decisions, companies do exhibit technical and business cooperation, merger negotiations, and contract discussions to reach agreements. Political parties negotiate during an election to increase their seats but also work and talk with opposing parties to reach compromises on bills and amendments. Moreover, in international relations, countries at war may search for a better solution through international conferences, diplomatic negotiations and trade negotiations.

Game theory has also expanded widely to fields such as biology, information science, management engineering, social engineering, operations research, economics, political science, and sociology.

A System of Game Theory

Game theory is classified roughly into noncooperative games and cooperative games. In noncooperative games, a player makes decisions independently. Conversely, cooperative games permit strategic discussions between players and consider a cooperative action.

As previously mentioned, game theory began as decision-making based on the behavior analysis of two persons competing for victory or defeat in games such as chess. In the decision-making in such a game, each player's interest is opposed in the form of the victory or defeat and is thus termed a zero-sum game. Conversely, in a nonzero-sum game between two persons, if one player understands one's own gain, another player's gain could also be grasped naturally because of the complete opposition between two players.

In this way, when all players participating in a game completely understand the rules and the structure of the game, it is known as a game with complete information. Conversely, when all players do not share the rules and structure of the game among them, the game is considered a game with incomplete information. The game handling of one-time decision-making is called a one-stage game, whereas the game handling with several decisions is a multi-stage game. Games in which decisions are made simultaneously by plural players are called simultaneous games, whereas games in which decisions are made in turns are known as sequential games. In this research, we combined the one-stage game and the simultaneous game.

Expression of Gaming Situations

1. **Case of Non-Cooperative Game:**
   a. **Strategic Form:** The gaming situation is expressed by a method called "strategic form" when players act at the same time. In the strategic form, the gaming situation is expressed in the three elements of player, strategy and gain. First, a player is the main constituent of decision-making affecting a gaming situation. Next, the strategy is a plan that decided which option was taken previously when each player decides on an action. Finally, the gain is the result that occurs when each player acts according to each strategy.
   b. **Development Form:** The gaming situation is expressed by a method called development form when decisions occur according to time. In development form, a tree structure is used for expressing who decides, when to decide and how to decide.
2. **Cooperative Game:** In cooperative games, the game situation is expressed by means of a special function. This special function expresses how much the group of players who contracted a collaborative relationship gains. These are called special function form or partner form in contrast to the terms strategic form or development form in non-cooperative games.
3. **Cournot Model, Bertrand Model, and Stackelberg Model:** The strategy which a company adopts is linked to the production output or the price of a product and divides competition into two parts. The Cournot competition is an oligopoly market in which two or two or more rival firms exist, and each company is in competition to attain profit maximization through the adjustment of production output. That is, it is competition of production output. Each company predicts a partner's action from the market demand and supply, making decisions regarding production output used as the profit maximization of its company.

Unlike the Cournot competition, the point of competition of the Bertrand competition is not production output, but rather a price. Each company analyzes the price of a rival firm's product and makes price decisions from which the profits of their own company become the maximum corresponding to the partner's price. The Cournot competition and the Bertrand competition are examples of cases in which rival firms make decisions simultaneously. In the Stackelberg competition, decision-making occurs in turns (i.e., decisions are divided into leader and a Follower).

Figure 6. Room price system in yield management

A	B	C
Very early	Early	Normal
25%	50%	Normal price : 100%

One month ago One week ago Staying date

5. BERTRAND MODEL

In this research, two hotels employ yield management. The two hotels are in a duopolistic situation. The strategy employed is analyzed in a single year. First, the optimal profits are derived theoretically based on a mathematical model in the game environment. Second, the theoretical result is compared with the one obtained by Monte Carlo simulation. In the same area, two hotels X and Y are considered to be duopolistic. The two hotels have the same number of rooms, that is, 100 rooms each. Both the room price systems are also the same. Therefore, it is assumed that the two hotels are the same spec with the exception of the yield management strategies of the two hotels. In addition, each room price is treated by yield management as follows. Let Z denote the number of customers. Hotel sales and room price vary depending on the number of room.

Price Rate: $P: 25-1$.
A: Selling price rate 0.25 results in room price as $25\% \rightarrow 0.25$.
B: Selling price rate 0.5 results in room price as $50\% \rightarrow 0.50$.
C: Selling price rate 1.0 results in room price as $100\% \rightarrow 1.0$.

The general demand curve of Figure 4 represents two results. First, demand increases when price becomes cheaper. Second, demand decreases when price become more expensive. This theory supports the same schematic in the price decision model of hotels in the case of duopoly situation. Therefore, when the room price become cheaper compared to the competition hotel, the number of customers increases. Conversely, when the room price increases, the number of customers decreases.

We therefore suppose the following points.

1. In this area the average demand rate is 80 persons. The absolute value of the price gap is less than or equal to 2000 yen.
2. When the general price is the same at the two hotels, the number of people staying in the two hotels is 80 persons.
3. When the absolute value of the price gap between the two hotels is above 2000 yen, the less expensive hotel's demand is 100%. By contrast, the more expensive hotel's demand is 60 persons.

Number of customer (X or Y) $= b1, b2$ Demand factor - Regular price

Figure 7. Demand curve

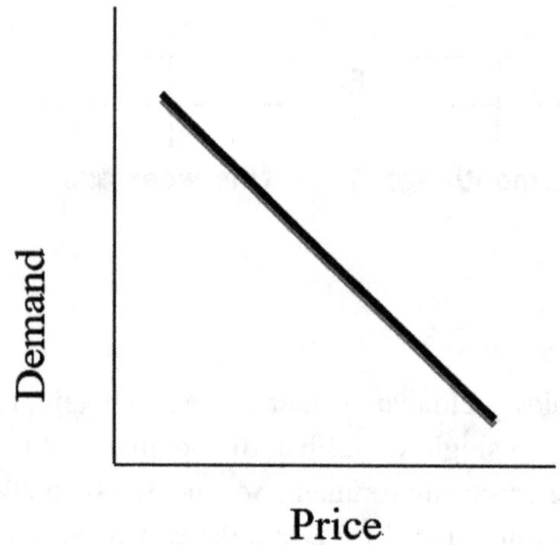

$$
\begin{aligned}
&Z_x = 80 - (P_x - P_y) \cdot M, \\
&Z_y = 80 - (P_y - P_x) \cdot M, \\
&0 \le Z \le 100, \\
&| P_x - P_y | \le 20
\end{aligned}
\tag{1}
$$

Z_x : The numbers of people staying at hotel X.
Z_y : The numbers of people staying at hotel Y.
P_x : The regular price of hotel X.
P_y : The regular price of hotel Y.
M : The variable for adjusting the dimension "From 100 JPY per 1 person".
80 : Assume that the average number of people staying at each hotel is 80 persons

The Detailed Explanation of Above-Mentioned Formula

Number Z_x of lodgers at the hotel X will decrease at once, if room unit price P_x which the hotel X determined goes up, and if a partner's price P_x goes up, it will increase at once.

Thus, in consideration of the causal relationship in the Bertrand competition, the strategies (decision variable) of each hotel are P_x and P_y and Z_x and Z_y define them as being decided according to it.

Each hotel selects a strategy out of three patterns a, b, and c as shown in Tables 1 and 2.

Table 1. Room allocation of Hotel X

X	Type of Strategy		
	a_1	b_1	c_1
A	10	10	10
B	30	20	40
C	60	70	50

Table 2. Room allocation of Hotel Y

Y	Type of Strategy		
	a_2	b_2	c_2
A	10	5	30
B	40	10	50
C	50	85	20

Figure 8. The relation between room price and demand

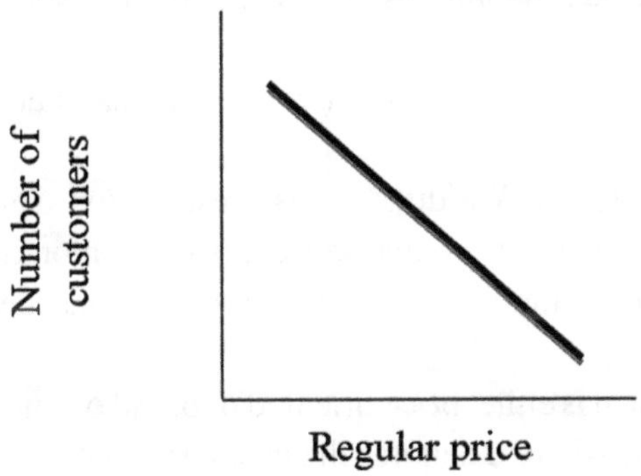

From previous data, we can classify customers into type (1) and type (2). A customer of type (1) focuses on room price as low as possible; conversely, a customer of type (2) emphasizes a hotel brand. In the type (2) case, customers decide on the hotel encountered first.

For example, in the case of type (1), people want to choose the lower-priced room. However, when rooms in period A_1 have already been fully booked at hotel X one month in advance, then this customer will book a room in period A_2 of hotel Y from hotel X, if the period A_2 of the hotel Y is available.

Conversely, in the case of type (2), people regard the hotel brand as important. When the rooms in period A_1 have already been fully booked in hotel X a month previously, even if there are vacant rooms of another type, the person will make a reservation in the same hotel X

The above results suggest that hotel profits will change depending on the number of rooms. Therefore, this research verifies which strategy is the best regarding yield management. In addition, each hotel does not know the room allocation of the other hotel.

From the above, hotel profits will change depending on the number of rooms. Therefore, this research verifies which strategy is the best in yield management. In addition, they do not know the room allocation of the other hotel each other.

Then, the strategy $c1$ of hotel Y and the strategy $a2$ of hotel Y are the same. These strategies for hotels X and Y have no difference other than the number of reservable rooms. Actually when both of the hotels select these strategies, the profit is the same in the simulation and evaluation function (see below). Even if it is impossible in real situations, it is important that this research is focused on the changes in the profits by using Bertrand model strategy for setting the room proportion strategy.

6. MONTE CARLO SIMULATION OF THE MODEL

Yield management is widely used as a hotel management strategy. In this section, the game theory of the Bertrand model is applied to validate the yield management of hotels.

Prerequisite

In this model, one of the purposes is to simulate the current situation using the Monte Carlo method. The following items are generated randomly by Monte Carlo simulation:

1. The gathering ratio of customers, meaning the sum of demand at two hotels by people staying in the region.
2. Reservation date by a customer. We divide these reservation dates into three groups, including reserved one month ago, reserved from one week ago to one month ago, and reserved from staying date to one week ago. The beginning of the reservation date has been set to be one year before the measurement date.
3. Feature of a customer, which is either price oriented or brand oriented. Both the room costs required per one room are the same where d denotes the cost per room.

Monte Carlo Simulation

Performing a Monte Carlo simulation to obtain the optimal strategy, we calculated the following:

Result of the Simulation

The simulation was conducted 1000 cycles for all the combination of strategies as shown in Table 3. Comparing the highest average values, different values were obtained each time because of the random Monte Carlo simulations.

Evaluation Function

We create and verify the evaluation function for the Bertrand model that selects the best for both the hotels in the game situation. The profits of both hotels depend on the method used to secure the number of sold rooms. In other words, the selected strategies of room combination influence the total profit.

Table 3. Result of the simulation

		Profit of X and Y		Price(x)			Price(y)		
		Hotel X (×10⁶)	Hotel Y (×10⁶)	A (×10⁴)	B (×10⁴)	C (×10⁴)	A (×10⁴)	B (×10⁴)	C (×10⁴)
1	a_1, a_2	0.462	0.451	0.299	0.598	1.196	0.301	0.603	1.205
2	a_1, b_2	0.421	0.442	0.292	0.584	1.168	0.287	0.574	1.148
3	a_1, c_2	0.540	0.461	0.312	0.624	1.249	0.328	0.655	1.310
4	b_1, a_2	0.460	0.440	0.295	0.590	1.180	0.299	0.599	1.197
5	b_1, b_2	0.418	0.431	0.288	0.576	1.152	0.285	0.570	1.140
6	b_1, c_2	0.537	0.450	0.308	0616	1.232	0.326	0651	1.302
7	c_1, a_2	0.464	0.464	0.304	0.607	1.214	0.304	0.607	1.214
8	c_1, b_2	0.423	0.455	0.296	0.593	1.186	0.289	0.579	1.157
9	c_1, c_2	0.543	0.452	0.317	0.633	1.266	0.330	0.660	1.319

Therefore, the method calculates an allocation of room assignments and selects the largest total profit for hotels X and Y. To maximize the total profit of each of the hotels, both the strategies are compared under consideration of the results of the Monte Carlo simulation. The evaluation function is written as follows. The one room price. The number of sold rooms Cost.

1. **Notations:**
    a. **Hotel Type:**
        i. X: Target hotel,
        ii. Y: Competitor hotel.
    b. **Reservation Timing:**
        i. A: Reservation period "before more than one month",
        ii. B: Reservation period "from one week to one month",
        iii. C: Reservation period "from one week previous to the present day".
    c. **Quantity of Staying:**
        i. Z: The total number of people who visit and stay in hotels $Z = (Z_x + Z_y)$,
        ii. Z_n: The total number of people who visit and stay in Hotel $n = X$ or Y.
    d. **Rate of Numbers of Sold Room on Reservation Timing against the Total Number of Rooms:**
        i. Z_A: The rate when the reservation is made during the period A,

ii. Z_B : The rate when the reservation is made during the period B,

iii. Z_C : The rate when the reservation is made during the period C.

e. **Price Rates:**

i. P_A : Price rate when the reservation is made during the period A,

ii. P_B : Price rate when the reservation is made during the period B,

iii. P_C : Price rate when the reservation is made during the period C.

f. **Official Price:**

i. P_n : Official price of Hotel $n = X$ or Y.

g. **Profit: π**

i. π_x : The profit of hotel X,

ii. π_y : The profit of hotel Y,

iii. π_x^* : The profit of hotel X in equilibrium,

iv. π_y^* : The profit of hotel Y in equilibrium.

2. Calculation of one room price in three periods, A, B and C for each hotel $n = X,Y$

$$R_{An} = P_A \times P_n \tag{2}$$

$$R_{Bn} = P_B \times P_n \tag{3}$$

$$R_{Cn} = P_C \times P_n \tag{4}$$

The number of hotel customers in three periods A, B and C as

$$A_n = A_X \text{ or } A_Y$$

$$B_n = B_X \text{ or } B_Y$$

$$C_n = C_X \text{ or } C_Y$$

where $Z_n = Z_X$ or Z_Y, $Z_X = A_X + B_X + C_X$, and $Z_Y = A_Y + B_Y + C_Y$.

The number of hotel customers "A_n" "B_n" or "C_n" in period A, B or C is written as

$$A_n = Z_A \times Z_n \tag{5}$$

$$B_n = Z_B \times Z_n \tag{6}$$

$$C_n = Z_C \times Z_n \tag{7}$$

The total room cost $(n = x, y)$ is written as

$$D = d \times Z_n \tag{8}$$

where d denotes a cost for one room.

Therefore, the total profit is derived using equations (2) to (8) as follows:

$$\Pi = R_{An} \times A_n + R_{Bn} \times B_n + R_{Cn} \times C_n - d \times Z_n \tag{9}$$

$$Z = Z_x + Z_y \text{ and } Z_n = Z_x, Z_y \tag{10}$$

Next, we compute the theoretical values of the evaluation function.

Price Rate:

$$P_A = 0.25\,,\ P_B = 0.5\,,\ P_C = 1.0\,, \tag{11}$$

Room cost is denoted as $d = 3{,}000 JPY$.

In this case, the hotel X's strategy $c1$ is same to the hotel Y's strategy $a2$. Therefore, this simulation focuses on a combination of $c1$ and $a2$. The evaluation function is calculated under $c1$ and $a2$. After that, these results are compared.

As a result, π_x and π_y result in the following equation:

$$\pi_x(P_x, P_y) = (-0.725 \cdot P_x^2 + 58 \cdot P_x + 0.725 \cdot P_x \cdot P_y) - 30 \cdot Z_x \tag{12}$$

$$\pi_y(P_x, P_y) = (-0.725 \cdot P_y^2 + 58 \cdot P_y + 0.725 \cdot P_x \cdot P_y) - 30 \cdot Z_y \tag{13}$$

From the above computation, we calculated the value of P_x and P_y.

$$P_y^* = P_y^* = 121.38 \tag{14}$$

P_y^*: The price of hotel X in equilibrium

Table 4. Results of simulation and evaluation function

Profit	$X : c_1(\times 10^6)$	$Y : a_2(\times 10^6)$
Evaluation function	0.464	0.464
Simulation	0.464	0.464

P_y^* : The price of hotel Y in equilibrium

From the above computation, when hotel X's strategy is c and hotel Y's strategy is a, the hotel total profit of the evaluation function is as follows:

$$\therefore \pi_x^*(P_x, P_y) = \pi_y^*(P_x, P_y) = 0.464 \times 10^6 \quad (15)$$

This evaluation function maximizes the total profit of both hotels when hotel X uses the strategy of $c1$ and hotel Y adopts the strategy of $a2$. When comparing the results between the Monte Carlo simulation and the theoretically evaluation function, hotel X with strategy $c1$, hotel Y with strategy $a2$, both hotels obtained the highest profits using the evaluation function as shown in Table 4.

The Simulation Results When the Environments of These Hotels Are Changed

In the simulation, the total number of customers is determined using random variables. The evaluation function derives the Bertrand model-Nash equilibrium for both hotels. The number of sold rooms in hotels X and Y is derived by the same result of Bertrand model-Nash equilibrium in an evaluation function. In the simulation, the same Bertrand model-Nash equilibrium is used with normal demand=80 and the value calculated as shown in Table 5. Based on this finding, this simulation has proven to be correct. When these two hotels executed strategy $c1$ and $c2$, this became the optimal strategy for these hotels.

1. **Demand:** Hotels X and Y competed for demand from people who would stay in the region. To increasingly gain this demand, hotels X and Y must select one of the yield management strategies by considering the strategy and needs of each hotel. At this time, the price-setting of rooms was decided by a strategy at each hotel.
    a. **Normal Demand:** Approximately 80%.
    b. **High Demand:** Approximately 90%.
    c. **Low Demand:** Approximately 70%.
2. **Simulation of Normal Demand Situation:** In the 9th combination of strategy, we could get the most profits for both hotels. Considering the hotel Y only, the 5th strategy from the top could gain the most profits. However, since it will not be concerned with the strategy of the hotel X and 500,000 yen or more profits can be gained if the strategy c2 is taken, the combination of the 9th strategy will be taken every year if the demand stays as usual.

Table 5. Simulation of normal demand situation

		Profit of X and Y		Price(x)			Price(y)		
		Hotel X ($\times10^6$)	Hotel Y ($\times10^6$)	A ($\times10^4$)	B ($\times10^4$)	C ($\times10^4$)	A ($\times10^4$)	B ($\times10^4$)	C ($\times10^4$)
1	$a1, a2$	0.483	0.477	0.299	0.598	1.196	0.301	0.603	1.205
2	$a1, b2$	0.457	0.442	0.292	0.584	1.168	0.287	0.574	1.148
3	$a1, c2$	0.610	0.535	0.312	0.624	1.249	0.328	0.655	1.310
4	$b1, a2$	0.462	0.487	0.295	0.590	1.180	0.299	0.599	1.197
5	$b1, b2$	0.480	0.581	0.288	0.576	1.152	0.285	0.570	1.140
6	$b1, c2$	0.577	0.530	0.308	0616	1.232	0.326	0651	1.302
7	$c1, a2$	0.482	0.482	0.304	0.607	1.214	0.304	0.607	1.214
8	$c1, b2$	0.474	0.507	0.296	0.593	1.186	0.289	0.579	1.157
9	$c1, c2$	0.618	0.541	0.317	0.633	1.266	0.330	0.660	1.319

Table 6. The optimal strategy of Table 5 for these hotels

Type of Hotel	Strategy of the Highest Profit
Hotel X	9: $c1, c2$
Hotel Y	5: $b1, b2$
Nash equilibrium of hotel X & Y	$9:\ c1, c2$

The result of situation when the hotel environment changed is recorded below.

3. **The Case during a Period of High Demand Situation (Booming Economy):** The optimal strategy is discussed based on the high demand situation given in Table 7. The result is explained in Table 8.
4. The case during a period of high demand "when there are many customers who reserve early, the sightseer would increase."

The optimal strategy is discussed based on the earlier reserving situation given in Table 9. The result is explained in Table 8.

Table 7. Simulation of high demand situation

		Profit of X and Y		Price(x)			Price(y)		
		Hotel X ($\times 10^6$)	Hotel Y ($\times 10^6$)	A ($\times 10^4$)	B ($\times 10^4$)	C ($\times 10^4$)	A ($\times 10^4$)	B ($\times 10^4$)	C ($\times 10^4$)
1	$a1, a2$	0.615	0.604	0.324	0.647	1.306	0.326	0.653	1.305
2	$a1, b2$	0.576	0.555	0.317	0.634	1.267	0.312	0.624	1.248
3	$a1, c2$	0.745	0.545	0.337	0.674	1.348	0.353	0.705	1.410
4	$b1, a2$	0.599	0.630	0.320	0.640	1.280	0.324	0.649	1.297
5	$b1, b2$	0.628	0.747	0.313	0.626	1.252	0.310	0.620	1.240
6	$b1, c2$	0.778	0.562	0.333	0667	1.333	0.351	0.701	1.402
7	$c1, a2$	0.620	0.620	0.329	0.657	1.131	0.329	0.657	1.131
8	$c1, b2$	0.605	0.704	0.321	0.643	1.286	0.314	0.629	1.257
9	$c1, c2$	0.691	0.572	0.342	0.683	1.366	0.355	0.710	1.419

Table 8. The optimal strategy of Table 7 for these hotels

Type of Hotel	Strategy of the Highest Profit
Hotel X	$6 : b1, c2$
Hotel Y	$5 : b1, b2$
Nash equilibrium of hotel X & Y	$5 : b1, b2$

5. **The Case during a Period of Low Demand Situation (Depressed Economy):** The optimal strategy is discussed based on the low demand situation given in Table 9. The result is explained in Table 11.
6. **The Case during a Period of Low Demand (Decreasing of Sightseer):** The element: Yen appreciation, holding of the main competitions like the Olympic Games in other place and so on.

7. THE FUSION OF REAL OPTIONS

When introducing a real option, it is important whether it is based on a CRR model or a Black-Scholes model.

The most fundamental model of a real option is the optimal halting problem adapting the option theory of the finance. The model which chooses the timing which the fountainhead of uncertainty is given by geometric Brownian motion, looks at the state, and takes action. It is also the most famous. For

Table 9. Simulation of many customers who reserve early

		Profit of X and Y		Price(x)			Price(y)		
		Hotel X ($\times 10^6$)	Hotel Y ($\times 10^6$)	A ($\times 10^4$)	B ($\times 10^4$)	C ($\times 10^4$)	A ($\times 10^4$)	B ($\times 10^4$)	C ($\times 10^4$)
1	$a1, a2$	0.515	0.556	0.324	0.647	1.306	0.326	0.653	1.305
2	$a1, b2$	0.489	0.469	0.317	0.634	1.267	0.312	0.624	1.248
3	$a1, c2$	0.651	0.450	0.337	0.674	1.348	0.353	0.705	1.410
4	$b1, a2$	0.501	0.540	0.320	0.640	1.280	0.324	0.649	1.297
5	$b1, b2$	0.628	0.728	0.313	0.626	1.252	0.310	0.620	1.240
6	$b1, c2$	0.717	0.466	0.333	0667	1.333	0.351	0.701	1.402
7	$c1, a2$	0.554	0.554	0.329	0.657	1.131	0.329	0.657	1.131
8	$c1, b2$	0.533	0.474	0.321	0.643	1.286	0.314	0.629	1.257
9	$c1, c2$	0.691	0.486	0.342	0.683	1.366	0.355	0.710	1.419

Table 10. The optimal strategy of Table 9 for these hotels

Type of Hotel	Strategy of the Highest Profit
Hotel X	$6 : b1, c2$
Hotel Y	$5 : b1, b2$
Nash equilibrium of hotel X & Y	$5 : b1, b2$

example, in an early stage, if the profits which a project gives are small, investment cannot be performed in the project. However, when the profits change with geometric Brownian motion and there is a rising trend, Whether investing in a project where timing becomes the optimal decision-making and how much value the right option that it can invest in the future to a project by the basis which wove in the optimal decision-making brings a company (timing of option use) pose a problem. On the contrary, although the project is already started, when the profits are in a downtrend, withdrawing from investment to which timing is the problem whether to become the optimal decision-making. These correspond to the theory of the call option (Black-Scholes model) of a financial option theory. In the problem of decision-making of such investment or entry, not only one company is facing this problem. Similarly in entry and research and development to a new product market, the competitor aims at research and development of the market and the field in many cases. Making decision according to the case of a partner company and decision-making of the partner company is also subject to the influence of a self-company. Of course,

Table 11. Simulation of low demand situation

		Profit of X and Y		Price(x)			Price(y)		
		Hotel X ($\times 10^6$)	Hotel Y ($\times 10^6$)	A ($\times 10^4$)	B ($\times 10^4$)	C ($\times 10^4$)	A ($\times 10^4$)	B ($\times 10^4$)	C ($\times 10^4$)
1	$a1, a2$	0.370	0.359	0.274	0.548	1.096	0.276	0.553	1.105
2	$a1, b2$	0.352	0.409	0.267	0.534	1.067	0.262	0.524	1.048
3	$a1, c2$	0.437	0.499	0.287	0.574	1.149	0.302	0.605	1.210
4	$b1, a2$	0.357	0.362	0.270	0.540	1.080	0.274	0.549	1.097
5	$b1, c2$	0.327	0.331	0.263	0.526	1.052	0.260	0.520	1.040
6	$b1, c2$	0.418	0.493	0.283	0567	1.133	0.301	0.601	1.202
7	$c1, a2$	0.372	0.372	0.279	0.557	1.114	0.279	0.557	1.114
8	$c1, b2$	0.382	0.354	0.271	0.543	1.086	0.264	0.529	1.057
9	$c1, c2$	0.476	0.504	0.292	0.583	1.166	0.304	0.610	1.219

Table 12. The optimal strategy of Table 11 for these hotels

Type of Hotel	Strategy of the Highest Profit
Hotel X	$9 : c1, c2$
Hotel Y	$9 : c1, c2$
Nash equilibrium of hotel X & Y	$9 : c1, c2$

this is applied not only to a company but a hotel. Therefore, the problem of decision-making of two companies which thought of the competitor can say a real option to the problem of the above-mentioned simple real option as the most typical model of game theory. In Dixit and Pindyck (1994) already mentioned above, the fountainhead of uncertainty is given by the stochastic process by geometric Brownian motion, and the problem two companies which compete according to the state consider the timing of investment to be is explained roughly. Grenadier (1996) which applied this to early entry of real estate development is a typical paper which tied up this model and actual proof. The research which applied this model through research of the first stages, such as Huisman (2001) and Shibata and Kijima (2002) now to others is very prosperous.

Then, Cox, Ross, Rubinstein and others rewrote the option theory of the finance based on the probability differential equation which made Black-Scholes model first in a roll to the model based on 2 clause model. Even if the feature or merit by 2 clause model do not overcome the difficult wall of a

Table 13. Simulation of decreasing of sightseers

		Profit of X and Y		Price(x)			Price(y)		
		Hotel X ($\times 10^6$)	Hotel Y ($\times 10^6$)	A ($\times 10^4$)	B ($\times 10^4$)	C ($\times 10^4$)	A ($\times 10^4$)	B ($\times 10^4$)	C ($\times 10^4$)
1	$a1, a2$	0.462	0.486	0.274	0.548	1.096	0.276	0.553	1.105
2	$a1, b2$	0.413	0.387	0.267	0.534	1.067	0.262	0.524	1.048
3	$a1, c2$	0.522	0.544	0.287	0.574	1.149	0.302	0.605	1.210
4	$b1, a2$	0.428	0.480	0.270	0.540	1.080	0.274	0.549	1.097
5	$b1, b2$	0.381	0.383	0.263	0.526	1.052	0.260	0.520	1.040
6	$b1, c2$	0.493	0.538	0.283	0567	1.133	0.301	0.601	1.202
7	$c1, a2$	0.500	0.500	0.279	0.557	1.114	0.279	0.557	1.114
8	$c1, b2$	0.441	0.400	0.271	0.543	1.086	0.264	0.529	1.057
9	$c1, c2$	0.554	0.550	0.292	0.583	1.166	0.304	0.610	1.219

Note: The element: Yen appreciation, holding of the main competitions like the Olympic Games in other place and so on.

Table 14. The optimal strategy of Table 13 for these hotels

Type of Hotel	Strategy of the Highest Profit
Hotel X	$9: c1, c2$
Hotel Y	$9: c1, c2$
Nash equilibrium of hotel X & Y	$9: c1, c2$

probability differential equation, it is easy to understand by the model of one period at the two times -- the price increases about the concept of pricing using the conditions of not deciding, and completeness, and volatility. Moreover, it is set to one of the concrete numerical computation to the option price of continuous time by increasing the number of periods.

From this, game theory is united with real option analysis, and the two-step option game model at the two times of two companies, which can analyze decision-making regarding the strategic investment of a company in uncertainty and market competition.

Binomial Cox, Ross, Rubinstein Model

Binomial Cox, Ross, Rubinstein model (CRR model) is one of the option value evaluation techniques. After subdividing the exercise period of an option, it is an effective method of predicting the transition

Table 15. Coefficient setting of binomial CRR model

Probability of Changing	q
Upward Profit Rate in Market Demand	u
Downward Profit Rate in Market Demand	d
Interest-Rate	r
Risk Neutral Probability	p

or change of a stock price (underlying assets price) and presumes the present option value. It is also called a "binary tree model."

According to the principle figure shown in Figure 9, a certain value (stock price of property) should pass a period when the project of V is fixed. It will increase u rate with the possibility of q, and the value will be uV. With the possibility of $1-q$ and d rate, it sometimes decreases and the value will be dV. This model computes each value from the current value V of a project in consideration of certain future possibilities from the fluctuating value in the future. A formula is shown below.

$$NPV = \frac{p \times V_{+} + (1-p) \times V_{-}}{1+r} \tag{16}$$

where $V_{+} = uV$ and $V_{-} = dV$

$$p = \frac{(1+r)-d}{u-d} \tag{17}$$

Option Game Model

The result of the competition expected from the strategy of one company and the strategy of the other company is considered. If one company and the other company invest in equipment simultaneously, it will become the Nash equilibrium by the Cournot competition, and the investment value will turn into

Figure 9. Principle figure of binomial CRR model

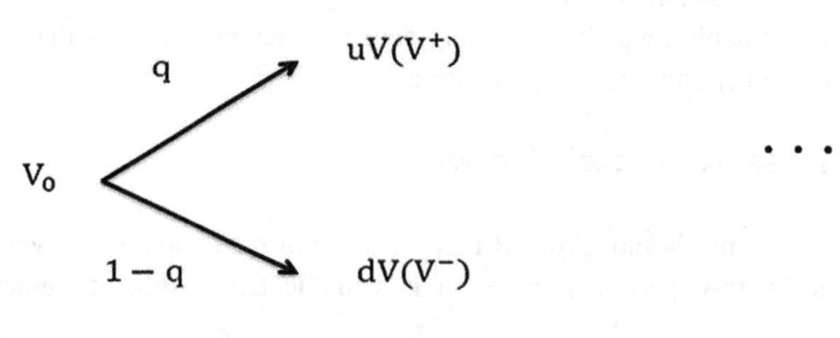

Figure 10. The structure of games

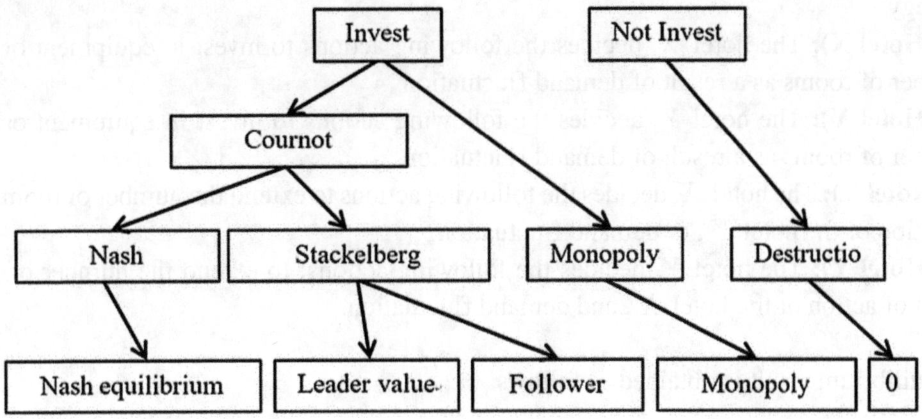

the Nash equilibrium value. When one company enters into a market, it reflects the Stackelberg balance and the investment value serves as leader value. The other company is set to Follower value. When its company is late entering a market, it becomes the Stackelberg balance and the investment value serves as Follower value. The other company is set to leader value. When only one company invests, it is monopolized and the investment value turns into monopoly value. When a company does not invest, it is abandoned and the value is 0. The result predicted from the action of one company and the other company is shown in Figure 10.

It is considered two companies by the case at the two times, and an option game model is built. Moreover, to consider this model as strategic, one company precedes and takes into consideration whether strategic-research investment for development is carried out.

Figure 11 shows the actions taken by each of hotels X and Y.

Figure 11. Option game model

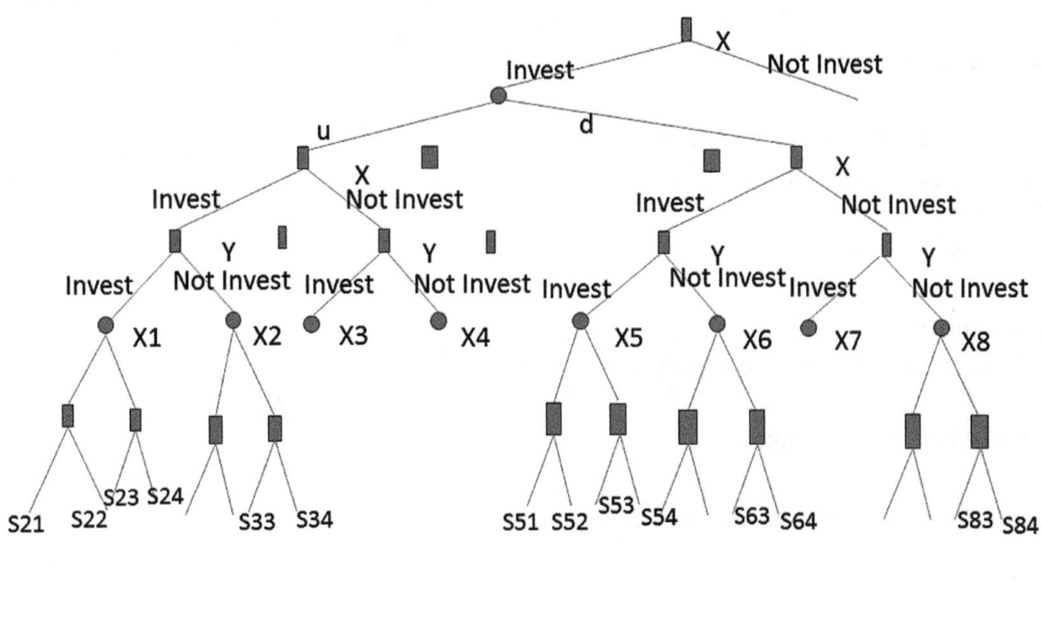

Phase 0 (Hotel X): The hotel X decides the following actions to precede or to invest towards a new strategy.
Phase 1 (Hotel X): The hotel X decides the following actions to invest in equipment or extend the number of rooms as a result of demand fluctuation.
Phase 1 (Hotel Y): The hotel Y decides the following actions to invest in equipment or extend the number of rooms as a result of demand fluctuation.
Phase 2 (Hotel X): The hotel X decides the following actions to extend the number of rooms as a result of action of the hotel Y, or demand fluctuation.
Phase 2 (Hotel Y): The hotel Y decides the following actions: to extend the number of rooms as a result of action of the hotel X, and demand fluctuation.

The equilibrium result is obtained as follows.

- **Hotel X:**
    - **Nash Equilibrium:** s1, s41, s45, s81, s85
    - **Stackelberg Equilibrium (Leader):** s21, s23, s31, s33
    - **Stackelberg Equilibrium (Follower):** s61, s63, s71, s73
    - **Monopoly:** s22, s24, s42, s46, s62, s64, s82, s86
    - **Destruction:** s32, s34, s43, s44, s47, s48, s72, s74, s83, s84, s87, s88
- **Hotel Y:**
    - **Nash Equilibrium:** s1, s41, s45, s81, s85
    - **Stackelberg Equilibrium (Leader):** s61, s63, s71, s73
    - **Stackelberg Equilibrium (Follower):** s21, s23, s31, s33
    - **Monopoly:** s32, s34, s43, s47, s72, s74, s83, s87
    - **Destruction:** s22, s24, s42, s44, s46, s48, s62, s64, s82, s84, s86, s88

As described above, the result is obtained when each of hotels X and Y took actions by using a real option-game model. Next, we calculate the value of the profit from the result of the competition.

The figure is a result of the prediction made from the actions taken by our company and other company

Parameter Settings

Q : The number of rooms (Production output),
I : Additional cost of the room,
C : Conception cost,
c : Operation unit cost,
q : Secondary cost coefficient,
P : Room price,
Z: Market demand rate with probability q (u rate).

With probability $1-q$ (d rate):

K_a : Strategic-research investment for development,

r : Interest rate,
k : Discount rate,
π : Operation profit,
V : Profit,
p : Risk neutral probability.

Then, the function that is used for the model is explained. Two competitive hotels are set to $i(i = X, Y)$ (Hotel X is the leading company) and j is a competitor of i.

Cost function:

$$C(Q_1) = c_1 Q_1 + \frac{1}{2} q_1 Q_1^2 \tag{18}$$

Management profit:

$$\pi_i(Q_i, Q_j, Z_n) = PQ_i - C(Q_i) \tag{19}$$

Item value:

$$V = \frac{\pi_i}{k} \tag{20}$$

Market demand function:

$$P(Q, Z_n) = Z_n - (Q_i + Q_j) \tag{21}$$

NPV value of each node:

$$NPV_i = V_i - I \tag{22}$$

Formula of Cournot Competition Equilibrium

The equilibrium value acquired from the result of competition is formulized.

Nash Equilibrium Value

Based on (22).

$$NPV_i = V_i - I = \frac{\pi_i}{k} - I$$

Based on (19).

$$PQ_i - C(Q_i) = \{Z_n - (Q_i + Q_j)\}Q_i - \left(c_iQ_i + \frac{1}{2}q_iQ_i^2\right) = -\left(1 + \frac{1}{2}q_i\right)Q_i^2 + \{(Z_n - c_i) - Q_j\}Q_i$$

This formula is a quadratic function. When $\frac{\partial \pi_i}{\partial Q_i} = 0$, the NPV value serves as the maximum with the value of Q_i.

If this is calculated, the optimum production output of Hotel i to the production output of Hotel j can be denoted by the following formula.

$$Q_i = \frac{(Z_n - c_i) - Q_j}{2 + q_i}$$

$$Q_j = \frac{(Z_n - c_j) - Q_i}{2 + q_j} = \frac{(Z_n - c_j) - \frac{(Z_n - c_i) - Q_j}{2 + q_i}}{2 + q_j}$$

The above calculation gives the following results:.

$$Q_i = \frac{(Z_n - c_j)(2 + q_j) - (Z_n - c_i)}{(2 + q_i)(2 + q_j) - 1}$$

$$Q_j = \frac{(Z_n - c_j)(2 + q_i) - (Z_n - c_i)}{(2 + q_i)(2 + q_j) - 1}$$

To simplify the calculation, let us assume $q_i = q_j = 0$,

$$Q_i = \frac{Z_n - 2c_i + c_j}{3} \tag{23}$$

$$Q_j = \frac{Z_n - 2c_j + c_i}{3} \tag{24}$$

Therefore, the maximum profits in the Nash equilibrium of the Cournot competition are as follows.

$$\pi_i(Q_i, Q_j, Z_n) = -\left(1 + \frac{1}{2}q_i\right)Q_i^2 + \left\{(Z_n - c_i) - Q_j\right\}Q_i$$

$$= \frac{(Z_n - 2c_i + c_j)^2}{9} \tag{25}$$

and the final profit can be expressed as follows.

$$V_i = \frac{\pi_i}{k} = \frac{(Z_n - 2c_i + c_j)^2}{9k} \tag{26}$$

As mentioned above, the optimum output and the final profit were calculated by Nash equilibrium point.

Stackelberg Equilibrium Value

In the case of a leader. If hotel i is made into a leader and hotel j is made into a follower, the follower j will adjust the production output such that profit maximization may be attained corresponding to the leader's i production output. Therefore, the production output of hotel j is denoted by the following formula.

$$Q_j = \frac{(Z_n - c_j) - Q_i}{2}$$

Corresponding to production output Q_i of hotel j, if hotel i is optimized for profit maximization, it will become the following formula.

$$\pi_i = -\left(1 + \frac{1}{2}q_i\right)Q_i^2 + \left\{(Z_n - c_i) - \frac{(Z_n - c_j) - Q_i}{2}\right\}Q_i$$

Here, the partial differentiation of the upper formula will be carried out as $\frac{\partial \pi_i}{\partial Q_i} = 0$ to give profit maximization. Production output of leading Hotel i is shown as follows:

$$Q_i = \frac{Z_n - 2c_i + c_j}{2} \tag{27}$$

The production output of Follower Hotel j is shown as

$$Q_j = \frac{(Z_n - c_i) + \frac{Z_n - 2c_i - c_j}{2}}{2} \tag{28}$$

By using the above, the final profit evaluation of the leader hotel and the follower hotel can be calculated in the following:

$$V_i = \frac{(Z_n - 2c_i + c_j)^2}{8k} \tag{29}$$

$$V_j = \frac{(Z_n - 3c_j + 2c_i)^2}{16k} \tag{30}$$

From the calculations above, the optimum production output and the final profit of the leader and the follower in Stackelberg equilibrium position were obtained.

Value of Monopoly Case and Abandonment Case

The forerunner hotel can monopolize the market until the competition hotel develops a new product as the late-coming hotel. Therefore, forerunner hotel $i(i = X \text{ or } Y)$ and determine that the production output will become the maximum about profits. Thus, the production output of hotel $j(j = X \text{ or } Y)$ is 0. $(Q_j = 0)$

The monopoly production output of hotel $i(i = X \text{ or } Y)$ is written as.

$$Q_i = \frac{(Z_n - c_i) - Q_j}{2 + q_i} = \frac{(Z_n - c_i)}{2 + q_i} \tag{31}$$

and the monopoly price is shown in the following:

$$P(Q_i, Z_n) = Z_n - (Q_i + Q_j) = \frac{Z_n(1 + q_i) - c_i}{2 + q_i} = \frac{Z_n + c_i}{2} \tag{32}$$

A monopolistic final profit becomes the following formula.

$$\pi_i = PQ_i - C(Q_i) = \frac{(Z_n + c_i)}{2}\frac{(Z_n - c_i)}{2} - c_i\frac{(Z_n - c_i)}{2} = \frac{(Z_n - c_i)^2}{4} \tag{33}$$

The monopoly final profit is shown as

$$V_i = \frac{(Z_n - c_i)^2}{4k} \tag{35}$$

It is set to 0 in the cases of abandonment.

Table 16. Formula of the Cournot competition

Market Competition Structure	Q_i	π_i	NPV_i
Nash Equilibrium	$\frac{Z_n - 2c_i + c_j}{3}$	$\frac{(Z_n - 2c_i + c_j)^2}{9}$	$\frac{(Z_n - 2c_i + c_j)^2}{9k} - I$
Stackelberg Equilibrium (Leader)	$\frac{Z_i - 2c_i + c_j}{3}$	$\frac{(Z_i - 2c_i + c_j)^2}{8}$	$\frac{(Z_i - 2c_i + c_j)^2}{8} - I$
Stackelberg Equilibrium (Follower)	$\frac{Z_i - 3c_i + 2c_j}{4}$	$\frac{(Z_i - 3c_i + 2c_j)^2}{16}$	$\frac{(Z_i - 3c_i + 2c_j)^2}{16} - I$
Monopoly	$\frac{Z_i - c_i}{2}$	$\frac{(Z_i - c_i)^2}{4}$	$\frac{(Z_i - c_i)^2}{4k} - I$
Destruction	0	0	0

Summary of Formula

An investment value can be evaluated, if a formula is substituted for each last item and it solves by Binomial CRR model.

- **Hotel X:**
  - **Nash Equilibrium:** $s1, s41, s45, s81, s85$
  - **Stackelberg Equilibrium (Leader):** $s21, s23, s31, s33$
  - **Stackelberg Equilibrium (Follower):** $s61, s63, s71, s73$
  - **Monopoly:** $s22, s24, s42, s46, s62, s64, s82, s86$
  - **Destruction:** $s32, s34, s43, s44, s47, s48, s64, s82, s86$
- **Hotel Y:**
  - **Nash Equilibrium:** $s1, s41, s45, s81, s85$
  - **Stackelberg Equilibrium (Leader):** $s61, s63, s71, s73$
  - **Stackelberg Equilibrium (Follower):** $s21, s23, s31, s33$
  - **Monopoly:** $s32, s34, s43, s47, s72, s74, s83, s87$
  - **Destruction:** $s22, s24, s42, s44, s46, s48, s62, s64, s82, s84, s86, s88$

An Option Game Model in the Application

Precondition

Hotel attempts to extend the room. Simultaneously, it is assumed that rival hotel has the same idea. The cost equivalent to a plan, the investment of a simulation, or capital investment begins the extension of the room. We would like to make decisions carefully. Moreover, considering market demand, we also must take the rival's hotel strategy into consideration. We would like to make investment decisions such that the profits of one hotel will serve as the maximum.

Q : The number of rooms (Production output),
I : Cost of additional of the room,
C : Conception cost,
c: Operation unit cost $(c_i = 20, c_j = 10)$,
P : Room price,
Z : Market demand; Increase $(u = 9/8)$; Decrease $(d = 7/8)$,
k : Discount rate $(= 1.05)$,
π : Operation profit,
V : Item value.

The Case When the Outside Environment Is Usual ($Z_x = 85, Z_y = 75$)

Becoming the Nash equilibrium from the above is a case in which hotel X chooses strategy c_1, and hotel Y chooses strategy c_2. Management unit cost ($c_i = 20, c_j = 10$) (see Table 18).

The Case When the Demand Is High ($Z_x = 89, Z_y = 91$)

Becoming the Nash equilibrium from the above is a case in which hotel X chooses strategy b_1, and hotel Y chooses strategy b_2. Management unit cost ($c_i = 20, c_j = 10$) (see Table 19).

The Case When the Demand Is Low ($Z_x = 75, Z_y = 65$)

See Table 20.

Table 17. The case when Hotel X is leading hotel

Structure of Market Competition	$Q_i(\times 10^2)$	$\pi_i(\times 10^5)$	$NPV_i(\times 10^6)$
Nash Equilibrium	0.183	0.336	0.320-I
Stackelberg Equilibrium (Leader)	0.175	0.378	0.360-I
Stackelberg Equilibrium (Follower)	0.087	0.076	0.073-I
Monopoly	0.325	1.056	1.006-I
Destruction	0	0	0

Table 18. The case when Hotel Y is leading hotel

Structure of Market Competition	$Q_i(\times 10^2)$	$\pi_i(\times 10^5)$	$NPV_i(\times 10^6)$
Nash Equilibrium	0.150	0.225	0.214-I
Stackelberg Equilibrium (Leader)	0.125	0.253	0.241-I
Stackelberg Equilibrium (Follower)	0.113	0.127	0.121-I
Monopoly	0.275	0.756	0.720-I
Destruction	0	0	0

Table 19. The case when Hotel X is leading hotel

Structure of Market Competition	$Q_i(\times 10^2)$	$\pi_i(\times 10^5)$	$NPV_i(\times 10^6)$
Nash Equilibrium	0.197	0.387	0.368-I
Stackelberg Equilibrium (Leader)	0.195	0.435	0.414-I
Stackelberg Equilibrium (Follower)	0.127	0.163	0.155-I
Monopoly	0.345	1.190	1.134-I

Table 20. The case when Hotel Y is leading hotel

Structure of Market Competition	$Q_i(\times 10^2)$	$\pi_i(\times 10^5)$	$NPV_i(\times 10^6)$
Nash Equilibrium	0.203	0.413	0.394-I
Stackelberg Equilibrium (Leader)	0.205	0.465	0.443-I
Stackelberg Equilibrium (Follower)	0.245	0.150	0.143-I
Monopoly	0.355	1.260	1.200-I
Destruction	0	0	0

Destruction	0	0	0

8. CONCLUSION

The optimum strategy has been developed for hotel yield management in the Bertrand situation. In this study, the number of customers varies randomly using exponential distribution. If we can determine the number of guests who stay in a particular hotel, we can obtain the optimal decision-making strategies by using the Bertrand simulation of our research. In the case of the Bertrand model, hotel managers also must accurately select strategies to estimate price. Moreover, the timing of a new capital budgeting decision and investment value evaluation can be solved from two companies at two points in time by introducing the real option. If the variation of a company's action is increased, it will be a proposal as a more practical model.

1. Although this option game model is under the two companies and two times, it is required to solve the optimum answer in the case of three or more companies and two or more points. This result occurs because there are few situations for which it fights actually in two companies.
2. When calculating by Section 7, management cost applied the value assumed while it had been ambiguous. Then, it is necessary to an ambiguous and indefinite value to introduce the fuzzy variable which a more suitable value can compute.

REFERENCES

Amram, M., & Kulatilaka, N. (1999). *Real options: managing strategic investment in an uncertain world.* Oxford Press.

Angelou, G. N., & Economides, A. A. (2009). A multi-criteria game theory and real-options model for irreversible ICT investment decisions. *Telecommunications Policy*, *33*(10-11), 686–705. doi:10.1016/j.telpol.2009.07.005

Baldursson, F. M. (1998). Irreversible Investment under Uncertainty in Oligopoly. *Journal of Economic Dynamics & Control*, *22*(4), 627–644. doi:10.1016/S0165-1889(97)00070-5

Cachon, G. P. (2001). Stock wars: Inventory competition in a two-echelon supply chain with multiple retailers. *Operations Research*, *49*(5), 658–674. doi:10.1287/opre.49.5.658.10611

Choi, S. C. (1991). Price competition in a channel structure with a common retailer. *Marketing Science*, *10*(4), 271–296. doi:10.1287/mksc.10.4.271

Cox, J. C., Ross, S., & Rubinstein, M. (1979, September). Option pricing: A simplified approach. *Journal of Financial Economics*, *7*(3), 229–263. doi:10.1016/0304-405X(79)90015-1

Dixit, A. K. (1989). Entry and Exit Decisions under Uncertainty. *Journal of Political Economy*, *97*(3), 620–638. doi:10.1086/261619

Dixit, A. K. (1994). *Investment under Uncertainty*. Princeton, NJ: Princeton University Press.

Doi, H. (2003). *Discounted Cash Flow: Business Analysis and Valuation. Toyo Keizai Inc.* (in Japanese)

Ertek, G., & Griffin, P. M. (2002). Supplier- and buyer-driven channels in a two-stage supply chain. *IIE Transactions*, *34*(8), 691–700. doi:10.1080/07408170208928905

Fudenberg, D., & Tirole, J. (1985, July). Pre-emption and Rent Equalisation in the Adoption of New Technology. *The Review of Economic Studies*, *52*(3), 383–401. doi:10.2307/2297660

Grenadier, S. R. (1996). The Strategic Exercise of Options: Development Cascades and Overbuilding in Real Estate Markets. *The Journal of Finance*, *51*(5), 1653–1679. doi:10.1111/j.1540-6261.1996.tb05221.x

Herath, H. S. B., & Barth, J. R. (2001). *A Nested Option Model for Multi-stage Capital Investment Decisions*. Eighth Asia Pacific Finance Association (APFA) Annual Conference 2001, Bangkok, Thailand.

Huisman, K. J. M. (2001). *Technology and Investment: A Game Theoretic Real Options Approach*. Kluwer Academic Publishers. doi:10.1007/978-1-4757-3423-2

Huisman, K. J. M., & Kort, P. M. (1999). *Effects of Strategic Interactions on the Option Value of Wating*. Discussion Paper 1999-92.

James, W. N. (2006). *Game Theory Decisions, Interaction and Evolution*. Springer-Verlag.

Kijima, M., & Shibata, T. (2002). *Real Options in a Duopoly Market with General Volatility Structure*. Discussion Paper, Kyoto University

Kobayashi, K., Matsushima, K., & Hishida, K. (2008). Economic benefit evaluation of reservation system. *JapanSociety of Civil Engineering, Transaction D*, *64*(2), 299-318. (in Japanese)

Kong, J. J., & Kwok, Y. K. (2007). Real options in strategic investment games between two asymmetric firms. *European Journal of Operational Research*, *181*(2), 967–985. doi:10.1016/j.ejor.2006.07.006

Kulatilaka, N., & Perotti, E. C. (1998, August). Strategic Growth Options. *Management Science*, *44*(8), 1021–1031. doi:10.1287/mnsc.44.8.1021

Lau, A., & Lau, H.-S. (2002). Effects of a Demand-Curve"s Shape on The Optimal Solutions of a MultiEchelon Inventory/Pricing-model. *European Journal of Operational Research*, *147*(3), 530–548. doi:10.1016/S0377-2217(02)00291-6

Leahy, J. V. (1993). Investment in Competitive Equilibrium: The Optimality of Myopic Behavior. *The Quarterly Journal of Economics*, *108*(4), 1105–1133. doi:10.2307/2118461

McDonald, R., & Siegel, D. (1986). The Value of Waiting to Invest. *The Quarterly Journal of Economics*, *101*(4), 707–728. doi:10.2307/1884175

Messinger, P. R., & Narasimhan, C. (1995). Has power shifted in the grocery channel? *Marketing Science*, *14*(2), 189–223. doi:10.1287/mksc.14.2.189

Murto, P. (2004). Exit in Duopoly under Uncertainty. *The Rand Journal of Economics*, *35*(1), 111–127. doi:10.2307/1593732

Murto, P., Nasakkala, E., & Keppo, J. (2004). Timing of investments in oligopoly under uncertainty: A framework for numberical analysis. *European Journal of Operational Research*, *157*(2), 486–500. doi:10.1016/S0377-2217(03)00234-0

Myers, S. C. (1977, November). Determinants of corporate borrowing. *Journal of Financial Economics*, *5*(2), 147–175. doi:10.1016/0304-405X(77)90015-0

Nash, J. F. Jr. (1950, January). Equilibrium points in n-person games. *Procs.of the National Academy of Sciences*, *36*(1), 48–49. doi:10.1073/pnas.36.1.48 PMID:16588946

Noro, R., Dobashi, Y., Wang, Z. Y., Takahashi, T., Okamoto, T., Tanaka, T., & Watada, J. (2006, April). Real Options Approach to Life Cycle Management of Power Equipments. *Proceedings, CMD 2006, International Conference on Condition Monitoring and Diagnosis Changwon.*

Parlar, M., & Weng, Z. K. (2006). Coordinating Pricing and Production Decisions in The Presence of Price Competition. *European Journal of Operational Research, 170*(1), 211–227. doi:10.1016/j.ejor.2004.06.026

Pawlina, G., & Kort, P. M. (2006). Real options in an asymmetric duopoly: Who benefits from your competitive disadvantage? *Journal of Economics & Management Strategy, 15*(1), 1–35. doi:10.1111/j.1530-9134.2006.00090.x

Reinganum, J. F. (1981). On the diffusion of new technology: A game theoretic approach. *The Review of Economic Studies, 48*(3), 395–405. doi:10.2307/2297153

Sarkar, S. (2000). On the investment–uncertainty relationship in a real options model. *Journal of Economic Dynamics & Control, 24*(2), 219–225. doi:10.1016/S0165-1889(99)00005-6

Shigeo, M. (2001). *Introduction to Game Theory*. Nikkei Publishing Inc.

Smit, H. T. J. (2003). Infrastructure investment as a real options game: The case of European airport expansion. *Financial Management, 32*(4), 27–57. doi:10.2307/3666135

Smit, H. T. J., & Ankum, L. A. (1993). A Real Options and Game-Theoretic Approach to Corporate Investment Strategy Under Competition. *Financial Management*, 223.

Smit, H. T. J., & Trigeorgis, L. (2006, January). Real options and games: Competition, alliances and other applications of valuation and strategy. *Review of Financial Economics, 15*(2), 95–112. doi:10.1016/j.rfe.2005.12.001

Stackelberg, H. V. (1934). Marktform und Gleichgewicht. Springer Verlag.

Stenbacka, R., & Tombak, M. M. (1994, September). Strategic Timing of Adoption of New Technologies under Uncertainty. *International Journal of Industrial Organization, 12*(3), 387–411. doi:10.1016/0167-7187(94)90040-X

Suzuki, K. (Ed.). (1999). *MBA Game Theory*. Diamond Ltd. (in Japanese)

Trigeorgis, L. (1991). Anticipated Competitive Entry and Early Preemptive Investment in Deferrable Projects. *Journal of Economics and Business, 43*(2), 143–156. doi:10.1016/0148-6195(91)90014-N

Trigeorgis, L. (1996). *Real Options: Managerial Flexibility and Strategy n Resource Allocation*. The MIT Press.

Tsai, W. H., & Hung, S. J. (2009, April). Dynamic pricing and revenue management process in Internet retailing under uncertainty: An integrated real options approach. *Omega, 37*(2), 471–481. doi:10.1016/j.omega.2007.07.001

von Neumann, J. (1928, December). Zur Theorie der Gesellschaftsspiele. *Mathematische Annalen, 100*(1), 295–320. doi:10.1007/BF01448847 von Neumann, J. (1928, June). Sur la theorie des jeux. *Comptes Rendus delAcademie des Sciences, 186*(25), 1689-1691.

von Neumann, J., & Morgenstern, O. (1947). *Theory of Games and Economic Behavior* (2nd ed.). Princeton, NJ: Princeton University Press.

Walker, P. (2005, October). A Chronology of Game Theory. *History of Game Theory.*

Wang, B., Wang, S. M., & Watada, J. (2009, July). Improved real option analysis based on fuzzy random variables. *Machine Learning and Cybernetics, 2009 International Conference on*, (pp. 694- 699). doi:10.1109/ICMLC.2009.5212448

Wang, B., Wang, S. M., & Watada, J. (2010, June). Real Options Analysis Basedon Fuzzy Random Variables. *International Journal of Innovative Computing, Information, & Control, 6*(4), 1689–1698.

Watada, J. (2008, May). *A Real Options Approach to Project Evaluation for MIS.* The 19th International Conference on Information Management.

Watada, J., & Song, W. (2010, December). Strategy Building for Foreign Exchange Exposure. *ICIC Express Letters, 1*(1), 1–17.

Watada, J., Waripan, T., & Wu, B. (2014). Optimal decision methods in two-echelon logistic models. *Journal of Management History, 52*(7), 1273 – 1287.

Williams, J. T. (1993). Equilibrium and Options on Real Assets. *Review of Financial Studies, 6*(4), 825–850. doi:10.1093/rfs/6.4.825

Yaakob, S.B., & Watada, J. (2010, July). *Power System Equipments Investment Decision-Making under Uncertainty: A Real Options Approach.* KES-IDT2010, Baltimore, MD.

Yang, S. L., & Zhou, Y.-W. (2006). Two-echelon supply chain models: Considering duopolistic retailers different competitive behaviors. *International Journal of Production Economics, 103*(1), 104–116. doi:10.1016/j.ijpe.2005.06.001

Zhu, M. Y., & Watada, J. (2008). A Real Option Approach to Pharmaceutical R&D Project Valuation. *Proceedings of, Int. Symposium on Management Engineering 2008.*

ADDITIONAL READING

Dixit, A. K., & Pindyck, R. S. (1994). *Investment Under Uncertainty.* Princeton University Press.

Grenadier, S. R. (Ed.). (2000). *Game Choices: The Interaction of Real Options and Game Theory.* Risk Books.

Maeland, I. (1999), "Valuation of irreversible investments and agency problems," Working Paper presented at the 3rd Annual Real Option Conference.

Sick, G. (1995), "Real Options," in Jarrow R.A., V.Maksimovic and W.T.Ziemba eds., Handbook in Operations Research and Management Science, Vol.9, Finance, chap.21, Elsevier.

Trigeorgis, L. (1996). *Real Options: Managerial Flexibility and Strategy in Resource Allocation*. MIT Press.

Wang, B., Wang, S. M., & Watada, J. (2009, July), "Improved real option analysis based on fuzzy random variables," *Machine Learning and Cybernetics, 2009 International Conference on*, 694- 699 doi:10.1109/ICMLC.2009.5212448

Wang, B., Wang, S. M., & Watada, J. (2010, June). Real Options Analysis Basedon Fuzzy Random Variables. *International Journal of Innovative Computing, Information, & Control*, *6*(4), 1689–1698.

Watada, J., & Song, W. (2010, December). Strategy Building for Foreign Exchange Exposure. *ICIC Express Letters*, *1*(1), 1–17.

Watada, J., Waripan, T., & Wu, B., (2014) "Optimal decision methods in two-echelon logistic models," *Journal of Management History (Archive) merged into Management Decision*, 52(7), 1273 - 1287

Chapter 3
Tourism Search and Metasearch Engines for Online Booking:
What Do They Offer?

Trinidad Domínguez
University of Vigo, Spain

Jose Antonio Fraiz
University of Vigo, Spain

Noelia Araújo
University of Vigo, Spain

Elisa Alén
University of Vigo, Spain

ABSTRACT

This chapter aims to analyze the different tourism search and metasearch engines for online booking based on supplier perspective (accommodation, flights, leisure and package deals) and taking into account the possible relations generated by the structure and content variables of web offers. A literature review is carried out to study user experience and provide greater in-depth knowledge which will, in turn, in addition to observation of online booking, trends, data, and profile and user preferences. With this information, the study establishes the main attributes of strucuture and offer content of webs focused on the user preferences. The empirical analysis is based on a representative sample of e-commerce of tourism websites, its main characteristics and possible correlations according to the structure and content of their offers. Based on this, the work identifies competitive advantages that will set the trends for the sector and future short-term strategies.

INTRODUCTION

The emergence of the Internet and the ease with which it can be used and accessed has no doubt changed many aspects of our lives. Certainly, one of these aspects is the way we buy. This has led many companies to review the way they market their products, focusing their efforts on the management of online marketing. Within this context, two main concepts arise:

- E-commerce, and
- E-business.

DOI: 10.4018/978-1-5225-1054-3.ch003

E-business is the use of information technologies in various business operations, while e-commerce is the latest step in the evolution of business transactions process. It comprises the sale of goods or services through computer networks, i.e., it uses internet as a channel. As a matter of fact, e-commerce may be considered as a set of e-business applications used in commercial transactions (Kalakota & Robinson, 2001).

Among the various activities that can be found within e-commerce, we are particularly interested those of businesses serving end consumers with products and / or services (Kumar & Raheja, 2012). B2C e-commerce sales include all products and services ordered or booked throughout the Internet on any search. B2C e-commerce sales worldwide reached $ 1,471 trillion in 2014; this represents an increase of nearly 20% with respect to 2013. However, e-commerce growth will slow down and settle at around 10% by the end 2018, with sales reaching $ 2,356 trillion (eMarket, 2014; United Nations, 2015). With the clear growth trend of e-commerce in the forthcoming years supported by an ever-increasing Asia-Pacific market and the continued presence of North America and Western Europe, B2C becomes of strategic value to the company. One of the sectors in which it is increasingly becoming relevant is the tourism sector. According to the report of the United Nations (2015), travel-related items (such as tours, hotel reservations and airline tickets) rank among the top five items consumers wished to buy online in 2014 throughout all regions; that is to say,

- Asia and Oceania (the Pacific Islands),
- Europe,
- Africa and the Middle East,
- Latin America, and
- North America.

Furthermore, according to the World Tourism Organization (UNWTO), tourism in the world in 2014 generated 9% of GDP (direct, indirect and induced impact), 1 in 11 jobs, US $ 1.5 trillion in exports, 6% of the world exports, 1,133 million international tourists and 6 billion domestic tourists. Moreover, tourism is expected to generate 1.8 billion international tourists by 2030 (UNWTO, 2015). A market opportunity will thus be created and supported by the continued medium and long-term growth of this sector along with the ever-growing e-commerce activity itself.

As in other sectors, tourism is also undergoing many changes and internet is to blame for most of them. Internet is the main source for travel planning. Travelers are turning to internet for travel inspiration in the same way they would turn to their own family, friends or colleagues for advice (Ipsos MediaCT & Google Travel Study, 2014). Supporting this statement, the gross bookings of Online Travel Agencies (OTAs) exceeded $ 150 billion in 2013. This represents 38% of the global online market and 13% of the overall travel market for that year. The online travel sales figure is estimated to be growing at 12% per year. Overall, global leaders of online travel include Priceline and Expedia which each sell more than 22 million and 12 million hotel room nights per month, respectively. The world's largest online travel platform, TripAdvisor, receives 315 million monthly unique visitors (Forbes, 2015).

Drawing on data from Nielsen (2014), we may see that 42% of the survey respondents make searches on tours and hotel reservations in the upcoming six months and 44% of them make purchases; in the case of airline tickets and reservations, this percentage increases to 48%. The most significant markets include Latin America, where 52% of the consumers search for tours/hotel reservations and 32% make online purchases; 44% of the consumers carry out these type of searches in the Asian - Pacific area and

53% of them make online purchases. This trend is the same for airline tickets, in which the propensity to purchase is greater than the tendency to only search, 59% vs. 43%. From a global perspective, we can see a substantial increase from 2013 to 2014 in terms of tours and hotel reservations which moved from a global intent-to-buy-online average of 27% in the following 6 months to 44%. The trend in airline tickets and reservations, akin to that of tours and hotel reservations, increased from 33% to 48%. One of the most striking features concerning online buyers of travel services is that over 70% of them start their search online before even deciding where or how they want to travel (Ipsos MediaCT & Google Travel Study, 2014). A remarkable 59% of search engines will become go-to sources for travel information and ideas as well as a tool to compare. This will particularly be the case of airlines, hotels and car rentals (Ipsos MediaCT & Google Travel Study, 2014).

These new consumer habits create new role models and intermediaries in the online travel market which lead to a reduction in the distribution chain and cause processes of "disintermediation" or "reintermediation", that is, online agents take the place of traditional intermediaries (Tourism Business Portal, 2014). Internet offers have grown, but so has the need for an intermediary who can assist consumers in finding products at the price they want and help them understand the offers targeted (PhoCusWright's, 2012). The information available on the net is vast and ever-growing; these intermediaries especially search and metasearch engines help the consumer "visualize" what matters and is most useful to them (Xiang et al., 2008). The tourism sector has already revealed that most travelers make use of these tools to plan their holidays (TIA, 2009). This is why general search engines have created specific tourism sites (Yahoo and Google among others) and many new specific ones have arisen (Trivago, Tripadvisor, Booking, etc.). Nowadays, search and meta-search engines are the most popular online planning source for travelers; 57% of leisure travelers and 64% of business travelers start their travel booking and shopping processes with the use of search or meta-search engines (Ipsos Media CT & Google Travel Study, 2014).

In light of the above, this work aims to analyze the different tourism search and metasearch engines for online booking based on what they actually offer (accommodation, flights, leisure and package deals) and taking into account the possible relations generated by the structure and content variables of web offers. A literature review is carried out to study user experience and provide greater in-depth knowledge which will, in turn, in addition to observation of online booking, trends, data, and profile and user preferences. With this information, the study establishes the attributes like online booking, trends, data, and profile and user preferences. The empirical analysis is based on a representative sample of e-commerce of tourism websites, its main characteristics and possible correlations according to the structure and content of their offers. Among other things, this work would identify competitive advantages that will set the trends for the sector and future short-term strategies. Possible relationships that may arise among these trends and usage patterns are analyzed in order to improve the operation of the web accordingly.

BACKGROUND

Trends and Reality of E-Commerce in the Tourism Sector: Booking Online

Research extensively deals with general issues associated to e-commerce (Lee et al., 2007; Ngai & Wat, 2002; Urbaczewski et al., 2002, Wareham et al., 2005; Wang & Chen, 2010). However, the interest of this paper focuses on specific matters such as:

- Accepting and adopting online shopping (Cao & Mokhtarian, 2005; Casanovas, 2010; Chen & Holsapple, 2012, 2013; Chitura et al., 2008; Mohamad & Ismail, 2009; Parker & Castleman, 2006; Zhou et al, 2007),
- Marketing and personalization or co-creation (Adolphs & Winkelmann, 2010; Paredes & Barrutia, 2014; Schibrowsky et al., 2007).

In some cases this activity is analyzed from the point of view of the user. This requires taking into account user purchasing attitude (Al-Debei, Akroush & Ashouri, 2015), and experience (Carlson & O'Cass, 2011; Pallud & Straub, 2014; Visinescu et al., 2015) or behaviour (Floh & Madlberger, 2013; Lin & Huang, 2014; Rondán et al., 2014). Among many other studies on e-commerce in tourism are those of Fesenmaier et al. (2011), Kamis and Frank (2014) and Ip, Lee and Law (2012) concerning online travel planning or the work of Crnojevac et al., (2010) and Andrés et al., (2014) on online hotel booking as well as the work that analyses e-commerce tourism websites principally from the perspective of the distribution, accessibility and quality of their interfaces and content (Akgül & Vatanserver, 2016; Davidavičienė & Tolvaišas, 2011, Granados et al., 2008).

Initially, the main objective of e-commerce was the exchange of goods or services between two parts through electronic means. This relationship has currently expanded and it continually implements new elements to generate ongoing added value. Nowadays, according to recent data from Eurostat, approximately half of the regular internet users in the EU have made some sort of online purchase. Internet users in the UK, with 82% of internet users buying online regularly (Tourism Economics, 2013), are reportedly more likely to make online purchases than are users in any other EU country.

Although e-commerce has different exchange modalities in function of the parties involved (see Table 1), for our analysis we will take e-commerce business to consumer (B2C) as a reference.

The most relevant features or advantages in B2C may be summed up in the fact that consumers are provided with more information and they may easily compare markets and products, participate in product design and comfort and they may save time and effort on purchases. Also, the company may improve the distribution and sale of their products, interact more with customers, and they may increase the speed of their management processes, simplify tasks, be more flexible, adaptable and gain operational profits by reducing errors, time, etc. (De Pablo, 2004).

Therefore, Thakran and Verma (2013) advise companies to invest resources to improve their web position on search engines. This is the best way to directly reach consumers and it generates a new role for online tourism intermediaries. Starting from the premise that one of the main functions of tourism intermediaries is disclosure, the emergence of the Internet has boosted the expansion and dissemination of online channels (Vallespín & Molinillo, 2014) which has spurred the emergence of new tourism intermediaries such as:

- **Search Engines:** Intermediaries are limited to providing information and contacting the members of electronic channels (Rodriguez, 2002).
- **Metasearch Engines:** Mega search engines that allow web users to browse various search engines and web directories at a time (Ho, Lin & Chen, 2012). That is to say, this system locates information in the most commonly used search engines. It lacks its own database and uses other search engines to show a combination of the best pages returned by each one of them.

Table 1. Types of e-commerce

Type	Components
Business To Business - B2B	Between companies that operate in the net/network. The following modalities can be distinguished: • The market controlled by a seller in search of buyers. • The market controlled by the buyer in search of suppliers. • The market controlled by intermediaries seeking an agreement among buyers and sellers.
Customer To Customer - C2C	Among consumers, who may use email correspondence or *peer to peer* technology.
Business To Customer - B2C	Among companies and the consumer. The most notable among them are: • **On-line Intermediaries:** Facilitate transactions among buyers and sellers in exchange of a share of economic compensation. • **Models Based on Publicity:** Publicity based on the system in which the companies have inventory websites to sell to interested parties. • **Models Based on the Community:** The companies let users all over the world interact in similar areas of interest. • **Models Based on Tariffs:** A payment system based on the system. In this case the company charges a fee to allow access to its content.
Business To Administration - B2A	Between the administration and the firms (It may also happen with citizens) for the realization of administrative tasks.
Business To Employee - B2E	Among the company and its employees.
Citizen To Government - C2G	Among consumers and the government for telematic interchange.
Business To Government - B2G	Among companies and the government seeking to optimize negotiation processes among them.

Source: Authors based on Montane (2010), Valverde (2011) and the Spanish Association of Electronic Commerce and Relationship Marketing (AECEM) (2009).

- **Online Travel Agencies (OTA):** Only available on the Internet with no physically located establishment (Vázquez-Casielles et al., 2009). In addition, customers may find a single tourist product, such as a flight or a set or combination of flights, as well as last-minute offers (Rensmann, 2012).
- **Destination Marketing Organizations (DMOs):** Entities responsible for promoting tourism and attracting tourists to a specific geographical area. Their ultimate goal is to develop the local economy and improve its image in coordination with other tourism industry agents (Presenza, Moreno & Lobo, 2007).
- **Collective Purchase Portals, Online Outlets, and Flash Sale Sites:** Outlet websites offering discounted products are also breaking into the tourism sector, although they are still in the consolidation phase. Apart from major generalist portals (e.g. Groupon, Planeo, Lets Bonus, Groupalia, etc.), some travel agencies are also based on the same outlet model and they work like a private travel club. Access to these travel agencies is given through subscription through which the site sends emails with special offers (Voyages Privé or Hotelminute) (Núñez, 2012).

This scenario evidences the great transformation of the tourism sector which is further heightened and marked by the massive use of social media, interactive television and mobile devices that provide a greater and more immediate experience exchange (Vallespín & Molinillo, 2014). According to the latest study by Ipsos MediaCT and Google Travel Study (2014), friends, family or colleagues and online sources are critical for travel inspiration with a total of 56% of users and 74% of leisure travellers and 77% of business travellers using internet as a planning source. The top online sources of inspiration are social networking, video or photo sites with 83% (YouTube 42%) and search engines with 61%. Search engines are the most popular online planning source for 60% of travellers and the second most popular for 55% of business travellers. Smart phones are used throughout the travel process just like computers and tablets but nearly half of the people who use their smart phone for leisure travel inspiration ultimately book through some other method. Those booking online represent 87% (70% do so via computer, 38% via smart phone and 17% via tablet) and 29% do so via phone call. Once at their destinations, leisure travellers decide on their activities:

- Heavily relying on smart phones (56%),
- Followed by the computer brought from home / work (49%),
- Tablets (35%),
- Computer at accommodation (29%), and
- Other means (4%).

Finally it is worth noting that four in ten travellers watch no commercials even when they are watching live TV. However, they do engage heavily with online video and they watch more than just travel content. Online videos are viewed throughout the travel journey, and this is particularly so prior to making decisions.

Based on the above, one may appreciate the importance of e-commerce in tourism as well as its tendency to keep growing, not only as a tool to seek information and compare bids but also as a way to manage the purchase of tourism products and services. If we briefly compare tourist sites performing e-commerce B2C worldwide, we may observe that the following online travel booking sites were the most popular in May 2014:

Booking.com with 210.5 milions estimated desktop visits,

- TripAdvisor sites with 159.6,
- Expedia sites with 63.5,
- Hotels.com with 34.5,
- Airbnb sites with 32.5M,
- Agoda.com with 31.6,
- Priceline.com with 28,
- Skyscanner sites with 23.4,
- Kayak sites with 23.3,
- Makemytrip.com with 18.4,
- Cheapoair.com with 18, and
- Trivago sites with 17.5 (Skift, 2014).

This ranking below does not account for direct sites of travel brands (airlines/hotels) themselves, which we have done in separate lists previously and all brands of different countries are combining. But these results are different if the analysis is focused on the companies:

- Priceline Group, which includes Booking.com, Priceline.com, Agoda.com and Kayak sites, is the leader by a huge margin.
- Expedia brand (all country domain sites together) is third largest, Expedia Group, which includes Hotels.com, Trivago, Hotwire and Venere, is the second most popular online travel booking company.

- TripAdvisor is surely the biggest media-plus-booking site globally, and the second most popular brand if you add up all country-specific sites.
- Remarkably, Airbnb is now the fifth most popular online travel booking site, and is insanely addictive when you count the time spent on site. Part of it has to do with the design of the site, meant to be browsed more leisurely; the other part is more endemic, where users have to browse more listing in sharing/rentals to come to the right choice.
- Skyscanner is now among the top 10 most popular booking brands globally, and it is now making a big push in U.S. as well.
- MakeMyTrip is now among the top 10 most popular online booking sites globally, and is the only local site to reach that high level, showing the potential and size of online travel market in India and beyond. Other Indian sites Cleartrip and Yatra are also in the Top 30 list globally.
- It seems that Trivago's various country sites are now seeing the network effects from being part of Expedia, and its traffic is rising fast.

Main Elements for the Success of E-Commerce Tourist Sites According to User Preferences

Since e-tourism website user preferences will be taken as the basis of our analysis, the work of Muhtaseb et al. (2012) serves as a reference to analyze the experience of website users. We also follow the work of Park and Gretzel (2007), which carries out a literature review to identify the success factors for destination marketing web sites. Both of these studies focus on the variables related to the structure of the web.

Muhtaseb et al. (2012) collect four broad categories with their respective factors, which include:

- Contents,
- Usability,
- Functionality, and
- Branding.

These categories coincide with other proposals made by different authors in terms of quality or intended web use. Thus, concerning contents, Santos (2003) states similar coverage factors:

- Content,
- Linkage, and
- Appearance.

David et al. (1989), Loiacono et al. (2007) and Vijayasarathy (2004) speak of ease of use; Loiacono also includes the entertainment value; and finally Calero et al. (2005) point out reliability and portability. In terms of usability terms like use (Santos, 2003), usefulness (David et al., 1989; Loiacono et al., 2007; Vijayasarathy, 2004) and usability (Calero et al., 2005) appear. Calero et al. (2005) and Santos (2003) also refer to functionality in their work using structure. Finally, in branding we find terms such as search and information products or brand recognition and reputation (Abeles et al., 1999; Santos, 2003). Moreover, Park and Gretzel (2007) include a total of nine factors:

- Information quality,
- Ease to use,
- Responsiveness,
- Security / privacy,
- Visual appearance,
- Trust,
- Interactivity,
- Personalization and fulfilment.

These factors either totally or partially coincide with the aforementioned categories. Taking into account all the factors and categories relative to the structure of the website based on user experience, we propose to establish a set of variables that collect the key analytical attributes and categories (see Table 2).

A problem arises when attempting to determine the dimensions to be analyzed for each of these factors. As already indicated, research carried out in other fields has been taken as a reference in order to perform a broader review given the small number of papers on e-commerce in tourism. This can help establish the most relevant attributes or dimensions when analyzing e-commerce sites taking into account the factors listed above (see Table 2). Likewise, given that the study focuses on tourism, it collects specific variables related to tourism, i.e., the offers made through tourist e-commerce sites. Thus it analyzes the main components of online tourist products (UNWTO, 2014, 2015).

Table 2. User preferences attributes of websites

Structure Attributes	Categories	Authors	Offer Content Attributes	Categories	Author
Access	Accessibility needs, design satisfies	Hanson (2000), Henry (2005), Jamshidi (2008), Muhtaseb et al., (2012), Zaharias and Poylymenakou (2009)	Loyalty system	Mailing, chat, promotions, discount bond, gift card, gift certificate, field collection of questions and/or suggestions	Ab Hamid et al., (2010), Lawson- Body and Limayem (2004),
Composition of contents	Organization, layout, heading and style, readability, current and timely information, complexity, informative product page	Banati et al., (2006), Barnes and Vidgen (2003), Cao et al., (2005), Calero et al., (2005), Carlshamre (2001), Chakraborty et al., (2003), Eighmey (1997), Floh and Madlberger (2013), Loiacono et al., (2007), Mich et al., (2003), Muhtaseb et al., (2012), Parasuraman et al., (2007), Santos (2003), Zaharias and Poylymenakou (2009)	Offer type	Offer by category, offer by product	Arsal et al., (2010), Crnojevac at al., (2010), Fresenmaier et al., (2011), Hu et al., (2014), Steinmetz et al., (2013)

continued on following page

Table 2. Continued

Structure Attributes	Categories	Authors	Offer Content Attributes	Categories	Author
Design and structure	Website design, homepage, graphical interface, page background, color consistency,	Banati et al., (2006), Barnes and Vidgen (2003), Cao et al., (2005), Carlshamre (2001), Dix et al., (2003), Floh and Madlberger (2013), Loiacono et al., (2007), Mich et al., (2003), Muhtaseb et al., (2012), Nielsen (1993), Parasuraman et al., (2007), Park and Gretzel (2007), Santos (2003), Zaharias and Poylymenakou (2009)	Operative CRM	Customer tracking, campaign tracking, RSS, alarm, mailing, blog, membership card	Choudhury and Harrigan (2014), Hamid et al., (2010), Jayachandran et al., (2015), Lawson- Body and Limayem (2004), Hu et al., (2014)
Interactivity (real time, help)	Computer-mediated, communication, real-time communication, dialogue, interactive decision aids, transaction-related vs. non-transaction related interactivity	Barnes and Vidgen (2003), Cao et al., (2005), Calero et al., (2005), Deighton (1996), Dix et al., (2003), Ghose and Dou (1998), Loiacono et al., (2007), Marlow (2010), Mich et al., (2003), Muhtaseb et al., (2012), Parasuraman et al., (2007), Park and Gretzel (2007), Santos (2003), Zaharias and Poylymenakou (2009)	Package components	Flight, accommodation, leisure, flight + hotel, flight + hotel + tour, hotel + tour + holidays, holidays, cruises, tour, breaks, destinations, cars, coaches, trains, apartments	Arsal et al., (2010), Crnojevac at al., (2010), Fresenmaier et al., (2011), Kamis and Frank (2014), Steinmetz et al., (2013)
Learning	Familiarity, consistency, generalizability, predictability, simplicity	Banati et al., (2006), Barnes and Vidgen (2003), Cao et al., (2005), Calero et al., (2005), Dix et al., (2003), Eighmey (1997), Loiacono et al., (2007), Louis (2003), Mich et al., (2003), Muhtaseb et al., (2012), Nielsen (1993), Parasuraman et al., (2007), Santos (2003), Vijayasarathy (2004),	Payment method	Visa, Mastercard, Maestro, Visa Electron, American Express, Paypal, Company Card	Barnes and Vidgen (2003), Cao et al., (2005), Calero et al., (2005), Loiacono et al., (2007), Park and Gretzel (2007), Santos (2003),
Memorability	Interface memorability	Banati et al., (2006), Muhtaseb et al., (2012), Nielsen (1993),	Product browser	Filtering by destination, product, category, use browsers, use filters, use some languages	Arsal et al., (2010), Crnojevac at al., (2010), Fresenmaier et al., (2011), Niemann et al., (2008)
Navegation	Know where they are, easily return, search	Barnes and Vidgen (2003), Cao et al., (2005), Calero et al., (2005), Carlshamre (2001), Floh and Madlberger (2013), Loiacono et al., (2007), Matsatsinis et al., (2003), Mauro (2002), Mich et al., (2003), Muhtaseb et al., (2012), Parasuraman et al., (2007), Qin and Zhang (2001), Park and Gretzel (2007), Santos (2003), Zaharias and Poylymenakou (2009)	Product type	Accommodation, journey, leisure, package holidays	Arsal et al., (2010), Crnojevac at al., (2010), Fresenmaier et al., (2011), Hu et al., (2014), Steinmetz et al., (2013)

continued on following page

Table 2. Continued

Structure Attributes	Categories	Authors	Offer Content Attributes	Categories	Author
Performance and availability (easy search)	Efficiency, product differentiation and comparisons	Banati et al., (2006), Barnes and Vidgen (2003), Cao et al., (2005), Calero et al., (2005), Dix et al., (2003), Fazlollahi (2002), Loiacono et al., (2007), Mich et al., (2003), Muhtaseb et al., (2012), Nielsen (1993), Parasuraman et al., (2007), Park and Gretzel (2007), Santos (2003), Vijayasarathy (2004), Zaharias and Poylymenakou (2009)	Purchase management method	Offline, bill, returns, return costs, administrative costs, premium service, possibility of reimbursement	Barnes and Vidgen (2003), Loiacono et al., (2007), Park and Gretzel (2007)
Personal-ization	Recognizing visitors, customization of content, tailoring, explicit and implicit personalization	Abbattista et al., (1999), Carlshamre (2001), Chang (2007), Iacobucci (2000), Muhtaseb et al., (2012), Parasuraman et al., (2007), Park and Gretzel (2007)	Social networking	Facebook, Instagram, Twitter, Pinterest, LinkedIn, G+, updated social media	Bilgihan et al., (2016), Choudhury and Harrigan (2014), Jayachandran et al., (2015)
Privacy and security	Personally identifiable information, transmission of transactional information, lack of control, trust, assurance sales	Banati et al., (2006), Barnes and Vidgen (2003), Cao et al., (2005), Calero et al., (2005), Loiacono et al., (2007), Mich et al., (2003), Muhtaseb et al., (2012), Parasuraman et al., (2007), Park and Gretzel (2007), Santos (2003), Srinivasan (2004), Vijayasarathy (2004),	Type of campaign	Daily campaigns, fixed period campaigns, season campaigns, registration possibility	Fresenmaier et al., (2010), Crnojevac at al., (2010)
Social media	Social networking, blogs, collaborative projects, virtual, content communities, applications	Bernabé et al, (2015), Choudhury and Harrigan (2014), Jayachandran et al., (2015)			

METHODOLOGY AND SAMPLE ANALYSIS

The works of Akügl and Vatanserver (2016), Granados et al. (2008) and Xiang et al. (2008; 2014) are used as a reference to design the methodology. From them we take the steps to fulfil our objective. We specifically do so by applying "a method that mimics travellers' use of a search engine to search for destination related information" (Xiang et al., 2008, p. 142). That is, we use a series of keywords combined with the selected destinations for the Google search to obtain the resulting study sample.

DeVellis (1991) establishes two critical considerations for a valid sampling method. First, predefined keywords should represent the object of study in the tourism sector as consistently and broadly as possible. To this end, our paper uses web sites and metasearch engines that perform B2C e-commerce. Yet this already implies a number of exceptions. That is to say, websites that do not develop or make e-commerce, reservations or direct purchases possible are unaccounted.

This paper follows several studies to set keywords. Xiang et al. (2008) presented a list of 90 predefined words grouped into nine categories:

- Accommodation,
- Activities,
- Areas,
- Attractions,
- Events,
- Information,
- Places,
- Restaurants, and
- Shopping.

The taxonomy for keyword aggregation of Mitsche (2005) also consists in nine blocks:

- Accommodation,
- Attractions,
- Destinations (places),
- Transport,
- Information,
- Hospitality,
- Activities,
- Events, and
- Target groups.

The results of this study highlight destination, accommodation, attractions, information and transport among them. Finally, Wöber (2003) takes data from the European Tourism City website as a starting point and establishes the following top ten suggested keywords:

- Hotels,
- Map, hotel,
- Information,
- Guide, events,
- English,
- Pension,
- Shopping,
- Points of interest.

The keywords chosen by the user are similar:

- Hotel,
- Events,

- Hotels,
- Map,
- Information,
- Shopping,
- Guide,
- English,
- Pension, and
- Guides.

Google is then used as a framework for data search at the expense of other search engines. The decision to do so rests on fact that Google Sites led the US explicit core search market in September of 2015, during which 17.8 billion explicit core searches were conducted. Google Sites ranked first with 11.4 billion (up 1 percent), Microsoft Sites ranked second with 3.7 billion searches (up 2 percent), followed by Yahoo Sites with 2.2 billion, Ask Network with 300 million and AOL Inc. with 210 million (Reston, 2015) searches. Google is considered one of the best available search engines (Bertolucci, 2007; Bring and Page, 1998).

Based on the above, we established three phases/stages to perform data collection:

1. *Selection of the destinations that will be used for search in combination with keywords.* Since this paper focuses on e-commerce, it looks at the main recipient destinations for international tourism by income rather than by number of tourist arrivals. It then selects the top ten destinations in 2014 of the UNWTO. Please note that Macao and Hong Kong are directly linked to China, so they are included as a part of China. The destinations selected in order of relevance are: USA, Spain, China, France, Italy, UK, Germany and Thailand. This sample is justified on the basis that in 2014 they represented $616,503 million out of $1,245,446.00 in total. This represents 49.5% of the total income for international tourism which arises from 3.77% of the countries, i.e. it arises from 8 out of a total of 212 countries (UNWTO, 2015).
2. *Selection of the search keywords used in conjunction with the chosen destinations.* As previously discussed, the works of Mitsche (2005), Xiang et al., (2008) and Wöber (2003) are taken as a reference. A series of criteria associated to the research is applied to determine these words. First, the selection must include terms referring to e-commerce processes, i.e., processes enabling reservations or service purchases. Thus other options such as information or map information, among others, are not taken into account. Secondly, these terms must refer to or display domains directly linked to the tourism sector (Xiang et al., 2008). Third and last, the terms must cover the activity as a whole and not a specific item, in order to standardize/homogenize the search as much as possible. That is to say, instead of using hotels in detriment of other types of accommodation, we should opt for the term that encompasses the whole category, accommodation (Mistche, 2005). Based on the keywords in the cited papers and the application of these criteria, we obtain the following six concepts: tourism, accommodation, trip, travel, transport and booking.
3. *Search for the combination of destinations and selected keywords.* A total of 48 search terms were generated. All of these terms fulfilled the listed criteria. A final filtering was applied to eliminate websites offering a unique product or service, i.e., unique brands. It is understood that even if these sites perform B2C e-commerce, there is no real search or metasearch function for tourism

products and services. Such is the case of airline or hotel websites that are only linked to what they themselves offer.

First, considerating the eight most popular destinations based on the International Tourism Receipts and the six search keywords, 48 search terms were generated resulting in a total amount of 3,025,000,000 websites (see Appendix 1). Due to the high number of results and their variety (including a wide range of sources, from articles to blogs and others), it was used as a search filter that the websites were search or meta-search engines, limiting/restricting the results to 887,130 outputs. In order to focus on the object of the study (B2C websites), only the websites that fullfill that requirement were considered, 1,402 websites. Finally, a serie of filters or delimitations were applied to prevent website repetition within the sample or not accomplishment with the object of the study. Following that, duplicate websites, same websites but with different domains depending on country, public administration websites or specific business websites (only search or meta-search) were excluded from the sample.

A sample of 99 search engines and metasearch tourist sites performing e-commerce (see Appendix 1) resulted from the implementation of these phases. The data used in our database was collected between March and June 2015.

In the final sample we may observe that the country with the most tourism e-commerce webs in both search and metasearch engines is the USA, followed by the United Kingdom, Spain and China (see Table 3). This fact is very interesting because if we consider the ranking of countries according to international

Table 3. Sample description

Attribute	Categories	N	%
Browser type	Search	82	83.2
	Meta search	17	16.8
Country	Germany	6	5.9
	USA	21	20.8
	Singapoure	3	3
	China	9	8.9
	United Kingdom	15	14.9
	Australia	6	5.9
	Spain	14	13.9
	Thailand	5	5
	France	7	6.9
	Netherlands	1	1
	Italy	6	5.9
	Ireland	2	2
	Brasil	1	1
	Canada	1	1
	India	1	1
	Denmark	1	1
	Russia	1	1
	Israel	1	1

tourism incomes, USA would come in first place followed by Spain, China and the United Kingdom would come in seventh place (UNWTO, 2015). There is a clear relationship between the number of sites and the main tourist income countries with the exception of two cases. It should be pointed out that France, despite being number one in number of tourists and number four in revenue, holds no leading position in e-commerce tourism websites. This sharply contrasts with the United Kingdom, which is in eighth place in the world ranking for tourist arrivals and seventh place in revenue but holds a strong position in e-commerce tourism and it is second only to the USA.

When talking about the type of browser, search engines clearly dominate over metasearch engines. This reflects the current market situation. Throughout the first half of 2012, metasearch visitors increased 13% in the United States but they only generated 2% of traffic for airlines and 1% for hotels. Actually, these are the search engines that tend to accumulate the highest traffic from searches (Vargas, 2014).

Once the most characteristic sample features were established, we analyzed the data to address our main objective, i.e., to conduct an analysis of several tourist website search and metasearch engines for online booking based on their main typologies (accommodation products, journey products -flights-, leisure products and pack products). In so doing, we took into account the relations likely to be generated by the variables of structure and content of the web offer determined by user preferences.

The attributes listed in Table 2 were used and measured according to the items identified in each category. Similar studies focusing on supply attributes have been taken as a reference (Crouch, 2007; Domínguez et al, 2015). Thus, in the case of structure we got 11 attributes with their corresponding categories for analysis:

- Access,
- Composition of contents,
- Design and structure,
- Interactivity,
- Learning,
- Memorability,
- Navigation,
- Performance and availability,
- Personalization,
- Privacy and security and social media.

And for the offer content, 10 attribute and their categories:

- Loyalty system,
- Offer type,
- Operative crm,
- Package components,
- Payment method,
- Product browser,
- Product type,
- Purchase management method,
- Social networking, and
- Type of campaign.

The values for the different categories were set based on a Likert scale.

SOLUTIONS AND RECOMMENDATIONS

When analyzing the results, we followed the scheme proposed in the theoretical context. We analyzed all the selected variables for the structure and scope of the web offer and their impact on the different types of web pages by product offered, i.e., websites that mainly offered accommodation products, journey products, leisure products and pack products. Given their importance, specific flight websites were analyzed and discussed separately within the block of journey products.

Table 4 shows the correlations as well as the percentage of the different websites possessing the variables of the different attributes that form part of the web structure analysis.

In light of the results, it seems noteworthy to point out how structure variables impact on the websites offering accommodation products which contain around 67-68% of all of the following personalization sub-categories: recognition of visitors, customization of content, tailoring and explicit and implicit personalization. However, the rest of the sub-categories show higher correlations and percentages; all of them above 90%, and at least half of the categories of each attribute maintain this relationship.

The websites offering the next greatest number of correlations with the different categories and attributes are pack products. They are outstanding in terms of interactivity, performance and availability, and their percentages are generally over 90%. Most importantly, along with the websites offering flights, 59.6%, they are the only ones that have assurance sales. More specifically speaking, 61.8% of them have assurance sales. Although simplicity is an attribute characterizing 94.1% of these sites, this attribute forms part of the different typologies so these sites are surprisingly the ones containing the most product information; 94.1% of them have an informative product page.

Websites offering journey products, including flights, are reportedly present in social media, 52.3% of the websites that offer journey products generally have blogs and 47.7% contain community content. Half of the specific flight webs have collaborative projects. Additionally, these sites, along with the websites offering pack products, are the only ones that offer the chance to differentiate/distinguish and compare products.

The impact of websites offering leisure products is low and these impacts are essentially transaction and blog related. This may be due to the variety and diversity of their offers as well as the fact that the offers are usually linked to other more relevant tourist offers that determine patterns and structures.

Once the structure is analyzed, as may be seen in Table 5, the same approach is performed for the offer content of the sites by product type.

The first thing that surprises us is the conflicting behaviour of the webs offering accommodation products. Structure accommodation presented the most correlations among the different variables. However in offer they present the fewest correlations. In fact only two do so and both of them are related to the payment method attribute. This shows a correlation based on the possibility of payment through Visa Electron and PayPal but it only happens in 11.3% and 13.4% of the webs. For the rest of the webs the correlations are not strong enough to establish clear patterns.

The websites that offer journey products, specifically flights, are associated to nearly all of the package component categories to different extents. They are closely associated to trip (93.2%), destination (97.7%) and leisure activities (75%) and they are associated to a lesser extent to hotel + tour (6.8%), flight + hotel (29.5%) or cruise (31.8%). Note that 88.5% of the websites offering flights presented trip

Table 4. Correlations of structure variables and the different web typologies

Structure	Accommodation Products		Journey Products				Leisure Products		Pack Products	
			All of Them		Flights					
Sub-Category	χ^2	%	χ^2	%	χ^2	%	χ^2	%	χ^2	%
Interactivity										
Interactive decision aids									0.007	94.1
Transaction related			0.022	90.9			0.007	94.6	0.048	91.2
Personalization										
Recognizing visitors	0.043	68								
Customization of content	0.048	67								
Tailoring	0.048	67								
Explicit and implicit personalization	0.048	67								
Privacy and Security										
Lack of control	0.009	94.8								
Assurance seals					0.011	59.6			0.039	61.8
Composition of Contents										
Layout heading and styles	0.000	97.9								
Readability	0.017	93.8								
Current and timely information	0.000	97.8								
Informative product page									0.048	94.1
Learning										
Familiarity	0.009	94.8								
Simplicity									0.027	88.2
Design and Structure										
Home page	0.042	91.8								
Graphical interface	0.003	95.9								
Color consistency	0.001	96.9								
Performance and Availability										
Efficiency									0.048	94.1
Product differentiation and comparison			0.004	63.6	0.000	75			0.039	61.8
Social Media										
Blogs			0.011	52.3			0.013	54.1	0.031	52.9
Collaborative project					0.007	45.5				
Content community			0.036	47.7						

package component options and 61.5% of them presented car rentals. 86.4% of the journey product offer websites are usually determined by category. Those that offer flights are determined by category (90.4%) as well as by destination (86.5%). The websites that offer flights are the only ones related to the attribute of purchase management method. In them we may observe the possibility of returns cost, return claims and promotions. As opposed to this, however, hardly any of them offer premium service

Table 5. Correlations of offer content variables and the different web typologies

Offer Content	Accommodation Products		Journey Products				Leisure Products		Pack Products	
			All of Them		Flights					
Sub-Category	χ^2	%	χ^2	%	χ^2	%	χ^2	%	χ^2	%
Search Typology										
Meta-search					0.030	76.5				
Offer Type										
Offer by destination					0.039	86.5			0.012	82.4
Offer by category			0.000	86.4	0.000	90.4	0.000	83.8	0.000	91.2
Package Components										
Flight + hotel			0.009	29.5	0.000	36.4	0.005	32.4	0.000	44.1
Flight + hotel + tour									0.048	5.9
Hotel + tour			0.049	6.8			0.023	8.1	0.015	8.8
Trip			0.000	93.2	0.000	88.5	0.000	94.6	0.000	97.1
Cruise			0.002	31.8	0.000	32.7	0.000	37.8	0.000	41.2
Tours and routes			0.000	43.2	0.024	28.8	0.000	51.4	0.001	38.2
Break			0.000	45.5	0.003	36.5	0.000	51.4	0.001	44.1
Destination			0.012	97.7			0.040	97.3		
Car renting			-	-	0.000	61.5	0.001	56.8	0.000	67.6
Bus			-	-			0.043	10.8		
Train			-	-			0.003	21.6		
Leisure activities			0.000	75	0.002	51.9	-	-	0.000	66.7
Type of Campaing										
Daily campaigns									0.025	88.2
Period campaigns									0.025	88.2
Registration in campaigns									0.004	73.5
Payment Method										
Payment with Visa Electron	0.027	11.3								
Payment with Paypal	0.037	13.4								
Purchase Management Method										
Management cost					0.017	5.9				
Returns cost					0.010	64.7				
Return claim possibility					0.006	67.3				
Premium service					0.011	2				
Promotions					0.006	82.7				
Social Networking										
Facebook					0.040	73.1				
Instagram									0.050	8.8

and management cost. Moreover, they are the only websites affected by the search typology variable given that 76.5% of them are meta-searches. In other words, they act as an information system that locates the most commonly used search engines. They have no database of their own so they use other search engines to display a combination of the best pages returned by each of them.

Finally, the webs that offer leisure products and those that offer pack products behave similarly although the second presents a higher variable relation. Nearly all of them present the trip categories (94.6% and 97.1%, respectively) followed by car rental (56.8% and 67.6%) and break (51.4% and 44.1%). At large, almost all the websites offering pack products structure the offer by destination and category. They make a special effort on campaign issues, which they carry out regularly giving the user the chance to register for follow-ups.

FUTURE RESEARCH DIRECTIONS

Many possible lines of research arise given the increasing importance of this topic both in terms of trend as well as consumer use in the short-term. Perhaps one of the most interesting lines of research would consist in conducting a cross-sectional analysis with a larger site sample. This would help establish structure and content patterns which could be immediately be applied to the sector. It also seems vital to conduct research focusing on social media, specifically on social networks, to study user perception before and after enjoying a product. This would make it easier to establish implementation strategies to improve the offered product.

In order to study the web structure and the user preferences, will be interesting to establish a model to identify the main factors and their attributes, and the relevance of these factors for the user experience in tourism e-commerce websites. The model resulting could make possible to determinate a common pattern for the structure and offer content of the web sites based on the user preferences and needs, enough to find the best market strategies.

CONCLUSION

It is obvious that e-commerce plays a central role throughout all the economic sectors and the tourism sector is no exception. New tourist trends and motivations associated to experience and product specialization make it possible for internet to act as an almost imperishable resource. One of the channels most used for this purpose are the exchanges from business to consumer. Thanks to the commercialization of flights and low-cost fares, for example, new habits and behaviour have emerged in the market.

In addition to what is enumerated herein we must also consider the ever-increasing use of mobile devices not only to answer queries about destinations but also to manage previous bookings and on-site adds. This along with an increased availability of applications to this end (Ipsos MediaCT & Google Travel Study, 2014) has created a market opportunity which forms part of real consumer behaviour.

The breadth and variety of websites offering tourism products in all sorts of shapes and sizes clearly reflects the new reality this paper has aimed to study. The type of web search, search or meta-search, has not been observed to determine web behaviour regardless of the range of products offered with the exception of websites offering flights. These sites are clearly distinct from the others; this may be explained by their comparatively extensive experience in the market or the fact that they have the highest rate of search and purchase intent (Nielsen, 2014). More than three quarters of these searches are meta- searches. That is to say, they act as information systems with no data base of their own that locate information employing the most commonly used search engines to display a combination of the best pages returned from each of them.

Since the type of browser included in a web has not been determined to be critical for most of the websites analyzed by type of product offered, it follows that this may also be said of the structure of all the webs except for those offering accommodation products. These have a common pattern, especially in terms of personalization and, to a lesser extent, in terms of privacy and security, composition of content, design and structure and learning. Surprisingly, websites offering accommodation products are observed to behave absolutely heterogeneously in terms of offer content despite their homogeneous behaviour in terms of structure.

A common set of variables in the structure of webs offering pack products seems to give rise to a similar behaviour in relation to some attributes. However, results obtained from analyzing the offer content associated to package components display more relevance and commonality. A similar situation befalls websites that offer leisure products. They display greater homogeneity in offer content attributes but not in structure as is the case of journey product webs.

Finally, let us highlight the singularly distinct behaviour of websites offering flights as compared to other sites in terms of the above-referenced characteristics, especially in the areas of purchase and management method and social networks.

In conclusion, issues of structure and offer generally make no display of clearly defined structure and offer content patterns associated to the different types of web pages by type of search engine, search and metasearch. This is even more so depending on the type of product offered even though the patterns are not clearly defined. The offer content of the website presents great heterogeneity but major shortcomings are observed in social media and customer loyalty tools in which there is no clear common strategy. Thus, different webs must make an effort to carry out strategies to differentiate these attributes based on facilitating a space where users may share their experiences. This will no doubt contribute to increasing the recognition and perceived value of the brand.

ACKNOWLEDGMENT

The authors would like to thank the anonymous reviewers for their helpful and constructive comments that greatly contributed to improving the final version of the chapter. They would also like to thank the Editor for his generous comments and support during the review process.

REFERENCES

Ab Hamid, N. R., Ahmad, F., Tahir Shas, S. N., & Arshad, N. H. (2010). Evaluating the adoption level of electronic customer relationship management by telecommunication companies. *WSEAS Transactions on Systems*, *9*(2), 115–124.

Abbatista, F., Licchelli, M. D. O., Lops, P., Semerao, G., & Zambetta, F. (1999). Improving the usability of an e-commerce web site through personalization. In *Proceedings of theWorkshop on Recommendation and Personalization in Electronic Commerce, 2nd International Conference on Adaptative Hypermedia and Adaptative Web Based Systems.* University of Bari.

Abeles, E. G., White, M. D., & Hahn, K. (1999). A user-based design process for web sites. *OCLC Systems & Services, 15*(1), 35–44. doi:10.1108/10650759910257850

Adolphs, C., & Winkelmann, A. (2010). Personalization research in e-commerce a state of the art review (2000-2008). *Journal of Electronic Commerce Research, 11*(4), 326–341.

Akgűl, Y., & Vatansever, K. (2016). Web accessibility evaluation of government websites for people with disabilities in Turkey. *Journal of Advance Management, 4*(3), 201–210.

Al-Debei, M., Ashouri, M., & Akroush, M. (2015). Consumer attitudes towards online shopping: The effects of trust, perceived benefits, and perceived web quality. *Internet Research, 25*(5), 707–733. doi:10.1108/IntR-05-2014-0146

Andrés, M. E., Gómez, M. A., & Mondéjar, J. A. (2014). A model to evaluate the effects of price fairness perception in online hotel booking. *Electronic Commerce Research, 14*(2), 171–187. doi:10.1007/s10660-014-9137-4

Arsal, I., Woosnam, K. M., Baldwin, E. D., & Backman, S. J. (2010). Residents as travel destination information providers: An online community perspective. *Journal of Travel Research, 49*(4), 400–413. doi:10.1177/0047287509346856

Banati, H., Bedi, P., & Grover, P. S. (2006). Evaluating web usability from the users perspective. *Journal of Computer Science, 2*(4), 314–317. doi:10.3844/jcssp.2006.314.317

Barnes, S. J., & Vidgen, R. (2003). Measuring website quality improvements: A case study of the Forum on Strategic Management Knowledge Exchange. *Industrial Management & Data Systems, 103*(5), 297–309. doi:10.1108/02635570310477352

Bertoluci, J. (2007). Search Engine Shoot-out. *PC World, 25*(6), 86–96.

Bring, S., & Page, L. (1998). The anatomy of a large-scale hypertextual web search engine. *Computer Networks, 30*, 107–117.

Calero, C., Ruiz, J., & Piattini, M. (2005). Classifying web metrics using the web quality model. *Online Information Review, 29*(3), 227–248. doi:10.1108/14684520510607560

Cao, X., & Mokhtarian, P. L. (2005). *The intended and actual adoption of online purchasing a brief review of recent literature.* Working paper Institute of Transportation Studies (ITS). 2007. Department of Civil and Environmental Engineering, University of California. Retrieved from http://www.tc.umn.edu/~cao/UCD-ITS-RR-05-07.pdf

Carlshamre, P. (2001). *A usability perspective on requirements engineering – From methodology to product development.* Linköping: Linköping University Electronic Press.

Carlson, J., & OCass, A. (2011). Developing a framework for understanding e-service quality, its antecedents, consequences, and mediators. *Managing Service Quality: An International Journal, 21*(3), 264–286. doi:10.1108/09604521111127965

Casanovas, I. (2010). Exploring the current theoretical background about adoption until institutionalization of online education in universities: Needs for further research. *Electronic Journal of E-learning, 8*(2), 73–84.

Chakraborty, G., Lala, V., & Warren, D. (2003). What do customers consider important in B2B websites. *Journal of Advertising Research, 43*(1), 50–61. doi:10.2501/JAR-43-1-50-61

Chen, L., & Holsapple, C. (2012). E-business adoption research: Analysis and structure. *Proceedings of AMCIS, 2012*(August), 9–12.

Chen, L., & Holsapple, C. (2013). E-Business adoption research: State of the art. *Journal of Electronic Commerce Research, 14*(3), 261–286.

Chitura, T., Mupemhi, S., Dube, T., & Bolongkikit, J. (2008). Barriers to electronic commerce adoption in small and medium enterprise: A critical literature review. *Journal of Internet Banking and Commerce, 13*(2), 1–13.

Choudhury, M. M., & Harrigan, P. (2014). RM to social CRM: The integration of new technologies into customer relationship management. *Journal of Strategic Marketing, 22*(2), 149–176. doi:10.1080/0965254X.2013.876069

Crnojevac, I. H., Gugié, J., & Karlovčan, S. (2010). eTourism: A comparison of online and offline bookings and the importance of hotel attributes. *Journal of Information and Organizational Sciences, 34*(1), 41–54.

Davidavičienė, V., & Tolvaišas, J. (2011). Measuring quality of e-commerce web sites: Case of Lithuania. *Economics and Management, 16*, 723–729.

Davis, F. D., Bagozzi, R. P., & Warshaw, P. R. (1989). User acceptance of computer technology: A comparison of two theoretical models. *Management Science, 35*(8), 982–1003. doi:10.1287/mnsc.35.8.982

De Pablo, R. (2004). *Las nuevas tecnologías aplicadas al turismo*. Madrid: Editorial Universitaria Ramón Areces.

De Villes, R. F. (1991). *Scale development: theory and applications. Applied Social Research Series 26*. Newbury Park: Sage.

Deighton, J. (1996). The future of interactivity marketing. *Harvard Business Review, 74*(6), 151–160.

Dix, A., Finlay, J. E., Abowd, G. D., & Beale, R. (2003). *Human-computer interaction* (3rd ed.). New York: Prentice Hall.

Economics, T. (2013). The impact of online content on European Tourism. Tourism Economics and Oxford Economics Company.

Eighmey, J. (1997). Profiling user responses to commercial websites. *Journal of Advertising Research, 37*(3), 59–66.

eMarket. (2014, July 23). Worldwide Ecommerce Sales to Increase Nearly 20% in 2014. *eMarket*.

Fazlollahi, B. (2002). *Strategies for e-commerce success*. IRM Press. doi:10.4018/978-1-93177-708-7

Fesenmaier, D. R., Xiang, Z., Pan, B., & Law, R. (2011). A framework of search engine use for travel planning. *Journal of Travel Research*, *50*(6), 587–601. doi:10.1177/0047287510385466

Floh, A., & Madlberger, M. (2013). The role of atmospheric cues in online impulse-buying behavior. *Electronic Commerce Research and Applications*, *12*(6), 425–439. doi:10.1016/j.elerap.2013.06.001

Forbes. (2015, January 5). Competition is shaking up the online travel market. *Forbes*.

Granados, N. F., Kauffman, R. J., & King, B. (2008). The emerging role of vertical search engines in travel distribution: a newly-vulnerable electronic markets perspective.*Proceedings of the 41st Hawaii International Conference on System Sciences*. doi:10.1109/HICSS.2008.435

Hanson, W. (2000). *Principles of internet marketing*. Cincinnati, OH: South-Western Collage Publishing.

Henry, S. L. (n.d.). *Introduction to web accessibility*. World Wide Web Consortium. Retrieved from www.w3.org/WAI/intro/accessibility.php

Ho, C. I., Lin, M. H., & Chen, H. M. (2012). Web users behavioural patterns of tourism information search: From online to offline. *Tourism Management*, *33*(6), 1468–1482. doi:10.1016/j.tourman.2012.01.016

Hu, T., Marchiori, E., Kalbaska, N., & Cantoni, L. (2014). Online representation of Switzerland as a tourism destination: An exploratory research on a Chinese microblogging plataform. *Studies in Communication Sciences*, *14*(2), 136–143. doi:10.1016/j.scoms.2014.12.001

Ip, C., Lee, H. A., & Law, R. (2012). Profiling the users of travel wesbites for planning and online experience sharing. *Journal of Hospitality & Tourism Research (Washington, D.C.)*, *36*(3), 418–426. doi:10.1177/1096348010388663

Ipsos Media CT & Google Travel Study. (2014). *The 2013 traveler's road to decision: affluent insights*. Thinkinsights Google. Retrieved from https://think.storage.googleapis.com/docs/travelers-road-to-decision-affluent-insights_research-studies.pdf

Jamshidi, B. (2008). *Web usability in B2B websites: Users' perspective* (M. S. Thesis). Lulea University of Technology, Sweden.

Jayachandran, S., Sharma, S., Kaufman, P., & Raman, P. (2005). The role of relational information processes and technology use in customer relationship management. *Journal of Marketing*, *69*(4), 177–192. doi:10.1509/jmkg.2005.69.4.177

Kalakota, R., & Robinson, M. (1999). *E-business: roadmap for success*. Addison Wesley Publishing Company.

Kamis, A., & Frank, J. (2014). Advice shearing between paired users in online travel planning. *Journal of Theoretical and Applied Electronic Commerce Research*, *9*(1), 1–15. doi:10.4067/S0718-18762014000100002

Kumar, V., & Raheja, G. (2012). Business To Business (B2b) and Business to Consumer (B2c) Management. *International Journal of Computer & Technology*, *3*(3), 447–451.

Lawson-Body, A., & Limayem, M. (2004). The impact of customer relationship management on customer loyalty: The moderating role of web site characterisitcs. *Journal of Computer-Mediated Communication*, *9*(4), 153–174.

Lee, S. M., Hwang, T., & Kim, J. (2007). An analysis of diversity in electronic commerce research. *International Journal of Electronic Commerce*, *12*(1), 31–67. doi:10.2753/JEC1086-4415120102

Lin, T. C., & Huang, S. L. (2014). Understanding the determinants of consumers switching intentions in a standards war. *International Journal of Electronic Commerce*, *19*(1), 163–189. doi:10.2753/JEC1086-4415190105

Loiacono, E. T., Watson, R. T., & Goodhue, D. L. (2007). WebQual: An instrument for consumer evaluation of web sites. *International Journal of Electronic Commerce*, *11*(3), 51–87. doi:10.2753/JEC1086-4415110302

Mich, L., Franch, M., & Cilione, G. (2003). The 2QCV3Q quality model for the analysis of web site requirements. *Journal of Web Engineering*, *2*(1), 105–127.

Mitsche, N. (2005). Understanding the information search process within a tourism domain specific search engine. In *Information and communication technologies in tourism* (pp. 183–193). Springer. doi:10.1007/3-211-27283-6_17

Mohamad, R., & Ismail, N. A. (2009). Electronic commerce adoption in SME: The trend of prior studies. *Journal of Internet Banking and Commerce*, *14*(2), 1–16.

Muhtaseb, R., Lakiotaki, K., & Matsatsinis, N. (2012). Applying a multicriteria satisfaction analysis approach based on user preferences to rank usability attributes in e-tourism websites. *Journal of Theoretical and Applied Electronic Commerce Research*, *7*(3), 28–48. doi:10.4067/S0718-18762012000300004

Ngai, E. W. T., & Wat, F. K. T. (2002). A literature review and classification of electronic commerce research. *Information & Management*, *39*(5), 415–429. doi:10.1016/S0378-7206(01)00107-0

Nielsen, J. (1993). *Usability engineering*. San Francisco: Academic Press.

Niemann, M., Mochol, M., & Tolksdorf, R. (2008). Enhancing hotel research with semantic web technologies. *Journal of Theoretical and Applied Electronic Commerce Research*, *3*(2), 82–96.

Núñez, J. D. (2012). Centrales de reservas: Sentido y sensibilidad. *Tecnohotel*, *454*, 8–10.

Pallud, J., & Straub, D. W. (2014). Effective Website Design for Experience-Influenced Environments: The Case of High Culture Museums. *Information & Management*, *51*(3), 359–373. doi:10.1016/j.im.2014.02.010

Parasuraman, A., Zeithaml, V. A., & Malhotra, A. (2007). E-S-QUAL: A multiple-item scale for assessing electronic servisse quality. *Journal of Service Research*, *7*(3), 13–33.

Paredes, M. R., Barrutia, J. M., & Echebarria, C. (2014). Resources for value co-creation in e-commerce: A review. *Electronic Commerce Research*, *14*(2), 111–136. doi:10.1007/s10660-014-9135-6

Park, Y. A., & Gretzel, U. (2007). Success factors for destination marketing websites: A Qualitative Meta-Analysis. *Journal of Travel Research*, *46*(1), 26–35. doi:10.1177/0047287507302381

Parker, C. M., & Castleman, T. A. (2006). A preliminary meta-analysis of SEM ebusiness journal publications: current trends and future research opportunities moving beyond adoption factor studies. In *Collecter 2006.Proceedings of the 2006 Collecter E-Commerce Conference.*

PhoCusWright's. (2012, September 22). The global travel e-commerce. *PhoCusWright's.*

Quin, Z., & Zhang, L. (2001). *Customizable e-commerce: introduction, application and proposition.* Xi'an Jiaotong University.

Rodríguez, I. (2002). *Marketing.com y comercio electrónico en la sociedad de la información.* Madrid: Pirámide.

Rondán, F. J., Arenas, J., & Ramierez, P. (2015). Travel buying behavior in Social Network site users: To buy online vs. offline. *Journal of Theoretical and Applied Electronic Commerce Research*, *10*(1), 49–62. doi:10.4067/S0718-18762015000100005

Santos, J. (2003). E-service quality: A model of virtual service quality dimensions. *Managing Service Quality: An International Journal*, *13*(3), 233–246. doi:10.1108/09604520310476490

Schibrowsky, J. A., Peltier, J. W., & Nill, A. (2007). The state of internet marketing research: A review of the literature and future research directions. *European Journal of Marketing*, *41*(7/8), 722–733. doi:10.1108/03090560710752366

Skift. (2014, June 17). The most popular online booking sites in travel, 2014 Edition. *Skift.*

Srinivasan, R. (2004). Role of trust in e-business success. *Information Management & Computer Security*, *12*(1), 66–72. doi:10.1108/09685220410518838

Stienmetz, J. L., Maxcy, J. G., & Fesenmaier, D. R. (2015). Evaluating destination advertising. *Journal of Travel Research*, *54*(1), 22–35. doi:10.1177/0047287513514295

Taft, M. (2014, October 13). The definition of online booking tool in business travel. *ClickTravel.*

Thakran, K., & Verma, R. (2013). The emergence of hybrid online distribution channels in travel, tourism and hospitality. *Cornell Hospitality Quarterly*, *54*(3), 240–247. doi:10.1177/1938965513492107

Tourism Business Portal. (2014). EU Virtual tourism observatory. *European Tourism Forum.* Retrieved from https://ec.europa.eu/growth/tools-databases/vto/

United Nations. (2015). *Information economy report 2015. Unlocking the potential of e-commerce for developing countries.* United Nations Publication, UNCTAD/IER/2015.

Urbaczewski, A., Jessup, L. M., & Wheeler, B. (2002). Electronic commerce research: A taxonomy and synthesis. *Journal of Organizational Computing and Electronic Commerce*, *12*(4), 263–305. doi:10.1207/s15327744joce1204_1

Vallespín, M., & Molinillo, S. (2014). El futuro de la intermediación en el sector turístico. *Revista de Análisis Turístico*, *17*, 13–25.

Vargas, A. (2014, April 5). Los metabuscadores aumentan su influencia en 2014. *Hosteltur*.

Vijayasarathy, L. R. (2004). Predicting consumer intentions to use on-line shopping: The case for an augmented technology acceptance model. *Information & Management*, *41*(6), 747–762. doi:10.1016/j.im.2003.08.011

Visinescu, L. L., Sidorova, A., Jones, M. C., & Prybutok, V. R. (2015). The influence of website dimensionality on customer experiences, perceptions and behavioral intentions: An exploration of 2D vs. 3D web design. *Information & Management*, *52*(1), 1–17. doi:10.1016/j.im.2014.10.005

Wang, C. C., & Chen, C. C. (2010). Electronic commerce research in latest decade: A literature review. *International Journal of Electronic Commerce Studies*, *1*(1), 1–14.

Wareham, J., Zheng, J. G., & Straub, D. (2005). Critical themes in electronic commerce research: A meta-analysis. *Journal of Information Technology*, *20*(1), 1–19. doi:10.1057/palgrave.jit.2000034

Wöber, K. W. (2003). Domain-specific search engines. In Destination recommendation systems: behavioral foundations and applications. CABI.

World Tourism Organization (UNWTO). (2014). *Tourism highlights 2014 Edition*. UNWTO.

World Tourism Organization (UNWTO). (2015). *Tourism highlights 2015 Edition*. UNWTO.

Xiang, Z., Wang, D., & Fesenmaier, D. R. (2014). Adaptive strategies to technological change: understanding travellers using the internet for trip planning. In Information and communication technologies in toursm (pp. 411–424).

Xiang, Z., Wöber, K., & Fesenmaier, D. R. (2008). Representation of the online tourism domaine in search engines. *Journal of Travel Research*, *47*(2), 137–150. doi:10.1177/0047287508321193

Zaharias, P., & Poylymenakou, A. (2009). Developing a usability evaluation method for e-learning applications: Beyond functional usability. *International Journal of Human-Computer Interaction*, *25*(1), 75–98. doi:10.1080/10447310802546716

Zhou, L., Dai, L., & Zhang, D. (2007). Online shopping acceptance model-A critical survey of consumer factors in online shopping. *Journal of Electronic Commerce Research*, *8*(1), 41–62.

ADITIONAL READING

Aiello, L. (2012). Role of meta-search engine in the relationship tourist / e-retailer. *Information Science and Digital Content Technology (ICIDT),8th International Conference*, 3: 749-756.

Bernabè, M. (2015). The booking challenge: tourism and sharing economy. *Tafter Journal*, 2015.

Bogdanovych, A., Berger, H., Simoff, S., & Sierra, C. (2006). Travel agents vs. online booking: tackling the shortcomings of nowadays online tourism portals. In: Information and Communication Technologies in Tourism, pp 418-428. doi:10.1007/3-211-32710-X_55

Buhalis, D., & Licata, M. C. (2002). The future etourism intermediates. *Tourism Management*, *23*(3), 207–220. doi:10.1016/S0261-5177(01)00085-1

Condratov, I. (2013). E-tourism: Concept and evolution. *Ecoforum*, *2*(1), 58–61.

Dorn, J., Hrastnik, P., Rainer, A., & Starzacher, P. (2009). Web Service based Meta-search for Accommodations. *Information Technology & Tourism*, *10*(2), 147–159. doi:10.3727/109830508784913095

European Commission (2011). User language preferences online. Analytical report. Eurobaremeter Flash EB No 313.

Fernández, S., Sebastia, L., & Fdez-Olivares, J. (2005). Planning tourist adapted to user preferences. In Castillo, L. et al.: Planning, scheduling and constraint satisfaction: from theory to practice, pp. 119-128.

Kabir, M. A., Jahan, K., Adnan, N., & Khan, N. (2012). Business model of e-Tourism for developing countries. *International Journal of Computer and Information Technology*, *3*(1), 30–34.

Littlehotelier (2015). Top Trends in Travel 2015 - for Bed & Breakfasts, Guesthouses and other small hospitality providers. http://www.littlehotelier.com/ebook/eBook_Top-Travel-Trends.pdf

Longhi, Ch. (2008). Usages of the internet and e-tourism. Towards a new economy of tourism *Archives – Ouvertes.fr*, May 2008.

Marzouk, T. (2015). *ITB World travel trends report 2015/16*. ITB Academy and IPH Intertnational.

Stiakakis, E., & Georgadies, Ch. K. (2009). Drives of a tourism e-business strategy: The impact of information and communication technologies. *Operations Research*, *11*(2), 149–169. doi:10.1007/s12351-009-0046-6

Tourism e kit (2011). Online distribution: online travel agents. Australia Tourism Data Warehouse. http://www.wikitourism.co.za/images/9/9e/OTA_Queensland.pdf

Werthner, H. (2011). e-Tourism: impact of new technologies. Vienna University of Technology & Electronic Commerce Center. http://www.minervaeurope.org/events/michael/materialiintconf06120405/Presentations/werthner.PDF

Xiang, Z., Pan, B., & Fesenmaier, D. R. (2013). Search engine marketing for tourist destinations. The Handbook of Internet and Marketing.

KEY TERMS AND DEFINITIONS

Booking Online: Process throughout customer's concluded arrangement with a goods or services supplier (carrier, hotel, restaurant, publisher, etc.) representing a completed sale, using networks. Also called reservation or transaction. An online booking tool is a tool used by organisations to manage their business travel online (Taft, 2014).

Business Commerce or E-Business: The use of information technologies in various business operations, while e-commerce is the latest step in the evolution of business transactions process. It comprises the sale of goods or services through computer networks, i.e., it uses internet as a channel (Kalakota and Robinson, 2001).

Business to Customer or B2C: Electronic commerce between company that operate in the net/ network and customer (Montane, 2010).

Electronic Commerce or E-Commerce: May be considered as a set of e-business applications used in commercial transactions (Kalakota and Robinson, 2001). There are different types as result of transaction combinations of business, consumer and agoverment.

Metasearch Engines: Mega search engines that allow web users to browse various search engines and web directories at a time (Ho, Lin and Chen, 2012). That is to say, this system locates information in the most commonly used search engines. It lacks its own database and uses other search engines to show a combination of the best pages returned by each one of them.

Online Travel Agencies (OTA): Only available on the Internet with no physically located establishment (Vázquez-Casielles et al., 2009). In addition, customers may find a single tourist product, such as a flight or a set or combination of flights, as well as last-minute offers (Rensmann, 2012).

Search Engines: Intermediaries are limited to providing information and contacting the members of electronic channels (Rodriguez, 2002).

APPENDIX

Table 6. Data collected process: number of results search

	Accommodation	Trip	Travel	Transport	Booking
USA	195,000,000	327,000,000	574,000,000	695,000,000	224,000,000
Spain	118,000,000	257,000,000	309,000,000	234,000,000	161,000,000
China	122,000,000	233,000,000	470,000,000	397,000,000	110,000,000
France	161,000,000	378,000,000	491,000,000	543,000,000	173,000,000
Italy	133,000,000	235,000,000	362,000,000	282,000,000	189,000,000
UK	332,000,000	400,000,000	843,000,000	694,000,000	297,000,000
Germany	147,000,000	216,000,000	416,000,000	551,000,000	167,000,000
Thailand	72,300,000	110,000,000	162,000,000	134,000,000	97,800,000
Total	1,119,300,000	2,156,000,000	3,627,000,000	3,530,000,000	1,418,000,000
Filter 1			**Tourism**		
USA	30,100,000	82,800,000	98,700,000	44,700,000	43,000,000
Spain	31,000,000	87,500,000	110,000,000	30,300,000	42,300,000
China	31,300,000	63,400,000	127,000,000	53,000,000	30,500,000
France	36,200,000	107,000,000	153,000,000	48,000,000	37,900,00
Italy	31,100,000	77,400,000	123,000,000	32,500,000	45,900,000
UK	138,000,000	259,000,000	271,000,000	130,000,000	195,000,000
Germany	30,400,000	73,400,000	110,000,000	37,900,000	30,600,000
Thailand	20,700,000	35,600,000	52,400,000	19,100,000	24,300,000
Total	348,800,000	786,100,000	1,045,100,000	395,500,000	449,500,000
Filter 2			**Search or Meta-Search Engine**		
USA	11,100	14,700	29,500	17,700	16,800
Spain	12,500	11,600	24,700	17,500	13,700
China	13,600	9,730	26,200	17,000	12,600
France	43,300	14,600	56,300	48,900	45,000
Italy	11,500	10,900	25,000	17,000	13,600
UK	13,000	16,200	33,100	18,200	16,300
Germany	11,800	10,900	24,700	17,500	13,500
Thailand	37,500	10,500	47,500	42,800	38,600
Total	154,300	99,130	267,000	196,600	170,100
Filter 3			**B2C Webs**		
Total	246	160	426	298	272
Filter 4	• Without duplícate websites • Without different websites depending on country or domain • Without public administration websites • Without specific business websites (only search or metasearch)				
Total			99		

Table 7. Search and meta-search engines websites

Search and Meta-Search Engines	Websites
9flats	http://www.9flats.com/
Affordable Tours	http://www.webjet.com/
Agoda	http://www.agoda.com/
Airbnb	https://www.airbnb.com/
Airchina	http://www.airchina.com/
Asia Rooms	http://www.asiarooms.com/
Asia Travel	http://www.asiatravel.com/
Asia Web Direct	http://www.asiawebdirect.com/
Atrápalo	http://www.atrapalo.com/
Back Yard Travel	http://www.backyardtravel.com/
Bahn	http://www.bahn.com/
B&b Hotels	http://www.hotel-bb.com/
Bed and brekfast	http://www.bedandbreakfast.com/
Booking	http://www.booking.com
Cheapflight	http://www.venere.com/
Cheapoair	http://www.cheapoair.com/
China Airline Travel	http://chinaairlinetravel.com/
China Discount Hotels	http://www.china-discount-hotels.com/
China Eastern	http://hk.ceair.com/
China Hotels Reservation	http://www.chinahotelsreservation.com/
China Connection Tours	http://www.china-tour.com/
Ctrip	http://english.ctrip.com/
Despegar	http://www.us.despegar.com/
Destinia	http://destinia.com/
Easy Viajar	http://www.easyviajar.com/
Easyjet	http://www.easyjet.com/
Ebookers	http://www.ebookers.com/
Edreams	http://www.edreams.es/
Elong	http://www.china-discount-hotels.com/
Emma Villas	http://www.emmavillas.com/
Eurolines	http://www.eurolines.co.uk/
Evaneos	http://www.evaneos.com/
Expedia	http://www.expedia.com/
Fare Compare	http://www.farecompare.com/
German Tourist Hotels	http://www.germantouristhotels.com/
Hihostels	https://www.hihostels.com/

continued on following page

Table 7. Continued

Search and Meta-Search Engines	Websites
Hipmunk	https://www.hipmunk.com/
Homelidays	http://www.homelidays.com/
Hostel Bookers	http://www.hostelbookers.com/
Hostel Word	http://www.hostelworld.com/
Hot Wire	http://www.hotwire.com/
Hotel Book	http://www.hotelbook.com/
Hotel Club	http://www.hotelclub.com/
Hotel Conbined	http://www.hotelscombined.com/
Hotel Search	http://www.hotelsearch.com/
Hotel Thailand	http://www.hotelthailand.com/
Hotels 2 Thailand	http://www.hotels2thailand.com/
Hotel Travel	http://www.hoteltravel.com/
Hotel Urbano	http://www.hotelurbano.com/
Hotels.com	http://www.hotels.com/
Hrs	http://www.hrs.com/
Ibis Budget	http://ibisbudgethotel.ibis.com/
In Italia	http://www.initalia.it/
Jetcost	http://www.jetcost.com/
Kayak	http://www.kayak.com/
Last Minute	http://www.es.lastminute.com/
Late Rooms	http://www.laterooms.com/
Liligo	http://www.liligo.com/
Logitravel	http://www.logitravel.com/
Lowcost Holydays	http://www.lowcostholidays.com/
Luxury Retreats	http://www.luxuryretreats.com/
Make my Trip	http://www.makemytrip.com/
Mejores Vuelos	http://www.mejores-vuelos.com/
Minube	http://www.minube.com/
Mobissimo	http://www.mobissimo.com/
Momondo	http://www.momondo.com/
Muchoviaje	http://vacaciones.muchoviaje.com/
Omega Travel	http://uk.omegatravel.net/
Onetwotrip	http://www.onetwotrip.com/
Opodo	http://www.opodo.co.uk/
Orbitz	http://www.orbitz.com/
Park Inn	http://www.parkinn.com/
Price Line	http://www.priceline.com/

continued on following page

Table 7. Continued

Search and Meta-Search Engines	Websites
Phuket	http://www.phuket.com/
Qunar	http://www.qunar.com/
Raileurope	http://www.raileurope.com/
Ratestogo	http://www.ratestogo.com/
Room 77	https://www.room77.com/
Routeperfect	http://www.routeperfect.com/
Rumbo	http://www.rumbo.com/
Sawadee	http://www.sawadee.com/
Sleep in Italy	http://www.sleepinitaly.com/
Skyscanner	http://www.skyscanner.com/
Statravel	http://www.statravel.co.uk/
Travel Europe	http://www.traveleurope.com/
Travel Jungle	http://www.traveljungle.fr/
Travelocity	http://www.travelocity.com/
Travelsupermarket	http://www.travelsupermarket.com/
Travelzoo	http://www.travelzoo.com/
Tripadvisor	http://www.tripadvisor.com/
Trivago	http://www.trivago.com/
Venere	http://www.venere.com/
Viggiare	http://www.viaggiare.it/
Voyages-SNCF	http://uk.voyages-sncf.com/en/
Vueling	http://www.vueling.com/es/reserva-tu-vuelo/busca-tu-vuelo
Vuelos Baratos	http://www.vuelosbaratos.es/
Web Jet	http://www.webjet.com.au/
Wego	http://www.wego.com/
Wimdu	http://www.wimdu.es/
Wotif	http://www.wotif.com/
You Vuelos	http://www.youvuelos.com/
Zingarate	http://www.zingarate.com/

Chapter 4
Optimizing Investment Decisions Using DCF, Decision Tree Analysis, and Real Options Analysis:
The Case of Hotel Expansions

Ramya Rajajagadeesan Aroul
Ecole Hôtelière de Lausanne, HES-SO – University of Applied Sciences Western Switzerland, Switzerland

ABSTRACT

Large scale infrastructure expansions in hotels are exposed to uncertainty. Since the costs involved in these expansion projects are high and often irreversible, hotels would benefit from analyses that incorporate uncertainty along with traditional valuation techniques like the discounted cash flow (DCF) method. Decision tree analysis (DTA) and real options analysis (ROA) have been in use for the past couple of decades to handle uncertainties and optimize investment decisions. DTA provides a distinct approach to strategic investments that quantitatively takes into account the uncertainties involved in the investments. Under uncertainty, the decision about whether to expand is analogous to the decision about whether to exercise an American call option. By using ROA to the hotel expansion scenario, managers can incorporate and quantify, flexibility and timing in their analysis. The objective of this chapter is to detail the DCF, DTA and ROA methodologies and their applications specific to hotel expansion investments.

INTRODUCTION

The discounted cash flow (DCF) technique is the most widely used valuation technique in practice. In this technique, a project is valued by discounting its expected cash flows at its cost of capital. The cost of capital, usually the weighted average cost of capital (WACC) is used to estimate the worth of future cash flows today. This process of estimating how much these future cash flows are worth

DOI: 10.4018/978-1-5225-1054-3.ch004

today is called discounting and hence the name for the technique. The net present value (NPV) of the project is then calculated as the difference between the total of all discounted future cash flows and the initial investment. A project is valuable to be undertaken if the NPV is not negative. Accepting projects with NPV less than zero decrease the value of the firm. Also, the higher the NPV of the project is, the higher it contributes to the overall value of the firm.

However, DCF works best when the certainty of future cash flows is high. But valuation using DCF is shaky when it's difficult to predict and estimate future cash flows with much confidence. Also, this analysis does not take into account the flexibility inherent in the investment decision. Also, past real estate studies suggest that DCF may be insufficient for evaluating real estate investment projects (Hayes & Abernathy 1980; Hodder & Riggs, 1985; Sirmans, 1997).

The complexity of the investment process is that the firm can take dynamic decisions which is not accounted for by DCF. When there is a lot of complex information that needs to be accounted for, decision trees (Gertner & Rosenfield, 1999) are often used as techniques for investment decision making. They provide an effective configuration in which different decisions and the consequences of taking those decisions are portrayed in a diagram and evaluated. Eventually they help to picture an accurate and balanced opinion of the risks and rewards that evolve from making a specific investment choice. In this regard, decision tree analysis is superior to the DCF analyses in the context of inclusion of uncertainty in the investment decision process. Although decision-tree analysis tends to pay a lot of attention to the cash flow uncertainties, it rarely considers dynamic decision making.

In 1970s, stochastic calculus was introduced in the finance literature leading to theories and models in pricing financial options. Many studies then applied these financial option-pricing models to investment projects. This new area of 'real option' analysis received a lot of attention in several areas especially in real estate valuation (Quigg 1993, 1995; Buetow & Albert 1998; Hendershott & Ward 2000; Holland et al. 2000). Real options analysis (ROA) has been in use for the past couple of decades to handle uncertainties and optimize investment decisions. It provides a distinct approach to strategic investments that quantitatively takes into account the risks and uncertainties. Similar to DCF, ROA also focuses on the goal of value maximization. However, ROA acknowledges that managers use available information to make decisions while DCF makes managers to choose years in advance and does not accommodate the changes that may happen in the future. ROA uses the "backward induction" approach like the decision tree analysis and combines this with a valuation model.

In the following sections, we evaluate a hypothetical hotel expansion project using DCF, decision tree analysis and real options analysis and explain the fundamental differences among these techniques. This chapter is divided into the following distinct sections. The first section provides the background of literature. Section 2 discusses the motivation for the book chapter. Section 3 discusses the DCF valuation framework, the basic decision tree valuation approach that is used in both single period and multi-period cash flow scenarios and introduces the ROA analysis. This section also discusses the advantages and disadvantages of each valuation technique. Section 4 introduces the case's basic framework and applies all three valuation techniques on this mini case. The chapter also provides ROA application to the hypothetical case study with illustrations of the real options optimization technique step by step in Excel.

MOTIVATION

The service sector has grown dramatically in both developed and developing countries alike during the past couple of decades. In contrast to most other service sectors, the hotel industry is usually characterized by capital-intensive investments with a complex supply chain (Chen & Dimou, 2005; Contractor & Kundu, 1998). The high capital requirement also raises the risk of hotel expansion activities. Therefore, it is important that an analyst accounts for all possible factors while valuing expansion projects.

There are many advantages to the DCF analysis, perhaps one of the best reasons is that it uses cash flows. Cash flows are more trust worthy to use since it overcomes most of the inaccuracy against reported earnings. Cash flows are discounted to their present value by the firm's cost of capital or a rate that appropriately reflects project risk. The theoretical rationale for utilizing DCF techniques is sound and the techniques are well accepted (Haley & Goldberg, 1995).

DCF analysis cannot be applied to scenarios involving dynamic decision making. In reality, a manager observes indicators or signals of the state of the world and act in ways that affect the project`s future cash flows. Therefore, we introduce the readers to the importance of incorporating uncertainty in determining optimal investment decisions.

BACKGROUND

Most hotels do not explicitly use real option techniques in evaluating investments. Yet, real option perspectives can be a significant component of value. Decisions about large scale infrastructure expansions in hotels are often taken while future developments are uncertain. As these projects take many years to be implemented, the hotels are exposed to multiple uncertainties surrounding these investments like economic changes, socio-political shifts etc. Since the costs involved in these expansion projects are quite high and most often irreversible, hotels would benefit from analyses that incorporate uncertainty in addition to traditional valuation techniques like the DCF.

The extant literature in hospitality applies real options analysis in various contexts. Because of the implicit option value in the product, Quan (2002) and Lu (2012) have introduced ROA into research on revenue management. Lai, Ng & Zhou (2001) presents a model framework that documents a theoretical explanation for overbuilding in the hotel industry through the derivation of optimal construction strategies using real option pricing methodology. Gallego, Kou & Phillips (2008) present the concept of callable products, which connotes an option, to reduce the loss from denied booking. Anderson, Davison & Rasmussen (2004) apply a real options approach to develop a car rental revenue management model. Miller & Huggins (2010) use real options binomial lattice structure to evaluate pricing strategies for service industries.

Alvarez, Ruiz & Solana (2008) present a real options model to optimal decision making when assigning rooms to hotel customers. Cunill, Forteza & Miralles (2008) suggests a way for valuing growth strategies by large hotel chains, particularly when the project is under high degree of uncertainty and highlights that ignoring the real options inherent in hotel investment strategies where uncertainty is involved will lead to undervaluation. Garcia de Soto Camacho (2015) documents the use of real options analysis in the internationalization of firms in the hotel industry. García de Soto-Camacho & Vargas-Sánchez (2015) presents a dynamic descriptive model for the international strategy in hotel chains using the real options theory. Yang & Chang (2011) uses a real option pricing model to create time-axis profiles of restaurant menu items and to examine their profitability assuming temporally varying constraints. Li (2014) shows a method to find the best overbooking level vector by modeling a series of real options contained in the joint overbooking decision.

In this chapter we present a case of real options along with other valuation models such as NPV and decision trees in making optimal investment decisions. By doing so, we contribute to the vast literature that incorporates uncertainty and flexibility into optimal decision making using real options analysis.

VALUATION TECHNIQUES

DCF Valuation: Simple Case of "Now or Never"

Discounted cash flow is the most common capital investment appraisal technique used by businesses. The fundamental theory behind this method was suggested by Fisher (1907) and popularized by Hirschleifer (1958). In order to evaluate the potential for investment, discounted cash flow (DCF) analysis uses future estimated cash flows and discounts them to the present by using a relevant discount rate (most often, the weighted average cost of capital).

Let us assume that the investment costs I during this period and will generate cash flows in the future. The current market value of the hotel is V0 and if the hotel manager rejects the investment opportunity, the hotel's value V0 will comprise of debt D0 and equity E0.

$$V0 = D0 + E0$$

$$E0 = V0 - D \quad (1)$$

If the hotel manager undertakes this project, the value of the hotel will increase to V' comprising of original debt D0, original equity E0 and the new securities issued by the hotel to raise the initial investment for the new project that equals I.

$$V1 = D1 + E1 + I$$

$$E1 = V1 - D1 - I \quad (2)$$

The manager's objective is to maximize the value of the shareholders and hence he would undertake the project only if E1≥ E0. Subtracting (1) from (2) we get

$$V1 - D1 - I \geq V0 - D0$$

$$\left(V1 - V0\right) - I \geq D1 - D0$$

Assuming that the current new investment does not affect the market value of debt (this assumption is made even in traditional capital budgeting analyses like NPV), we get

$$V1 - V0 - I \geq 0$$

Here, (V'-V0)-I is the NPV of the project. If the NPV is positive, the project is viable. A negative NPV implies that the corporation is better off not making the investment. Therefore, in the "now or never" investment situation, the NPV analysis is consistent with the value maximization. However, there are other situations where assumptions are relaxed. ROA provides optimal decisions in all cases while NPV is inconsistent and provides optimal decisions only in a relatively small set of special cases like the one discussed above. DCF is very popular among managers, with its limitations being often ignored. Other complementary methods, as proposed in this chapter, offer valuable contribution to decision making in investments.

$$Net\ Present\ Value\left(NPV\right) = Present\ value\ of\ inflows - Present\ value\ of\ outflows$$
$$Invest\ if\ NPV > 0$$

Decision Tree Analysis

DCF analysis cannot be applied to scenarios involving dynamic decision making. In reality, a manager observes indicators or signals of the state of the world and act in ways that affect the project`s future cash flows. For instance, if the manager of a hotel observes steady increase in the demand for laundry services, he/she may be motivated to include laundry facilities which would impact the future cash flows. The decision tree analysis recognizes that the cash flows in the future would depend on future managerial actions. The decision tree analysis maps out the sequence of actions that the manager could take representing them in a decision tree and then works through the tree backwards to identify the best action at every point where the tree branches out called the node.

The decision tree model has two crucial aspects:

1. The capacity of managers to learn from what is happening around (the choice of the state variable).
2. Managerial actions (their willingness and the ability to adapt behavior based upon that learning).

The state variable is independent of the managerial actions. The state variable describes the broader risks the project is subjected to. The manager takes actions based on his observation of the state variable. The actions along with the state variable determines the cash flows.

The first step in decision tree analysis is to differentiate between root nodes, decision nodes, event nodes and end nodes. The root node is the starting point of the decision tree at t=0, where a manager is faced with a decision choice. The final goal of the decision tree analysis is to estimate the value of the risky investment at this point or node. Event nodes signify the possible outcomes; whether a company's first stage expansion to a market pans out or not is a good example at t=1 in the two stage decision tree. We have to identify the possible outcomes and the probabilities of the outcomes, based upon the current information available. Decision nodes denote choices that can be made by the decision maker. End nodes typically characterize the final outcomes of previous risky outcomes and decisions made. Decision tree analysis uses one cost of capital to discount cash flows in each branch to the present and the probabilities used to compute an expected value need not be the same as accounting for uncertainty using the real options analysis.

Real Options Analysis

The theory behind options is quite old, as Bachelier (1900) already discussed the basics of options in his doctoral thesis more than a century back. The basic method for valuing financial options was however suggested by Black & Scholes (1973), and is named after them. Financial options (call and put options) give the holder the right, but not the obligation, to buy (call option) or sell (put option) shares at a pre-determined price, known as the exercise or strike price, during an already decided period of time. These financial options are exposed to market volatility and exhibit an asymmetric return distribution as a result of this right exercising the option (Amran & Kulatilaka, 1999; Trigeorgis, 1996).

A "real option" is an application of the financial option concept to real capital. The underlying asset in a financial option is the company's stock while in a real option it is the real assets. Real options presents a firm's manager the opportunity to increase the project/investment's gains or reduce its losses (Calle & Tamayo, 2009). Real options were first developed by Myers (1977) and can be used in evaluating investment projects.

Under uncertainty, the decision about whether to expand is analogous to the decision about whether to exercise a call option, and therefore the hotel's expansion option is considered a real option. Cunill et al. (2008) highlights the problem that failure to recognize the real options inherent in hotel investment strategies where uncertainty is involved will lead to undervaluation of the project. By using real options optimization techniques to the hotel expansion scenario, managers can incorporate and quantify flexibility and timing in their analysis which were amiss in traditional DCF technique. The objective of this section is to detail the real options methodology's application specific to hotel expansion investments.

The real option concept was derived from the continuous option pricing theory developed by Black & Scholes (1973). A call option gives the holder the right not an obligation to buy shares of a company at a specified time at a specific day. The Black-Scholes formula given below calculates the price of a call option for a non-dividend paying stock.

$$C\left(S,t\right) = N\left(d1\right)S - N\left(d2\right)Ke^{-r\left(T-t\right)}$$

$$d1 = \frac{1}{\sigma\sqrt{T-t}}\left[ln\left(\frac{S}{K}\right) + \left(r + \frac{\sigma^2}{2}\right)\left(T-t\right)\right]$$

$$d2 = d1 - \sigma\sqrt{T-t}$$

where

1. C(S,t) is the price of call option at a given spot price S and given time t.
2. N(.) is the cumulative distribution function (CDF) of the standard normal distribution.
3. T-t is the time to maturity.
4. S is the spot price.
5. K is the strike price.

Figure 1. Call option

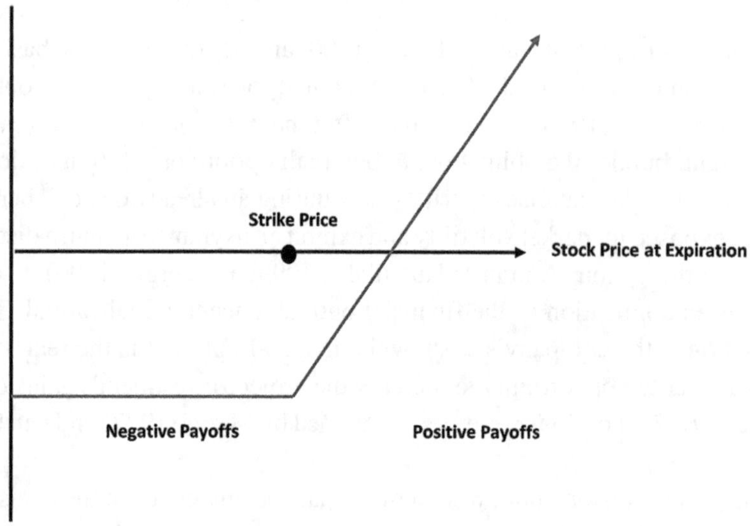

6. R is the risk free rate.
7. σ is the volatility of stock returns.

The replicating portfolio is embedded in the Black-Scholes model. To replicate this call, you would need to buy N (d1) shares of stock and borrow N (d2) Ke-r(T-t)

Although the Black – Scholes formula provides a helpful way to value options, it has few drawbacks as well. Some of the constraints of the Black – Scholes formula are:

- It can only value European options (You can exercise an European option only at maturity).
- The income follows a normal distribution.
- There has to be only one source of uncertainty.
- The exercise price has to stay constant.

MINI CASE: HELVETICA HOTEL DEVELOPMENT CORPORATION'S EXPANSION TO SOUTH AMERICA

Helvetica Hotel Development Corporation is a hypothetical company involved in hotel development and operations in the United States. Managers at Helvetica are currently evaluating their expansion plans to the Brazilian market. The managers have proposed to expand their operations (build hotels) to many locations in South America if this venture is successful. This expansion plan would require major allocation of resources and has a huge strategic impact. The company is currently the leader in the US operations and if this South American venture pans out well, it would be a major move towards evolving a market leader in the Americas. However, by this scale of expansion Helvetica is exposed to market risks that it does not fully comprehend.

In the following sections, we evaluate the project using NPV, Decision Trees and Real Options Modeling and explain the fundamental differences among the models.

1. DCF Valuation

In this section we look at a special situation where Helvetica has a now or never investment opportunity i.e. either the hotel invests now or the opportunity of investment vanishes forever. In our hypothetical case, managers at Helvetica estimate that the initial investment to enter Brazilian market is $200 million. At the end of the first year, the expected cash flows is $50 million. The expected cash flows in the following four years are $40 million, $32 million, $30 million and $25 million respectively. Suppose Helvetica's discount rate is 5%. The cash flow is then discounted by a factor of 1/ (1+5%) every year. As shown below, the computed NPV turns out to be negative.

The managers will have to reject the hotel expansion project in Brazil since the NPV is negative. Because of negative NPV situation, Helvetica would not consider expanding into the South American market using DCF analysis. Since the DCF analysis does not consider uncertainty and it is too rigid in its "now or never" framework, big investment decisions with strategic importance based only on DCF analysis could be costly.

DCF is very popular among managers, with its limitations being often ignored. Other complementary methods, as proposed in this chapter, offer valuable contribution to decision making in investments.

2. Decision Tree Analysis

Consider the case example where Helvetica, an all-equity hotel development company intends to buy a new hotel in Brazil. The manager is deciding whether to invest in this project. There are two possibilities of the state variable (up market and down market). The managerial decision will depend on the probabilities and values associated with both up and down markets. If Helvetica decides to get the property the value of the hotel could go up ($80 million) or down (-$100 million). Assuming equal probabilities

Figure 2. Discounted cash flow analysis

DCF Analyisis	
Initial Investment	$200.00
Cashflows	
Year 1	$50.00
Year 2	$40.00
Year 3	$32.00
Year 4	$30.00
Year 5	$25.00
Discount Rate	0.05
PV of Future Cash Flows	$155.81
Net Present Value	-$44.19

of up and down movements, and given the larger potential loss, the expected value for this investment is negative. The expected value Ve is

$$Ve = \left(0.5 * -100\right) + \left(0.5 * 80\right) = -\$10 \ million$$

Now let us extend this to a slightly more complicated two-phase decision tree. This decision tree analysis includes time t=2 as well where Helvetica makes its South American expansion project decision. The payoffs and the probabilities have to adjust such a way that the potential profits and losses over the two phases would be similar to the ones from the single stage.

$$The\ expected\ value\ Ve$$
$$= 0.33 * \left[10 + \left(0.67 * 70\right) + \left(0.33 * -90\right)\right] + 0.67 * -10 = \$21.88 \ million$$

This $21.88 million value reflects all of the potential outcomes that can develop over time and indicates the choices at each node that are not optimal and therefore should be rejected. The overall value of this project is positive while considering the embedded option the hotel company has in addition to the current Brazilian expansion project. In spirit, the value of real options comes from the notion that while investing in risky assets, a manager can observe what happens in the real world and incorporate that learning in adapting his/her behavior to increase the project's potential upside return and to decrease the possible downside risk. When a manager decides to make a risky investment, there are three possible actions that s/he could take based upon the new information that s/he observes from the state variables.

1. Option to expand if the outlook is positive
2. Option to abandon if the outlook is negative, and
3. Option to delay if the outlook is inconclusive.

Figure 3. Decision tree analysis: single period

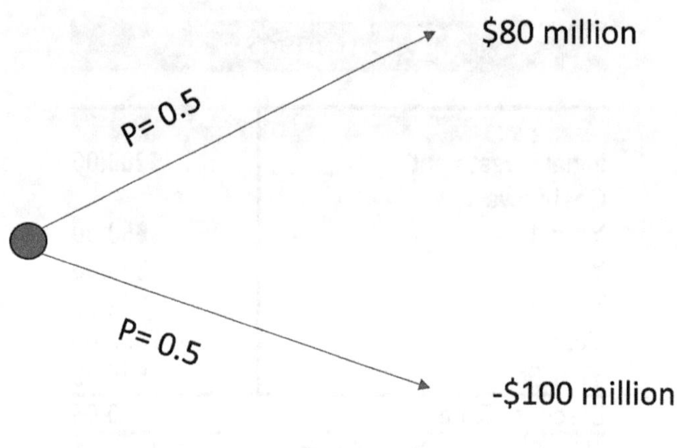

Figure 4. Decision tree analysis: two period

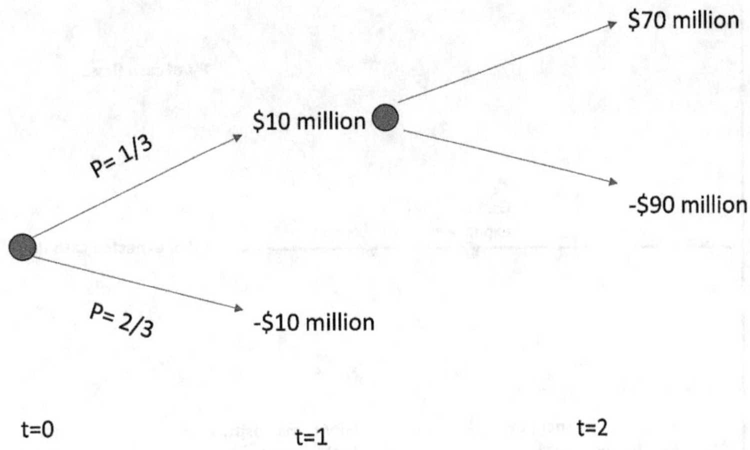

In the following section, we will discuss how a hotel manager could add more value to an investment by considering one of the most important options embedded in investment decision – option to expand.

3. Real Options Analysis

Consider a simple example of an option to expand for a hotel. Helvetica Hotel Development Corporation is currently expanding its operation to Brazil and estimates it net present value to be -44 million. The negative net present value traditionally suggests rejecting the investment. However in Helvetica's case, it has plans to expand the hotel to the rest of South America at a cost of $500 million if the Brazilian venture pans out well. Helvetica estimates that based on current assessment, the expansion project's present value is $400 million which is again a negative net present value. The $400 million estimate is an approximate estimate and Helvetica is not confident about the estimate because it does not have a firm grasp of the market. A Monte Carlo simulation performed by their analysts show that the volatility of this project's value is 50%. Assuming Helvetica has to make this decision within 5 years of the Brazilian investment, what is the value of this option to expand given the five-year risk free rate is 4%. The steps to solve the case are provided below:

Step 1: Inputs.

First we need to get all the required inputs to solve the real options valuation case.

1. S (Spot price) = Enter the assumed value of expansion cash flows at the time of option acquisition or the present value of hotel expansion cash flows = $400 million
2. K (Strike price) = Enter the exercise price of the option to expand which is equal to the cost of expansion =$500 million
3. σ (volatility) = Enter the standard deviation of returns for the underlying asset, stated as the annualized standard deviation (got from Monte Carlo simulation) = 50%

Figure 5. The option to expand

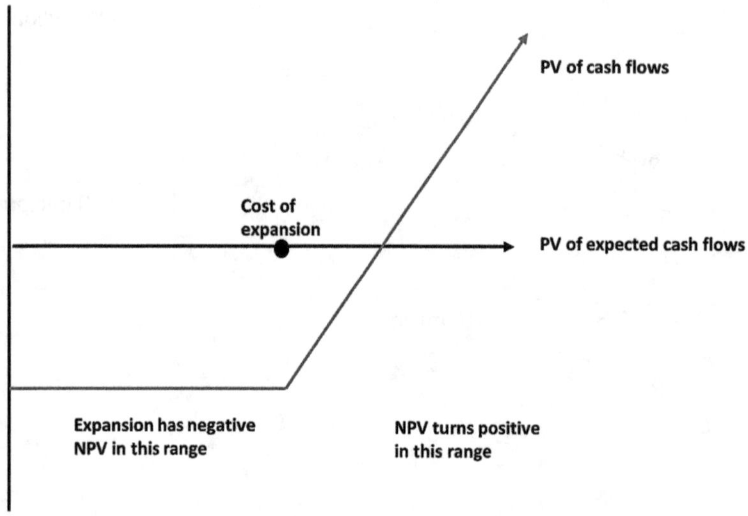

4. value t (time) = Enter the time to option expiration in years or decimal fractions of years = 5 years
5. r (risk free rate) = Enter the risk-free rate at the time of acquisition for the duration of the option, stated as a compound annual rate = 4%

Step 2: Calculate d1.

$$d1 = \frac{1}{\sigma\sqrt{T-t}}\left[\ln\left(\frac{S}{K}\right) + \left(r + \frac{\sigma^2}{2}\right)(T-t)\right]$$

Step 3: Calculate d2.

$$d2 = d1 - \sigma\sqrt{T-t}$$

Step 4: Calculate call option price.

Figure 6. Step 1: inputs

Real Option Value using Black-Scholes Option Pricing		
Input Data		
Stock Price now (P)	Enter the assumed value of expansion cash flows at the time of option acquisition	400
Exercise Price of Option (EX)	Enter the exercise price of the option to expand which is the cost of expansion	500
Number of periods to Exercise in years (t)	Enter the time to option expiration in years or decimal fractions of years.	5
Compounded Risk-Free Interest Rate (rf)	Enter the risk-free rate at the time of acquisition for the duration of the option, stated as a compound annual rate.	4.00%
Standard Deviation (annualized σ)	Enter the standard deviation of returns for the underlying asset, stated as the annualized standard deviation.	50.00%

Figure 7. Step 2: calculate dl

	A	B	C
1	Real Option Value using Black-Scholes Option Pricing		
2			
3	**Input Data**		
4	Stock Price now (P)	Enter the assumed value of expansion cash flows at the time of option acquisition	400
5	Exercise Price of Option (EX)	Enter the exercise price of the option to expand which is the cost of expansion	500
6	Number of periods to Exercise in years (t)	Enter the time to option expiration in years or decimal fractions of years.	5
7	Compounded Risk-Free Interest Rate (rf)	Enter the risk-free rate at the time of acquisition for the duration of the option, stated as a compound annual rate.	4.00%
8	Standard Deviation (annualized σ)	Enter the standard deviation of returns for the underlying asset, stated as the annualized standard deviation.	50.00%
9			
10			
11	**Output Data**		
12	Present Value of Exercise Price (PV(EX))	C5*EXP(-C7*C6)	409.3654
13	σ*t^.5	C8*C6^0.5	1.1180
14	d1	(LN(C4/C5)+(C7+C8*C8/2)*C6)/(C8*C6^0.5)	0.5383

Figure 8. Step 3: calculate d2

	A	B	C
1	Real Option Value using Black-Scholes Option Pricing		
2			
3	**Input Data**		
4	Stock Price now (P)	Enter the assumed value of expansion cash flows at the time of option acquisition	400
5	Exercise Price of Option (EX)	Enter the exercise price of the option to expand which is the cost of expansion	500
6	Number of periods to Exercise in years (t)	Enter the time to option expiration in years or decimal fractions of years.	5
7	Compounded Risk-Free Interest Rate (rf)	Enter the risk-free rate at the time of acquisition for the duration of the option, stated as a compound annual rate.	4.00%
8	Standard Deviation (annualized σ)	Enter the standard deviation of returns for the underlying asset, stated as the annualized standard deviation.	50.00%
9			
10			
11	**Output Data**		
12	Present Value of Exercise Price (PV(EX))	C5*EXP(-C7*C6)	409.3654
13	σ*t^.5	C8*C6^0.5	1.1180
14	d1	(LN(C4/C5)+(C7+C8*C8/2)*C6)/(C8*C6^0.5)	0.5383
15	d2	C14-C13	-0.5797

$$C(S,t) = N(d1)S - N(d2)Ke^{-r(T-t)}$$

- NPV of Limited Brazilian market expansion: -$44 million.
- Value of Option to Expand to full South American market: $167 million.
- NPV of Project with option to expand: -$44 million + $167 million = $123 million.

Taking an expansion project today may allow a hotel to consider and take other valuable expansion projects in the future. Thus, even though the expansion project may have a negative NPV, it may be a project worth taking if the option it provides the hotel (to take other projects in the future) provides a more-than-compensating value. These are the options that firms often call “strategic options” and use as a rationale for taking on “negative NPV” projects.

Figure 9. Step 4: calculate call option price

	A	B	C
1	Real Option Value using Black-Scholes Option Pricing		
2			
3	***Input Data***		
4	Stock Price now (P)	Enter the assumed value of expansion cash flows at the time of option acquisition	400
5	Exercise Price of Option (EX)	Enter the exercise price of the option to expand which is the cost of expansion	500
6	Number of periods to Exercise in years (t)	Enter the time to option expiration in years or decimal fractions of years.	5
7	Compounded Risk-Free Interest Rate (rf)	Enter the risk-free rate at the time of acquisition for the duration of the option, stated as a compound annual rate.	4.00%
8	Standard Deviation (annualized σ)	Enter the standard deviation of returns for the underlying asset, stated as the annualized standard deviation.	50.00%
9			
10			
11	***Output Data***		
12	Present Value of Exercise Price (PV(EX))	C5*EXP(-C7*C6)	409.3654
13	σ*t^.5	C8*C6^0.5	1.1180
14	d1	(LN(C4/C5)+(C7+C8*C8/2)*C6)/(C8*C6^0.5)	0.5383
15	d2	C14-C13	-0.5797
16	Delta N(d1) Normal Cumulative Density Function	NORMDIST(C14;0;1;TRUE)	0.7048
17	Bank Loan N(d2)*PV(EX)	NORMDIST(C15;0;1;TRUE)*C12	115.0532
18			
19	*Value of Call*	C16*C4-C17	166.8751
20			

DISCUSSIONS

DCF analysis is easy to understand and apply. It can be applied to simple and complex situations alike. DCF analysis can be used for valuing the entire business and also for valuing separate business units of a company. Like any valuation tool, it relies heavily on the quality of inputs used. Therefore, if inputs are modified slightly there can be large deviations in valuation. The success of DCF analysis is related to the accuracy of future cash flows predictions. Traditional DCF analysis attempts to optimize capital budgeting decisions when capital resources are limited and existing investment alternatives are known. DCF analysis assumes fixed scenario for outlays and operations. With uncertainty and irreversibility, DCF rule is often wrong. While valuing important real options like land, offshore oil reserves, or patent that provides an option to invest, DCF analysis fails. DCF analysis cannot determine the value of flexibility, like delaying construction, installing small solar panel units in the hotel instead of building a large turbine etc. DCF does not provide the tools for managers to incorporate uncertainty in any form.

A main benefit of decision tree analysis is that it provides a complete outline for alternative scenarios of a decision. Simple decision trees are usually simple to use, easy to understand, and offer many advantages compared to DCF analysis. As with DCF, the consistency of the decision tree analysis rests on the precision of the input information. Also, formulating large complex decision tree with many branches and nodes, is difficult and time-consuming. Calculating probabilities of different probable branches at every node, evaluating and selecting optimal split of each node in the decision tree are complex tasks that require expertise and experience in quantitative and statistical analysis. Among the major disadvantages of a decision tree analysis is its inherent limitations in predicting values of a continuous variable. Another major limitation includes the chance of replication of the same sub-branch in different branches.

ROA provides optimal decisions predominantly while NPV is inconsistent and provides optimal decisions only in a relatively small set of special cases like the one discussed above in DCF analysis. The helpfulness of a real options approach to managerial decision making depends entirely on how close are the attributes of the capital budgeting proposal being evaluated to the assumptions of the option valuation model being used. One should note that the similarity between financial and real options is limited, making the use of financial option valuation models sometimes challenging for real options. Strategic options many a times do not include the characteristics of financial options. These variances in fact determine whether traditional option valuation models (such as the Black-Scholes model) are useful for managerial decision making. As with DCF and decision tree analysis, ROA also suffers from its dependencies on precise input information.

FUTURE RESEARCH DIRECTIONS

Real options analysis takes a more realistic, understanding and interpretation of the investment decision. It incorporates in its analysis, the option that the investments may be delayed or altered as conditions change or as new information becomes available. We strongly believe that integrating real options analysis to traditional valuation approaches thereby complementing DCF analysis will provide more insightful decisions that incorporates uncertainty and flexibility. Future research should be aimed at creating combined valuation approaches that specifically address the valuation challenges in the hospitality industry.

CONCLUSION

Discounted cash flow analysis is the most common method for valuing projects in a company, but an increasing number of practitioners are now using options methodology to get a better picture of the real value of a project. In this book chapter, we introduce the readers to the importance of incorporating uncertainty in determining optimal investment decisions using a hypothetical case. We explain the rationale as to why it is important to go beyond the traditional valuation techniques like discounted cash flow analysis. We then describe the decision tree analysis technique and the real options analysis method in detail and its application in hotel expansion investment decisions. In the hypothetical case, we focus on one investment decision-hotel expansion and elaborate on the assumptions and considerations required by managers to use the real options analysis in this context. We also provide the step by step estimations of various investment decisions like DCF and decision tree analysis and contrast it with real options optimization technique step by step in Excel.

Value of the project, derived from the valuation of real options, is the best way of capturing the value of flexibility and uncertainty. However, this valuation method is still not in widespread use, for similar reasons that earlier deferred the use of the DCF method — mainly due to the lack of understanding of the underlying tools and comprehending the benefits of options valuation (Anderson, 2002).

ACKNOWLEDGMENT

I would like to thank Thai Trang, a Ph.D. candidate in Finance at the University of Texas at Arlington for her insights. I also thank the anonymous reviewers for their comments and suggestions.

REFERENCES

Alvarez, L., Ruiz, F., & Solana, P. (2008, September). *Yield management under the real options scheme for optimal decision making in hotels*. Paper presented at the II International Conference on Industrial Engineering and Industrial Management, Burgos, Spain.

Amran, M., & Kulatilaka, N. (1999). *Real options—Managing strategic investment in an uncertain world*. Boston, MA: Harvard Business School Press.

Andersen, L. (2002). How options analysis can enhance managerial performance. *European Management Journal*, *20*(5), 505–511. doi:10.1016/S0263-2373(02)00093-2

Anderson, C. K., Davison, M., & Rasmussen, H. (2004). Revenue management: A real options approach. *Naval Research Logistics*, *51*(5), 686–703. doi:10.1002/nav.20026

Bachelier, L. (1900). *The'orie de la Speculation*. Paris: Gauthier-Villars.

Black, F., & Scholes, M. S. (1973). The pricing of options and corporate liabilities. *Journal of Political Economy*, *81*(3), 637–659. doi:10.1086/260062

Calle, A. M., & Tamayo, V. M. (2009). Decisiones de inversión a través de opciones reales. *Estudios Gerenciales*, *25*(111), 107–126. doi:10.1016/S0123-5923(09)70073-7

Chen, J. J., & Dimou, I. (2005). Expansion strategy of international hotel firms. *Journal of Business Research*, *58*(12), 1730–1740. doi:10.1016/j.jbusres.2004.11.001

Contractor, J. F., & Kundu, K. S. (1998). Modal choice in the world of alliances: Analysing organizational forms in the international hotel sector. *Journal of International Business Studies*, *29*(2), 325–358. doi:10.1057/palgrave.jibs.8490039

Cunill, O. M., Forteza, C. M., & Miralles, M. R. (2008). Valuing growth strategy management by hotel chains based on the real options approach. *Tourism Economics*, *14*(3), 511–526. doi:10.5367/000000008785633640

Fisher, I. (1907). *The Rate of Interest*. New York: The Macmillan Company.

Gallego, G., Kou, S. G., & Phillips, R. (2008). Revenue management of callable products. *Management Science*, *54*(3), 550–564. doi:10.1287/mnsc.1070.0782

Garcia de Soto Camacho, E. (2015). Mode of entry, real options and strategic flexibility in internationalization processes of hotel chains. Doctoral Dissertation Summary. *European Journal of Tourism Research, 10*, 131-135.

García de Soto-Camacho, E., & Vargas-Sánchez, A. (2015). Choice of entry mode, strategic flexibility and performance of international strategy in hotel chains: An approach based on real options. *European Journal of Tourism Research*, *9*, 92–114.

Gertner, R., & Rosenfield, A. (1999). *How Real Options Lead to Better Decisions*. Retrieved from http://www.ftmastering.com/risk/sr3.html

Haley, G., & Goldberg, S. (1995). Net Present Value Techniques and Their Effects on New Product Research. *Industrial Marketing Management*, *24*(3), 177–190. doi:10.1016/0019-8501(94)00044-W

Hayes, R., & Abernathy, W. (1980). Managing our way to economic decline. *Harvard Business Review*, *58*(4), 66–77.

Hendershott, P. H., & Ward, C. W. R. (2000). Incorporating option-like features in the valuation of shopping centers. *Real Estate Finance*, (Winter), 31–36.

Hirschleifer, J. (1958, August). On the theory of optimal investment decision. *Journal of Political Economy.*

Hodder, J., & Riggs, H. (1985). Pitfalls in evaluating risky projects. *Harvard Business Review*, *63*(1), 128–135.

Holland, A. S., Ott, S. H., & Riddough, T. J. (2000). The role of uncertainty in investment: An examination of competing investment models using commercial real estate data. *Real Estate Economics*, *28*(1), 33–64. doi:10.1111/1540-6229.00792

Lai, R. N., Ng, L. H. M., & Zhou, Y. (2001, January). *Analysis of Hotel Overbuilding Based on Real Options Exercise Strategies*. Paper presented at the sixth annual meeting of the Asian Real Estate Society, Tokyo, Japan.

Li, B. (2014). A Cruise Line Dynamic Overbooking Model with Multiple Cabin Types from the View of Real Options. *Cornell Hospitality Quarterly*, *55*(2), 197–209. doi:10.1177/1938965513507126

Lu, P. (2012). *Optimizing hotel pricing: A new approach to hotel reservations. Cornell Hospitality Reports, 12(10).*

Myers, S. C. (1977). Determinants of corporate borrowing. *Journal of Financial Economics*, *5*(2), 147–175. doi:10.1016/0304-405X(77)90015-0

Myers, S. C. (1984). Finance Theory and Financial Strategy. *Interfaces*, *14*(1), 126–137. doi:10.1287/inte.14.1.126

Quan, D. C. (2002). The price of a reservation. *The Cornell Hotel and Restaurant Administration Quarterly*, *43*(3), 77–86. doi:10.1016/S0010-8804(02)80021-0

Quigg, L. (1993). Empirical Testing of Real Option-Pricing Models. *The Journal of Finance*, *48*(2), 621–641. doi:10.1111/j.1540-6261.1993.tb04730.x

Quigg, L. (1995). Optimal land development. In L. Trigeorgis (Ed.), *Real Options in Capital Investment: Models, Strategies and Applications* (pp. 265–280). Westport, CT: Praeger.

Rappaport, A. (1986). Creating Shareholder Value. The Free Press.

Sirmans, C. F. (1997). Research on discounted cash flow models. *Real Estate Finance*, *13*(4), 93–95.

Slater, S. F., Reddy, V. K., & Zwirlein, T. J. (1998). Evaluating strategic investments: Complementing discounted cash flow analysis with options analysis. *Industrial Marketing Management*, *27*(5), 447–458. doi:10.1016/S0019-8501(97)00097-7

Trigeorgis, L. (1996). *Real options: Managerial flexibility and strategy in resource allocation*. Cambridge, MA: MIT Press.

Yang, C., & Chang, T. (2011). Binomial real option pricing for restaurant menu analysis. *Cornell Hospitality Quarterly*, *52*(3), 273–282. doi:10.1177/1938965511410054

KEY TERMS AND DEFINITIONS

Call Option: A call option provides a buyer the right not an obligation to buy a security (stock, bond, commodity etc.) at a specified price during a specified time.

DCF: Discounted Cash Flow (DCF) analysis estimates the value of an asset based on its expected future cash flows, by discounting them to the present using an appropriate discount rate.

Decision Tree: A decision tree is a tool that uses a tree like graph to model decisions and consequences to help managers incorporate uncertainty in valuations.

Real Option: A real option provides the manager the right not an obligation to undertake certain business initiatives like expansion, abandonment, deferment, etc.

Valuation: The process of determining the present value of all potential future cash flows of an asset.

Chapter 5
Mastering Customer Service, Customer Experience, and Customer Orientation in the Hospitality and Tourism Industry

Kijpokin Kasemsap
Suan Sunandha Rajabhat University, Thailand

ABSTRACT

This chapter presents the utilization of optimization techniques in the hospitality and tourism industry; the perspectives on customer service and customer service teams; customer service and service leadership; customer service behavior in the travel industry; the overview of customer experience; customer experience, customer satisfaction, and service quality in the airline industry; customer experience and budget hotel in the hospitality industry; the issues with customer orientation; and the relationship between customer orientation and service innovation in the hotel industry. Optimization techniques are the essential approaches to many real-world problems, and can be applied to promote the performance of hospitality and tourism industry regarding customer service, customer experience, and customer orientation. The chapter argues that enhancing customer service, customer experience, and customer orientation through optimization techniques has the potential to increase organizational performance and reach strategic goals in the hospitality and tourism industry.

INTRODUCTION

In the service sector, high service quality effectively promotes customer satisfaction, and customer satisfaction has a positive relationship with business revenue (Kasemsap, 2016a). Customer service can enable the company to differentiate its offers from competition (Melovic, Mitrovic, Djokaj, & Vatin, 2015). Regarding effective customer service, staying close to customers, serving them productively,

DOI: 10.4018/978-1-5225-1054-3.ch005

and dealing effectively with complaints are the essential capabilities that contribute to business success (O'Reilly & Paper, 2012). It is important for service personnel to have high ethical standards, be committed to the business, and be willing to serve customers well, as this perspective can help their employer gain sustainable competitive advantage (Lin, 2014).

Focusing on customer experience can help to maintain competitive advantage (Chen & Lin, 2015). Creating a positive customer experience enables a company to differentiate itself from the competition (MacGillavry & Wilson, 2014). Customer experience can be incorporated into the life contexts of the customer to produce values that are consistent with those of the customer or create stunning thinking patterns, attracting the interest of the customer and leaving a positive impression, creating tangible revenue and intangible reputation for the brand enterprise (Bruhn, Schnebelen, & Schäfer, 2014).

In modern business, customers expect sellers not only to respond effectively to their expressed needs, but also to understand their business sufficiently well to proactively address their latent and future needs regarding customer orientation (Blocker, Flint, Myers, & Slater, 2011). Customer orientation means creating products that should meet customers' desires (Badea, 2014). Customer orientation and the development of long-term relationships with customers are the major elements for business growth and profit sustainability (Valenzuela, Mulki, & Jaramillo, 2010). In the travel industry, employees' customer orientation depends on employees' organizational identification and their leaders' acting as the role models of customer orientation (Wieseke, Ullrich, Christ, & van Dick, 2007).

The chapter aims to bridge the gap in the literature on the thorough literature consolidation of customer service, customer experience, and customer orientation. The extensive literature of customer service, customer experience, and customer orientation provides a contribution to practitioners and researchers by describing the current issues and trends with customer service, customer experience, and customer orientation in the hospitality and tourism industry.

BACKGROUND

Customer service is a direct result of the connection between sales activities and the delivery process, which begins with the order and ends with the delivery, but in some cases it continues with the post-purchase stage of the product usage (Fawcett, Mcleish, & Ogden, 1992). Brohman et al. (2009) indicated that customer service is a key component of a firm's value proposition and a fundamental driver of differentiation and competitive advantage in nearly every industry. Managing both technologies and customer service teams, multichannel customer support creates an operational challenge in modern business (Froehle, 2006).

Customer experience refers to the sensory, emotional, and cognitive impacts generated by customers following participatory and observational experiences, which elevates their interest, motivations, and recognition, and subsequently adds value to the brand communities (Kim & Perdue, 2013). Customer experience can be incorporated into customers' lives to produce perspectives consistent with those of the customer and citizen, or create the modes of thinking, thus creating the tangible revenue for enterprises and intangible reputation for governments (Bruhn et al., 2014). Customer experience is essential because experience is the psychological perception in the experiences of users, and affects usage behaviors (Hsu & Tsou, 2011).

As a strategic orientation, a customer orientation provides the firm with the strategic direction to encourage appropriate behaviors that not only focus on creating superior customer value, but also promote a culture that is an approach to building innovation competencies (Day, 1994). Research on customer orientation has been conducted at two levels of analysis:

- The individual level (Brown, Boya, Humphreys, & Widing, 1993), and
- The organizational level (Kennedy, Goolsby, & Arnould, 2003).

Research at the individual level relates to the interpersonal contact between employees and customers (Maurya, Mishra, Anand, & Kumar, 2015). Research at the organizational level has focused on the perspectives of organizational behavior with respect to a firm's customers and competitors (Maurya et al., 2015).

ASPECTS OF CUSTOMER SERVICE, CUSTOMER EXPERIENCE, AND CUSTOMER ORIENTATION

This section emphasizes:

- The utilization of optimization techniques in the hospitality and tourism industry;
- The perspectives on customer service and customer service teams;
- Customer service and service leadership;
- Customer service behavior in the travel industry;
- The overview of customer experience;
- Customer experience, customer satisfaction, and service quality in the airline industry;
- Customer experience and budget hotel in the hospitality industry;
- The issues with customer orientation; and
- The relationship between customer orientation and service innovation in the hotel industry.

Utilization of Optimization Techniques in the Hospitality and Tourism Industry

Optimization is the collection of mathematical principles and methods used for solving quantitative problems in many disciplines, including:

- Business,
- Economics,
- Engineering,
- Physics, and
- Biology (Vora & Mirnalinee, 2015).

Optimization problems are typically defined in terms of a set of decision variables (i.e., an assignment of a value to each of these constitutes a solution to the problem), a set of constraints (which specify the allowed combinations of values for these variables), and an objective function defined on the set of decision variables (Ghose & Billiau, 2011).

Major soft computing-related optimization techniques include:

- Genetic algorithms,
- Simulated annealing,
- Ant colony optimization,
- Particle swarm optimization,
- Fuzzy logic,
- Neural networks, and
- Meta-heuristics (Xie & Zhang, 2012).

Due to the presence of fuzzy or linguistic and dynamic variables, the modeling of dynamic systems (e.g., social system and humanistic system) is a challenge to many decision makers (Mutingi & Mbohwa, 2014). Hospitality and tourism companies today have to make difficult decisions about targeting the right customers with the right offers while staying within budget and channel capacities, all without disassembling future sales or saturating customers with too many messages toward enhancing customer service, customer experience, and customer orientation.

Dostál (2014) indicated that soft computing-related optimization techniques can be utilized to:

- Address the potential customer (fuzzy logic),
- Which kind of customer could be provided by a loan or a mortgage (neural networks),
- The sorting of products according to the kind of customers (genetic algorithms),
- Or solving the traveling salesman problem (evolutionary algorithms).

The benefits of applying soft computing-related optimization techniques to the hospitality and tourism industry include the enhanced return on investment (ROI), better adherence to customer contact policies, and improved overall operational efficiency.

Through the applications of mobile device and optimization techniques, consumers (e.g., guests and tourists) in the hospitality and tourism industry are able to receive suitable announcements, alerts, promotions, discounts, and location-relevant offerings along with access to the reports of purchasing history, loyalty rewards status, and a host of concierge services. Hotels, restaurants, clubs, and casinos recognize the unparalleled competitive advantages that can be gained from the location-based service (LBS) applications, including guest check-ins, word of mouth broadcasting, and shared experiences. LBS technology involves the use of device location coordinates to determine what directional or promotional information to transmit.

There are two broad categories of optimization-related LBS applications (i.e., push and pull applications). Push applications deliver information requested by the user in response to the user opting-in or triggering transmission based on entry into a specific location. Push applications rely on pre-set content developed for the guest by the host (e.g., hotel, restaurant, club, and casino). A push service is activated by an event (e.g., guest arrives in a targeted area or a time-dependent setting expires). In the pull applications, information is delivered as a result of a direct query by the user. In other words, the user requests specific information (e.g., hours of operation, nearest location, and reservation information) and the application provides the reactive response toward improving customer service, customer experience, and customer orientation in the hospitality and tourism industry.

Perspectives on Customer Service and Customer Service Teams

Customer service represents a measure of the efficiency of the logistics system in creating time and space value for the product, including the after-sales activities of the company (Melovic et al., 2015). Customer service is recognized as a supportive activity to the basic activities of the company-production or sale of products or services (Berman & Evans, 1995). The policy of the customer service is not only created for current customers, but also it is a process which has the task to attract as many new customers as possible (Lambert & Stocs, 1993).

As long as the products or services remain in use, customers tend to retain the high levels of expectation related to service provider (Wu & Lo, 2012). The provision of logistics customer services is a crucial point in supply chain management that can affect customers' brand perceptions (Oflaç, Sullivan, & Baltacioğlu, 2012). Superior logistics customer service leads to improve firm performance (Leuschner, Charvet, & Rogers, 2013). Developing quality relationships with customers can improve opportunities for future sales (Ernst, Hoyer, Krafft, & Krieger, 2011). Rayport et al. (2005) indicated that in the perspective of demanding customers and a shortage of skilled workers to serve them, companies are facing a crisis in customer interaction and customer relationship management (CRM). CRM becomes one of the most important business strategies in the digital age, thus involving organizational capability of managing business interactions with customers in an effective manner (Kasemsap, 2015a).

Customer service teams are increasingly developed because the selling process is beyond the capabilities of individual and may call for an integrated effort (Workman, Homburg, & Jensen, 2003). Customer service teams are thought able to combine diverse information to analyze clients' needs and influence buying decisions effectively (Guenzi, Pardo, & Georges, 2007). Effective customer service teams are characterized by in-role performance (Banker, Field, Schroeder, & Sintia, 1996), customer orientation (Flaherty, Dahlstrom, & Skinner, 1999), and the capability to recover from service errors (De Jong, Wetzels, & de Ruyter, 2008).

Effective customer service teams are highly oriented toward customers (Homburg, Müller, & Klarmann, 2011). Customer orientation occurs when customers believe that the service team helps them make purchasing decisions that are highly beneficial and fitting for the customers themselves (Schwepker, 2003). As the key characteristic of customer orientation is focused on satisfying the customer's long-term needs, customer service team has to expend the greater effort in customer-related interactions by working with customers (Thakor & Joshi, 2005).

Customer Service and Service Leadership

Companies are increasingly relying on customer service teams to deliver quality service critical for repeat business and profitability (Wong, Liu, & Tjosvold, 2015). But these teams need leadership to be effective (Wong et al., 2015). Customer service teams are expected to have the potential to combine expertise and effort so that they can flexibly respond to changing requirements, recover from service errors, and help customers feel satisfied and committed to the company (Greer & Lei, 2012).

Effective leadership lays the foundation for teamwork that results in the productive action in the service industry (Morgeson, DeRue, & Karam, 2010). Transformational leadership, empowerment, and innovation support significantly lead to organizational innovation in modern business (Kasemsap, 2014). Leadership development enables individuals to initiate qualities, which can make them a leader, and helps managers to develop their leadership qualities for becoming more productive and managerial

(Kasemsap, 2016b). Leaders can help develop the environment in which customer service personnel beneficially exceed (Wong et al., 2015).

Customer service leaders can strengthen the values of caring for customers and delivering high-quality service to them that encourage customer service teams to engage in adaptive selling where they understand customers and utilize a wide range of approaches to meet customers' needs (Gronfeldt & Strother, 2006). This adaptive selling, as it is both a flexible and a committed approach to customer service, results in the productive customer service teams that are oriented toward serving customers (Saxe & Weitz, 1982) and effectively recover from customer service deficiencies (Ashill, Rod, & Carruthers, 2008).

Customer Service Behavior in the Travel Industry

Customer service in the retail settings can be considered as an important factor in order to maintain satisfied and loyal customers (Molina, Martín, Santos, & Aranda, 2009). In the travel industry, retail travel agencies are the high-contact service businesses, with travel agents serving as the main route for interaction with customers (Chow, Lai, & Loi, 2015). Cheyne et al. (2006) recognized travel agents as the key intermediary between travel suppliers and consumers. As the front line service employees, travel agents are responsible for delivering customer service, which plays a critical role in the success of the travel agency (Chow et al., 2015).

Two types of travel agent performances are crucial for sustaining service quality: customer service behavior and organizational citizenship behavior (OCB). Customer service behavior refers to both role-prescribed and extra-role service behavior directed toward customers (Bettencourt & Brown, 1997), and OCB refers to discretionary or voluntary behavior directed toward the organization that is not explicitly specified in the employee's formal job description or recognized in the formal reward structure (Organ, 1988). Leader-member exchange (LMX), organizational justice, and job satisfaction are positively related to OCB (Kasemsap, 2013). Increasing evidence shows a positive relationship of customer service behavior and OCB to customers' perception of service quality (Yoon, Seo, & Yoon, 2004). When travel agents experience a low level of LMX, they have little access to resources and receive less supervisor support (Bolino & Turnley, 2009).

Tourism marketing promotes the dynamic evolutions and the demands of tourism industry concerning the diversity of tourism marketing topics (Kasemsap, 2015b). Market orientation as a business strategy has recently been adopted in the travel and tourism industry, as well as other industries, to enhance performance (Chen & Myagmarsuren, 2013). A number of studies demonstrated the positive impact of internal marketing orientation (IMO) on employees' satisfaction (Tortosa, Moliner, & Sanchez, 2009) and the success of an organization (Bitner, Booms, & Mohr, 1994) in the service industry.

Lings and Greenley (2009) provided empirical evidence on the positive relationship between the level of IMO and employee motivation to provide the excellent customer service. Employees consider IMO as the organization's commitment to understanding and responding to their needs (Lings & Greenley, 2009). In the travel service context, a travel agency needs to train, motivate, and provide emotional support to its travel agents to interact with customers (Chow et al., 2015). When travel agents perceive high IMO, they will be strongly motivated to reciprocate with customer service behavior to generate high-quality service and encourage customer satisfaction (Chow et al., 2015).

Overview of Customer Experience

Customer experience is defined as the cognitive acknowledgment or perception that follows from stimulated motivation of a customer who observes or participates in an event (Chen & Lin, 2015). Customer experience is about the conscious and unconscious sides of the human mind, expressed as rational and emotional decisions and actions toward the company's offering (Vanharanta, Kantola, & Seikola, 2015). Klaus and Maklan (2013) viewed customer experience as the customer's cognitive and affective evaluation of all direct and indirect encounters with the firm relating to their purchasing behavior. Lemke et al. (2011) defined customer experience as the customer's subjective response to the holistic direct and indirect encounter with the firm. Meyer and Schwager (2007) recognized customer experience as the internal and subjective response customers have to contact with a company. Shaw (2005) described customer experience as an interaction between an organization and a customer.

The purpose of customer experience is to drive brand success by creating a brand-based customer experience which is differentiated from its competitors and for which consumers are willing to pay (Laming & Mason, 2014). Customer experience has been theoretically and empirically clarified in terms of its components that create value (Gentile, Spiller, & Noci, 2007), as well as its creation, determinants, and dynamics (Verhoef et al., 2009). Customer experience incorporates company-customer interaction that includes both functional and emotional clues developed through different touch points (Khan, Garg, & Rahman, 2015). By delivering customer satisfaction in its execution, good customer experience significantly delivers revenue, profitability, and growth (Frow & Payne, 2007).

The domain of customer experience is effectively developed in the face-to-face context (Rose, Hair, & Clark, 2011). Physical environment contacts offer experiences to the customer (Pullman & Gross, 2004). The effective interactions with human or physical elements are important in shaping the customer experience (Ren, Qiu, Wang, & Lin, 2016). The proper management of customer experience can turn customers into advocates, upon which profitability can be achieved (Smith & Wheeler, 2002). It is an intentional effort on the part of the company to develop and maintain good experience which is differentiated from the competition, consistent at every touch point and most importantly valued by the customer (Joshi, 2014).

Improving customer experience can help marketers indicate niche markets than does maintaining competitive prices or superior product functions and features (Kao, Huang, & Yang, 2007). The effective customer-oriented strategies can promote the enhancement of customer experience (Ojiako & Maguire, 2009). Schmitt (1999) developed an integrated customer experience framework to help marketers evaluate the different types of experiences (e.g., strategic experiential modules). Customer experience is an important method that generates a link between customer and product and provides a memorable experience for customer, thus adding value to product (Chen & Lin, 2015). The effective customer experience refers to a positive interaction between a customer and the experience (Christodoulides, de Chernatony, Furrer, Shiu, & Abimbola, 2006), which enhances customer satisfaction (Kim, 2005).

Through customer experience, consumers implicitly generate various psychological assessments to measure metrics for assessing the value that can be gained from experiential services that are provided by enterprises (Vera & Trujillo, 2013). Perceived value is the perception that is perceived by customers following a customer-experience interaction, presented in terms of assessment items. Perceived value-related assessment items include a customer's return on investment,

- Determined by comparing resources invested and benefits gained, superiority of service,
- Determined by whether the commitments made by the service provider were met, esthetic value,
- Determined by subjectively judging the esthetic conception, and the pleasure gained in the pattern of experienced enjoyment and happiness (Vera & Trujillo, 2013).

Customer experience is an important factor for the service brands (Hartman, Meyer, & Scribner, 2009). Hospitality and tourism industry has been considered as the more experience-based service industry (Yuan & Wu, 2008). Regarding online marketing, customer experience greatly affects online browsing and purchase intentions of users in the online shopping environments (Luo, Chen, Ching, & Liu, 2011). It is important to better understand shoppers' online experiences when shopping with an approach to retaining them in the online marketing channel or at least in the retailer's multi-channels for the longer term (Martin, Mortimer, & Andrews, 2015). Regarding the increasing advances in technology, information technology and knowledge management effectively improve the strategic tools for providing the direct link between customers and tourism organizations, thus encouraging the communication channels in global tourism (Kasemsap, 2016c). Sorooshian et al. (2013) indicated that the application of online infrastructure can facilitate the process of purchasing and selling of products toward increasing customer experience.

Customer experience refers to the sensory, emotional, and cognitive impacts of participatory experiences on customers, and specifically on their interest, motivations, and recognition, and significantly adds perceived value to the online communities (von Wallpach & Kreuzer, 2013). Customer experience is the perception that is generated by customers after they have in an experience, and customer satisfaction is the beneficial outcome when customers generate a positive perception and recognition (Janda & Ybarra, 2005). Creating positive customer experiences across all stages of a product's life cycle drives higher satisfaction among current customers, which builds customer loyalty (Kasemsap, 2016d). Customer experience is favorably related to customer satisfaction (Torres, Fu, & Lehto, 2014). Enterprises can employ customer experiences to improve both customer satisfaction and customer loyalty (Brakus, Schmitt, & Zarantonello, 2009).

Customer experience incorporates service quality definition along with the objective to enhance the static measurement of service quality (Berry, Carbone, & Haeckel, 2002). Customer experience is an observation of a journey that precedes service encounters, passes through service deliver, and continues even after delivery of services (Berry et al., 2002). Before service delivery or purchase, customer experiences direct or indirect communications made by a firm (Brakus et al., 2009) and receives word of mouth from other customers (Kwortnik & Ross, 2007). Experiential marketing strategy is designed to attract customer to be involved with the product or service, thus making the customers practically experience favorable responses, either through direct contact or indirect contact with the company through the intermediaries that connect them (Gopalani & Shick, 2011). These interactions will result in customer experiences with the provider (Nasution, Sembada, Miliani, Resti, & Prawono, 2014).

Customer Experience, Customer Satisfaction, and Service Quality in the Airline Industry

The major field of utilization of optimization techniques in tourism research is the airline industry. Areas where optimization techniques have been introduced include:

- Personnel scheduling,
- Ground holiday delays,
- Airline network design,
- Flight frequency determination, and
- Optimization problems in yield management systems in the airline industry toward enhancing customer experience, customer satisfaction, and service quality.

Introducing a customer experience function is considered as a major foundation of success for future profitability and growth for airline industry. Managing various complex elements of the airline customer experience is a distinct challenge for airlines, made more challenging as many of airline staff (e.g., airports, security, and customs) are outside an airline's direct control and complicate the customer journey (Verhoef et al., 2009). Smith and Wheeler (2002) indicated that this perspective is one of the major elements in achieving great customer experience within airlines, citing the examples of Virgin Atlantic and Southwest, who embed their staff within the customer experience, and empower their team to deliver.

Service quality is the difference between performance and expectations (Sorooshian et al., 2013). Parasuraman et al. (2005) stated that any attempt to develop the face-to-face services into online services can affect the validity of convergence. More widespread in the literature is the research that focuses on service quality in the airline industry, particularly the application of the SERVQUAL model to the airline industry (e.g., Liou, Hsu, Yeh, & Lin, 2011; Nadiri, Hussain, Ekiz, & Erdoğan, 2008). SERVQUAL is an effective service quality measurement of how an experience matches an individual's expectation (Laming & Mason, 2014).

In order to measure customer experience in terms of its purpose toward delivering customer satisfaction to increase loyalty and advocacy in the airline industry (Johnston & Kong, 2011), an individual customer satisfaction rating is required at all the airline-related touch points, ranging from booking, to check-in, lounge, boarding, in-flight, transfer, and baggage. The elements of the customer journey can be measured to ascertain which are the important approaches to enhance customer satisfaction and delivering customer experience in the airline industry (Strombeck & Wakefield, 2008). Driver (1999) indicated that achieving differentiation in the airline industry is difficult, with essentially common environments, virtually identical airplane and similar airports, as they are not in control of essential elements of their customer journey.

Customer Experience and Budget Hotel in the Hospitality Industry

At the micro level, customer experience is influenced by the sensory effects of sight, sound, touch, smell, and taste (Pine & Gilmore, 2011). Hemmington (2007) proposed five dimensions of hospitality customer experience (i.e., host-guest relationship, generosity, theater and performance, numerous small surprises, and safety and security). Knutson et al. (2009) indicated four dimensions of hospitality customer experience (i.e., environment, accessibility, driving benefit, and incentive). Walls (2013) described two dimensions of hospitality customer experience (i.e., physical environment and human interaction).

The nature of budget hotels determines the prerequisites for the survival and development of this hotel sector in the competitive market (Ren et al., 2016). Providing no-frills accommodations with no-frills service and creating value for money are top priorities for budget hotel-related business (Ren et al., 2016). Clemes et al. (2011) proposed three primary dimensions that affect customer experience in the budget hotel sector. These customer experience-related dimensions are:

- Interaction,
- Physical environment, and
- Outcome quality, with the price as the moderating factor between service quality and customer satisfaction (Clemes et al., 2011).

Brotherton (2004) investigated the success factors for budget hotels through a quantitative analysis of 239 survey questionnaires in the United Kingdom. The study identified seven dimensions:

- Customer service emerged on top, and was followed by
- Core product,
- Strategic control,
- Hygiene,
- Quality,
- Consistency,
- Pricing, and
- Location.

The 11 value dimensions identified by Brotherton (2004) significantly reflect the holistic nature of customer experience with budget hotels. In the context of China, Zhang et al. (2008) explored the dimensions of brand experience with budget hotels by interviewing 15 budget hotel managers and executives. Three customer experience-related dimensions (i.e., theme and activities, social interactions, and physical environment) affect the cognitive, affective, and behavioral aspects of budget hotel guests (Zhang et al., 2008). Hua et al. (2009) indicated that service quality is the primary concern for budget hotel customers based in China.

Issues with Customer Orientation

Customer orientation can enhance the presence of organizational factors that are instrumental in putting ideas into action (Racela, 2014). Customer orientation includes all the activities that are involved in acquiring information about customers in a market and in disseminating the information in the firm (Narver & Slater, 1990). These behaviors include gathering market intelligence about the needs of customers and sharing that information in the firm (Gatignon & Xuereb, 1997). Customer orientation is positively linked to the both learning behaviors and innovation capabilities (Han, Namwoon, & Srivastava, 1998). Customer-oriented attitudes have a direct effect on customer satisfaction (Stock & Hoyer, 2005).

The organizational processes that are associated with a customer orientation are the dynamic capabilities that facilitate innovation (Blocker et al., 2011). Customer-oriented processes of market sensing, customer relating, and customer response are aimed at acquiring and transforming customer information into knowledge that can be utilized so that the firm can effectively respond to customer requirements (Racela, 2014). De Luca et al. (2010) stated that customer orientation is only valuable to research and development (R&D) effectiveness when there is a high level of knowledge integration within the firm. Learning orientation is positively related to customer orientation, while performance orientation is positively related to selling orientation (Harris, Mowen, & Brown, 2005).

Im and Workman (2004) stated that customer orientation has a positive influence on new products and marketing programs. Internal marketing, job satisfaction, and customer orientation have significant influences on relationship marketing (Wu, Tsai, & Fu, 2013). Effective customer orientation has a positive impact on the introduction of radical innovations, but a negative impact on disruptive innovation, while emerging customer orientation has a positive effect on disruptive innovation and is unrelated to radical innovations (Govindarajan, Kopalle, & Danneels, 2011). Transformational leadership not only directly increased employee customer orientation, but also indirectly enhanced employee customer orientation through employee-perceived supervisor support (Liaw, Chi, & Chuang, 2010). Employee orientation has an indirect relationship with both performance outcomes through customer orientation (Zhang, 2010).

Relationship between Customer Orientation and Service Innovation in the Hotel Industry

As hotels face greater competitive pressure from globalization and customization, they need to meet customer demand for the memorable experiences (Chathoth, Altinay, Harrington, Okumus, & Chan, 2013), and provide high-quality service to customers (Wang, Chen, & Chen, 2012) to survive in the hotel industry. In the hotel industry, customer requirements for continue improving service quality, service production and consumption occur simultaneously, and customer needs are diverse and heterogeneous (Orfila-Sintes, Crespí-Cladera, & Martínez-Ros, 2005). These service characteristics have led to the development of new services and to the improvement of existing hotel services, both of which are crucial for hotel operations (Chang, Gong, & Shum, 2011).

Customer orientation is the main factor that expedites hotels to become the ambidextrous firms (Tang, 2014). In response to inquiries regarding customer service, ambidextrous hotels can create new services and improve the quality of current services (Tang, 2014). Customer orientation enhances a hotel's understanding of customers, and helps it design new or improved service solutions to meet customer needs (Grissemann, Plank, & Brunner-Sperdin, 2013); thus, customer orientation enhances managerial and financial performance (Tajeddini, 2010). In order to reduce the risk of new service development (Grissemann et al., 2013) and develop ideas for improving existing services (Lages & Piercy, 2012), hospitality companies need to understand customer needs; consequently, hotels can balance the capabilities of service creation and improvement for satisfying customer needs regarding effective customer orientation.

Service innovation involves the utilization of new ideas, methods, and technologies (Tang, 2014). Service innovation is utilized to establish new service delivery processes or unique sensory experiences (Shaw, Bailey, & Williams, 2011). By indicating new methods to create enhanced value, service innovation can meet future customer needs (Olsen & Sallis, 2006) and enable revisit customers enjoy distinct service experience. Radical and incremental innovation results from the application of these capabilities by hotels. The application of the service innovation capability leads to radical innovation; by contrast, the application of the service improvement capability leads to incremental innovation (Tang, 2014).

Regarding service innovation in the hotel industry, the development of new services can offer both unique consumption experiences and excellent core values to customers (Paswan, D'Souza, & Zolfagharian, 2009). These core values include impressive life experiences (Gilmore & Pine, 2002), indoor recreational facilities, and a private business center. The development of core values allows a hotel to differentiate itself from competitors and improve the service quality and reputation of the hotel (Tajeddini & Trueman, 2012).

Customer orientation encourages hotels to use a unique approach to respond to current customer requirements (Grissemann et al., 2013). Hotels that apply customer orientation normally place an emphasis on collecting, analyzing, and disseminating knowledge concerning current customers (Sin, Tse, Heung,

& Yim, 2005). By exploring potential demand and closing the gap between hotel services and potential customer requirements, customer orientation helps hotels satisfy customer requirements through service innovation (Zhou, Brown, & Dev, 2009). Customer orientation enables hotels to understand potential customer requirements and guides hotels toward creating effective check-out systems, establishing attractive architectural designs, and providing leisure activities (Ottenbacher & Gnoth, 2005) in order to deliver a memorable accommodation experience (Berry, Shankar, Parish, Cadwallader, & Dotzel, 2006).

FUTURE RESEARCH DIRECTIONS

The classification of the extensive literature in the domains of customer service, customer experience, and customer orientation will provide the potential opportunities for future research. With the advent of the World Wide Web, marketing for electronic commerce (e-commerce) can achieve a better customer acquisition cost (CAC) ratio as putting an advertisement on the Internet is comparably much cheaper than putting up a roadside banner or filming a television commercial (Kasemsap, 2016e). The use of social media improves effective communication, increases thought leadership, enhances sales, and provides better customer service (Kasemsap, 2016f). Social media marketing programs usually center on efforts to create the content that attracts attention and encourages readers to share it across their social networks toward competition in the global marketplace (Kasemsap, 2016g). Facilitating the activities of hospitality and tourism industry through the applications of e-commerce and social media platforms should be further studied.

Consumer attitude is a composite of a consumer's beliefs, feelings, and behavioral intentions toward specific objects within the context of marketing, usually a brand or retail store. The study of consumer attitude is essential to understand the motivation and decision strategies employed by consumers. Regarding consumer attitude, the combination of beliefs, attitudes, and behaviors considerably affect how a consumer reacts to the product or service. Consumer behavior is the study of how a consumer thinks, feels, and selects between competing products. Cross-cultural understanding and cultural knowledge for global marketing approaches are the key issues to be considered for a favorable integration into the cross-cultural organizations (Kasemsap, 2015c). The relationships among customer service, customer experience, customer orientation, and consumer attitude, and consumer behavior in cross-cultural settings will be the important topics for future research directions.

CONCLUSION

This chapter highlighted the utilization of optimization techniques in the hospitality and tourism industry; the perspectives on customer service and customer service teams; customer service and service leadership; customer service behavior in the travel industry; the overview of customer experience; customer experience, customer satisfaction, and service quality in the airline industry; customer experience and budget hotel in the hospitality industry; the issues with customer orientation; and the relationship between customer orientation and service innovation in the hotel industry. Optimization techniques are the essential approaches to many real-world problems, and can be applied to promote the performance of hospitality and tourism industry regarding customer service, customer experience, and customer orientation. Customer service is important to an organization because it is the only contact a customer has with a company and is one of the most important elements of the marketing mix for products and services. The high-quality customer service helps to create customer loyalty and increase customer satisfaction in the hospitality and tourism industry.

Customer experience is the practice of designing and reacting to customer interactions for the purpose of exceeding customer expectations, thus increasing customer satisfaction, loyalty, and advocacy. Positive customer experience is important in that customers have the power choose between competing companies, which offer the varying levels of customer service from poor to excellent. Good customer experience management can strengthen brand preference through differentiated experiences, boost revenue with incremental sales from existing customers and new sales from word of mouth, improve customer loyalty through valued and memorable customer interactions, and lower costs by reducing customer churn in the hospitality and tourism industry.

Customer orientation is of ultimate importance to deliver the value-added products and is a business strategy that requires management and employees to focus on the changing needs of its customers. Customer orientation emphasizes on understanding customers' real needs, and satisfying them better than any competitor in the hospitality and tourism industry. Promoting customer service, customer experience, and customer orientation is necessary for modern organizations that seek to serve suppliers and customers, increase business performance, strengthen competitiveness, and achieve continuous success in the hospitality and tourism industry. Enhancing customer service, customer experience, and customer orientation through optimization techniques has the potential to increase organizational performance and reach strategic goals in the hospitality and tourism industry.

REFERENCES

Ashill, N. J., Rod, M., & Carruthers, J. (2008). The effect of management commitment to service quality on frontline employees' job attitudes, turnover intentions and service recovery performance in a new public management context. *Journal of Strategic Marketing*, *16*(5), 437–462. doi:10.1080/09652540802480944

Badea, M. (2014). Customer orientation in the local media. Case study: The audiovisual in Dambovita county. *Procedia: Social and Behavioral Sciences*, *149*, 76–80. doi:10.1016/j.sbspro.2014.08.333

Banker, R. D., Field, J. M., Schroeder, R. G., & Sintia, K. K. (1996). Impact of work teams on manufacturing performance: A longitudinal field study. *Academy of Management Journal*, *39*(4), 867–890. doi:10.2307/256715

Berman, B., & Evans, J. (1995). *Principles of marketing*. Englewood Cliffs, NJ: Prentice Hall.

Berry, L. L., Carbone, L. P., & Haeckel, S. H. (2002). Managing the total customer experience. *MIT Sloan Management Review*, *43*(3), 85–89.

Berry, L. L., Shankar, V., Parish, J. T., Cadwallader, S., & Dotzel, T. (2006). Creating new markets through service innovation. *MIT Sloan Management Review*, *47*(2), 56–63.

Bettencourt, L. A., & Brown, S. W. (1997). Contact employees: Relationships among workplace fairness, job satisfaction and prosocial service behaviors. *Journal of Retailing*, *73*(1), 39–61. doi:10.1016/S0022-4359(97)90014-2

Bitner, M. J., Booms, B. H., & Mohr, L. A. (1994). Critical service encounters: The employee's viewpoint. *Journal of Marketing*, *58*(4), 95–106. doi:10.2307/1251919

Blocker, C. P., Flint, D. J., Myers, M. B., & Slater, S. F. (2011). Proactive customer orientation and its role for creating customer value in global markets. *Journal of the Academy of Marketing Science*, *39*(2), 216–233. doi:10.1007/s11747-010-0202-9

Bolino, M. C., & Turnley, W. H. (2009). Relative deprivation among employees in lower-quality leader-member exchange relationships. *The Leadership Quarterly*, *20*(3), 276–286. doi:10.1016/j.leaqua.2009.03.001

Brakus, J. J., Schmitt, B. H., & Zarantonello, L. (2009). Brand experience: What is it? How is it measured? Does it affect loyalty? *Journal of Marketing*, *73*(3), 52–68. doi:10.1509/jmkg.73.3.52

Brohman, M. K., Piccoli, G., Martin, P., Zulkernine, F., Parasuraman, A., & Watson, R. T. (2009). A design theory approach to building strategic network-based customer service systems. *Decision Sciences*, *40*(3), 403–430. doi:10.1111/j.1540-5915.2009.00242.x

Brotherton, B. (2004). Critical success factors in the UK budget hotel operations. *International Journal of Operations & Production Management*, *24*(9/10), 944–969. doi:10.1108/01443570410552135

Brown, G., Boya, U. O., Humphreys, N., & Widing, R. E. (1993). Attributes and behaviors of salespeople preferred by buyers: High socializing vs. low socializing industrial buyers. *Journal of Personal Selling & Sales Management*, *13*(1), 25–33.

Bruhn, M., Schnebelen, S., & Schäfer, D. (2014). Antecedents and consequences of the quality of e-customer-to-customer interactions in B2B brand communities. *Industrial Marketing Management*, *43*(1), 164–176. doi:10.1016/j.indmarman.2013.08.008

Chang, S., Gong, Y., & Shum, C. (2011). Promoting innovation in hospitality companies through human resource management practices. *International Journal of Hospitality Management*, *30*(4), 812–818. doi:10.1016/j.ijhm.2011.01.001

Chathoth, P., Altinay, L., Harrington, R. J., Okumus, F., & Chan, E. S. (2013). Co-production versus co-creation: A process based continuum in the hotel service context. *International Journal of Hospitality Management*, *32*, 11–20. doi:10.1016/j.ijhm.2012.03.009

Chen, C. F., & Myagmarsuren, O. (2013). Exploring the moderating effects of value offerings between market orientation and performance in tourism industry. *International Journal of Tourism Research*, *15*(6), 595–610. doi:10.1002/jtr.1900

Chen, S. C., & Lin, C. P. (2015). The impact of customer experience and perceived value on sustainable social relationship in blogs: An empirical study. *Technological Forecasting and Social Change*, *96*, 40–50. doi:10.1016/j.techfore.2014.11.011

Cheyne, J., Downes, M., & Legg, S. (2006). Travel agent vs. Internet: What influences travel consumer choices? *Journal of Vacation Marketing*, *12*(1), 41–57. doi:10.1177/1356766706059307

Chow, C. W. C., Lai, J. Y. M., & Loi, R. (2015). Motivation of travel agents' customer service behavior and organizational citizenship behavior: The role of leader-member exchange and internal marketing orientation. *Tourism Management, 48*, 362–369. doi:10.1016/j.tourman.2014.12.008

Christodoulides, G., de Chernatony, L., Furrer, O., Shiu, E., & Abimbola, T. (2006). Conceptualising and measuring the equity of online brands. *Journal of Marketing Management, 22*(7/8), 799–825. doi:10.1362/026725706778612149

Clemes, M. D., Gan, C., & Ren, M. (2011). Synthesizing the effects of service quality, value, and customer satisfaction on behavioral intentions in the motel industry: An empirical analysis. *Journal of Hospitality & Tourism Research (Washington, D.C.), 35*(4), 530–568. doi:10.1177/1096348010382239

Day, G. S. (1994). The capabilities of market-driven organizations. *Journal of Marketing, 58*(4), 37–52. doi:10.2307/1251915

De Jong, A., Wetzels, M., & de Ruyter, K. (2008). Linking employee perceptions of collective efficacy in self-managing service teams with customer-perceived service quality: A psychometric assessment. *International Journal of Service Industry Management, 19*(3), 353–378. doi:10.1108/09564230810875011

De Luca, L. M., Verona, G., & Vicari, S. (2010). Market orientation and R&D effectiveness in high-technology firms: An empirical investigation in the biotechnology industry. *Journal of Product Innovation Management, 27*(3), 299–320. doi:10.1111/j.1540-5885.2010.00718.x

Dostál, P. (2014). The use of soft computing in management. In P. Vasant (Ed.), *Handbook of research on novel soft computing intelligent algorithms: Theory and practical applications* (pp. 294–326). Hershey, PA: IGI Global. doi:10.4018/978-1-4666-4450-2.ch010

Driver, J. (1999). Developments in airline marketing practice. *Journal of Marketing Practice: Applied Marketing Science, 5*(5), 134–150. doi:10.1108/EUM0000000004571

Ernst, H., Hoyer, W. D., Krafft, M., & Krieger, K. (2011). Customer relationship management and company performance: The mediating role of new product performance. *Journal of the Academy of Marketing Science, 39*(2), 290–306. doi:10.1007/s11747-010-0194-5

Fawcett, P., Mcleish, R. E., & Ogden, I. D. (1992). *Logistics management*. London, UK: Pitman Publishing.

Flaherty, T. B., Dahlstrom, R., & Skinner, S. J. (1999). Organizational values and role stress as determinants of customer-oriented selling performance. *Journal of Personal Selling & Sales Management, 19*(2), 1–18.

Froehle, C. M. (2006). Service personnel, technology, and their interaction in influencing customer satisfaction. *Decision Sciences, 37*(1), 5–38. doi:10.1111/j.1540-5414.2006.00108.x

Frow, P., & Payne, A. (2007). Towards the "perfect" customer experience. *Journal of Brand Management, 15*(2), 89–101. doi:10.1057/palgrave.bm.2550120

Gatignon, H., & Xuereb, J. M. (1997). Strategic orientation of the firm and new product performance. *JMR, Journal of Marketing Research, 34*(1), 77–90. doi:10.2307/3152066

Gentile, C., Spiller, N., & Noci, G. (2007). How to sustain the customer experience: An overview of experience components that co-create value with the customer. *European Management Journal*, *25*(5), 395–410. doi:10.1016/j.emj.2007.08.005

Ghose, A. K., & Billiau, G. (2011). The optimizing WEB: A green ICT research perspective. In B. Unhelkar (Ed.), *Handbook of research on green ICT: Technology, business and social perspectives* (pp. 184–196). Hershey, PA: IGI Global. doi:10.4018/978-1-61692-834-6.ch012

Gilmore, J. H., & Pine, B. J. (2002). Differentiating hospitality operations via experiences: Why selling services is not enough. *The Cornell Hotel and Restaurant Administration Quarterly*, *43*(3), 87–96. doi:10.1016/S0010-8804(02)80022-2

Gopalani, A., & Shick, K. (2011). The service-enabled customer experience: A jump-start to competitive advantage. *The Journal of Business Strategy*, *32*(3), 4–12. doi:10.1108/02756661111121947

Govindarajan, V., Kopalle, P. K., & Danneels, E. (2011). The effects of mainstream and emerging customer orientations on radical and disruptive innovations. *Journal of Product Innovation Management*, *28*(s1), 121–132. doi:10.1111/j.1540-5885.2011.00865.x

Greer, C. R., & Lei, D. (2012). Collaborative innovation with customers: A review of the literature and suggestions for future research. *International Journal of Management Reviews*, *14*(1), 63–84. doi:10.1111/j.1468-2370.2011.00310.x

Grissemann, U., Plank, A., & Brunner-Sperdin, A. (2013). Enhancing business performance of hotels: The role of innovation and customer orientation. *International Journal of Hospitality Management*, *33*, 347–356. doi:10.1016/j.ijhm.2012.10.005

Gronfeldt, S., & Strother, J. (2006). *Service leadership: The quest for competitive advantage*. London, UK: Sage Publications.

Guenzi, P., Pardo, C., & Georges, L. (2007). Relational selling strategy and key account managers' relational behaviors: An exploratory study. *Industrial Marketing Management*, *36*(1), 121–133. doi:10.1016/j.indmarman.2005.03.014

Han, J. K., Namwoon, K., & Srivastava, R. K. (1998). Market orientation and organizational performance: Is innovation a missing link? *Journal of Marketing*, *62*(4), 30–45. doi:10.2307/1252285

Harris, E. G., Mowen, J. C., & Brown, T. J. (2005). Re-examining salesperson goal orientations: Personality influencers, customer orientation, and work satisfaction. *Journal of the Academy of Marketing Science*, *33*(1), 19–35. doi:10.1177/0092070304267927

Hartman, K. B., Meyer, T., & Scribner, L. L. (2009). Retail and service encounters: The intercultural tourist experience. *Journal of Hospitality Marketing & Management*, *18*(2/3), 197–215. doi:10.1080/19368620802591934

Hemmington, N. (2007). From service to experience: Understanding and defining the hospitality business. *Service Industries Journal*, *27*(6), 747–755. doi:10.1080/02642060701453221

Homburg, C., Müller, M., & Klarmann, M. (2011). When should the customer really be king? On the optimum level of salesperson customer orientation in sales encounters. *Journal of Marketing*, *75*(2), 55–74. doi:10.1509/jmkg.75.2.55

Hsu, H. Y., & Tsou, H. T. (2011). Understanding customer experiences in online blog environments. *International Journal of Information Management*, *31*(6), 510–523. doi:10.1016/j.ijinfomgt.2011.05.003

Hua, W., Chan, A., & Mao, Z. (2009). Critical success factors and customer expectation in budget hotel segment: A case study of China. *Journal of Quality Assurance in Hospitality & Tourism*, *10*(1), 59–74. doi:10.1080/15280080802713702

Im, S., & Workman, J. P. Jr. (2004). Market orientation, creativity, and new product performance in high-technology firms. *Journal of Marketing*, *68*(2), 114–132. doi:10.1509/jmkg.68.2.114.27788

Janda, S., & Ybarra, A. (2005). Do product and consumer characteristics affect the relationship between online experience and customer satisfaction? *Journal of Internet Commerce*, *4*(4), 133–151. doi:10.1300/J179v04n04_09

Johnston, R., & Kong, X. (2011). The customer experience: A road-map for improvement. *Managing Service Quality*, *21*(1), 5–24. doi:10.1108/09604521111100225

Joshi, S. (2014). Customer experience management: An exploratory study on the parameters affecting customer experience for cellular mobile services of a telecom company. *Procedia: Social and Behavioral Sciences*, *133*, 392–399. doi:10.1016/j.sbspro.2014.04.206

Kao, Y. F., Huang, L. S., & Yang, M. H. (2007). Effects of experiential elements on experiential satisfaction and loyalty intentions: A case study of the super basketball league in Taiwan. *International Journal of Revenue Management*, *1*(1), 79–96. doi:10.1504/IJRM.2007.011195

Kasemsap, K. (2013). Innovative human resource practice: A synthesized framework and causal model of leader-member exchange, organizational justice, job satisfaction, and organizational citizenship behavior. *International Journal of e-Education, e-Business, e-Management Learning*, *3*(1), 13–17. doi:10.7763/IJEEEE.2013.V3.185

Kasemsap, K. (2014). Developing a unified framework and a causal model of transformational leadership, empowerment, innovation support, and organizational innovation. In N. Erbe (Ed.), *Approaches to managing organizational diversity and innovation* (pp. 280–303). Hershey, PA: IGI Global. doi:10.4018/978-1-4666-6006-9.ch014

Kasemsap, K. (2015a). The role of customer relationship management in the global business environments. In T. Tsiakis (Ed.), *Trends and innovations in marketing information systems* (pp. 130–156). Hershey, PA: IGI Global. doi:10.4018/978-1-4666-8459-1.ch007

Kasemsap, K. (2015b). The role of marketing strategies in the tourism industry. In N. Ray (Ed.), *Emerging innovative marketing strategies in the tourism industry* (pp. 174–194). Hershey, PA: IGI Global. doi:10.4018/978-1-4666-8699-1.ch010

Kasemsap, K. (2015c). The roles of cross-cultural perspectives in global marketing. In J. Alcántara-Pilar, S. del Barrio-García, E. Crespo-Almendros, & L. Porcu (Eds.), *Analyzing the cultural diversity of consumers in the global marketplace* (pp. 37–59). Hershey, PA: IGI Global. doi:10.4018/978-1-4666-8262-7.ch003

Kasemsap, K. (2016a). Promoting service quality and customer satisfaction in global business. In U. Panwar, R. Kumar, & N. Ray (Eds.), *Handbook of research on promotional strategies and consumer influence in the service sector* (pp. 247–276). Hershey, PA: IGI Global. doi:10.4018/978-1-5225-0143-5.ch015

Kasemsap, K. (2016b). Promoting leadership development and talent management in modern organizations. In U. Aung & P. Ordoñez de Pablos (Eds.), *Managerial strategies and practice in the Asian business sector* (pp. 238–266). Hershey, PA: IGI Global. doi:10.4018/978-1-4666-9758-4.ch013

Kasemsap, K. (2016c). The roles of information technology and knowledge management in global tourism. In A. Nedelea, M. Korstanje, & B. George (Eds.), *Strategic tools and methods for promoting hospitality and tourism services* (pp. 109–138). Hershey, PA: IGI Global. doi:10.4018/978-1-4666-9761-4.ch006

Kasemsap, K. (2016d). Encouraging supply chain networks and customer loyalty in global supply chain. In N. Kamath & S. Saurav (Eds.), *Handbook of research on strategic supply chain management in the retail industry* (pp. 87–112). Hershey, PA: IGI Global. doi:10.4018/978-1-4666-9894-9.ch006

Kasemsap, K. (2016e). Implementing electronic commerce in global marketing. In I. Lee (Ed.), *Encyclopedia of e-commerce development, implementation, and management* (pp. 591–602). Hershey, PA: IGI Global. doi:10.4018/978-1-4666-9787-4.ch043

Kasemsap, K. (2016f). Utilizing social media in modern business. In I. Lee (Ed.), *Encyclopedia of e-commerce development, implementation, and management* (pp. 2171–2182). Hershey, PA: IGI Global. doi:10.4018/978-1-4666-9787-4.ch156

Kasemsap, K. (2016g). The roles of social media marketing and brand management in global marketing. In W. Ozuem & G. Bowen (Eds.), *Competitive social media marketing strategies* (pp. 173–200). Hershey, PA: IGI Global. doi:10.4018/978-1-4666-9776-8.ch009

Kennedy, K. N., Goolsby, J. R., & Arnould, E. J. (2003). Implementing a customer orientation: Extension of theory and application. *Journal of Marketing*, *67*(4), 67–81. doi:10.1509/jmkg.67.4.67.18682

Khan, I., Garg, R. J., & Rahman, Z. (2015). Customer service experience in hotel operations: An empirical analysis. *Procedia: Social and Behavioral Sciences*, *189*, 266–274. doi:10.1016/j.sbspro.2015.03.222

Kim, D., & Perdue, R. R. (2013). The effects of cognitive, affective, and sensory attributes on hotel choice. *International Journal of Hospitality Management*, *35*(1), 246–257. doi:10.1016/j.ijhm.2013.05.012

Kim, H. R. (2005). Developing an index of online customer satisfaction. *Journal of Financial Services Marketing*, *10*(1), 49–64. doi:10.1057/palgrave.fsm.4770173

Klaus, P., & Maklan, S. (2013). Towards a better measure of customer experience. *International Journal of Market Research*, *55*(2), 227–246. doi:10.2501/IJMR-2013-021

Knutson, B. J., Beck, J. A., Kim, S., & Cha, J. (2009). Identifying the dimensions of the guests' hotel experience. *Cornell Hospitality Quarterly*, *50*(1), 44–55. doi:10.1177/1938965508326305

Kwortnik, R. J. Jr, & Ross, W. T. Jr. (2007). The role of positive emotions in experiential decisions. *International Journal of Research in Marketing*, *24*(4), 324–335. doi:10.1016/j.ijresmar.2007.09.002

Lages, C. R., & Piercy, N. F. (2012). Key drivers of frontline employee generation of ideas for customer service improvement. *Journal of Service Research*, *15*(2), 215–230. doi:10.1177/1094670511436005

Lambert, D., & Stocs, J. (1993). *Strategic logistics management*. New York, NY: McGraw–Hill.

Laming, C., & Mason, K. (2014). Customer experience: An analysis of the concept and its performance in airline brands. *Research in Transportation Business & Management*, *10*, 15–25. doi:10.1016/j.rtbm.2014.05.004

Lemke, F., Clark, M., & Wilson, H. (2011). Customer experience quality: An exploration in business and consumer contexts using repertory grid technique. *Journal of the Academy of Marketing Science*, *39*(6), 846–869. doi:10.1007/s11747-010-0219-0

Leuschner, R., Charvet, F., & Rogers, D. S. (2013). A meta-analysis of logistics customer service. *Journal of Supply Chain Management*, *49*(1), 47–63. doi:10.1111/jscm.12000

Liaw, Y. J., Chi, N. W., & Chuang, A. (2010). Examining the mechanisms linking transformational leadership, employee customer orientation, and service performance: The mediating roles of perceived supervisor and coworker support. *Journal of Business and Psychology*, *25*(3), 477–492. doi:10.1007/s10869-009-9145-x

Lin, C. F. (2014). Exploring the hotel service personnels' cognitive implications toward service attributes and ethics. *Human Factors and Ergonomics in Manufacturing & Service Industries*, *24*(1), 14–28. doi:10.1002/hfm.20349

Lings, I. N., & Greenley, G. E. (2009). The impact of internal and external market orientations on firm performance. *Journal of Strategic Marketing*, *17*(1), 41–53. doi:10.1080/09652540802619251

Liou, J. J. H., Hsu, C., Yeh, W., & Lin, R. (2011). Using a modified grey relation method for improving airline service quality. *Tourism Management*, *32*(6), 1381–1388. doi:10.1016/j.tourman.2011.01.013

Luo, M. M., Chen, J. S., Ching, R. K., & Liu, C. C. (2011). An examination of the effects of virtual experiential marketing on online customer intentions and loyalty. *Service Industries Journal*, *31*(13), 2163–2191. doi:10.1080/02642069.2010.503885

MacGillavry, K., & Wilson, A. (2014). Delivering loyalty via customer experience management at DHL freight. *Global Business and Organizational Excellence*, *33*(6), 6–20. doi:10.1002/joe.21569

Martin, J., Mortimer, G., & Andrews, L. (2015). Re-examining online customer experience to include purchase frequency and perceived risk. *Journal of Retailing and Consumer Services*, *25*, 81–95. doi:10.1016/j.jretconser.2015.03.008

Maurya, U. K., Mishra, P., Anand, S., & Kumar, N. (2015). Corporate identity, customer orientation and performance of SMEs: Exploring the linkages. *IIMB Management Review*, *27*(3), 159–174. doi:10.1016/j.iimb.2015.05.001

Melovic, B., Mitrovic, S., Djokaj, A., & Vatin, N. (2015). Logistics in the function of customer service: Relevance for the engineering management. *Procedia Engineering*, *117*, 802–807. doi:10.1016/j.proeng.2015.08.255

Meyer, C., & Schwager, A. (2007). Understanding customer experience. *Harvard Business Review*, *85*(2), 117–126. PMID:17345685

Molina, A., Martín, V. J., Santos, J., & Aranda, E. (2009). Consumer service and loyalty in Spanish grocery store retailing: An empirical study. *International Journal of Consumer Studies*, *33*(4), 477–485. doi:10.1111/j.1470-6431.2009.00780.x

Morgeson, F. P., DeRue, D. S., & Karam, E. P. (2010). Leadership in teams: A functional approach to understanding leadership structures and processes. *Journal of Management*, *36*(1), 5–39. doi:10.1177/0149206309347376

Mutingi, M., & Mbohwa, C. (2014). Fuzzy system dynamics: An application to supply chain management. In P. Vasant (Ed.), *Handbook of research on novel soft computing intelligent algorithms: Theory and practical applications* (pp. 234–257). Hershey, PA: IGI Global. doi:10.4018/978-1-4666-4450-2.ch008

Nadiri, H., Hussain, K., Ekiz, E. H., & Erdoğan, Ş. (2008). An investigation on the factors influencing passengers' loyalty in the North Cyprus national airline. *Total Quality Management Journal*, *20*(3), 265–280.

Narver, J. C., & Slater, S. F. (1990). The effect of a market orientation on business profitability. *Journal of Marketing*, *54*(4), 20–35. doi:10.2307/1251757

Nasution, R. A., Sembada, A. Y., Miliani, L., Resti, N. D., & Prawono, D. A. (2014). The customer experience framework as baseline for strategy and implementation in services marketing. *Procedia: Social and Behavioral Sciences*, *148*, 254–261. doi:10.1016/j.sbspro.2014.07.041

Oflaç, B. S., Sullivan, U. Y., & Baltacioğlu, T. (2012). An attribution approach to consumer evaluations in logistics customer service failure situations. *Journal of Supply Chain Management*, *48*(4), 51–71. doi:10.1111/j.1745-493X.2012.03280.x

Ojiako, U., & Maguire, S. (2009). Seeking the perfect customer experience: A case study of British Telecom. *Strategic Change*, *18*(5/6), 179–193. doi:10.1002/jsc.846

Olsen, N. V., & Sallis, J. (2006). Market scanning for new service development. *European Journal of Marketing*, *40*(5/6), 466–484. doi:10.1108/03090560610657796

O'Reilly, K., & Paper, D. (2012). CRM and retail service quality: Front-line employee perspectives. *International Journal of Retail & Distribution Management*, *40*(11), 865–881. doi:10.1108/09590551211267610

Orfila-Sintes, F., Crespí-Cladera, R., & Martínez-Ros, E. (2005). Innovation activity in the hotel industry: Evidence from Balearic Islands. *Tourism Management*, *26*(6), 851–865. doi:10.1016/j.tourman.2004.05.005

Organ, D. W. (1988). *Organizational citizenship behavior: The good soldier syndrome*. Lexington, MA: Lexington Books.

Ottenbacher, M., & Gnoth, J. (2005). How to develop successful hospitality innovation. *Cornell Hospitality Quarterly*, *46*(2), 205–222. doi:10.1177/0010880404271097

Parasuraman, A., Zeithaml, V., & Malhotra, N. (2005). E-S-QUAL: A multiple-item scale for assessing electronic service quality. *Journal of Service Research*, *7*(3), 213–233. doi:10.1177/1094670504271156

Paswan, A., D'Souza, D., & Zolfagharian, M. A. (2009). Toward a contextually anchored service innovation typology. *Decision Sciences*, *40*(3), 513–540. doi:10.1111/j.1540-5915.2009.00239.x

Pine, B. J., & Gilmore, J. H. (2011). *The experience economy*. Boston, MA: Harvard Business Review Press.

Pullman, M. E., & Gross, M. A. (2004). Ability of experience design elements to elicit emotions and loyalty behaviours. *Decision Sciences*, *35*(3), 551–578. doi:10.1111/j.0011-7315.2004.02611.x

Racela, O. C. (2014). Customer orientation, innovation competencies, and firm performance: A proposed conceptual model. *Procedia: Social and Behavioral Sciences*, *148*, 16–23. doi:10.1016/j.sbspro.2014.07.010

Rayport, J. F., Jaworski, B. J., & Kyung, E. J. (2005). Best face forward: Improving companies' service interfaces with customers. *Journal of Interactive Marketing*, *19*(4), 67–80. doi:10.1002/dir.20050

Ren, L., Qiu, H., Wang, P., & Lin, P. M. C. (2016). Exploring customer experience with budget hotels: Dimensionality and satisfaction. *International Journal of Hospitality Management*, *52*, 13–23. doi:10.1016/j.ijhm.2015.09.009

Rose, S., Hair, N., & Clark, M. (2011). Online customer experience: A review of the business-to-consumer online purchase context. *International Journal of Management Reviews*, *13*(1), 24–39. doi:10.1111/j.1468-2370.2010.00280.x

Saxe, R., & Weitz, B. A. (1982). The SOCO scale: A measure of the customer orientation of salespeople. *JMR, Journal of Marketing Research*, *19*(3), 343–351. doi:10.2307/3151568

Schmitt, B. H. (1999). *Experiential marketing: How to get customers to sense, feel, think, act and relate to your company and brand*. New York, NY: The Free Press.

Schwepker, C. H. Jr. (2003). Customer-oriented selling: A review, extension, and directions for future research. *Journal of Personal Selling & Sales Management*, *23*(2), 151–171.

Shaw, C. (2005). *Revolutionize your customer experience*. New York, NY: Palgrave Macmillan. doi:10.1057/9780230513457

Shaw, G., Bailey, A., & Williams, A. (2011). Aspects of service-dominant logic and its implications for tourism management: Examples from the hotel industry. *Tourism Management*, *32*(2), 207–214. doi:10.1016/j.tourman.2010.05.020

Sin, L. Y. M., Tse, A. C. B., Heung, V. C. S., & Yim, F. H. K. (2005). An analysis of the relationship between market orientation and business performance in the hotel industry. *International Journal of Hospitality Management*, *24*(6), 555–577. doi:10.1016/j.ijhm.2004.11.002

Smith, S., & Wheeler, J. (2002). *Managing the customer experience: Turning customers into advocates*. Upper Saddle River, NJ: Prentice Hall.

Sorooshian, S., Salimi, M., Salehi, M., Nia, N. B., & Asfaranjan, Y. S. (2013). Customer experience about service quality in online environment: A case of Iran. *Procedia: Social and Behavioral Sciences*, *93*, 1681–1695. doi:10.1016/j.sbspro.2013.10.101

Stock, R. M., & Hoyer, W. D. (2005). An attitude-behavior model of salespeoples' customer orientation. *Journal of the Academy of Marketing Science*, *33*(4), 536–552. doi:10.1177/0092070305276368

Strombeck, S. D., & Wakefield, K. L. (2008). Situational influences on service quality evaluations. *Journal of Services Marketing*, *22*(5), 409–419. doi:10.1108/08876040810889175

Tajeddini, K. (2010). Effect of customer orientation and entrepreneurial orientation on innovativeness: Evidence from the hotel industry in Switzerland. *Tourism Management*, *31*(2), 221–231. doi:10.1016/j.tourman.2009.02.013

Tajeddini, K., & Trueman, M. (2012). Managing Swiss hospitality: How cultural antecedents of innovation and customer-oriented value systems can influence performance in the hotel industry. *International Journal of Hospitality Management*, *31*(4), 1119–1129. doi:10.1016/j.ijhm.2012.01.009

Tang, T. W. (2014). Becoming an ambidextrous hotel: The role of customer orientation. *International Journal of Hospitality Management*, *39*, 1–10. doi:10.1016/j.ijhm.2014.01.008

Thakor, M. V., & Joshi, A. W. (2005). Motivating salesperson customer orientation: Insights from the job characteristics model. *Journal of Business Research*, *58*(5), 584–592. doi:10.1016/j.jbusres.2003.10.002

Torres, E. N., Fu, X., & Lehto, X. (2014). Examining key drivers of customer delight in a hotel experience: A cross-cultural perspective. *International Journal of Hospitality Management*, *36*, 255–262. doi:10.1016/j.ijhm.2013.09.007

Tortosa, V., Moliner, M. A., & Sanchez, J. (2009). Internal market orientation and its influence on organisational performance. *European Journal of Marketing*, *43*(11/12), 1435–1456. doi:10.1108/03090560910989975

Valenzuela, L. M., Mulki, J. P., & Jaramillo, J. F. (2010). Impact of customer orientation, inducements and ethics on loyalty to the firm: Customers' perspective. *Journal of Business Ethics*, *93*(2), 277–291. doi:10.1007/s10551-009-0220-z

Vanharanta, H., Kantola, J., & Seikola, S. (2015). Customers' conscious experience in a coffee shop. *Procedia Manufacturing*, *3*, 618–625. doi:10.1016/j.promfg.2015.07.283

Vera, J., & Trujillo, A. (2013). Service quality dimensions and superior customer perceived value in retail banks: An empirical study on Mexican consumers. *Journal of Retailing and Consumer Services*, *20*(6), 579–586. doi:10.1016/j.jretconser.2013.06.005

Verhoef, P. C., Lemon, K. N., Parasuraman, A., Roggeveen, A., Tsiros, M., & Schlesinger, L. A. (2009). Customer experience creation: Determinants, dynamics and management strategies. *Journal of Retailing*, *85*(1), 31–41. doi:10.1016/j.jretai.2008.11.001

von Wallpach, S., & Kreuzer, M. (2013). Multi-sensory sculpting (MSS): Eliciting embodied brand knowledge via multi-sensory metaphors. *Journal of Business Research*, *66*(9), 1325–1331. doi:10.1016/j.jbusres.2012.02.032

Vora, M., & Mirnalinee, T. T. (2015). From optimization to clustering: A swarm intelligence approach. In P. Vasant (Ed.), *Handbook of research on artificial intelligence techniques and algorithms* (pp. 594–619). Hershey, PA: IGI Global. doi:10.4018/978-1-4666-7258-1.ch019

Walls, A. R. (2013). A cross-sectional examination of hotel consumer experience and relative effects on consumer values. *International Journal of Hospitality Management*, *32*, 179–192. doi:10.1016/j.ijhm.2012.04.009

Wang, C. H., Chen, K. Y., & Chen, S. C. (2012). Total quality management, market orientation and hotel performance: The moderating effects of external environmental factors. *International Journal of Hospitality Management*, *31*(1), 119–129. doi:10.1016/j.ijhm.2011.03.013

Wieseke, J., Ullrich, J., Christ, O., & van Dick, R. (2007). Organizational identification as a determinant of customer orientation in service organizations. *Marketing Letters*, *18*(4), 265–278. doi:10.1007/s11002-007-9021-2

Wong, A., Liu, Y., & Tjosvold, D. (2015). Service leadership for adaptive selling and effective customer service teams. *Industrial Marketing Management*, *46*, 122–131. doi:10.1016/j.indmarman.2015.01.012

Workman, J. P. Jr, Homburg, C., & Jensen, O. (2003). Intraorganizational determinants of key account management effectiveness. *Journal of the Academy of Marketing Science*, *31*(1), 3–21. doi:10.1177/0092070302238599

Wu, C. C., & Lo, Y. H. (2012). Customer reactions to encountering consecutive service failures. *Journal of Consumer Behaviour*, *11*(3), 217–224. doi:10.1002/cb.1376

Wu, W. Y., Tsai, C. C., & Fu, C. S. (2013). The relationships among internal marketing, job satisfaction, relationship marketing, customer orientation, and organizational performance: An empirical study of TFT-LCD companies in Taiwan. *Human Factors and Ergonomics in Manufacturing & Service Industries*, *23*(5), 436–449. doi:10.1002/hfm.20329

Xie, S., & Zhang, Z. (2012). Models and optimization techniques of machining parameters in turning operations. In J. Davim (Ed.), *Computational methods for optimizing manufacturing technology: Models and techniques* (pp. 162–192). Hershey, PA: IGI Global. doi:10.4018/978-1-4666-0128-4.ch007

Yoon, M. H., Seo, J. H., & Yoon, T. S. (2004). Effects of contact employee supports on critical employee responses and customer service evaluation. *Journal of Services Marketing*, *18*(5), 395–412. doi:10.1108/08876040410548311

Yuan, Y. H. E., & Wu, C. K. (2008). Relationships among experiential marketing, experiential value, and customer satisfaction. *Journal of Hospitality & Tourism Research (Washington, D.C.)*, *32*(3), 387–410. doi:10.1177/1096348008317392

Zhang, J. (2010). Employee orientation and performance: An exploration of the mediating role of customer orientation. *Journal of Business Ethics*, *91*(S1s1), 111–121. doi:10.1007/s10551-010-0570-6

Zhang, J., Cai, L. A., & Kavanaugh, R. R. (2008). Dimensions in building brand experience for economy hotels: A case of emerging market. *Journal of China Tourism Research*, *4*(1), 61–77. doi:10.1080/19388160802099782

Zhou, K. Z., Brown, J. R., & Dev, C. S. (2009). Market orientation, competitive advantage, and performance: A demand-based perspective. *Journal of Business Research*, *62*(11), 1063–1070. doi:10.1016/j.jbusres.2008.10.001

ADDITIONAL READING

Ali, F., Hussain, K., & Ragavan, N. A. (2014). Memorable customer experience: Examining the effects of customers experience on memories and loyalty in Malaysian resort hotels. *Procedia: Social and Behavioral Sciences*, *144*, 273–279. doi:10.1016/j.sbspro.2014.07.296

Arnold, T. J., Fang, E., & Palmatier, R. W. (2011). The effects of customer acquisition and retention orientations on a firm's radical and incremental innovation performance. *Journal of the Academy of Marketing Science*, *39*(2), 234–251. doi:10.1007/s11747-010-0203-8

Brach, S., Walsh, G., Hennig-Thurau, T., & Groth, M. (2015). A dyadic model of customer orientation: Mediation and moderation effects. *British Journal of Management*, *26*(2), 292–309. doi:10.1111/1467-8551.12049

Cheng, C. C., & Krumwiede, D. (2012). The role of service innovation in the market orientation: New service performance linkage. *Technovation*, *32*(7/8), 487–497. doi:10.1016/j.technovation.2012.03.006

Eggers, F., Hansen, D. J., & Davis, A. E. (2012). Examining the relationship between customer and entrepreneurial orientation on nascent firms' marketing strategy. *The International Entrepreneurship and Management Journal*, *8*(2), 203–222. doi:10.1007/s11365-011-0173-4

Feng, T., Sun, L., Zhu, C., & Sohal, A. S. (2012). Customer orientation for decreasing time-to-market of new products: IT implementation as a complementary asset. *Industrial Marketing Management*, *41*(6), 929–939. doi:10.1016/j.indmarman.2011.11.027

Fernandez, J. I. P., Cala, A. S., & Domecq, C. F. (2011). Critical external factors behind hotels' investments in innovation and technology in emerging urban destinations. *Tourism Economics*, *17*(2), 339–357. doi:10.5367/te.2011.0033

Halbesleben, J. R. B., & Stoutner, O. K. (2013). Developing customers as partial employees: Predictors and outcomes of customer performance in a services context. *Human Resource Development Quarterly*, *24*(3), 313–335. doi:10.1002/hrdq.21167

Harris, E. G., Brown, T. J., Mowen, J. C., & Artis, A. (2014). Exploring the role of productivity propensity in frontline employee performance: Its relationship with customer orientation and important outcomes. *Psychology and Marketing*, *31*(3), 171–183. doi:10.1002/mar.20685

Homburg, C., Müller, M., & Klarmann, M. (2011). When does salespeople's customer orientation lead to customer loyalty? The differential effects of relational and functional customer orientation. *Journal of the Academy of Marketing Science*, *39*(6), 795–812. doi:10.1007/s11747-010-0220-7

Joshi, S., Majumdar, A., & Malhotra, A. (2014). Enhancing customer experience using business intelligence tools with specific emphasis on the Indian DTH industry. *Procedia Economics and Finance*, *11*, 289–305. doi:10.1016/S2212-5671(14)00197-X

Komarov, M., & Avdeeva, Z. (2015). Customer experience management for smart commerce based on cognitive maps. *Procedia Computer Science*, *55*, 970–979. doi:10.1016/j.procs.2015.07.106

Kwan, H. K., Yim, F. H. K., & Zhou, X. (2015). Effects of mentoring on customer orientation: The moderating role of gender. *Asia Pacific Journal of Human Resources*, *53*(1), 124–140. doi:10.1111/1744-7941.12029

Lam, C. F., & Mayer, D. M. (2014). When do employees speak up for their customers? A model of voice in a customer service context. *Personnel Psychology*, *67*(3), 637–666. doi:10.1111/peps.12050

Lee, C. H., Chou, C., & Sun, T. L. (2015). Evaluating presence for customer experience in a virtual environment: Using a nuclear power plant as an example. *Human Factors and Ergonomics in Manufacturing & Service Industries*, *25*(4), 484–499. doi:10.1002/hfm.20563

Li, J., & Moul, C. C. (2015). Who should handle retail? Vertical contracts, customer service, and social welfare in a Chinese mobile phone market. *International Journal of Industrial Organization*, *39*, 29–43. doi:10.1016/j.ijindorg.2015.01.001

Nguyen, H., Groth, M., Walsh, G., & Hennig-Thurau, T. (2014). The impact of service scripts on customer citizenship behavior and the moderating role of employee customer orientation. *Psychology and Marketing*, *31*(12), 1096–1109. doi:10.1002/mar.20756

Perry, S. J., Witt, L. A., Callison, K., & Malka, A. (2013). Developing dedicated service employees. *Journal of Psychological Issues in Organizational Culture*, *4*(2), 35–52. doi:10.1002/jpoc.21113

Shao, R., & Skarlicki, D. P. (2014). Service employees' reactions to mistreatment by customers: A comparison between North America and East Asia. *Personnel Psychology*, *67*(1), 23–59. doi:10.1111/peps.12021

Shin, D. H. (2015). Effect of the customer experience on satisfaction with smartphones: Assessing smart satisfaction index with partial least squares. *Telecommunications Policy*, *39*(8), 627–641. doi:10.1016/j.telpol.2014.10.001

Spiess, J., T'Joens, Y., Dragnea, R., Spencer, P., & Philippart, L. (2014). Using big data to improve customer experience and business performance. *Bell Labs Technical Journal*, *18*(4), 3–17. doi:10.1002/bltj.21642

Terho, H., Eggert, A., Haas, A., & Ulaga, W. (2015). How sales strategy translates into performance: The role of salesperson customer orientation and value-based selling. *Industrial Marketing Management*, *45*, 12–21. doi:10.1016/j.indmarman.2015.02.017

Verhagen, T., van Nes, J., Feldberg, F., & van Dolen, W. (2014). Virtual customer service agents: Using social presence and personalization to shape online service encounters. *Journal of Computer-Mediated Communication, 19*(3), 529–545. doi:10.1111/jcc4.12066

Victorino, L., Verma, R., & Wardell, D. G. (2013). Script usage in standardized and customized service encounters: Implications for perceived service quality. *Production and Operations Management, 22*(3), 518–534. doi:10.1111/j.1937-5956.2012.01382.x

Zablah, A. R., Franke, G. R., Brown, T. J., & Bartholomew, D. E. (2012). How and when does customer orientation influence frontline employee outcomes? A meta-analytic evaluation. *Journal of Marketing, 76*(3), 21–40. doi:10.1509/jm.10.0231

Zhang, H., Lu, Y., Gupta, S., & Zhao, L. (2014). What motivates customers to participate in social commerce? The impact of technological environments and virtual customer experiences. *Information & Management, 51*(8), 1017–1030. doi:10.1016/j.im.2014.07.005

KEY TERMS AND DEFINITIONS

Customer Experience: The entirety of the interactions that a customer has with a company and its products.

Customer Orientation: A group of actions taken by a business to support its sales and service staff in considering customer requirements.

Customer Satisfaction: The degree of satisfaction provided by the products or services of a company as measured by the number of repeat customers.

Customer Service: All interactions between a customer and a product provider at the time of sales.

Marketing: The social and managerial processes by which individuals or companies obtain what they need or want through creating, offering, and exchanging the products with each other.

Optimization: The investigation of an alternative with the most cost effective or highest achievable performance under the given constraints, by maximizing desired factors and minimizing undesired ones.

Purchase Intention: A plan to purchase the particular products or services in the future.

Service Quality: An evaluation of how well the delivered service conforms to the customer expectations.

Chapter 6
Using Data Science to Predict Hotel Booking Cancellations

Nuno António
ISCTE Instituto Universitário de Lisboa, Portugal

Ana de Almeida
ISCTE Instituto Universitário de Lisboa, Portugal & Centro de Informática e Sistemas da Universidade de Coimbra, Portugal

Luis M. M. Nunes
ISCTE Instituto Universitário de Lisboa, Portugal & Instituto de Telecomunicações, Portugal

ABSTRACT

Booking cancellations in the hospitality industry not only generate revenue loss and affect pricing and inventory allocation decisions, but they also, in overbooking situations, have the potential to affect the hotel's online social reputation. By employing data sets from four resort hotels and addressing this issue as a classification problem in the scope of data science, the authors demonstrate that it is possible to build models for predicting booking cancellations with accuracy results in excess of 90%. This research also demonstrates that despite what was alleged by Morales and Wang (2010), it is possible to predict with high accuracy whether a booking will be canceled. Results allow hotel managers to act on bookings with high cancellation probability and contain the associated revenue losses, produce better net demand forecasts, improve overbooking/cancellation policies, and have more assertive pricing and inventory allocation strategies.

INTRODUCTION

Bookings represent a contract between a customer and a service provider (Talluri & Van Ryzin, 2004). This contract gives the customer the right to use the service in the future at a settled price, usually with an option to cancel the contract prior to the service provision. In the case of the hospitality industry, this option to cancel the booking puts the risk on the hotel. The hotel has to guarantee rooms to the customers who honor their bookings, but at the same time it has to bear with the opportunity cost of

DOI: 10.4018/978-1-5225-1054-3.ch006

vacant capacity when a customer cancels a booking or does not show up (Talluri & Van Ryzin, 2004). Cancellation rates vary from hotel to hotel. For the purpose of this chapter, no-shows will be treated as cancellations, even though there are some differences between them. A cancellation occurs when the customer terminates the contract prior to his or her arrival. A no-show occurs when the customer does not inform the hotel and fails to check in.

Canceled bookings can represent up to 20% of the total bookings (Morales & Wang, 2010). However, in airport/roadside hotels this number can rise as high as 60% (Liu, 2004). These cancellations can have a substantial affect on revenue, not only because of the revenue loss they represent themselves, but also because of the effect they can have on pricing and inventory allocation decisions (Morales & Wang, 2010).

To compensate for the potential revenue losses caused by cancellations, hotels often sell above their capacity (overbooking) (Ivanov & Zhechev, 2012; Mehrotra & Ruttley, 2006; Morales & Wang, 2010). However, overbooking can also generate costs (Hayes & Miller, 2011; Mehrotra & Ruttley, 2006): reallocation of customers to alternative hotels, cash compensations, or social reputation. Thus, classifying hotel bookings with high cancellation probability is relevant to enable hotels to act on those bookings to prevent or mitigate their cancellation effects. At the same time, this prediction facilitates an easier identification of cancellation patterns, hence allowing a better understanding of net demand and a better definition of overbooking/cancellation policies.

Using uncensored data from four hotel Property Management Systems (PMS) that represent this tendency for hotels to have increasingly higher booking cancellations rates (illustrated in Figure 1), the authors aim to demonstrate how data science can be applied in the scope of hotel revenue management to:

1. Identify features from PMS databases with predictive strength regarding a booking cancellation probability.
2. Build a model that could predict bookings with a high cancellation probability.
3. Understand if one prediction model fits all hotels or if a specific model should be built for each hotel.

BACKGROUND

Booking cancellations in the hospitality industry have a well-known body of knowledge in the scope of revenue management, particularly how the definition of a hotel's overbooking and cancellation policies can affect its revenue and inventory allocation (Hayes & Miller, 2011; Ivanov, 2014; Talluri & Van Ryzin, 2004). Revenue management is defined as "the application of information systems and pricing strategies to allocate the right capacity to the right customer at the right price at the right time" (Kimes & Wirtz, 2003, p. 125). Originally developed in 1966 by the aviation industry (Chiang, Chen, & Xu, 2007), revenue management was gradually adopted by other services industries, such as hotels, rental cars, golf courses, and casinos (Chiang et al., 2007; Kimes & Wirtz, 2003). For the hospitality industry (rooms division), the revenue management definition was adapted to "making the right room available for the right guest and the right price at the right time via the right distribution channel" (Mehrotra & Ruttley, 2006, p. 2). It requires constant monitoring of supply and demand, historical data analysis, forecasts, implementation of strategic and tactical restrictions/controls by market segments and distribution channels, detection of patterns, and tendencies and anomalies, among other processes that require data from multiple sources (Hayes & Miller, 2011; Ivanov & Zhechev, 2012; Serra, 2013).

Figure 1. Booking cancellation ratios over time per hotel

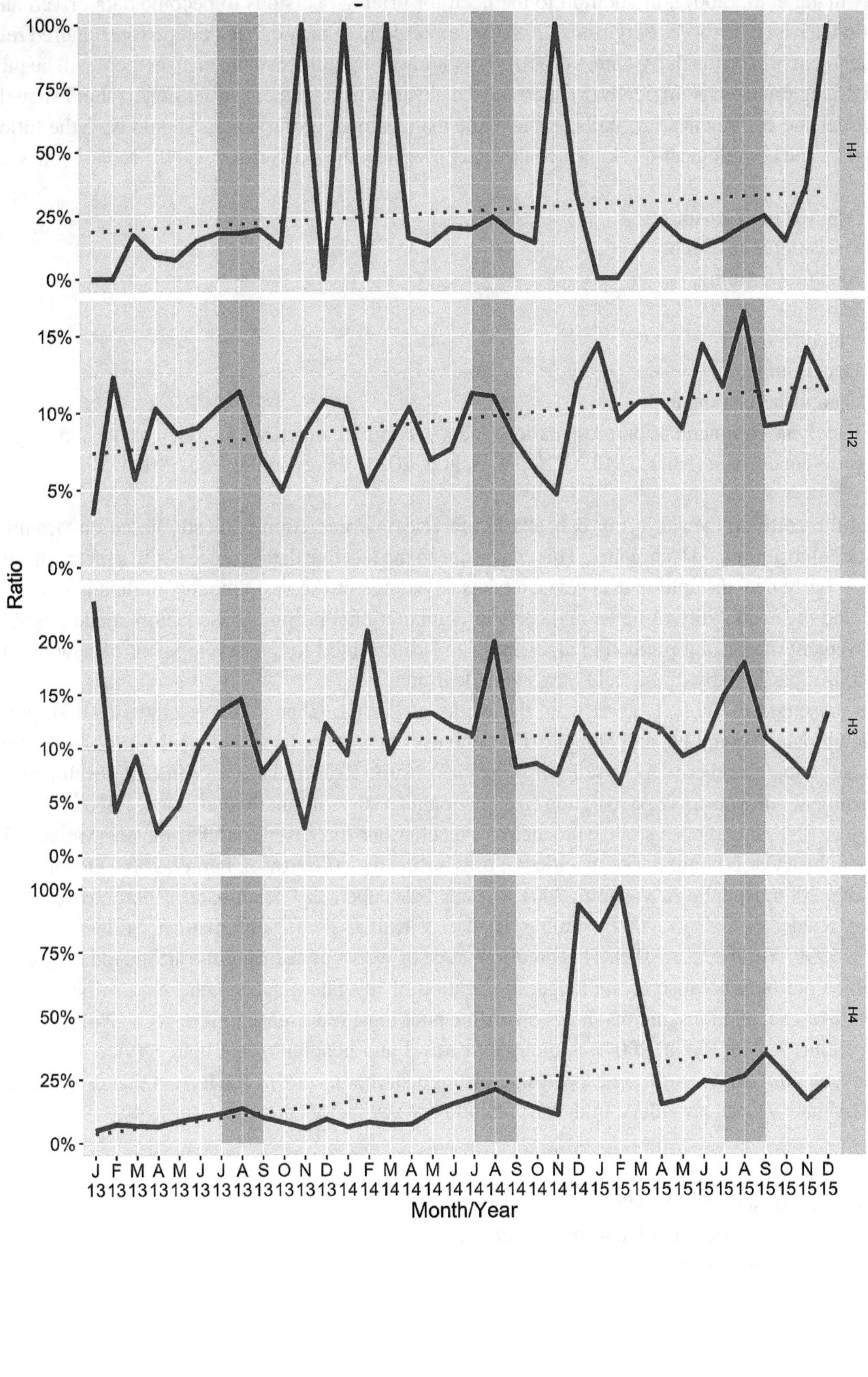

Use of mathematical and forecast models that take better advantage of the available data and technology (Chiang et al., 2007), in addition to the need for pricing decisions to become data driven and not intuition driven (Garrow & Ferguson, 2008), is supported in a study carried out on five hundred revenue management professionals by Kimes (2010). This study shows that revenue management will be increasingly strategic and technologically oriented and that revenue management professionals should have better analytical and communication skills. All revenue management professionals should have the following analytical and communication skills, and these are precisely the skills that are at the root of data science:

- Applied mathematics,
- Operational research,
- Machine learning,
- Statistics,
- Databases,
- Data mining,
- Data visualization, and
- Excellent communication/presentation fluency, complemented with a deep understanding of the problem domain (Dhar, 2013; O'Neil & Schutt, 2014; Yangyong & Yun, 2011).

Yet literature on the subject of cancellation forecasting/prediction is mostly based on "traditional" statistics algorithms and techniques. This can be due to the fact that data science is still a recent discipline.

As a fairly new discipline, data science takes advantage of the vast amounts of data at our disposal today and the availability of better and cheaper computational power. These factors made possible the improvement of existing prediction algorithms and contributed to the development of new and better algorithms, particularly in the field of machine learning.

As presented in Table 1, apart from the works of Huang, Chang, Ho, & others (2013) who uses restaurant data, Yoon, Lee, and Song (2012), who uses hotel simulated data, and Liu (2004), who uses real hotel data, all other works use PNR—Personal Name Record data—a standard developed by the International Air Transport Association (International Civil Aviation Organization, 2010).

Use of PNR data is not a strange practice as cancellations forecast research is mostly available for the airline industry or is nonindustry specific but uses airline data (Garrow & Parker, 2008; Gorin, Brunger, & White, 2006; Hueglin & Vannotti, 2001; Iliescu, Freisleben, & Gleichmann, 1993; Lawrence, 2003; Lemke, Riedel, & Gabrys, 2009; Neuling, Riedel, & Kalka, 2004; Subramanian, Stidham, & Lautenbacher, 1999; Yoon et al., 2012). This predominance of works concerning the airline industry could be explained not only because of the longer application of revenue management but also because of the high rate of cancellations and no-shows on airline bookings, representing from 30% (Phillips, 2005) to 50% (Talluri & Van Ryzin, 2004). Although air travel and hospitality are both service industries and have many similarities, there are aspects that distinguish them, such as the factors that drive customers to select their service providers. In the airline industry key factors are:

- Price,
- Quality of service (in flight),
- Airline image (specially in terms of safety),
- Loyalty programs, and

Table 1. Literature that addresses the subject of cancellation forecast (sorted by publication year)

Author(s)	Industry Incidence/Data	Algorithms/Techniques Used
Huang, Chang, Ho, & others (2013)	General, with restaurant data	Neural networks
Yoon, Lee, & Song (2012)	Airline	Linear programming
Morales & Wang (2010)	General, with hotel data	Logistic regression, probability estimation tree, random forest, minimum squared expected error tree, support vector machine and kernel logistic regression
Lemke, Riedel, & Gabrys (2009)	Airline	Traditional forecast combination methods
Iliescu, Garrow, & Parker (2008)	Airline	Discrete time proportional odds model
Gorin, Brunger, & White (2006)	Airline	Weighted averaging
Neuling, Riedel, & Kalka (2004)	Airline	Hybrid of decision tree and Naïve Bayes, and the adjusted probability model
Liu (2004)	Hotel, with censored data	Standard statistical methods
Lawrence (2003)	Airline	Hybrid of decision tree and Naïve Bayes, and the adjusted probability model
Hueglin & Vannotti (2001)	Airline	Decision tree and logistic regression
Subramanian, Stidham Jr, & Lautenbacher (1999)	Airline	Dynamic programming
Freisleben & Gleichmann (1993)	Airline	Neural networks

Source: (Extended from Morales & Wang, 2010, p. 556).

- Accessibility to transport hubs at the end destination (Chen, Peng, & Hackley, 2008; Park, Robertson, & Wu, 2006),

In the hospitality industry the importance of these factors changes and other factors, such as:

- Social reputation,
- Location,
- Cleanliness, and
- Price, come into play.

Although some of the previous works published on this subject consider booking cancellations prediction a classification problem, most consider it a regression problem. However, even some works that consider it a classification problem, such as Morales and Wang (2010), focused on global cancellation rate forecast and not on each booking cancellation probability. In fact, Morales and Wang stated that "it is hard to imagine that one can predict whether a booking will be canceled or not with high accuracy simply by looking at PNR information" (p. 556). However, as presented in the next sections, classification of whether a booking will be canceled is possible, especially if suitable PMS data is used.

Morales and Wang (2010) also stated that "in the revenue management context, the classification or even probability of cancellation of an individual booking is not important" (p. 556), which is not in accordance with what is frequently affirmed in revenue management theory. In revenue management, registering cancellations, at least by market segments or type of bookings, is an essential tool to identify

Figure 2. Documents found on Scopus database related to "booking cancellations" and "overbooking"

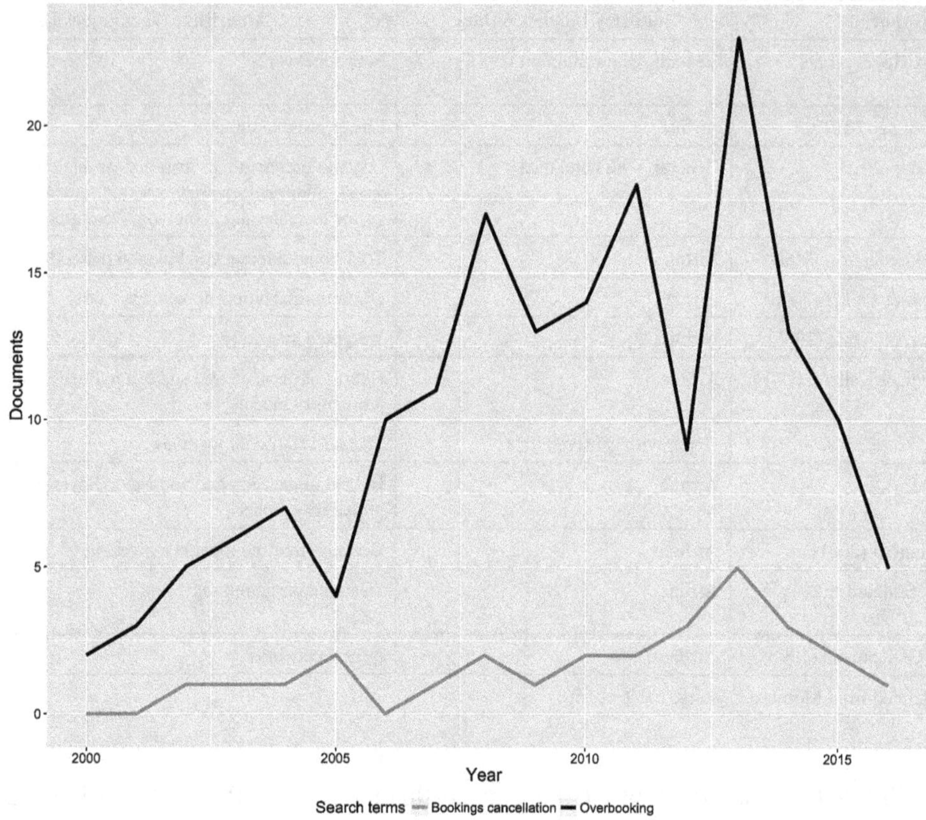

patterns and with it create better forecasts (Ivanov, 2014; Mehrotra & Ruttley, 2006) and better cancellation and overbooking policies.

Booking cancellations and overbooking research are highly related. Nonetheless, there seems to exist more publications on the topic of overbooking than on the topic of booking cancellations. As presented in Figure 2, in a search executed in the Scopus bibliographic database for publications from the years 2000 to 2016, of documents who used the terms "Bookings cancellation" and "Overbooking" on their title, abstract or keywords, related to the tourism/travel service industries, the number of documents published on overbooking is substantially superior to the number of documents published on bookings cancellation.

This predominance of publications on overbooking could be explained by the use given by hoteliers to overbooking. Overbooking helps hoteliers to reduce the risk associated with booking cancellations. However, overbooking has the potential to become a serious problem. The denial of service, by not allowing a customer to check in at a hotel he or she has previously booked, can be a terrible experience for the customer. This experience can have a negative affect on both the reputation and revenue of the service provider (Noone & Lee, 2011). On the other hand, not all revenue from canceled bookings is completely lost. Sometimes, canceled bookings are replaced by other bookings, which in some cases, such as last-minute bookings, can generate more revenue than the canceled bookings. To compensate for the uncertainty caused by booking cancellations, hotels implement cancellation policies, such as nonrefundable bookings. The downside of nonrefundable bookings is that they not only reduce the number of bookings, but they also diminish the average daily rate (ADR), due to their significant discounts on price (Smith, Parsa, Bujisic, & van der Rest, 2015).

CANCELLATION PREDICTION

Hotel Characterizations

This research is based on reservation data from four hotels located in the resort region of the Algarve, Portugal. Data span from years 2013 to 2015. Because these four hotels required anonymity, hereinafter they will be designated as H1 to H4. For the sake of better understanding the demand and market of these hotels, some details on facilities and services are provided. All four hotels are four- and five-star resort hotels, ranging in size from 86 to 180 rooms. All four hotels have at least one bar and one restaurant. H2 and H3 are mixed-ownership units—besides renting units owned by the hotels' management companies, these hotels also rent units that were sold in timeshare or factional ownership schemes. Summer months, from July to September, are considered high season in this region. H1 closes temporarily during low season, but not regularly.

Execution

The methodology for this research was based on CRoss-Industry Standard Process for Data Mining (CRISP-DM) (Chapman et al., 2000). CRISP-DM is one of the most-used process models in predictive analytics projects (Abbott, 2014). CRISP-DM provides a six-step process:

- Business understanding,
- Data understanding,
- Data preparation,
- Modeling,
- Evaluation, and
- Deployment.

The following sections offer a succinct description of the execution of each of these steps.

Business Understanding

As depicted in Figure 3, in the four studied hotels, except in H3 from 2014 to 2015, booking cancellations have been increasing since 2013. In 2015, booking cancellation rates ranged from 11.8% to 26.4%. From the point of view of canceled room nights, that rate reaches values as high as 27.5% in H4. These cancellation rates are in harmony with what was observed by Morales and Wang (2010). These high cancellation rates impose uncertainty in hotel managers, substantially influencing pricing and inventory allocation, especially in high-demand dates. If hotel managers do not "gamble" and overbook, they will certainly have room inventory, which eventually will not sell. On the contrary, if hotel managers overbook, they risk the costs related to accommodating guests in other hotels, cash or services compensations, and, worst of all in an age of social media, damaging the hotel's online social reputation. As a

Figure 3. Distribution of cancellations by bookings and room nights, per year and hotel

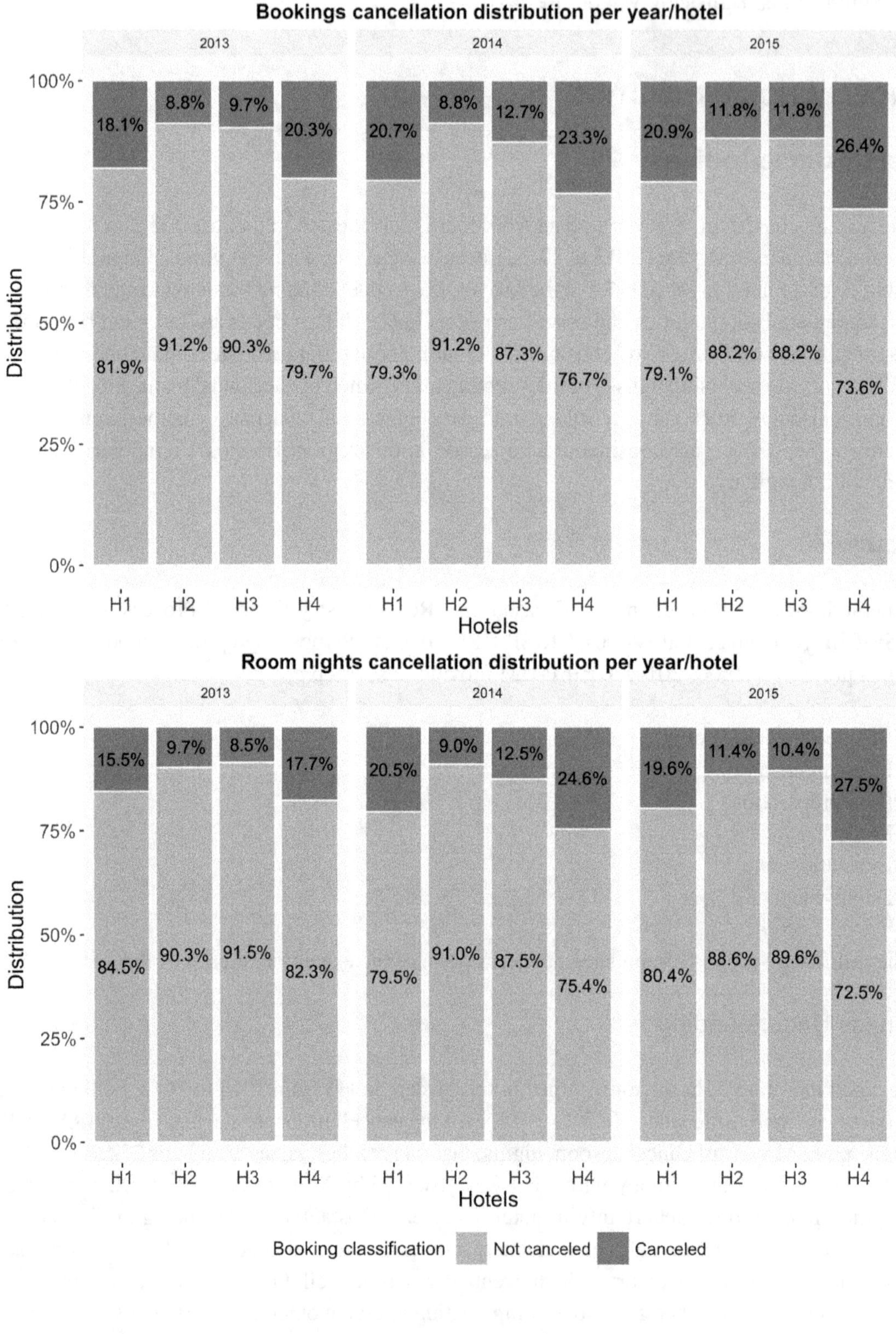

result of this high rate of cancellations, hotel managers need to forecast net demand (effective demand minus cancellations) to better forecast hotel occupancy.

To forecast demand long term (in a period much longer than the average lead time), forecast should be treated as a regression problem because it will most likely build a prediction model based on cancellation rates. This type of prediction modeling is where most research is focused. By contrast, to determine a more accurate forecast, one should forecast demand in the short- to midterm and view it as a classification prediction problem (i.e., Is this booking going to be canceled?). This means that it is possible to have better accuracy in predicting the overall cancellation ratio as well as identifying particular bookings with a high cancellation probability and, if possible, acting on those bookings.

Acting on bookings marked as having a high probably of being canceled can range from offering hotel services (e.g., spa treatments, free dinner or airport transfer) to discounts in certain services or entrances to local amusement parks. These actions could mitigate booking cancellations and therefore reduce the hotel's risk. These actions generate costs for the hotel, but by reducing the need to overbook, or at least enabling a better overbooking policy, the costs related to overbooking will also decrease. Moreover, by reducing uncertainty, pressure is taken from pricing and inventory allocation decisions. Acting on bookings marked as a high probability of being canceled should occur in the time gap between the time the booking is placed at the hotel and the expected arrival date. As illustrated in Figure 4, on average for each of the four hotels during the span of the 3 years, cancellations occurred, respectively, 25, 54, 33, and 55 days after bookings were made. Surprisingly, it is not during months of high demand (the high-season months highlighted in Figure 4) that lead time and cancellation time are higher. In fact, lead time and cancellation time are higher when there are special events in the region than during high season.

Understanding these cancellation patterns is one of the reasons why the development of a model for classifying bookings with high cancellation probability is important. To accomplish this and to help further a better understanding of net demand, model(s) should achieve a prediction accuracy above 0.8 and an area under the curve (AUC) also above 0.8, which is commonly considered a good prediction result (Zhu, Zeng, & Wang, 2010).

Data Understanding

Data were collected directly from the hotels' PMS databases using Microsoft SQL Server. All hotels used the same PMS application and therefore had the same database structure, but it was necessary to understand the database design and particularities of each hotel's data prior to building the data extraction queries. Selection of features to predict the probability of a booking being canceled started here. Table 2 presents a list of all variables extracted. These variables were chosen based on prior literature, but they also took into account the richness of data available in the PMS databases when compared to a PNR database. PNR database structure, as describer earlier, was built for the airline industry and for that reason it does not have fields or variables that are specific to the hotel industry. But, as recognized in data science and advocated by Guyon and Elisseeff (2003), extensive domain knowledge was fundamental to understand this data and to conduct a good variable selection.

As recognized in data science literature, domain knowledge is essential to select the best predictor variables and to avoid some prediction model "traps":

Figure 4. Lead time and cancellation time by month per hotel

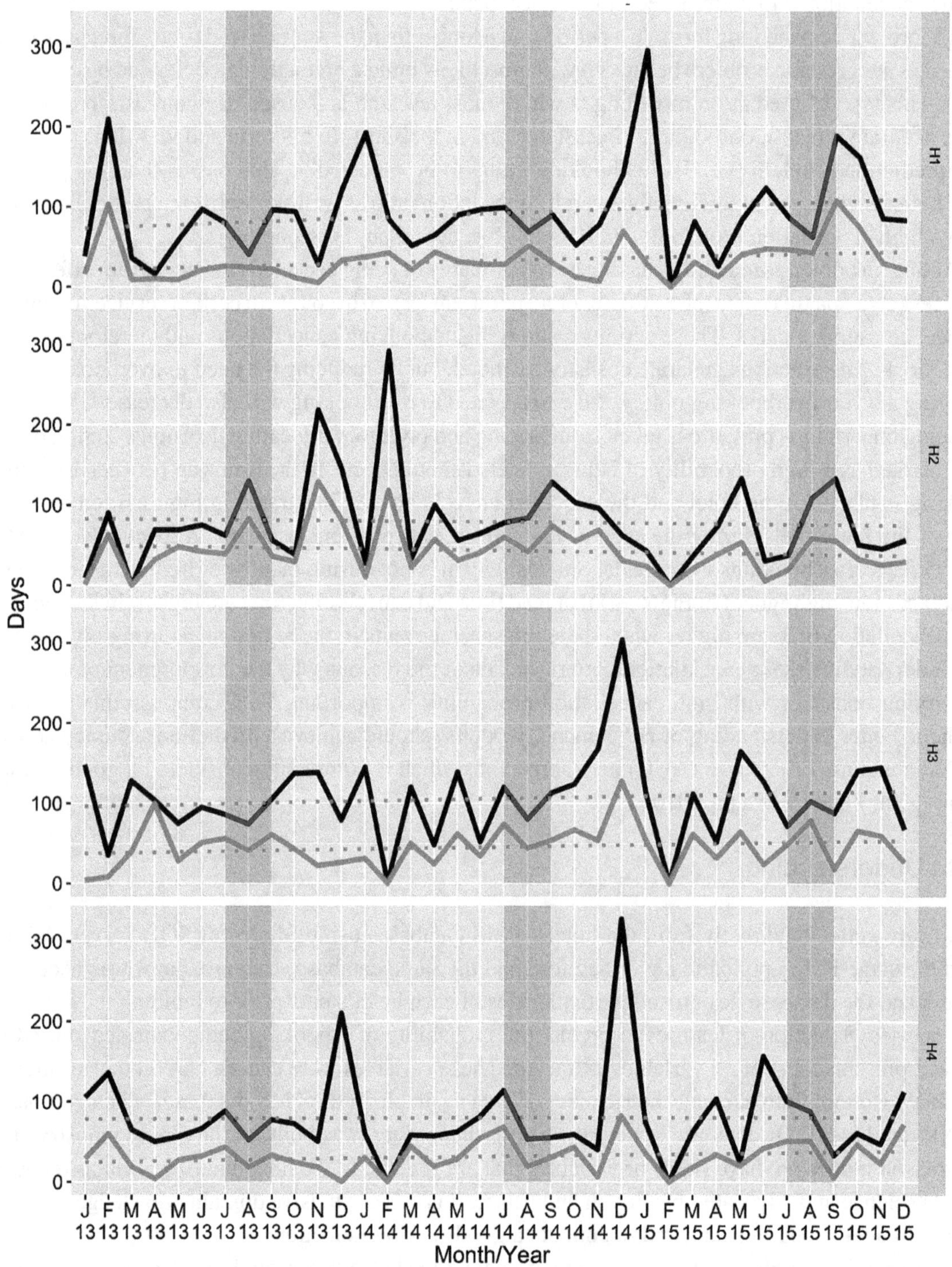

Table 2. Variables extracted from each booking from PMS databases

Name	Type	Description
ADR	Numeric	Average daily rate
Adults	Number	Number of adults
AgeAtBookingDate	Number	Age in years of the booking holder at the time of booking
Agent	Categorical	ID of agent (if booked through an agent)
ArrivalDateDayOfMonth	Numeric	Day of month of arrival date (1 to 31)
ArrivalDateDayOfWeek	Categorical	Day of week of arrival date (Monday to Sunday)
ArrivalDateMonth	Categorical	Month of arrival date
ArrivalDateWeekNumber	Numeric	Number of week in the year (1 to 52)
AssignedRoomType	Categorical	Room type assigned to booking
Babies	Numeric	Number of babies
BookingChanges	Numeric	Heuristic created by summing the number of booking changes (amendments) prior to arrival that could indicate cancellation intentions (arrival or departure dates, number of persons, type of meal, ADR, or reserved room type)
BookingDateDayOfWeek	Categorical	Day of week of booking date (Monday to Sunday)
CanceledTime	Numeric	Number of days prior to arrival that booking was canceled; when booking was not canceled it had the value of –1
Children	Numeric	Number of children
Company	Categorical	ID of company (if an account was associated with it)
Country	Categorical	Country ISO identification of the main booking holder
CustomerType	Categorical	Type of customer (group, contract, transient, or transient-party); this last category is a heuristic built when the booking is transient but is fully or partially paid in conjunction with other bookings (e.g., small groups such as families who require more than one room)
DaysInWaitingList	Numeric	Number of days the booking was in a waiting list prior to confirmed availability and to being confirmed as a booking
DepositType	Categorical	Because no specific field in the database existed with the type of deposit, based on how hotels operate, a heuristic was developed to define deposit type (nonrefundable, refundable, no deposit): payment made in full before the arrival date was considered a "nonrefundable" deposit, partial payment before arrival was considered a "refundable" deposit, otherwise it was considered as "no deposit"
DistributionChannel	Categorical	Name of the distribution channel used to make the booking
IsCanceled	Categorical	Outcome variable; binary value indicating if the booking was canceled (0: no; 1: yes)
IsRepeatedGuest	Categorical	Binary value indicating if the booking holder, at the time of booking, was a repeat guest at the hotel (0: no; 1: yes); created by comparing the time of booking with the guest history creation record
IsVIP	Categorical	Binary value indicating if the guest should be considered a Very Important Person (0: no; 1: yes)
LeadTime	Numeric	Number of days prior to arrival that the booking was placed in the hotel
LenghtOfStay	Numeric	Number of nights the guest stayed at the hotel
MarketSegment	Categorical	Market segmentation to which the booking was assigned
Meal	Categorical	ID of meal the guest requested
PreviousBookingsNotCanceled	Numeric	Number of previous bookings to this booking the guest had that were not canceled
PreviousCancellations	Numeric	Number of previous bookings to this booking the guest had that were canceled

continued on following page

Table 2. Continued

Name	Type	Description
PreviousStays	Numerical	Number of nights the guest had stayed at the hotel prior to the current booking
RequiredCarParkingSpaces	Numeric	Number of car parking spaces the guest required
ReservedRoomTypes	Categorical	Room type requested by the guest
RoomsQuantity	Numeric	Number of rooms booked
StaysInWeekendNights	Numeric	From the total length of stay, how many nights were in weekends (Saturday and Sunday)
StaysInWeekNights	Numeric	From the total length of stay, how many nights were in weekdays (Monday through Friday)
TotalOfSpecialRequests	Numeric	Number of special requests made (e.g., fruit basket, sea view, etc.)
WasInWaitingList	Categorical	Binary value indicating if the guest was in a waiting list prior to confirmed availability and to being confirmed as an effective booking (0: no; 1: yes)

- **The Curse of Dimensionality:** When the amount of data conjugated with a high number of predictor variables requires a high computational cost (Abbott, 2014). As a consequence, as advised by O'Neil and Schutt (2014), only variables that were considered relevant and useful for prediction were selected.
- **Correlation:** As described by Guyon and Elisseeff (2003): "Perfectly correlated variables are truly redundant in the sense that no additional information is gained by adding them" (p. 1164). Therefore, some variables were not included as they would be perfectly correlated (e.g., all studied hotels only assign room numbers to bookings on guest arrival; consequently, all noncanceled bookings would have a room number and canceled bookings would not). Even so, some known high correlated variables were selected because, in some cases, this high correlation "does not mean absence of variable complementarity" (Guyon & Elisseeff, p. 1164).
- **Leakage:** All created variables considered the possible leaking of future information. All created variables represent the value at the time of booking and not the value at the time of data extraction. For example, "IsRepeatedGuest," a binary indicator if a customer prior to a particular booking stayed at the hotel, should have the value of 0 (no) in the first booking of that specific guest. Only on subsequent bookings from the same guest should this variable assume the value of 1 (yes).

Date and time variables are particularly difficult to work with in predictive models (Abbott, 2014). For this reason, another consideration for the selection of variables to include in the prediction data sets was to decompose date variables into segments (month and week numbers). This is a common technique for employing fast and easy access to data (Yan, Ulanova, Ouyang, & Xu, 2014).

Exploration of the resulting data sets was executed in R. Resulting data sets had 20522, 9809, 9365, and 33445 observations, respectively, from H1 to H4. This exploration revealed both differences and similarities among the hotels:

- **Marketing Classification:** Each hotel had its own market segmentation and distribution channel classifications/categories.
- **Meal Types:** Although similar in three of the hotels, one offered more types of meals.

- **Cancellation Ratio (Bookings Canceled/Total Number of Bookings):** As depicted in Figure 1, all of the hotels show an upward tendency for this ratio. Peaks related to the temporary closing of hotels make it difficult to gain a more precise reading, but by the trend line it is possible to see that, over time, this cancellation rate is indeed rising.
- **Correlated Variables:** By plotting a correlation matrix for each hotel, it was possible to verify that, despite resemblances in some variable correlation among hotels, this correlation differs from hotel to hotel (e.g., the correlation matrix for H1 is presented in Figure 5). Resemblances appear in variables such as:
    - **"AgeAtBookingDate" and "CanceledTime" with "IsCanceled":** "AgeAtBookingDate" can be easy explained by the fact that birth dates are inserted on bookings, particularly for nonrepeated guests, only during check in. Thus, most times, canceled bookings do not have

Figure 5. Correlation matrix for H1 of numeric and categorical (binary) variables

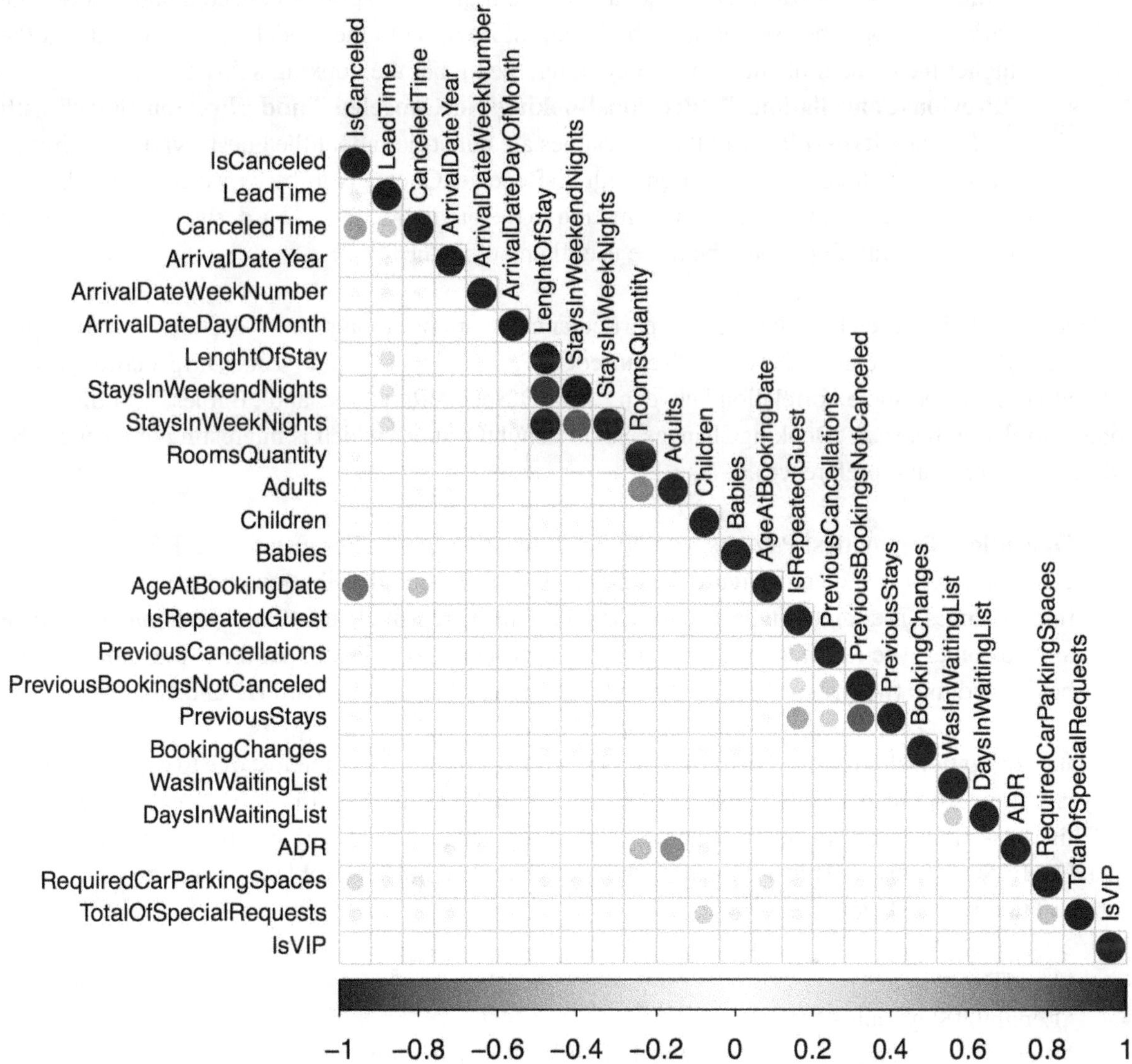

a birth date, which does not allow for the calculation of this variable. "CanceledTime" is even easier to explain because all noncanceled bookings have the same value of –1. As a consequence, "AgeAtBookingDate" and "CanceledTime" are two variables that should not be used in modeling.

 - **"StaysInWeekendNights" and "StaysInWeekNights" with "LenghtOfStay":** Because "LenghtOfStay" is the sum of the other two variables, it is redundant. Because the other two variables hold more information than "LenghtOfStay," it should not be used to model predictions.
 - **"RoomsQuantity" with "Adults":** In some cases, hotels create bookings with more than one room and a high quantity of adults simply to block rooms. Nevertheless, at check in, for the sake of registering extras and details about guests, each booking must have only one room assigned to it. Usually, when this number is different from 1 it means that the booking was canceled (as represented in Figure 7). This creates a "leak" problem; therefore, this variable should not be included in the modeling data set.
 - **"Adults" with "ADR":** This is a case where high correlation is expected and where both variables should be used because both add information to the model. It is expected that the higher the number of adults in one booking, the higher the booking's ADR.
 - **"PreviousCancellations," "PreviousBookingsNotCanceled," and "PreviousStays" with "IsRepeatedGuest":** As all three variables are quantities about the guests who have already stayed at the hotels, a correlation with "IsRepeatedGuest" is to be expected. Nevertheless, not all of these may add extra information to the prediction model. A derived feature created from these variables could be more useful and efficient.

Notwithstanding, each hotel has its own particular correlations among some variables. For example, H1 revealed some correlation between "RequiredCarParkingSpaces" and "TotalOfSpecialRequests." H2 and H3 revealed some correlation between "TotalOfSpecialRequests" and "Babies." H4 displayed some correlation between "BookingChanges" and "LenghtOfStay," which is interesting in what can be extracted toward guest preferences.

- **Operation Similarities:** Patterns were found in some variables. For instance, in H2 and H3, cancellation rates (as shown in Figure 3), markets segments, and distribution channels all presented resemblances. The same can be said for cancellations by repeated guest classification/deposit type and customer type (as presented in Figure 6). This seems to indicate that these four hotels could be divided into two groups: one group with H1 and H4 and another with H2 and H3.

Data exploration also revealed that, over all of the hotels, the total amount lost due to cancellations in the years from 2013 to 2015 was in excess of 1.3, 2.2, and 2.7 millions of Euros, per year, respectively.

This data-mining step was complemented with a data quality verification. To do that, R package "dataQualityR" was used. This verification showed some data quality problems, some of which were common to all of the hotels:

- "LeadTime,"
- "LenghtOfStay," and
- "PreviousStays" were positively skewed in terms of distribution.

Figure 6. Cancellations by repeated guest classification, deposit type, and customer type

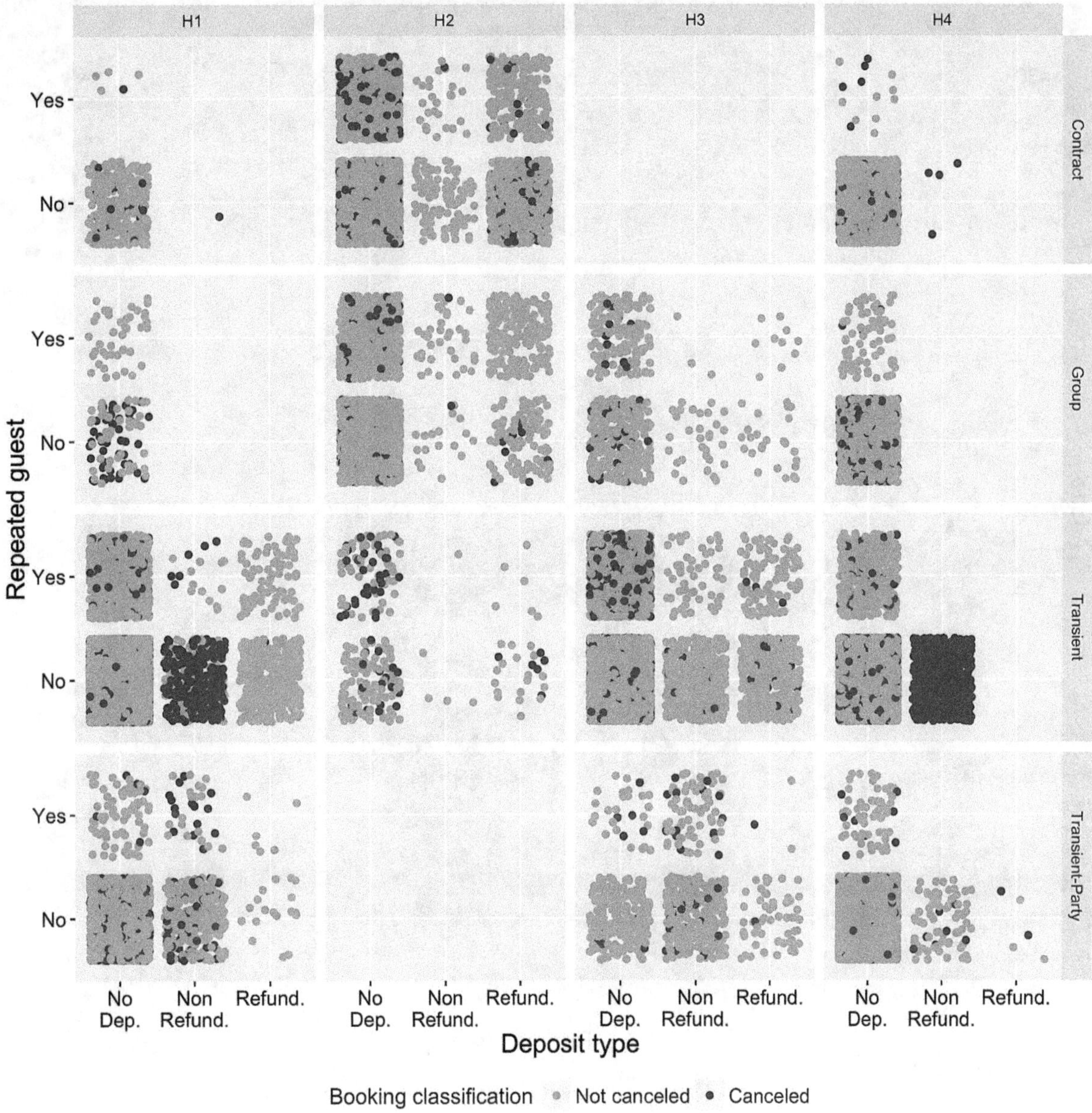

This is a problem usually solved by implementing a transformation function (e.g., Log10), but in this case the authors found that algorithms worked better without transformation. "ReservedRoomType" and "AssignedRoomType" did not seem to have any correlation to booking cancellations. "IsVIP" had very few observations. But, most important, "Agent" and "Company," although necessary features in terms of predicting a booking cancellation probability, both had too many categories, which make models computationally expensive.

There were also data quality problems specific to some of the hotels. "ADR," "Country," "Reserved-RoomType," and "AssignedRoomType" variables had missing values or outliers. Higher numbers of

Figure 7. Number of adults by room quantities per hotel

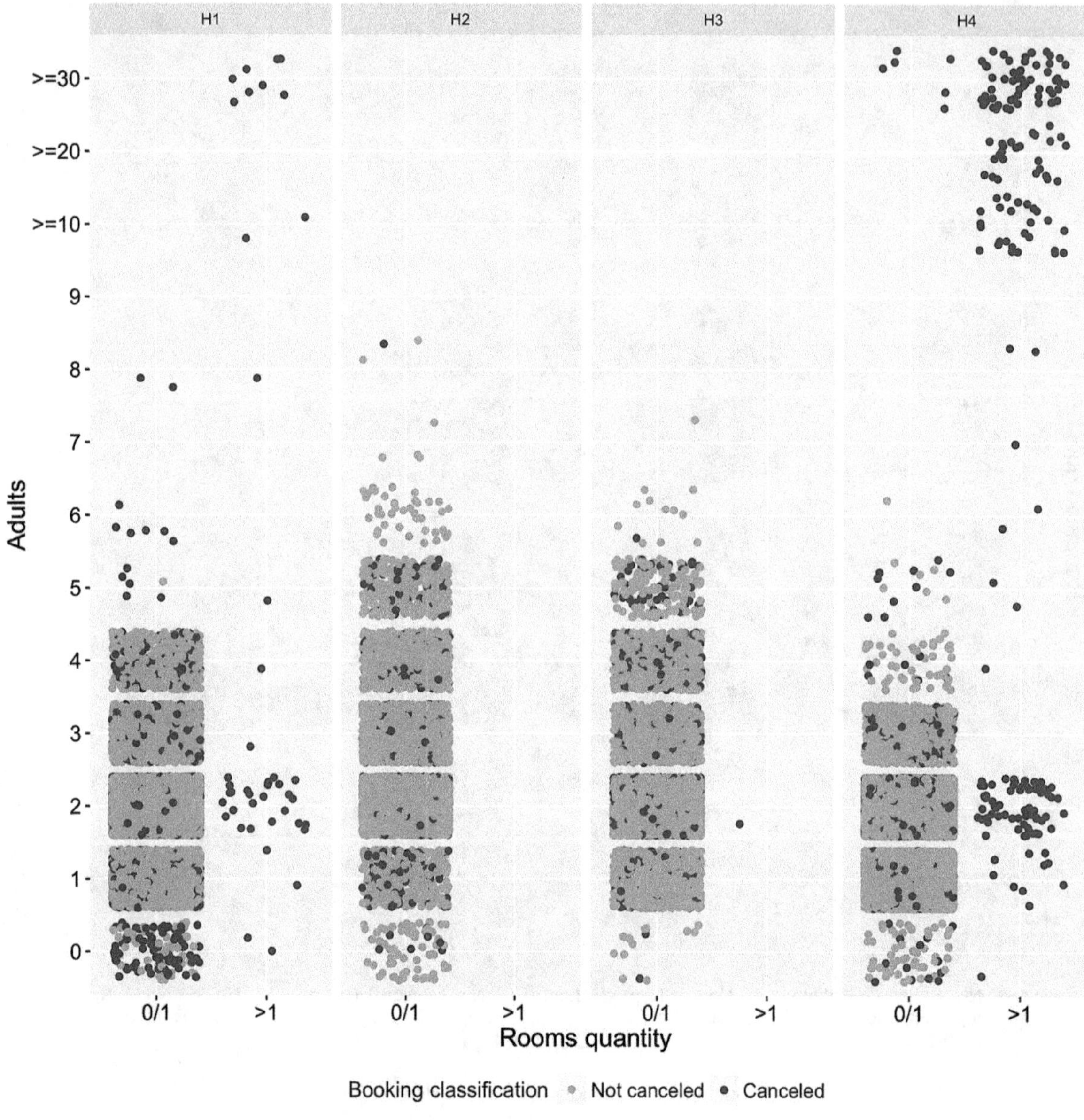

"Adults" and "Children" in some of the hotels, as illustrated in Figure 7, were related to higher numbers in "RoomsQuantity" and canceled bookings.

Finally, it was also verified that some variables were not used in some of the hotels, such as:

- "MarketSegment,"
- "RequiredCarParkingSpaces,"
- "TotalOfSpecialRequests," and
- "DaysInWaitingList."

Data Preparation

This step took advantage of the data exploration and quality verification made earlier to create the final data sets to be used in the prediction model development. It started with the removal from the original data sets of observations (rows) and variables (columns) based on the previous considerations. To validate this process for data selection, the "mutual information feature selection filter" was used. This filter measures the contribution of a predictor variable toward uncertainty about the value of the outcome variable; it is a very effective filter in the selection of features for nonlinear models (Tourassi, Frederick, Markey, & Floyd, 2001). Application of this filter in each hotel data set, as illustrated in Figure 8, demonstrated that the order and importance of booking cancellations for each predictor variable differ from hotel to hotel.

Figure 8. Mutual information feature selection filter results (order and value) per hotel

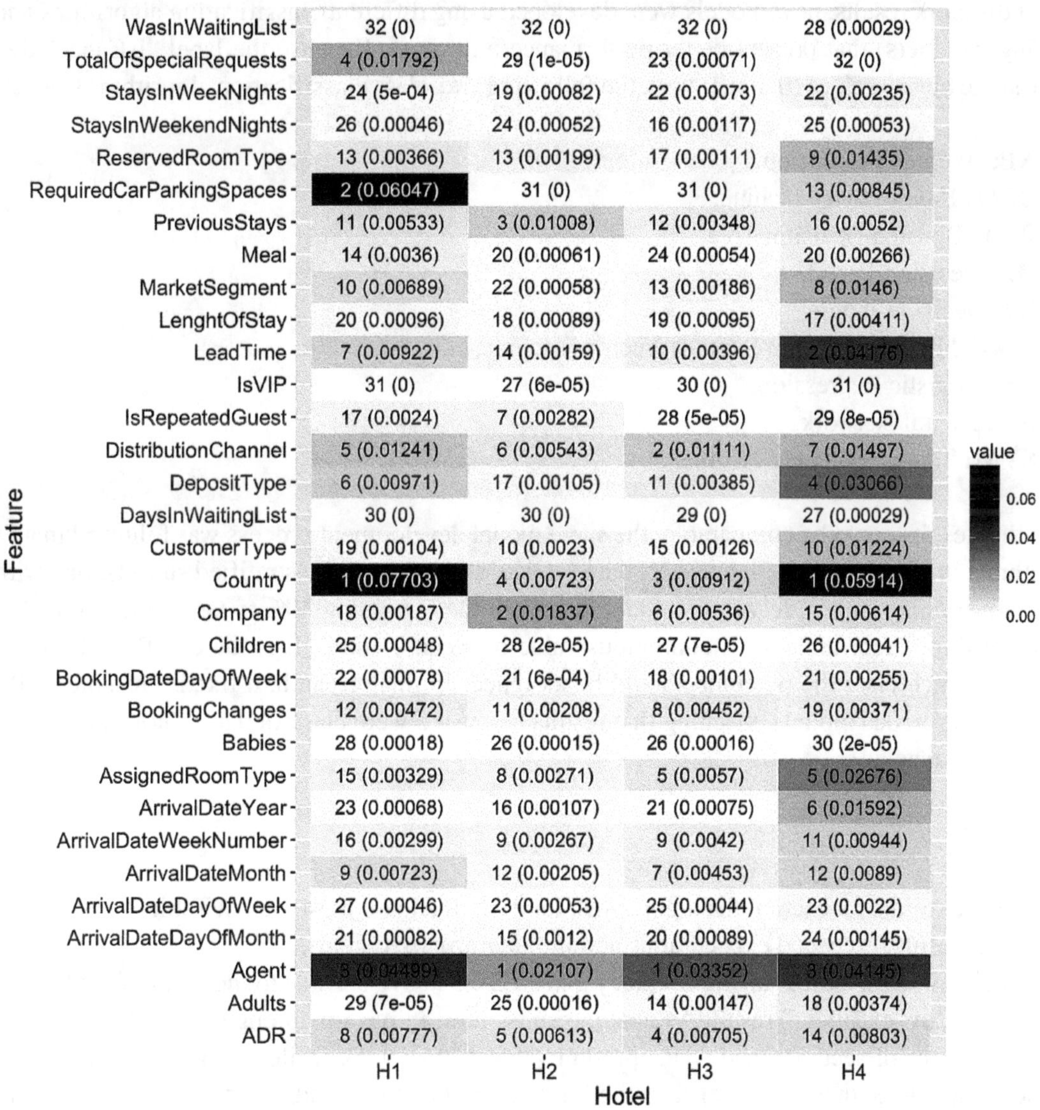

Based on the additional information provided by the application of the "mutual information feature selection filter" and the results of tests on the computational power required to create the model, it was decided to remove from all data sets the variables "BookingDateDayOfWeek" and "ArrivalDate-DayOfWeek." Lastly, the columns "PreviousBookingsNotCanceled," "PreviousCancellations," and "PreviousStays" were also removed from all data sets and substituted with a derived feature called "PREPPreviousCancellationRatio" that was calculated by dividing "PreviousCancellations" by the sum of "PreviousBookingsNotCanceled" with "PreviousCancellations."

Modeling

Due to the different order of contribution and the weight each feature presents for the four hotels (as presented in Figure 8), specific models were built for each hotel.

Microsoft Azure Machine Learning Studio was used to build these models. As different algorithms present different results, new models were developed using different classification algorithms and then selecting the one(s) that present better performance indicators. Because the label "IsCanceled" could only assume binary values (0: no; 1: yes), the following two-class classification algorithms were chosen:

- **AP:** Average Perceptron,
- **BPM:** Bayes Point Machine,
- **BDT:** Boosted Decision Tree,
- **DF:** Decision Forest,
- **DJ:** Decision Jungle,
- **LDSVM:** Locally Deep Support Vector Machine,
- **LR:** Logistic Regression,
- **NN:** Neural Network,
- **SVM:** Support Vector Machine.

So that results could be comparable, the same model development process was followed in all of the algorithms' implementations. Each hotel data set was divided into two stratified subsets, one with 70% of data for training (model learning) and another with the remaining 30% to test the developed model. Results of these algorithms are a continuous value between 0 and 1. It is the cutoff or threshold that defines to which class the outcome should be assigned. In this research, a fixed threshold of 0.5 was used in all developed models, meaning that results below 0.5 were classified as 0 (noncanceled) and all others as 1 (canceled).

Evaluation

Performance results for each of the algorithms, for each hotel, are presented in Table 3. For some of the algorithms, results exceeded expectations and initially proposed goals.

A prediction with more than 0.9 accuracy and an AUC above 0.9 was found in all of the hotels, what is considered an "excellent" result. As noted, results show a division among the hotels: Results for H2 and H3, in terms of accuracy and AUC, present values above 0.96, while for H1 and H4, although being excellent values, they are lower than the former. On the other hand, there were algorithms that did not perform to the level of the expected goals. For example, Bayes Point Machine did not reach 0.8 in

Table 3. Test set prediction performance indicators

Hotel	Algorithm	TP	FP	FN	TN	Accuracy	Recall	Precision	F1 Score	AUC
H1	AP	435	125	623	4 913	0.877	0.411	0.777	0.538	0.869
	BPM	412	94	646	4 944	0.879	0.389	0.814	0.527	0.878
	BDT	701	191	357	4 847	0.910	0.663	0.786	0.719	0.925
	DF	530	94	528	4 944	0.898	0.501	0.849	0.630	0.925
	DJ	375	26	683	5 012	0.884	0.354	0.935	0.514	0.907
	LDSVM	468	87	590	4 951	0.889	0.442	0.843	0.580	0.892
	LR	416	96	642	4 942	0.879	0.393	0.813	0.530	0.879
	NN	464	78	594	4 960	0.890	0.439	0.856	0.580	0.901
	SVM	323	57	735	4 981	0.870	0.305	0.850	0.449	0.851
H2	AP	255	13	35	2 627	0.984	0.879	0.951	0.914	0.971
	BPM	236	654	54	1 986	0.758	0.814	0.265	0.400	0.923
	BDT	260	15	30	2 625	0.985	0.897	0.945	0.920	0.972
	DF	251	6	39	2 634	0.985	0.866	0.977	0.918	0.979
	DJ	258	8	32	2 632	0.986	0.890	0.970	0.928	0.975
	LDSVM	258	9	32	2 631	0.986	0.890	0.966	0.926	0.975
	LR	259	19	31	2 621	0.983	0.893	0.932	0.912	0.977
	NN	257	13	33	2 627	0.984	0.886	0.952	0.918	0.977
	SVM	258	15	32	2 625	0.984	0.890	0.945	0.917	0.970
H3	AP	282	32	41	2 454	0.974	0.873	0.898	0.885	0.961
	BPM	260	553	63	1 933	0.781	0.805	0.320	0.458	0.920
	BDT	276	45	47	2 441	0.967	0.854	0.860	0.857	0.956
	DF	274	24	49	2 462	0.974	0.848	0.919	0.882	0.968
	DJ	277	24	46	2 462	0.975	0.858	0.920	0.888	0.967
	LDSVM	278	30	45	2 456	0.973	0.861	0.903	0.881	0.960
	LR	278	31	45	2 455	0.973	0.861	0.900	0.880	0.961
	NN	275	29	48	2 457	0.973	0.851	0.905	0.877	0.857
	SVM	279	37	44	2 449	0.971	0.864	0.883	0.873	0.956
H4	AP	719	276	831	8 147	0.889	0.464	0.723	0.565	0.893
	BPM	705	287	845	8 136	0.886	0.455	0.711	0.555	0.895
	BDT	1 145	330	405	8 093	0.926	0.739	0.776	0.757	0.946
	DF	1 020	222	530	8 201	0.925	0.658	0.821	0.731	0.948
	DJ	904	265	646	8 158	0.909	0.583	0.773	0.665	0.934
	LDSVM	885	265	665	8 158	0.907	0.571	0.770	0.656	0.902
	LR	719	270	831	8 153	0.890	0.464	0.727	0.566	0.895
	NN	855	266	665	8 157	0.907	0.571	0.769	0.655	0.916
	SVM	693	255	857	8 168	0.888	0.447	0.731	0.555	0.887

Legend: AP: Average Perceptron; BPM: Bayes Point Machine; BDT: Boosted Decision Tree; DF: Decision Forest; DJ: Decision Jungle; LDSVM: Locally Deep Support Vector Machine; LR: Logistic Regression; NN: Neural Network; SVM: Support Vector Machine

terms of accuracy in H2 and H3. This same algorithm in H2 and the Support Vector Machine algorithm also had results below 0.5 in terms of F1 score. Nevertheless, the results show that at least three of the algorithms have good performance results with three of the hotels:

- Boosted Decision Tree,
- Decision Forest, and
- Decision Jungle.

However, Decision Forest seems to be the algorithm that generates the more suitable model for the problem of booking cancellation prediction. This is aligned to the findings of Fernández-Delgado, Cernadas, Barro, and Amorim (2014). These authors tested 179 classifiers from 17 families and concluded that the best results are usually obtained with the random forest algorithms family.

The Decision Forest algorithm obtained the best AUC results in three of the four hotels. It also presented some of the best results in terms of accuracy, but, most important, in three of the hotels this algorithm also had the best performance results in terms of false positives. Good results in terms of the number of false positives are important due to the fact that if a hotel wants to act on bookings classified as "going to be canceled," then the least "false predicts" the model generates, the least the hotel will spend in cash/services with bookings that would not have been canceled.

Deployment

To deploy these models in a production environment was not in the scope of this research. In spite of that, development of these models brought attention to some considerations that should be taken into account in terms of model deployment:

1. Some predictor variables vary with time (e.g., "LeadTime") or can assume new values every day, as in the case of changes/amendments to bookings (e.g., "BookingChanges" or "Adults"). Thus, the model should be run every day so that all in-house bookings and results are evaluated on a daily basis.
2. As Abbott (2014) asserts, "Even the most accurate and effective models don't stay effective indefinitely. Changes in behavior due to new trends, fads, incentives, or disincentives should be expected" (p. 618). For example, when a hotel changes its marketing efforts and starts to capture more market from online travel agencies instead of traditional tour operators, this could influence many predictor variables, such as "MarketSegment," "DistribuitonChannel," and "LeadTime." If the model is not revised, its performance will not stay at the same level.

SOLUTIONS AND RECOMMENDATIONS

Despite what was alleged by Morales and Wang (2010), results unquestionably demonstrate that features extracted and derived from bookings in the hotels' PMS databases are a good source to predict, with high accuracy, if bookings are going to be canceled. Accuracy reached 98% in H2 and reached values above 91% in all other hotels.

These results confirm that it is possible to identify bookings with a high likelihood of being canceled. This makes it possible for hotel managers to take measures to avoid these potential cancellations, such as offering services, discounts, entrances to shows/amusement parks, or other perks. However, this cannot be applied to all customers because some are insensitive to these kinds of offers (e.g., corporate guests). Yet there is more to be gained from building and deploying this prediction model. By running the model each day against all in-house bookings, it is possible to obtain another important result: How many room nights are predicted to be canceled for each of the following days?

It is only necessary to add the number of bookings predicted as going to be canceled by night. Hotel managers can deduce this value of their demand, thus calculating their net demand. When provided with a more accurate value of net demand, hotel managers can develop better overbooking and cancellation policies, which would result in fewer costs and decreased risk.

Nevertheless, to build a booking cancellation model, suitable data and a good selection of features are crucial. As mentioned earlier and illustrated in Figure 8, not all features have the same order of importance, nor do they contribute the same way to predict if a booking is going to be canceled. This calls for a specific characterization from each hotel. Hotel location, services, facilities, nationality of guests, markets, and distribution channels are among the many features with different weight for predicting cancellation. One example of this is the feature "RequiredCarParkingSpaces." It is ranked in second place for H1 and 13th for H4 but had no importance in terms of H3 and H4. This is easily understandable if one knows these hotels' operations, as H2 and H3 do not have such limited car parking spaces as H1 and H4. Therefore, hotel revenue management and general business domain knowledge are not enough to undertake a good selection of features. It is also essential to understand each hotel's particular operation and characteristics. This can make a difference in terms of final model performance and adequacy.

As with any other predictive analytics problem, developing a model to predict booking cancellations requires that data meet all of the attributes of quality data: accurate, reliable, unbiased, valid, appropriate, and timely (Rabianski, 2003). As previously mentioned, some of the data sets had variables with inaccurate values (e.g., "ADR" and "Country" in the H1 data set). This lack of quality can affect model performance. For this reason, hotels that want to build prediction models must ensure that a data quality policy is in place.

FUTURE RESEARCH DIRECTIONS

This research employed data from four resort hotels that are using the same PMS application, which raises some questions that further research can help clarify:

1. Could similar results be obtained if data from other PMS databases were used, as not all applications record the same information?
2. If the number of observed hotels increases, would it be possible to maintain the same level of model performance?
3. Could similar model performance be attained in other types of hotels (e.g., city hotels)?

Consequently, further research with different PMS data, additional hotels, and different types of hotels could contribute to a better understanding of the topic of booking cancellation prediction.

Further research could also make use of features from additional data sources, such as weather information, competitive intelligence (prices and social reputation), and currency exchange rates, to improve model performance and measure the influence of these features in booking cancellations.

Finally, deployment of these predictive models in a production environment, in hotels, with the purpose of executing A/B testing could contribute to measuring the effect of having previous knowledge of which bookings have a high probability of cancellation.

CONCLUSION

Selecting PMS data from four hotels instead of the usually adopted PNR data, in conjunction with data science skills such as data visualization, data mining, and machine learning, demonstrates that it is possible to predict the cancellation probability of a hotel booking with high accuracy.

This prediction can be useful for hotel managers not only to take measures to avoid cancellations on bookings classified as "potentially going to be canceled" but also to produce more precise demand forecasts. Knowledge obtained from the built prediction models can enable hotel managers to mitigate the revenue loss derived from booking cancellations and the risks associated with overbooking (reallocation costs, cash or service compensations, and, particularly important today, social reputation costs) to enhance pricing and inventory allocation decisions.

In the case of the four studied hotels, cancellations from 2013 to 2015 represented a total amount of 6.2 million Euros. By reducing the risk associated with unsold rooms, even if only a small fraction, such as 10%, of this amount is retained/not lost by the application of prediction models, results could have a significant influence on the hotel's performance, meaning that prediction models can become an important tool in terms of yield management.

These prediction models also contribute to a better understanding of cancellation patterns and tendencies, thus empowering hotel managers to improve deposit policies and reduce the risk associated with unsold rooms because of booking cancellations.

Simultaneously, development of these models should contribute to the improvement of hotel revenue management, as its use of technology and mathematical/scientific models is in accordance with the works of Chiang et al. (2007), Ivanov and Zhechev (2012), and Kimes (2010).

ACKNOWLEDGMENT

The authors wish to express their gratitude to the board and general managers of the four studied hotels. Without their support in allowing access to their PMS's databases' this work could not be accomplished. The authors wish also to acknowledge their gratitude to Instituto de Telecomunicações/ISCTE-IUL for the financial support.

REFERENCES

Abbott, D. (2014). *Applied predictive analytics: Principles and techniques for the professional data analyst*. Indianapolis, IN: Wiley.

Chapman, P., Clinton, J., Kerber, R., Khabaza, T., Reinartz, T., Shearer, C., & Wirth, R. (2000). *CRISP-DM 1.0: Step-by-step data mining guide.* Retrieved from https://the-modeling-agency.com/crisp-dm.pdf

Chen, A. H., Peng, N., & Hackley, C. (2008). Evaluating service marketing in airline industry and its influence on student passengers purchasing behaviour using Taipei–London route as an example. *Journal of Travel & Tourism Marketing*, *25*(2), 149–160. doi:10.1080/10548400802402503

Chiang, W.-C., Chen, J. C. H., & Xu, X. (2007). An overview of research on revenue management: Current issues and future research. *International Journal of Revenue Management*, *1*(1), 97–128. doi:10.1504/IJRM.2007.011196

Dhar, V. (2013). Data science and prediction. *Communications of the ACM*, *56*(12), 64–73. doi:10.1145/2500499

Fernández-Delgado, M., Cernadas, E., Barro, S., & Amorim, D. (2014). Do we need hundreds of classifiers to solve real world classification problems? *Journal of Machine Learning Research*, *15*(1), 3133–3181.

Freisleben, B., & Gleichmann, G. (1993). Controlling airline seat allocations with neural networks. *Proceeding of the Twenty-Sixth Hawaii International Conference on System Sciences*. doi:10.1109/HICSS.1993.284243

Garrow, L., & Ferguson, M. (2008). Revenue management and the analytics explosion: Perspectives from industry experts. *Journal of Revenue and Pricing Management*, *7*(2), 219–229. doi:10.1057/rpm.2008.3

Gorin, T., Brunger, W. G., & White, M. M. (2006). No-show forecasting: A blended cost-based, PNR-adjusted approach. *Journal of Revenue and Pricing Management*, *5*(3), 188–206. doi:10.1057/palgrave.rpm.5160039

Guyon, I., & Elisseeff, A. (2003). An introduction to variable and feature selection. *Journal of Machine Learning Research*, *3*, 1157–1182.

Hayes, D. K., & Miller, A. A. (2011). *Revenue management for the hospitality industry*. Hoboken, NJ: Wiley.

Huang, H.-C., Chang, A. Y., & Ho, C.-C. (2013). Using artificial neural networks to establish a customer-cancellation prediction model. *Przeglad Elektrotechniczny*, *89*(1b), 178–180.

Hueglin, C., & Vannotti, F. (2001). Data mining techniques to improve forecast accuracy in airline business.*Proceedings of the seventh ACM SIGKDD international conference on knowledge discovery and data mining*, (pp. 438–442). doi:10.1145/502512.502578

Iliescu, D. C., Garrow, L. A., & Parker, R. A. (2008). A hazard model of US airline passengers refund and exchange behavior. *Transportation Research Part B: Methodological*, *42*(3), 229–242. doi:10.1016/j.trb.2007.10.005

International Civil Aviation Organization. (2010). *Guidelines on Passenger Name Record (PNR) data.* Retrieved from https://www.iata.org/iata/passenger-data-toolkit/assets/doc_library/04-pnr/New%20Doc%209944%201st%20Edition%20PNR.pdf

Ivanov, S. (2014). *Hotel revenue management: From theory to practice.* Varna, Bulgary: Zangador.

Ivanov, S., & Zhechev, V. (2012). Hotel revenue management—A critical literature review. *Turizam: Znanstveno-Strucnicasopis, 60*(2), 175–197.

Kimes, S. E. (2010). The future of hotel revenue management. *Cornell Hospitality Report, 10*(14), 6–15.

Kimes, S. E., & Wirtz, J. (2003). Has revenue management become acceptable? Findings from an international study on the perceived fairness of rate fences. *Journal of Service Research, 6*(2), 125–135. doi:10.1177/1094670503257038

Lawrence, R. D. (2003). A machine-learning approach to optimal bid pricing. In H. K. Bhargava & N. Ye (Eds.), *Computational modeling and problem solving in the networked world: Interfaces in computer science and operations research* (pp. 97–118). New York, NY: Springer. doi:10.1007/978-1-4615-1043-7_5

Lemke, C., Riedel, S., & Gabrys, B. (2009). Dynamic combination of forecasts generated by diversification procedures applied to forecasting of airline cancellations. In *IEEE Symposium on Computational Intelligence for Financial Engineering* (pp. 85–91). New York, NY: Institute of Electrical and Electronics Engineers. doi:10.1109/CIFER.2009.4937507

Liu, P. H. (2004). Hotel demand/cancellation analysis and estimation of unconstrained demand using statistical methods. In I. Yeoman & U. McMahon-Beattie (Eds.), *Revenue management and pricing: Case studies and applications* (pp. 91–101). London, England: Thomas Learning.

Mehrotra, R., & Ruttley, J. (2006). *Revenue management* (2nd ed.). Washington, DC: American Hotel & Lodging Association.

Morales, D. R., & Wang, J. (2010). Forecasting cancellation rates for services booking revenue management using data mining. *European Journal of Operational Research, 202*(2), 554–562. doi:10.1016/j.ejor.2009.06.006

Neuling, R., Riedel, S., & Kalka, K.-U. (2004). New approaches to origin and destination and no-show forecasting: Excavating the passenger name records treasure. *Journal of Revenue and Pricing Management, 3*(1), 62–72. doi:10.1057/palgrave.rpm.5170094

Noone, B. M., & Lee, C. H. (2011). Hotel overbooking: The effect of overcompensation on customers reactions to denied service. *Journal of Hospitality & Tourism Research (Washington, D.C.), 35*(3), 334–357. doi:10.1177/1096348010382238

O'Neil, C., & Schutt, R. (2014). *Doing data science: Straight talk from the frontline*. Sebastopol, CA: O'Reilly Media.

Park, J.-W., Robertson, R., & Wu, C.-L. (2006). Modelling the impact of airline service quality and marketing variables on passengers future behavioural intentions. *Transportation Planning and Technology, 29*(5), 359–381. doi:10.1080/03081060600917686

Phillips, R. L. (2005). *Pricing and revenue optimization*. Stanford, CA: Stanford University Press.

Rabianski, J. S. (2003). Primary and secondary data: Concepts, concerns, errors, and issues. *The Appraisal Journal, 71*(1), 43–55.

Serra, F. (2013). Estratégias de negócio: Os conceitos de yield e de revenue management e a sua aplicação às operações hoteleiras. In C. Henriques, I. Monteiro, F. Serra, J. Santos, & P. Águas (Eds.), *Inovação e Qualidade na Hotelaria* (pp. 161–172). Faro, Portugal: Universidade do Algarve, Escola Superior de Gestão, Hotelaria e Turismo.

Smith, S. J., Parsa, H. G., Bujisic, M., & van der Rest, J.-P. (2015). Hotel cancelation policies, distributive and procedural fairness, and consumer patronage: A study of the lodging industry. *Journal of Travel & Tourism Marketing*, *32*(7), 886–906. doi:10.1080/10548408.2015.1063864

Subramanian, J., Stidham, S. Jr, & Lautenbacher, C. J. (1999). Airline yield management with overbooking, cancellations, and no-shows. *Transportation Science*, *33*(2), 147–167. doi:10.1287/trsc.33.2.147

Talluri, K. T., & Van Ryzin, G. J. (2004). *The theory and practice of revenue management*. New York, NY: Springer. doi:10.1007/b139000

Tourassi, G. D., Frederick, E. D., Markey, M. K., & Floyd, C. E. (2001). Application of the mutual information criterion for feature selection in computer-aided diagnosis. *Medical Physics*, *28*(12), 2394–2402. doi:10.1118/1.1418724 PMID:11797941

Yan, T., Ulanova, L., Ouyang, Y., & Xu, F. (2014). Data mining in time series: Current study and future trend. *Journal of Computer Science*, *10*(12), 2358–2359. doi:10.3844/jcssp.2014.2358.2359

Yangyong, Z., & Yun, X. (2011). *Dataology and data science: Up to now*. Retrieved from http://iwdds.fudan.edu.cn/second/abstract/Dataology%20and%20DataScienceUp%20to%20Now_ZYYFudan.pdf

Yoon, M. G., Lee, H. Y., & Song, Y. S. (2012). Linear approximation approach for a stochastic seat allocation problem with cancellation & refund policy in airlines. *Journal of Air Transport Management*, *23*, 41–46. doi:10.1016/j.jairtraman.2012.01.013

Zhu, W., Zeng, N., & Wang, N. (2010). Sensitivity, specificity, accuracy, associated confidence interval and ROC analysis with practical SAS implementations. *NESUG Proceedings: Health Care and Life Sciences*.

ADDITIONAL READING

Chen, C.-C., & Schwartz, Z. (2013). On revenue management and last minute booking dynamics. *International Journal of Contemporary Hospitality Management*, *25*(1), 7–22. doi:10.1108/09596111311290192

Chen, C.-C., Schwartz, Z., & Vargas, P. (2011). The search for the best deal: How hotel cancellation policies affect the search and booking decisions of deal-seeking customers. *International Journal of Hospitality Management*, *30*(1), 129–135. doi:10.1016/j.ijhm.2010.03.010

Chen, C.-C., & Xie, K. (2013). Differentiation of cancellation policies in the U.S. hotel industry. *International Journal of Hospitality Management*, *34*, 66–72. doi:10.1016/j.ijhm.2013.02.007

DeKay, F., Yates, B., & Toh, R. S. (2004). Non-performance penalties in the hotel industry. *International Journal of Hospitality Management*, *23*(3), 273–286. doi:10.1016/j.ijhm.2003.11.003

Dong, Y., & Ling, L. (2015). Hotel overbooking and cooperation with third-party websites. *Sustainability, 7*(9), 11696–11712. doi:10.3390/su70911696

Guo, L. (2009). Service cancellation and competitive refund policy. *Marketing Science, 28*(5), 901–917. doi:10.1287/mksc.1080.0457

Guo, X., Dong, Y., & Ling, L. (2016). Customer perspective on overbooking: The failure of customers to enjoy their reserved services, accidental or intended? *Journal of Air Transport Management, 53*, 65–72. doi:10.1016/j.jairtraman.2016.01.001

Hastie, T., Tibshirani, R., & Friedman, J. (2009). *The elements of statistical learning* (2nd ed.). New York, NY: Springer. doi:10.1007/978-0-387-84858-7

Hong, L., Yan, J., & Panpan, L. (2013). Study on the model of hotel rooms overbooking. *Journal of Applied Sciences, 13*(15), 2994–2997. doi:10.3923/jas.2013.2994.2997

Hwang, J., & Wen, L. (2009). The effect of perceived fairness toward hotel overbooking and compensation practices on customer loyalty. *International Journal of Contemporary Hospitality Management, 21*(6), 659–675. doi:10.1108/09596110910975945

James, G., Witten, D., Hastie, T., & Tibshirani, R. (2013). *An introduction to statistical learning*. New York, NY: Springer. doi:10.1007/978-1-4614-7138-7

Koide, T., & Ishii, H. (2005). The hotel yield management with two types of room prices, overbooking and cancellations. *International Journal of Production Economics, 93-94*, 417–428. doi:10.1016/j.ijpe.2004.06.038

Law, R. (1998). Hospitality data mining myths. *Hospitality Review, 16*(1), 7.

Magnini, V. P., Honeycutt, E. D., & Hodge, S. K. (2003). Data mining for hotel firms: Use and limitations. *The Cornell Hotel and Restaurant Administration Quarterly, 44*(2), 94–105. doi:10.1016/S0010-8804(03)90022-X

Shalev-Shwartz, S., & Ben-David, S. (2014). *Understanding machine learning: from theory to algorithms*. New York, NY: Cambridge University Press. doi:10.1017/CBO9781107298019

Siegel, E. (2013). *Predictive analytics: The power to predict who will click, buy, lie, or die*. Hoboken, N.J: Wiley.

Sierag, D. D., Koole, G. M., van der Mei, R. D., van der Rest, J. I., & Zwart, B. (2015). Revenue management under customer choice behaviour with cancellations and overbooking. *European Journal of Operational Research, 246*(1), 170–185. doi:10.1016/j.ejor.2015.04.014

Smola, A., & Vishwanathan, S. V. N. (2010). *Introduction to machine learning*. Cambridge, UK: Cambridge University Press.

Sulistio, A., Kim, K. H., & Buyya, R. (2008). Managing cancellations and no-shows of reservations with overbooking to increase resource revenue. In *Cluster Computing and the Grid, 2008. CCGRID'08. 8th IEEE International Symposium on* (pp. 267–276). Institute of Electrical and Electronics Engineers
Toh, R. S., & Dekay, F. (2002). Hotel room-inventory management: An overbooking model. *The Cornell Hotel and Restaurant Administration Quarterly*, *43*(4), 79–90.

Xie, J., & Gerstner, E. (2007). Service escape: Profiting from customer cancellations. *Marketing Science*, *26*(1), 18–30. doi:10.1287/mksc.1060.0220

KEY TERMS AND DEFINITIONS

Accuracy: Measure of outcome correctness. Measures the proportion of true results (True Positives and True Negatives) among the total number of predictions.

AUC (Area Under the Curve): Measure of success calculated from the area under the plot of true positive rate against false positive rate.

Cancellation Time: Time (usually measured in days) between a booking's date of cancellation and the guest's expected arrival.

Classification Problem: When the outcome of prediction is a class/category (discrete value).

F1 Score: Measure of prediction accuracy, which is the harmonic means of precision and recall.

FN (False Negative): The outcome prediction was false and the actual value was true (e.g., the booking was predicted as "has not canceled" but in fact it was canceled).

FP (False Positive): The outcome prediction was true and the actual value was false (e.g., the booking was predicted as "has a cancellation" but in fact it was not canceled).

Lead Time: Time (usually measured in days) between a booking's date of placement in the hotel and the guest's expected arrival date.

Outcome: A variable which one's want to predict. Also known as response variable, dependent variable, or label.

PMS (Property Management System): A computerized system used to facilitate the management of hotels and other types of properties. Considered equivalent to Enterprise Resource Planning systems in other types of industries.

Precision: Measures the proportion of True Positives against the sum of all positive predictions (True Positives and False Positives).

Predictor: A variable that explains the potential reasons behind the outcome variable variations. Also known as independent variable, explanatory variable, or feature.

Recall: Measure of relevant predictions that are retrieved. It can be interpreted as the probability of a randomly selected prediction could be a True Positive.

Regression Problem: When the outcome of prediction is a continuous value.

TN (True Negative): The outcome prediction was false and the actual value was false (e.g., the booking was predicted as "has not canceled" and it was not canceled).

TP (True Positive): The outcome prediction was true and the actual value was true (e.g., the booking was predicted as "has a cancellation" and it was canceled).

Chapter 7
User-Generated Content and Its Influence on Tourists' Choices:
How Do the Spanish Make Hotel Reservations Online?

Giacomo Del Chiappa
University of Sassari, Italy

Giuseppe Melis
University of Cagliari, Italy

Marcello Atzeni
University of Cagliari, Italy

ABSTRACT

This study aims to analyse referral and conversion traffic from user-generated content (UGC) to other key online/offline information sources and when consumers use UGC along with their information-search and the related cross-visitation. Data were collected by sending a semi-structured questionnaire by e-mail to Spanish consumers aged 18 years and above. A snowball sampling technique was used, and a total of 825 complete questionnaires were returned. Manual coding of open-ended questions and descriptive statistics were applied. Findings revealed that UGC plays a relevant role, along with all the online searching and booking behaviour, particularly UGC within online travel agencies (OTAs), which appears to be significant in the early stage of the tourist information search process. Contributions to the body of knowledge and managerial implications are discussed and suggestions for further research are given.

INTRODUCTION

According to Internet World Stats (2014), in 2014 42.3% of the world's population (nearly 7.2 billion people) were internet users with significant penetration ratios in countries all over the world. The highest penetration ratio was in North America (87.7%), followed by Oceania (72.9%) and Europe (70.5%), while the weight of the digital divide was felt heavily in Africa (26.5%) and Asia (34.7%). Table 1 (Internet

DOI: 10.4018/978-1-5225-1054-3.ch007

Table 1. Internet penetration rates

Country	Penetration (% Population)		Country	Penetration (% Population)	
	2013	2011		2013	2011
India	15.8	8.4	Iran	55.7	46.9
Indonesia	21.7	16.1	Italy	58.5	49.2
Nigeria	38.0	28.3	Spain	74.8	62.2
Philippines	41.1	29.2	Egypt	49.6	35.6 (2012)
Mexico	43.5	30.7	France	83.3	69.5
Vietnam	43.9	32.3	United States	84.2.	78.2
China	45.8	36.3	Japan	(86.2)	78.4
Brazil	54.2	37.4	Germany	86.2	79.9
Russia	61.4	43.0	South Korea	84.8	80.9
Turkey	46.3	44.4	United Kingdom	89.8	82

Source: (Internet World Stats, 2013; Internet World Stats, 2011).

World Stats, 2013; Internet World Stats, 2011) shows the rate of internet penetration in some countries of the World in 2011 and 2013.

The rapid and increasingly pervasive spread of the internet has substantially changed how people buy and consume products and services (Buhalis & Law, 2008); tourism has become the foremost industry in terms of online market share (Werthner & Ricci, 2004). Recently, the most significant development in internet applications has been in the area of user-generated content (UGC) and peer-to-peer applications. UGC has been found to have a great influence on service expectations (Mauri & Minazzi, 2013) in generating the idea of travelling during the actual planning process and after tourists' decisions have been made (Gretzel & Yoo, 2008; Law, Buhalis, & Cobanoglu, 2014). This could be explained by the fact that online comments and reviews (UGC) are usually perceived to have a higher credibility and trustworthiness than traditional tourist information sources (Akehurst, 2009).

Further, the emergence of new e-intermediaries, in particular online travel agencies (OTAs) (e.g., Expedia.com, Booking.com), has drawn increasing attention to what has become known as cybermediation (Anckar, 2003), a term that refers to the utilisation of information and communication technologies (ICT) and internet tools for developing new intermediaries, or for enabling existing intermediaries to re-design tourism distribution channels (Bennet & Buhalis, 2003; Bennet & Lai, 2005). This explains why most industries and academic experts talk about 'e-reintermediation', a term that describes a situation where OTAs are replacing 'bricks and mortar' travel agencies. In this scenario, high street travel agents have an uncertain future and their competitiveness is becoming strongly linked to their ability to carry out their marketing and promotions operations via websites (e.g., Abou-Shuck, Megicks, & Lim, 2013; Cheyne, Downes, & Legg, 2006) and social media (Huang, 2011), focusing on improving their brand image, selling packages online, and exploiting opportunities of reintermediation in the retail market (e.g., Abou-Shouk et al., 2013).

All that said, it is evident that the role of traditional tourism intermediaries is currently being seriously challenged (Kracht & Wang, 2010) and even further compromised by the fact that tourists quite often change the accommodation suggested by a travel agency based on UGC (Del Chiappa, 2013; Del Chiappa, Lorenzo-Romero, & Gallarza, 2014; Prayag & Del Chiappa, 2014). Previous studies (Law,

Leung, & Wong, 2004) showed that the propensity to purchase online increases with age, education level, and income; conversely, gender generally has no influence on online behaviour (Beldona, Racherla, & Mundhra, 2011; Ip, Lee, & Law, 2012). Further, previous research showed that people usually buy convenience or low involvement products and services online (such as short-haul or short trips), while relying on the traditional intermediaries for the purchase of more complex products or high involvement (such as honeymoons, travel in the medium to long term, or intercontinental travel) (Chu, 2001; Beldona, Morrison, & O'Leary, 2005; Card, Chen, & Cole, 2003; Del Chiappa, 2011b; Del Chiappa et al., 2015; Prayag & Del Chiappa, 2014; Wertener & Klein, 1999). Therefore, for example, with reference to high-risk products, consumers collect and exchange information on the Internet. However, they do not buy online due to the lack of sufficient information to tangibilize services and a secure payment system (Jun, Vogt, & MacKay, 2010). Moreover, it must be added that business travellers use the internet less frequently than those who travel for leisure. This occurs for any and all types of travellers considered (Cobanoglu, Bill, & Patrick, 2001; Del Chiappa, 2011b; Kim & Kim, 2004). All that said, it should be noted that the current research suggests that, in reality, consumers mix online and offline information sources when searching for information and in their booking behaviour (Ho, Lin, & Chen, 2012). Hence, travel agencies seem to continue to play an important role in the marketing mix of a significant majority of hotels (Ku, Wu, & Lin, 2011), at least when they need to reach customers belonging to specific segments, and who are looking for complex, high-involvement products.

Despite the growing attention tourism literature has devoted to the analysis of the influence of the internet and UGC on tourists' decision making, surprisingly little research has attempted to gain an overall snapshot of the tourism information-search process that leads consumers to make hotel reservations through the internet. This results in a lack of knowledge about "when" consumers use UGC in their information-search, as well as poor analysis of how referral and conversion traffic from UGC moves to other key online and offline information sources. To the best of our knowledge, just one paper has been published on this topic on a sample of Italian consumers. This present study was therefore carried out with the aim of deepening the knowledge of the topic by examining the path followed by a sample of 825 Spanish tourists making hotel reservations.

BACKGROUND

The development of the internet and other ICTs applications introduced complexity to the tourism distribution system (Kracht & Wang, 2010), changed the structure of the value chain, and generated opportunities and threats for all tourism organisations and enterprises (Berne, García-González, & Múgica, 2012; Buhalis & Law, 2008; Tang & Jang, 2014). On the whole, it is evident that ICTs has been contributing to the growth of tourism and has been causing a radical re-organization of travel consumption patterns worldwide. However, undesirable impacts of ICTs and social media should not be neglected. As far as this latter point is concerned, we could refer, for example, to the fact that technologies have been creating a new cosmology of uncertainty and conflict, where a decline of oral communication could negatively affect the extent to which hotels are able to properly manage the no-shows rate in their accommodation (e.g. Korstanje & George, 2015). Nowadays, many hotel managers opt for overbooking as a strategy to absorb the no-shows cases. As Korstanje and George (2015) argued managers tend to estimate the percentage of no-shows with the support of ICTs; given that this kind of behaviour might lowers the trust that consumers have in the hotel, a high number of no show cases could occur thus suggesting that the advent and development of ICTs has been influencing social trust and has a huge influence on online consumer behaviour. One strategy that hotel marketers could adopt to reduce this kind of problem could be to prompt their customers to cancel the reservations if not really needed (Mohanty, 2008) without waiting for costumers to cancel their reservation last minute taking advantage of the free of charge cancellation policy, as usually happens when the reservation has been made using an OTA.

Internet, ICTs, social media, and in general any type of interactive technology, are changing the way tourists shape their connections to the urban space of places they visit producing a mixture of digital and physical geography of landscapes; nowadays, travellers' tales clearly reveal that blogging and flashpacking (i.e. packing digital devices in a back pack when travelling), are normal activities for people when travelling (Molz, 2012). Theoretically, the development of ICTs and the advent of augmented reality allow people to experience tourism and products and services even without travelling, thus questioning researchers and practitioners about the extent to which e-tourism could offer tourists with virtual experiences that could be accepted as substitutes for real visitation, thus entailing a significant alternation in the way tourism phenomenon occurs (Guttentag, 2010). Further, interactive travel changes sociality among individuals who can now be in touch with people who are far away even when on the move (Molz, 2012), thus completely changing the way togetherness can be produced and precluded. People are now connected in a physical and virtual mobile world where interactive travel conflates social, spatial and digital connectivity (Molz, 2012); in such a context sociality is no longer related to spatial proximities, rather it is related to business digital ecosystems (Baggio & Del Chiappa, 2014) where people and organization agents interact physically and virtually via technologies. Hence, social relations are far from declining; rather, they are evolving and adjusting to new patterns which are leading to more solipsistic models. This means that not only are people selecting their networks and relations but these new relations are switched on or off depending individual desires.

In current literature there exists a significant number of studies that highlight the overwhelming use of the internet for both searching for and booking hotel rooms (Lu & Stepchenkova, 2014). Fesenmaier, Xiang, Pan, and Law (2011), in their study on a sample of 2,508 Americans, showed that respondents considered:

- Search engines (65.4%),
- Tourism providers' websites (62%), and
- Online travel agencies (56%) as useful and essential tools in their travel planning.

A relatively recent study (Econsultancy, 2011) on a sample of 5,000 UK consumers showed that 71% of travel research takes place online, including mobile and tablet. According to the same survey, 71% were lookers (just searching online) and 56% were bookers (searching and booking online), thus suggesting that people are seeking reassurance via other channels before making the final decision, as well as the fact that travel research is not necessarily followed by an actual purchase behaviour. For example, Pizam, Jeong, Reichel, Van Boemmel, Lusson, Steynberg and Montmany's (2004) study on a sample of 1,429 students at 11 universities located in 11 different countries showed that 46.5% of respondents booked and purchased their trip using a travel agent, especially those respondents scoring low on risk-taking and sensation seeking.

According to PhoCusWright (2012a), the reasons why travellers book offline change very little year to year; the convenience of personal service is still the major driver for offline booking in Germany and the UK, whilst French tourists book offline both for personal service and simply out of habit. As far as the preference for a street travel agency is concerned, it should be noted that 20% of German consumers book offline by a travel agent, whilst in France and the UK this percentage is less than 15% (PhoCus-Wright, 2012a). According to prior studies around the influence of culture on organization and consumer behaviour (e.g. Hofstede, 1980), this suggests that differences still exist in the way tourists search for information and book their holiday based on culture of consumers (Del Chiappa, Alarcón-del-Amo, & Lorenzo-Romero, 2016; Nunkoo, Gursoy, & Ramkissoon, 2013; Lu & Chen, 2014) and need to be further investigated. For example, it has been argued that in countries which are characterized by collectivism culture (e.g. China), people tend to conform to group norms and their decisions are influenced by the opinion of others. As a matter of fact, in these types of countries word of mouth (both offline and online) influences consumer behaviour more that in countries characterized by a higher level of individualism such as Italy, the United Kingdom, and Spain (e.g. Michopoulou & Moisa, 2016). As far as this last country is concerned, in 2012, 67.9% of Spanish households had internet access (Instituto Nacional de Estadistica, 2012), and for 54.6% of online buyers holiday accommodation was the product/service most often purchased online. The tourism sector in Spain is leading all the sectors in terms of internet penetration, with 55.6% of the total e-commerce income coming from air transportation, travel agencies, tour operators, hotels, camping, and road transportation (Observatorio Nacional de las Telecomunicaciones y de la Sociedad de la Información, 2011). According to PhoCusWright (2012a), desktop web dominates during the destination selection phase; over 6 in 10 travellers reference websites on their desktops and laptops while searching where to go. The use of mobile devices, including smartphones and tablets, in search of the destination is, in percentage, small, but not entirely insignificant. French travellers are the most active in the use of smartphones (9%), while Germany and the UK are using mobile devices at a rate of around 8%. In the three markets considered in the research, tablets were used for destination searching in 2012 in the range of 6–7%. With only a few exceptions, all traditional or offline shopping channels attracted fewer shoppers during 2012.

More recently, several studies have been devoted to the rise of UGC and peer-to-peer applications and to the analysis of their influence on tourists' choices (Law et al., 2014) and on online room sales (Ye, Law, Gu, & Chen, 2011). Given the fact that tourism products are intangible goods (Lewis & Chambers, 2000) and are characterised by emotional, economic, and psychological involvement (Swarbrooke & Horner, 2007), UGC are in an ideal position to be considered by tourists as being more credible and trustworthy than market-provided information. As a consequence, they are more likely to drive and influence consumers' choices (Akehurst, 2009; Chung & Buhalis, 2008; Gretzel and Yoo, 2008; Herr, Kardes & Kim, 1991; Weiss, Lurie & MacInnis, 2008). As far as trustworthiness towards UGC is concerned, it is noteworthy that relatively recent research (e.g: Del Chiappa, 2011a; PhoCusWright, 2009; Yoo, Lee, Gretzel & Fesenmaier, 2009) has highlighted that tourists tend to trust differently UGC uploaded in different travel 2.0 applications, such as:

- Media and content syndication (RSSfeeds),
- Mashups,
- Tagging,
- Wikis,
- Web forums,
- Customer rating and evaluation systems,
- Podcasting,
- Blogs,

- Microblogging (Twitter, Pownce, Jaiku, Hictu, etc.),
- Photo sharing, and
- Video sharing (Akehurst, 2009; O'Connor, 2008; Schmalleger & Carson, 2008), with OTAs being considered the most credible and reliable information source.

This figure, coupled with the evidence that tourists' decision-making often starts by seeking information from OTAs (Anderson, 2011; Del Chiappa, 2012), confirms that OTAs play an important role in building hotel reputation (Inversini & Masiero, 2014) and, together with other social media platforms, in better positioning the official website of hotels in search engines (Xiang & Gretzel, 2010). Consequently, it is clear that a suitable development and adequate control of the online presence of the hotel on a UGC platform allows hotel managers to improve the success rate of their SEO (search engine optimisation), thus increasing the traffic on their website. According to data from PhoCusWright (2011), OTA's hotel review pages grew by 84% in 2010, and these online platforms were found the most popular place for travelers to post their reviews.

UGC has a great influence on service expectations (Mauri & Minazzi, 2013), in generating the idea of travelling, during the current planning process, and even after tourists' decisions have been made (Gretzel & Yoo, 2008; Law et al., 2014; Xiang, Magnini, & Fesenmaier, 2015), thus explaining the relevant impact that UGC has on online room sales (e.g., Ye et al., 2011). Indeed, many researches have shown the increasing influence that UGC and Travel 2.0 applications exert in stimulating travel, during the whole planning process (pre-planning phase, actual planning phase and post-travel phase) (Gretzel & Yoo, 2008). In particular, during the last phase tourists can use UGC to voice hotel service failures and complaints (Au, Law & Buhalis, 2010; Sparks & Browning, 2010). Moreover, sometimes UGC induces tourists to alter their decisions after obtaining further information online. This online interpersonal influence is referred to as electronic word-of-mouth (eWOM) and can be defined as "all informal communications directed at consumers through internet-based technology related to the usage or characteristics of particular goods and services, or their sellers" (Litvin, Goldsmith & Pan, 2008).

Specifically, according with eMarketer (2007), the percentages of travellers who changed their reservation based on the reviews posted online by other consumers are 25% and 33%, respectively, for infrequent and frequent leisure travellers. Similarly, Del Chiappa's (2011a) study on a sample of 823 Italian tourists reported respondents changing their hotel accommodations sometimes (64.8%), almost always (12%), or always (0.5%) based on UGC; the same study reported gender differences on this issue, with female respondents revising their bookings "always" or "almost always" more often (14.8%) than male respondents (9.4%). Recent research shows that UGC can sometimes even lead to tourists changing the accommodations already suggested by a travel agency (Del Chiappa, 2013; Prayag & Del Chiappa, 2014). Specifically, Del Chiappa et al.'s (2015) study on a sample of 986 Spanish consumers reported that 57.36% of them changed the accommodation suggested by a travel agency based on UGC. However, it should also be noted that the influence UGC has on tourists' choices differs based on:

- Age,
- Gender,
- Income, and
- Education level.

Users are more likely to be younger, college-educated, and employed full-time (Yoo & Gretzel, 2011); differences have also been found based on cultural variations (Volo, 2008) and type of travel (McCarthy, Stock, & Verma., 2010).

The future of high street travel agents is becoming highly uncertain because of an even higher number of consumers using the Internet to make their hotel reservations, to develop and purchase their own self-made itineraries (e.g., Law et al., 2014). As previously noted, this is further affected by travelers' modification of accommodation choices on the basis of UGC.

In 2007, about 75.2 million American online users used UGC applications, and almost half of European internet users made decisions about their travel plans using Travel 2.0 applications (eMarketer, 2007). Further, out of every three European tourists, two use the internet to upload their blogs and share reviews about their holidays with other people (Li & Bernoff, 2008). According to PhoCusWright (2011), the final choice of more than 40% tourists is influenced by UGC within their social networks. Skyscanner.com (2011) found that more than half (52%) of Facebook users are inspired to book a holiday in places that are somehow "advertised" by images posted by Facebook friends. Further, 46% of people were reported to have either organised or were invited on a trip through the same social networking site (TourismExchange, 2011). UK tourists continue to show great interest in the websites containing reviews: more than half (52%) visited one in 2012, against only 29% in France and 27% in Germany. Further, UK tourists were reported as being almost twice as likely to read reviews on dedicated review sites (72%) compared to French (35%) and German tourists (37%) (PhoCusWright, 2012a).

Amadeus (2011) reported that 71% of Spanish social media users make use of peer-to-peer travel applications for travel-related purposes. In such a scenario, OTAs and travellers review online websites, increasingly attracting consumers who are accustomed to buying online, with a consequent reduction in the use of general search engines (PhoCusWright, 2012b). In France, from 2011 to 2012, the percentage of those who use more than six websites in their research decreased (from 13% to 11%), as well as that of those using one or two websites (37–36%), while the percentage grew slightly for those who make use of three to five (51–53%). In Germany, however, the percentage of those who focus on one or two websites grew enormously (22–36%), while it decreased both for those that use three to five (59–53%) and, above all, that use more than six (from 20–13%). Finally, in the UK, the percentage of travellers who focus on one or two websites slightly increased (from 34–35%), grew a little more for those who make use of three to five (48–52%), and decreased for those who make more than six (18–13%) (PhoCusWright, 2012b).

Anderson (2011), based on a sample of 1,720 reservations for InterContinental Hotel Brands, reported that almost 75% of consumers who booked through the official website of Intercontinental had visited an OTA prior to making the reservation, with nearly 83% performing a search on Google, Yahoo, or Bing, and two-thirds conducting both a search and an OTA visit (Anderson, 2011). According to PhoCusWright (2011), for every 10 customers of an OTA, there are 7 that cross the information with the information on Facebook, while the same figure applies for Twitter accounts for about 25%, for buyers of both the OTA and hotel websites. Recent research reported how social media websites represent a considerable proportion of the online tourism domain within the context of planning trips using a search engine, even though there are considerable variations in the amount of social media displayed by Google (Xiang, 2011). Finally, previous studies showed that online social networks (OSNs) can influence the popularity and traffic of tourist operators' websites in different ways; for example, a study carried out in Italy shows a low level of influence exerted by Facebook and Twitter on popularity and traffic of tourist operators' websites (Milano, Baggio, & Piattelli, 2011).

With the aim of analysing referral and conversion traffic from user-generated content (UGC) to other key online/offline information sources, a recent study carried out on a sample of 623 Italian tourists (Del Chiappa, 2012) shows that the majority of respondents usually start their online booking by using OTAs (45.6%) or search engines (23.2%). Specifically, 45.6% of the respondents who start their research through OTAs proceed mostly by visiting the hotel's website (55.2%) or other travel 2.0 applications that are mainly represented by tourism related social networks (19.7%). Then, 69.4% of those visiting hotel websites at the second step of their search proceed via:

- Email (36.5%),
- Phone call (32.9%), or
- Other tourism-related social network (12.9%).

On the other hand, 19.7% of those visiting tourism-related social networks at the second step then:

- Visit the hotel's website (72.7%),
- Place a phone call (9.1%), or
- Visit photo-sharing websites (3%).

Of those starting their search through search engines (23.2%),

- 37.2% visit the hotel's website,
- 24.6% browse OTAs, and
- 8.7% visit tourism-related social networks.

Then, 22.6% of those visiting the hotel website at the second step of their search proceed via:

- Phone call (22.6%),
- Email (19.4%), or
- By visiting an OTA to search for comments and reviews and/or to look for more convenient rates.

There are also 12.9% of tourists who continue to look for other UGC in tourism-related social networks. Additionally, 24.6% of those visiting OTAs at the second step proceed to:

- Visit the official websites (50%),
- Other tourism-related social networks (19.2%), or
- Make a phone call to the hotel (11.5%).

STUDY METHODS

For the purpose of this study, a semi-structured questionnaire was developed and web administered exclusively to Spanish people aged more than 18 years. Specifically, an online survey using a snowball sampling technique was considered as the best choice to obtain a large sample of consumers who reside in different parts of Spain (thus hard to reach through face-to-face structured interviews), and to cope with the financial constraints we faced in managing the research project (Wrenn, Stevens, & Loudon, 2007). The initial subjects were generated from the list of 600 contacts owned by the research team. These 600 individuals, residing in different areas in Spain, received an e-mail inviting them to fill in the online questionnaire by clicking on a link provided in the e-mail. At the same time, they were invited to forward the survey to at least 7 of their friends, relatives and contacts (16 years of age or over). Using a specific tracking program, the forwarding behaviour of subjects in the original list was tracked. Tracking records indicated that 400 subjects in the original list forwarded the survey invitation. This process resulted in a total of 2,800 individuals receiving the e-mail invitation. After a three-week survey period (during November 2012), we obtained a total of 1,699 valid questionnaires, thus yielding a response rate of 24.26%.

The survey instrument comprised three sections. In the first part, respondents were asked to state if they had any previous experience with booking hotel rooms online. Only people who answered positively were allowed to complete the questionnaire. Respondents were then asked to give some general demographic information (gender, age, income, and education level). In the second part, respondents were asked to give us some information about the types of travel for which they usually make hotel reservations through the internet and by referring to UGC, the motivations (leisure or business), and the geographical scale (national and intercontinental). Finally, the researchers posed an open-ended question asking respondents to fully describe the way they usually search for information when making hotel reservations.

FINDINGS AND DISCUSSION

A total of 825 completed questionnaires were returned and available for data analysis. The analysis started by manually coding the final open answers that respondents gave to describe their information-search behaviour. This coding process considered the sequence by which each category of information (both online or offline) was cited (that is step 1, step 2, etc.). Finally, the researchers calculated the frequencies by which the different sources were found to be used along with the respondents' information-search using SPSS (version 21.0).

The sample (Table 2) was composed of 825 Spanish people who were mostly female (54.84%), under 25 years old (41.94%), and had a secondary school degree (35.28%). The monthly average household income was under 2,000 € for 58.17% of respondents (15.61% under 1,000 €, 42.56% from 1,001–2,000 €). When leisure trips were considered, respondents reported booking their accommodation online and using UGC mostly for domestic (87%) or European (60%) trips; for intercontinental trips this percentage dropped down to 26%. As far as business travellers are considered, the percentage of respondents using the internet and UGC is much lower, specifically 21% for domestic trips, 13% for European trips, and 6% for intercontinental trips. Table 3 shows that for the majority of the respondents, the online booking process begins through the OTAs (44.9%) or search engines (27.8%) (confirming the existence of the so-called billboard effect: Anderson, 2009; Del Chiappa, 2012), while 15.2% of them start the search from tourism-related social networks (such as Tripadvisor).

Further, Table 3 shows that the buying process starts online and often closes through direct contact with the hotel (phone or email) or through the traditional travel agencies, highlighting the importance of using a multi-channel strategy by the hotels. Further insights on tourists' booking behaviour can be obtained by focusing our attention on what happens once tourists start a search via search engines and

Table 2. The sample

Gender	
Male	45.16%
Female	54.84%
Age	
< 18	0%
18–25	41.94%
26–35	13.53%
36–45	15.92%
46–55	17.38%
56–65	7.91%
≥ 66	3.33%
Level of Education	
Primary School	8.95%
Middle school	42.77%
Secondary school	35.28%
University Degree	10.09%
Master's degree	-
Phd	-
No education	2.91%
Monthly Household Income (€)	
< 1.000	15.61%
1.001–2.000	42.56%
2001–3.000	22.06%
3.001–4.000	8.22%
4.001–5.000	4.68%
5.001–6.000	2.71%
6.001–7.000	0.62%
> 7.000	3.54%

OTAs, the information sources that influence the tourist information behaviour at the very early stage. On the whole, this confirms the overwhelming use of the Internet for both searching and booking hotel rooms (Lu & Stepchenkova, 2014), and the fact that

- Search engines (65.4%),
- Tourism providers' websites (62%), and
- Online travel agencies (56%) are useful and essential tools in their travel planning (Fesenmaier, Xiang, Pan, & Law, 2011).

Table 3. Buying process

	Media/Application	Step I	Step II	Step III	Step IV	Step V
1.	Search engines	27.8	1.3	0.4		
2.	OTAs	44.9	18.9	6.1	15	
3.	Website	5.1	39.3	17.1	10	12.5
4.	Tourism related social networks (TSN)	15.2	12.2	12.1		
5.	Travel agencies	4	4.4	5.4	3.3	
6.	E-mail	-	6	1.1		
7.	Phone call to the hotel	0.3	5.7	26.4	33.3	50
8.	Blogs	0.4	3.3	4.6	1.7	
9.	App	0.3				
10.	Website of destinations	0.4				
11.	Flash deals	1.5	0.7	0.4		

Further, results seem to confirm that tourists express higher levels of trustworthiness and credibility toward UGC uploaded over OTA than in tourism-related social networks (Del Chiappa, Lorenzo-Romero, & Alarcón-del-Amo, 2015); hence, consumers start to browse information using OTAs (Expedia, Booking.com, etc) and then use tourism related social networks to "reassure" and "redefine" their choice set.

Given the complexity and length of the visual representation that could arise from decomposing the online booking process step by step, for the purposes of the present study the attention will be focused on the first three steps, and for each of them only the most frequently cited information sources will be considered. As Figure 1 shows, 27.8% of respondents start the search through a search engine and then visit OTAs (49.7%) or the hotel's official website (33.1%). Then, those visiting OTAs at the second step book their accommodation via:

- A direct phone call to the hotel (24.6%),
- By visiting the hotel's website (26.2%), or
- Through a tourism-related social network (27.7%).

On the other hand, 32.4% of those who visit the official website at the second step proceed via:

- Website booking (32.4%),
- While 35.3% proceed via phone and
- 11.8% book via traditional travel agencies (Figure 1).

According to Figure 2, those who begin the search starting from the OTAs at the second step proceed by:

- Visiting the official website (42.5%) and then
- Book via the official website (35.9%) or
- By a phone call (29.7%), or
- Continue to seek information by visiting other tourism-related social networks (21%).

Figure 1. Online booking starting with search engines: how do tourists search for information?

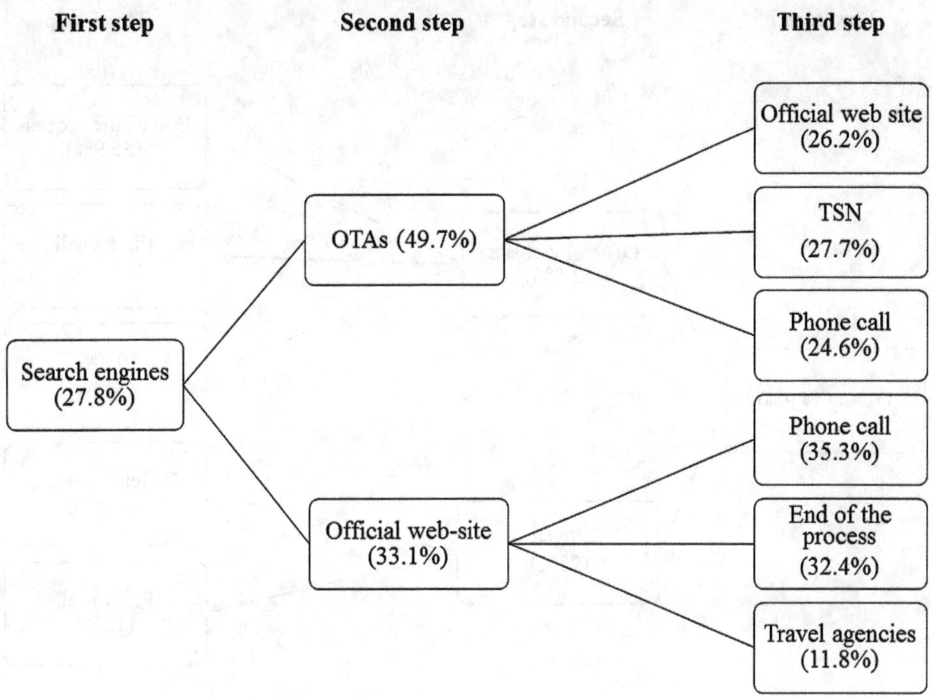

Then, those visiting tourism-related social networks at the second step (21%) proceed to book their final accommodation through the hotel's official website (37.5%) or via phone call (12%).

This confirms the existence of the so called "billboard effect", a term coined by Anderson (2009) to indicate the fact that consumers' decision making often starts by looking at Online Travel Agencies. Specifically, Anderson (2009) showed that traffic on a hotel's official website increased when the hotels are displayed in Expedia and highlighted the fact that the number of reservations made via their website, by phone or by email increased from 7.5% to 26%. This confirms that OTAs bring in hotel distribution and reach, which is particularly important for hotels of small and medium size that usually do not have enough budget or skills to effectively run search optimization activity (SEO).

CONCLUSION

This study, based on the answers given by a sample of 825 Spanish travellers, tries to explore tourists' buying decision processes regarding travel purchases and the role of UGC and traditional information sources. The study of tourists' buying decision processes is relevant both from academic and marketing points of view. In fact, a better knowledge of tourists' information-acquiring processes can enable marketers to better guide tourists' buying decisions.

Specifically, this chapter aimed at deepening knowledge about the information-search process that leads consumers to make hotel reservations through the Internet by analyzing the "when" consumers use UGC, as well as the referral and conversion traffic from peer-to-peer platforms to other online (i.e. hotel official web-site) and offline (phone, email, etc) information sources.

Figure 2. Online booking starting with OTAs: how do tourists search for information?

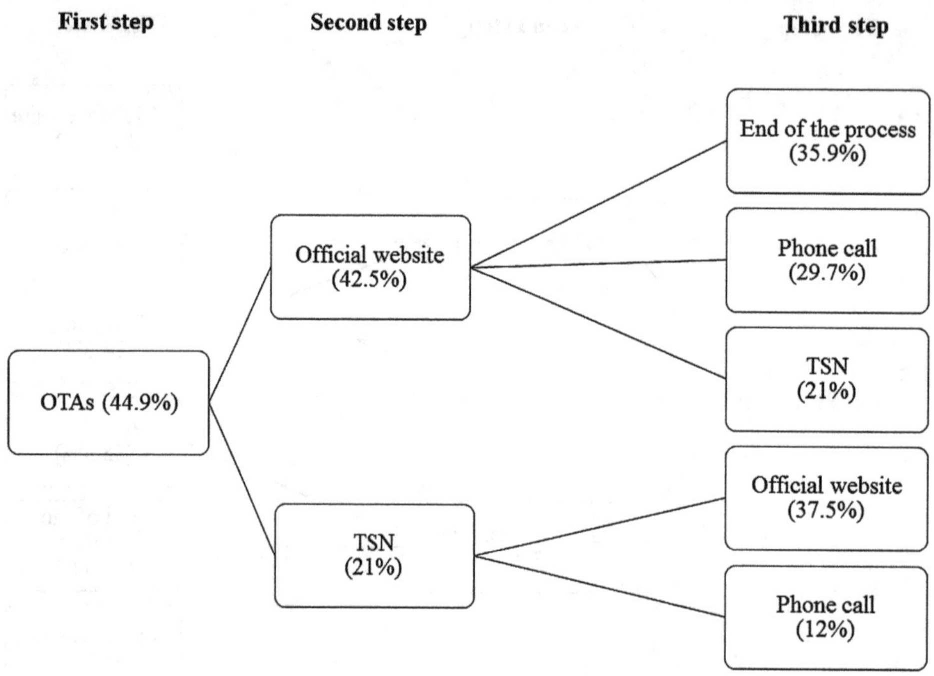

On the whole, our results show that UGC and "traditional" information sources (both online and off-line, such as emails, phone calls, official websites, etc.) play a relevant role in information-search and booking behaviour. In particular, they show that tourists are influenced by UGC uploaded in OTAs (e.g., Expedia, Booking.com) from the early stage of their information search, thus confirming that consumer decision-making frequently starts by looking at these sites (Anderson, 2009, 2011; Del Chiappa, 2012). This result suggests two main considerations. Firstly, it provides evidence of the fact that tourists trust UGC in OTAs more those uploaded in other Travel 2.0 applications (Del Chiappa, 2011a, 2011c; Del Chiappa, Lorenzo-Romero, & Alarcón-del-Amo, 2015; PhoCusWright, 2009; Yoo, et al., 2009). Secondly, it highlights that a hotel can be excluded by tourists from an early stage of the search on the basis of the UGC they read within OTA web pages. Hence, the less positive the posts about a given hotel on OTAs websites, the higher the likelihood that the same hotel will be not considered at all in the subsequent stages of the search. Conversely, the more positive the hotel rating/reviews posted on OTAs, the higher the likelihood that the same hotel will be considered as one of the potential accommodation options.

Our study also shows that UGC in OTA often acts as a referral platform for tourism-related social network (where people presumably go to read more comments and reviews about hotels in order to confirm or to redefine their choice set) or for the official web-site of the hotel (where consumers presumably go to search for further information about the hotel services/facilities and the surrounding area or for better rates); hence, in both cases consumers often use direct and offline channels (phone calls) to make their reservation.

These conclusions are relevant for both researchers and hospitality managers. From the academic point of view, they contribute to better understanding the tourist information-search process and behaviour. On the other hand, the findings provide relevant information for hospitality marketers. In particular, the findings suggest that hotel marketers should monitor and manage their corporate brand by concentrating their attention and resources on UGC within OTAs to better keep in contact whit tourists in the early stage of their information search processes. Further, hospitality marketers should consider the Internet as a complementary, rather than a substitute, source for information and booking (Beritelli, Bieger, & Laesser, 2007). At the same time, this study highlights the importance of effectively managing hotels' internet strategies (parity rate, effective booking engine, call-to-action, etc.) and customer service in order to convert as much traffic generated by OTAs (and other Travel 2.0 applications) or customer service as possible onto the website of the hotel. For this purpose, hotel marketers should provide extensive informational texts and photos relating to hotel services, assure that the reservation and client support systems on their website are working properly, and design their website in order to speed up the process of searching for information and booking as much as they can. Further, they could include virtual experiences and virtual tours in their website in order to be more attractive and effective in their promotion and/or to reduce consumers' travel anxiety by offering them psychological relief. Furthermore, with the aim of incentivizing consumers to book using hotel websites instead of travel agencies, hotel marketers should offer bonus loyalty points to repeaters who returning to hotel websites to book another stay; the same strategy could be used by a travel agency with the opposite aim. Broadly, hotel marketers should emphasise:

- Web usability,
- Security,
- Website functionality,
- Customer responsiveness and information quality, with information on reservations, facilities, and contacts being a crucial element of hotel website design.

LIMITATIONS AND FUTURE RESEARCH WORKS

Despite the theoretical and managerial contributions of the study, it does have limitations. Firstly, this study is country specific (e.g Spain) and it utilized a convenience sample derived from a web-based survey with a snowball sampling approach; further, like most online surveys, this study suffers from a coverage error. That said, it is evident that findings cannot be generalized. It would be interesting to repeat the study in other countries in order to analyze whether differences based on cultural background do exist and should be investigated via cross-cultural comparison. Secondly, this research did not investigate searching and booking behavior pertaining to groups, and it did not explicitly consider the effect that socio-demographic variables (such as gender, age, level of education, employment status, ecc.) or perceived risk and security exert on online tourist behavior (Del Chiappa, Lorenzo-Romero, & Alarcón-del-Amo, 2015; Chen, 2001); these aspects should be considered in the future. According to some pioneering research (Grønflaten, 2011), the issue of disintermediation should be analyzed both as a choice between two information sources (travel agents vs. service providers), and as a choice between two information channels (face-to-face vs. the Internet). Thus, researchers should consider more in detail the growing phenomenon of disintermediation in the online domain (hotel website versus online travel agencies; e.g., Liu & Zhang, 2014).

ACKNOWLEDGMENT

The authors would like to warmly thank the anonymous reviewers for their valuable help and suggestions in enhancing the quality of the manuscript.

REFERENCES

Abou-Shouk, M., Megicks, P., & Lim, W. M. (2013). Perceived benefits and e- commerce adoption by SME travel agents in developing countries: Evidence from Egypt. *Journal of Hospitality & Tourism Research (Washington, D.C.), 37*(4), 490–515. doi:10.1177/1096348012442544

Akehurst, G. (2009). User-generated content: The use of blogs for tourism organizations and tourism consumers. *Service Business, 3*(1), 51–61. doi:10.1007/s11628-008-0054-2

Amadeus. (2011). *The always-connected traveller: How mobile will transform the future of air travel.* Retrieved March 15, 2015 from http://www.amadeus.com/airlineit/the-always-connected-traveller/docs/amadeus-the-always-connected-traveller-2011-en.pdf

Anckar, B. (2003). Consumer intentions in terms of electronic travel distribution: Implications for future market structures. *e-Service Journal, 2*(2), 68–86. doi:10.2979/ESJ.2003.2.2.68

Anderson, C. K. (2009). The billboard effect: Online travel agent impact on non-OTA reservation volume. CHR Report. *The Center for Hospitality Research at Cornell University, 9*(16).

Anderson, C. K. (2011). Search, OTAs, and online booking: An expanded analysis of the billboard effect. CHR Report. *The Center for Hospitality Research at Cornell University, 11*(8).

Au, N., Law, R., & Buhalis, D. (2010). The impact of Culture on eComplaints: Evidence from Chinese Consumers in Hospitality Organizations. In U. Gretzel R. Law, & M Fuchs (Eds.), Information and Communication Technologies in Tourism 2010 (pp. 285-296). Vienna, Austria: Springer.

Baggio, R., & Del Chiappa, G. (2014). Real and virtual relationships in tourism digital ecosystems. *Information Technology & Tourism, 14*(1), 3–19. doi:10.1007/s40558-013-0001-5

Beldona, S., Morrison, A. M., & OLeary, J. (2005). Online shopping motivations and pleasure travel products: A correspondence analysis. *Tourism Management, 26*(4), 561–570. doi:10.1016/j.tourman.2004.03.008

Beldona, S., Racherla, P., & Mundhra, G. D. (2011). To buy or not to buy: Indian consumers choices of online versus offline channels for air travel purchase. *Journal of Hospitality Marketing and Management, 20*(8), 831–854. doi:10.1080/19368623.2011.577693

Bennet, M. M., & Buhalis, D. (2003). *Tourism and travel distribution, the travel agency perspective. Insights, D25-30.* English Tourism Council.

Bennett, M. M., & Lai, C. (2005). The impact of the internet on travel agencies in Taiwan. *Tourism and Hospitality Research, 6*(1), 8–23. doi:10.1057/palgrave.thr.6040041

Beritelli, P., Bieger, T., & Laesser, C. (2007). The impact of the Internet on information source portfolios: Insight from a mature market. *Journal of Travel & Tourism Marketing*, *22*(1), 63–80. doi:10.1300/J073v22n01_05

Berne, C., Garcia-Gonzalez, M., & Mugica, J. (2012). How ICT shifts the power balance of tourism distribution channels. *Tourism Management*, *33*(1), 205–214. doi:10.1016/j.tourman.2011.02.004

Buhalis, D., & Law, R. (2008). Progress in information technology and tourism management: 20 years on and 10 years after the Internet – The state of eTourism research. *Tourism Management*, *29*(4), 609–623. doi:10.1016/j.tourman.2008.01.005

Card, J., Chen, C., & Cole, S. (2003). Online travel product shopping: Differences between shoppers and non shoppers. *Journal of Travel Research*, *42*(2), 133–139. doi:10.1177/0047287503257490

Chen, S. L. (2001). Effects of value, affect, security, and web content on informa- tional and transactional usage of the Internet. *Asia Pacific Journal of Tourism Research*, *6*(81), 63–72. doi:10.1080/10941660108722089

Cheyne, J., Downes, M., & Legg, S. (2006). Travel agencies vs Internet: What influences travel consumer choices? *Journal of Vacation Marketing*, *12*(1), 41–57. doi:10.1177/1356766706059307

Chu, R. (2001). What online Hong Kong travelers look for on airline/travel website? *International Journal of Hospitality Management*, *20*(1), 95–100. doi:10.1016/S0278-4319(00)00046-3

Chung, J. Y., & Buhalis, D. (2008). Web 2.0: A study of online travel community. In P. O'Connor, W. Höpken and U. Gretzel (Eds.), Information and Communication Technologies in Tourism 2008 (pp. 70-81). Vienna, Austria: Springer.

Cobanoglu, C., Bill, W., & Patrick, J. M. (2001). A comparison of mail, fax and web-based survey methods. *International Journal of Market Research*, *43*(4), 441–452.

Del Chiappa, G. (2011a). Trustworthiness of Travel 2.0 applications and their influence on tourist behaviour: An empirical investigation in Italy. In *Proceedings of TTRA Europe and AFM Conference, Creativity and Innovation in Tourism*. doi:10.1007/978-3-7091-0503-0_27

Del Chiappa, G. (2011b). *Trustworthiness of Travel 2.0 applications and their influence on tourist behaviour*. Retrieved August 27, 2015, from http://tourismintelligence.ca/2011/06/17/trustworthiness-of-travel-2-0-applications-and-their-influence-on-tourist-behaviour/

Del Chiappa, G. (2012). User-generated content and its influence on tourists' choices: How do tourists make hotel reservations online? *Turistica: Italian Journal of Tourism*, *1*(1), 39–50.

Del Chiappa, G. (2013). Internet versus travel agencies: The perception of different groups of Italian online buyers. *Journal of Vacation Marketing*, *19*(1), 1–12.

Del Chiappa, G., Alarcón-del-Amo, M., & Lorenzo-Romero, C. (2016). Internet and user-generated content versus high-street travel agencies: A latent gold segmentation in the context of Italy. *Journal of Hospitality Marketing and Managament*, *25*(2), 197–217. doi:10.1080/19368623.2014.1001933

Del Chiappa, G., Lorenzo-Romero, C., & Alarcón-del-Amo, M. D. C. (2015). Profiling tourists based on their perceptions of the trustworthiness of different types of peer-to-peer applications. *Current Issues in Tourism*, 1–18. doi:10.1080/13683500.2015.1077208

Del Chiappa, G., Lorenzo-Romero, C., & Gallarza, M. (2014). Attitude toward disintermediation in hotel reservations: Spanish travellers profile. *European Journal of Tourism Research*, *9*(1), 129–143.

Econsultancy. (2011). *Three quarters of travel research takes place online: Survey*. Retrieved December 30, 2015, from http://econsultancy.com/uk/blog/8275-three-quarters-of-travel-research-takes-place-online-survey

Emarketer. (2007). *Niche sites invigorate online travel*. Retrieved April 11, 2015, from http://www.eMarketer.com

Fesenmaier, D. R., Xiang, Z., Pan, B., & Law, R. (2011). A framework of search engine use for travel planning. *Journal of Travel Research*, *50*(6), 587–601. doi:10.1177/0047287510385466

Gretzel, U., & Yoo, K. H. (2008). Use and impact of online travel reviews. In P. O'Connor, W. Höpken, & U. Gretzel (Eds.), *Information and communication technologies in tourism 2008 proceedings of the international conference in Innsbruck, Austria, 2008* (pp. 35–46). Vienna: Springer-Verlag. doi:10.1007/978-3-211-77280-5_4

Grønflaten, Ø. (2011). Predicting travelers choice of information sources and information channels. *Journal of Travel Research*, *48*(2), 230–244. doi:10.1177/0047287509332333

Guttentag, D. A. (2010). Virtual reality: Applications and implications for tourism. *Tourism Management*, *31*(5), 637–651. doi:10.1016/j.tourman.2009.07.003

Herr, P. M., Kardes, F. R., & Kim, J. (1991). Effects of word-of-mouth and product-attribute information on persuasion: An accessibility-diagnosticity perspective. *The Journal of Consumer Research*, *17*(4), 454–462. doi:10.1086/208570

Ho, C. I., Lin, M. H., & Chen, H. M. (2012). Web users behavioural patterns of tourism information search: From online to offline. *Tourism Management*, *33*(6), 1468–1482. doi:10.1016/j.tourman.2012.01.016

Hofstede, G. (1980). *Culture's Consequences*. Beverly Hills, CA: Sage Publications.

Huang, L. (2011). Social media as a new play in a marketing channel strategy: Evidence from Taiwan travel agencies blogs. *Asia Pacific Journal of Tourism Research*, *17*(6), 615–634. doi:10.1080/10941665.2011.635664

Instituto Nacional de Estadística. (2012). *Survey on equipment and use of information and communication technologies (ICT) in households year 2012*. Retrieved April 12, 2012, from http://www.ine.es/en/prensa/np738_en.pdf

International, I. P. K. (2011). *Global travel trends 2011/2012*. Retrieved September 20, 2013, from http://www.ipkinternational.com/uploads/media/IPK_PR_ITB_2012.pdf

Internet Word Stats. (2011). *Top 20 countries with the highest number of internet users*. Retrieved August 31, 2015, from http://www.internetworldstats.com/top20.htm

Internet Word Stats. (2013). *Top 20 countries with the highest number of internet users*. Retrieved December 31, 2014, from http://www.internetworldstats.com/top20.htm

Inversini, A., & Masiero, L. (2014). Selling rooms online: The use of social media and online travel agencies. *International Journal of Contemporary Hospitality Management*, *26*(2), 272–292. doi:10.1108/IJCHM-03-2013-0140

Ip, C., Lee, H. A., & Law, R. (2012). Profiling the users of travel websites for planning and online experience sharing. *Journal of Hospitality & Tourism Research (Washington, D.C.)*, *36*(3), 418–426. doi:10.1177/1096348010388663

Jun, S. H., Vogt, C. A., & MacKay, K. J. (2010). Online information search strategies: A focus on flights and accommodations. *Journal of Travel & Tourism Marketing*, *27*(8), 579–595. doi:10.1080/10548408.2010.507151

Kim, W. G., & Kim, D. J. (2004). Factors affecting online hotel reservation intention between online and non-online customers. *Hospital Management*, *23*(4), 381–395. doi:10.1016/j.ijhm.2004.02.001

Korstanje, M. E., & George, B. P. (2015). Understanding the Sociological Roots of'No-Shows' in Technology Mediated Transactions in Tourism. *E-Review of Tourism Research*, 11.

Kracht, J., & Wang, Y. (2010). Examining the tourism distribution channel: Evolution and transformation. *International Journal of Contemporary Hospitality Management*, *22*(5), 736–757. doi:10.1108/09596111011053837

Ku, E. C. S., Wu, W., & Lin, A. (2011). Strategic alignment leverage between hotels and companies: The buyer-supplier relationship perspective. *International Journal of Hospitality Management*, *30*(3), 735–745. doi:10.1016/j.ijhm.2010.10.006

Law, R., Buhalis, D., & Cobanoglu, C. (2014). Progress on information and communication technologies in hospitality and tourism. *International Journal of Contemporary Hospitality Management*, *26*(5), 727–750. doi:10.1108/IJCHM-08-2013-0367

Law, R., Leung, K., & Wong, J. (2004). The impact of the internet on travel agencies. *International Journal of Contemporary Hospitality Management*, *16*(2), 100–107. doi:10.1108/09596110410519982

Lewis, R. C., & Chambers, R. E. (2000). *Marketing leadership in hospitality, foundations and practices*. New York: Wiley.

Li, C., & Bernoff, J. (2008). *Groundswell: Winning in a world transformed by social technologies*. New York: Harvard Business Press.

Litvin, S. W., Goldsmith, R. E., & Pan, B. (2008). Electronic word-of-mouth in hospitality and tourism management. *Tourism Management*, *29*(3), 458–468. doi:10.1016/j.tourman.2007.05.011

Liu, J. N. K., & Zhang, E. Y. (2014). An investigation of factors affecting customer selection of online hotel booking channels. *International Journal of Hospitality Management*, *39*, 71–83. doi:10.1016/j.ijhm.2014.01.011

Lu, A. C. C., & Chen, B. T. (2014). Information search behavior of independent Guests' Complaints Online. *Journal of Hospitality Marketing & Management, 19*(7), 797–818.

Lu, W., & Stepchenkova, S. (2014). User-generated content as a research mode in tourism and hospitality applications: Topics, methods, and software. *Journal of Hospitality Marketing & Management.* doi:.10.1080/19368623.2014.907758

Mauri, A. G., & Minazzi, R. (2013). Web review influence on expectations and purchasing intentions of hotel potential customers. *International Journal of Hospitality Management, 34*(1), 99–107. doi:10.1016/j.ijhm.2013.02.012

McCarthy, L., Stock, D., & Verma, R. (2010). How travelers use online and social media channels to make hotel-choice decisions. *Cornell Hospitality Report, 10*(18), 4–18.

Michopoulou, E. E., & Moisa, D. (2016). The Role of Culture on Online Search Behaviour: A Comparative Study Between British and Chinese Travellers. In A. Inversini & R. Schegg (Eds.), Information and Communication Technologies in Tourism 2016 (pp. 765-777). Vienna: Springer International Publishing.

Milano, R., Baggio, R., & Piattelli, R. (2011, January). The effects of online social media on tourism websites. In R. Law, M. Fuchs, & F. Ricci (Eds.), *Information and Communication Technologies in Tourism 2011* (pp. 343–353). Vienna, Austria: Springer. doi:10.1007/978-3-7091-0503-0_38

Mohanty, P. (2008). *Hotel industry and Tourism in India.* APH Publishing.

Molz, J. G. (2012). *Travel connections: Tourism, technology, and togetherness in a mobile world.* New York: Routledge.

Nunkoo, R., Gursoy, D., & Ramkissoon, H. (2013). Developments in hospitality marketing and management: Social network analysis and research themes. *Journal of Hospitality Marketing & Management, 22*(3), 269–288. doi:10.1080/19368623.2013.753814

O'Connor, P. (2008). User-Generated Content and Travel: A Case Study on Tripadvisor.com. In P. O'Connor, W. Höpken & U. Gretzel (Eds.), Information and Communication Technologies in Tourism 2008 (pp. 47-58). Vienna, Austria: Springer.

Observatorio Nacional de las Telecomunicaciones y de la Sociedad de la Informaciốn. (2011). *Estudio sobre Comercio Electrónico B2C 2011 Edición 2012.* Retrieved March 12, 2013, from http://www.ontsi.red.es/ontsi/es/estudios-informes/estudio-b2c-2011-edición-2012

PhoCusWright. (2009). *Consumer Travel Report.* Report. Author.

PhoCusWright. (2011). *Social Media in Travel 2011: Traffic, Activity & Sentiment.* Report. Author.

PhoCusWright. (2012a). European Consumer Travel Report (3rd ed.). Report. Author.

PhoCusWright. (2012b). *Asia Pacific Online Travel Overview.* Report. Author.

Pizam, A., Jeong, G. H., Reichel, A., Van Boemmel, H., Lusson, J. M., Steynberg, L., & Montmany, N. (2004). The relationship between risk-taking, sensation-seeking, and the tourist behavior of young adults: A cross-cultural study. *Journal of Travel Research, 42*(3), 151–260. doi:10.1177/0047287503258837

Prayag, G., & Del Chiappa, G. (2014). Hotel disintermediation in France: Perceptions of students from Generation Y. *Anatolia – An International Journal of Tourism and Hospitality Research, 25*(3), 417-430.

Quan, D. C. (2002). The price of a reservation. *Cornell Hospitality Quarterly, 43*(3), 77–86. doi:10.1016/S0010-8804(02)80021-0

Schmallegger, D., & Carson, D. (2008). Blogs in tourism: Changing approaches to information exchange. *Journal of Vacation Marketing, 14*(2), 99-110.

Skyscanner.com. (2011). *Facebook factor inspires*. Retrieved April 22, 2014, from http://www.skyscanner.net/news/facebook-factor-inspires-52-book-holiday

Swarbrooke, J., & Horner, S. (2007). *Consumer behaviour in tourism*. Elsevier Ltd.

Tang, L., & Jang, S. (2014). Information value and destination image: Investigating the moderating role of processing fluency. *Journal of Hospitality Marketing & Management, 23*(7), 790–814. doi:10.1080/19368623.2014.883585

Tourism Exchange. (2011). *Friends' holiday pictures on Facebook inspires*. Retrieved August 31, 2014, from http://www.tourismexchange.com/exchange/en/newsroom/home/getArticle.jsp?articleID=13738

Volo, S. (2008). Bloggers reported tourist experiences: Their utility as a tourism data source and their effect on prospective tourists. *Journal of Vacation Marketing, 16*(4), 297–311. doi:10.1177/1356766710380884

Weiss, A. M., Lurie, N. H., & MacInnis, D. J. (2008). Listening to strangers: Whose responses are valuable, how valuable are they, and why? *JMR, Journal of Marketing Research, 45*(4), 425–436. doi:10.1509/jmkr.45.4.425

Wertener, H., & Klein, S. (1999). *Information technology and tourism: A challenging relationship*. Wien: Springer-Verlag. doi:10.1007/978-3-7091-6363-4

Werthner, H., & Ricci, F. (2004). E-commerce and tourism. *Communications of the ACM, 47*(12), 101–105. doi:10.1145/1035134.1035141

Wrenn, B., Stevens, R. E., & Loudon, D. L. (2007). *Marketing research: Text and cases*. Binghamton: Haworth Press.

Xiang, Z. (2011). *Dynamic social media in online travel information search: A preliminary analysis* (pp. 343–353). Vienna: Springer.

Xiang, Z., & Gretzel, U. (2010). Role of social media in online travel information search. *Tourism Management, 31*(2), 179–188. doi:10.1016/j.tourman.2009.02.016

Ye, Q., Law, R., Gu, B., & Chen, W. (2011). The influence of user-generated content on traveler behavior: An empirical investigation on the effects of e-word-of-mouth to hotel online bookings. *Computers in Human Behavior, 27*(2), 634–639. doi:10.1016/j.chb.2010.04.014

Yoo, K. H., & Gretzel, U. (2011). Influence of personality on travel-related consumer-generated media creation. *Computers in Human Behavior, 27*(2), 609–621. doi:10.1016/j.chb.2010.05.002

Yoo, K. H., Lee, Y., Gretzel, U., & Fesenmaier, D. R. (2009). Trust in travel-related consumer-generated media. In W. Höpken, U. Gretzel, & R. Law (Eds.), *Information and Communication Technologies in Tourism 2009* (pp. 49–59). Vienna, Austria: Springer. doi:10.1007/978-3-211-93971-0_5

ADDITIONAL READING

Gretzel, U., & Yoo, K. H. (2008). Use and impact of online travel reviews. In P. O'Connor, W. Höpken, & U. Gretzel (Eds.), Information and communication technologies in tourism 2008 proceedings of the international conference in Innsbruck, Austria, 2008 (35–46). Vienna, Austria: Springer-Verlag. doi:10.1007/978-3-211-77280-5_4

Gretzel, U., Yoo, K. H., & Purifoy, M. (2007). *Online travel review study: The role and impact of online travel reviews*. College Station, TX: Laboratory for Intelligent Systems in Tourism.

Guo, X., Ling, L., Dong, Y., & Liang, L. (2013). Optimal pricing strategy based on market segmentation for service products using online reservation systems: An application to hotel rooms. *International Journal of Hospitality Management*, *35*, 274–281. doi:10.1016/j.ijhm.2013.07.001

Gursoy, D. (2010). Chaotic changes in distribution channels: Implications for hospitality companies. *European Journal of Tourism. Hospitality and Recreation*, *1*(1), 126–137.

Gursoy, D., & Umbreit, T. (2004). Tourist information search behavior: Crosscultural comparison of European Union member states. *International Journal of Hospitality Management*, *23*(1), 55–70. doi:10.1016/j.ijhm.2003.07.004

Hernández, B., Jiménez, J., & Martín, M. J. (2011). Age, gender and income: Do they really moderate online shopping behavior? *Online Information Review*, *35*(1), 113–123. doi:10.1108/14684521111113614

Inversini, A., & Masiero, L. (2014). Selling rooms online: The use of social media and online travel agencies. *International Journal of Contemporary Hospitality Management*, *26*(2), 272–292. doi:10.1108/IJCHM-03-2013-0140

Law, R. (2009). Disintermediation of hotel reservations: The perception of different groups of online buyers in Hong Kong. *International Journal of Contemporary Hospitality Management*, *21*(6), 766–772. doi:10.1108/09596110910976007

Ng, E. (2009). An exploratory study on the use of information sources by consumers to select Australian regional travel agencies. *Journal of Hospitality Marketing and Management*, *18*(1), 26–43. doi:10.1080/19368620801988990

Pan, B., & Fesenmaier, D. R. (2006). Online information search: Vacation planning process. *Annals of Tourism Research*, *33*(3), 809–832. doi:10.1016/j.annals.2006.03.006

Pan, B., Zhang, L., & Law, R. (2013). The complex matter of online hotel choice. *Cornell Hospitality Quarterly*, *54*(1), 74–83. doi:10.1177/1938965512463264

Qi, S., Law, R., & Buhalis, D. (2013). Who booked five-star hotels in Macau? A study of hotel guests online booking intention. *Journal of Hospitality and Tourism Management, 20*, 76–83. doi:10.1016/j.jhtm.2013.05.002

Ramkissoon, H., & Nunkoo, R. (2012). More than just biological sex differences: Examining the structural relationship between gender identity and information search behavior. *Journal of Hospitality & Tourism Research (Washington, D.C.), 36*(2), 191–215. doi:10.1177/1096348010388662

Serra Cantallops, A., & Salvi, F. (2014). New consumer behavior: A review of research on eWOM and hotels. *International Journal of Hospitality Management, 36*, 41–51. doi:10.1016/j.ijhm.2013.08.007

Sigala, M., Lockwood, A., & Jones, P. (2001). Strategic implementation and IT: Gaining competitive advantage from the hotel reservations process. *International Journal of Contemporary Hospitality Management, 13*(7), 364–371. doi:10.1108/09596110110403956

Susskind, A. M., & Stefanone, M. A. (2010). Internet apprehensiveness. An examination of on-line information seeking and purchasing behavior. *Journal of Hospitality and Tourism Technology, 1*(1), 5–29. doi:10.1108/17579881011022990

Toh, R. S., Raven, P., & Dekay, F. (2011). Selling rooms: Hotel vs. third-party websites. *Cornell Hospitality Quarterly, 52*(2), 181–189. doi:10.1177/1938965511400409

Tsai, H., Huang, L., & Lin, C. (2005). Emerging e-commerce development model for Taiwanese travel agencies. *Tourism Management, 26*(5), 787–796. doi:10.1016/j.tourman.2004.04.009

Tse, T. S. M. (2013). The marketing role of the Internet in launching a hotel: The case of hotel ICON. *Journal of Hospitality Marketing and Management, 22*(8), 895–908. doi:10.1080/19368623.2013.734224

Volo, S. (2010). Bloggers reported tourist experiences: Their utility as a tourism data source and their effect on prospective tourists. *Journal of Vacation Marketing, 16*(4), 297–311. doi:10.1177/1356766710380884

Wen, I. (2009). Factors affecting the online travel buying decision: A review. *International Journal of Contemporary Hospitality Management, 21*(6), 752–765. doi:10.1108/09596110910975990

Xiang, Z., Magnini, V. P., & Fesenmaier, D. R. (2015). Information technology and consumer behavior in travel and tourism: Insights from travel planning using the internet. *Journal of Retailing and Consumer Services, 22*, 244–249. doi:10.1016/j.jretconser.2014.08.005

Yoon, S. J. (2002). The antecedents and consequences of trust in online-purchase decisions. *Journal of Interactive Marketing, 16*(2), 47–63. doi:10.1002/dir.10008

KEY TERMS AND DEFINITIONS

Disintermediation: The phenomenon that occurs when companies sell directly to their customers, for example trough their official website, rather than relying on tourism intermediaries.

E-Wom: All informal communications directed at consumers through internet-based technology related to the usage or characteristics of particular goods and services, or their sellers.

OTA: Websites offering comprehensive travel shopping and Reservations Solutions to consumers.

Peer-to-Peer Applications: Application architecture that partitions tasks or workloads between peers.

RSS Feeds: Standard web feed formats to publish frequently updated information online.

Snowball Sampling: Non-probability sampling technique where existing study subjects recruit future subjects from among their acquaintances.

UGC: Any form of content (comment, photo and/or video) created by users and uploaded in social peer-to-peer online platforms.

Chapter 8
Minimizing the Cost of Capital in Hotel Investments

Prashant Das
Ecole Hôtelière de Lausanne, HES-SO – University of Applied Sciences Western Switzerland, Switzerland

Gabrielle Bodenmann
Ecole Hôtelière de Lausanne, HES-SO – University of Applied Sciences Western Switzerland, Switzerland

ABSTRACT

In this book chapter, we introduce the readers to typical sources of hotel financing using a hypothetical case-study. First, we provide a commentary on various types of funding sources. We provide rationale for why a particular surplus unit specifies certain constraints to an (investment) manager. A discussion is offered on various factors that may lead to a certain mix of financing. We walk the readers through various steps of the optimization process. Finally, we provide a case study on optimizing the funding sources using the SOLVER function in MS Excel.

BACKGROUND

Investing in a hotel involves a big-ticket transaction. From a few million to over a billion US Dollars has been the range of hotel prices observed in recent years. The hotel investment community comprises of a large set of players: high net-worth individuals, boutique partnerships, REITs, pension funds, private equity funds and hotel brands to name a few. The amount of funding required in such projects is significant. Often, financing such assets becomes a complex business problem. Traditionally, a number of cash-surplus entities are selected to pool-in the capital necessary for the acquisition. Also, such investments tend to be highly leveraged. In other words, a large amount of debt capital is combined with equity to meet the amount required for acquisition.

Cash-surplus units (individuals, organizations, firms or governments) offer a broad range of financing alternatives to the investor with various maturity terms, conditions, priority for repayment and, of course costs. What a capital-surplus unit expects as return from her capital is a cost to the cash-deficit

DOI: 10.4018/978-1-5225-1054-3.ch008

unit which is raising the capital. A critical consideration is to minimize the overall cost of capital resulting from the return expectations of the surplus units. A non-optimal capital structure could lead to failure of an acquisition project whereas the acquisition with an optimal financing mix could increase the shareholder wealth. The investment management problem involves

1. Addressing a set of constraints imposed by the surplus units, and
2. Allocating the overall cash required towards various surplus units such that not only the constraints are met, but the weighted average cost of capital (WACC) is minimized as well.

Mathematically, solving the minimization problem requires knowledge of differential calculus. Most managers simplify the challenge by resorting to user-friendly, GUI-driven software such as SOLVER.

INTRODUCTION

In the discounted cash flow (DCF) method, an investment project is valued based on its cash flow generating potential. First, future cash flows (FV) are forecast. Second, for each future cash flow projection, an equivalent present value (PV) is calculated. The PV signifies how much cash the firm must invest today, at its expected rate of return (R), which will grow to the estimated FV in future. Finally net present value (NPV), the sum of all PV's in excess of the initial investment is calculated. If the NPV equals zero, the future cash flows exactly meet the expected return requirements. A positive NPV implies value-enhancement whereas a negative NPV is considered value-corroding. In the DCF method, a required rate of return is (R) estimated to "discount" the FV to a smaller, equivalent PV. Therefore, this rate is called the "discount rate[1]."

A firm can create value for its shareholders as long as the yield from a project exceeds costs. Therefore the discount rate, on average, should meet the return requirement of capital providers: creditors as well as owners. What the capital-providers expect as their "return" on the capital they finance is effectively the cost to the project. Therefore, the total R is also called "weighted average cost of capital," (WACC). Lower the WACC, higher the perceived value of a project, for a given set of cash flows[2]. This chapter describes various sources of capital available to hotel investment projects and how to structure the capital in order to minimize the WACC.

WACC is often based on an assumption that most hotel investors use a combination of debt or equity capital. Beyond the expected return ("cost of capital") of capital providers, the capital structure of the project (the proportion in which various types of capital are raised for a project) is critical to determining its WACC.

The expected return for any capital provider typically depends on current capital market conditions (e.g. stock returns and interest rates) and the capital provider's respective risk in the project. An investment manager signs different types of contract with different capital providers. Depending on the nature of contract, the same project may imply different risk-exposure to various capital providers. Thus, in order to estimate the cost of their capital, firms first need to determine the cost of each financing source they select to use. Second, they must identify the optimal financing mix of funding sources. A WACC which meets all constraints stemming from these contracts, but is minimized poses an optimization challenge.

MULTIVARIATE MINIMIZATION

Suppose, we are considering n sources of capital with their respective costs and allocations denoted as follows:

$$\begin{aligned} &Costs\ C = \begin{bmatrix} c_1 & c_2 & \dots & c_n \end{bmatrix} \\ &and\ allocations\ X = \begin{bmatrix} x_1 & x_2 & \dots & x_n \end{bmatrix} \\ &Then, WACC\ y = C * X' = c_1x_1 + c_2x_2 + \dots c_nx_n \end{aligned} \tag{1}$$

As the cost quoted by individual sources of capital are practically fixed, the investment manager tries various combinations of X, rendering x_i's as variable. Further, Equation (1) can be represented as $y = f(x_1, x_2, \dots, x_n)$ which is an equation of a surface.

Keeping other variables constant· the partial derivative of y with respect to x_i is (Hoy & Stengo, 2001)

$$f_{x_1} = \frac{\delta f}{\delta x_1} \tag{2}$$

It can be shown mathematically that

$$dy = \left[\frac{\delta y}{\delta x_1} + \frac{\delta y}{\delta x_2} + \dots + \frac{\delta y}{\delta x_n}\right] \begin{bmatrix} dx_1 \\ dx_2 \\ \vdots \\ dx_n \end{bmatrix} \tag{3}$$

$$PV_t = \frac{FV_t}{\left(1+R\right)^t};\ FV_t = CF_t$$

where CF signifies the cash flow at time t;

$$NPV_{t=0} = -Initial\ investment\ amount_{t=0} + \sum_{t=1}^{T} \frac{CF_t}{\left(1+R\right)^t};$$

where T denotes the life of the project[1]

The objective is

$$\text{MIN}\left[c_1x_1 + c_2x_2 + \ldots c_nx_n\right] \tag{4}$$

The first order condition is that a vector of small changes in capital allocation should not change the value of the WACC function:

$$\text{FOC: dy} = 0 \tag{5}$$

The second order condition is that the surface should be locally convex:

$$\text{SOC } d^2y \geq 0 \tag{6}$$

Further, for an optimal capital allocation, where the WACC is minimum, a specific set of X must be identified such that the FOC and SOC are met.

In a real world scenario, there are two major challenges to the mathematical approach described above. First, the surface may not be strictly convex and, therefore, may have multiple local minima. Second, beyond the FOC and SOC, the set of X may be subject to additional, business-specific constraints. These challenges make the minimization exercise computationally intense.

Computer algorithms such as the one used in SOLVER are based on brute-force iterative algorithm (Microsoft, n.d.-b). SOLVER tries several set of values for the variable cells (in our case, capital allocation) which must meet the constraints provided and observes the result (WACC) calculated. The algorithm performs "extensive analyses of the observed outputs and their rates of change as the inputs are varied to guide the selection of new trials." These analyses reduce the computation time and make the algorithm superior to naïve algorithms. Partial derivatives play a crucial role in SOLVER. However, such algorithms are still focused around finding the local minima (or maxima). As a result, the solution provided is sensitive to the initially specified allocation. According to Microsoft, "… (t)he only way to find the global optimum is to apply external knowledge of the problem…"

TWO MAJOR SOURCES OF CAPITAL: DEBT AND EQUITY

To finance an investment, a manager has a choice to use her firm's retained earnings or use external financing. Retained earnings are not costless. Since they signify equity holders' claim, they must generate the returns that equity holders expect. Besides, a firm may invite additional equity capital from new investors who will be offered ownership. All such funds generated from equity are technically a firm's own funds. In principle, equity providers are owners of the project themselves. Therefore, they have participation rights in the business unlike debt investors who, in theory, are passive capital providers. Equity capital providers are usually the last in the priority of claims on the cash flows. The manager has freedom to decide whether to distribute the cash flows (e.g. dividends) to equity holders or not (even if the project is cash surplus).

On the contrary, creditors are principally external capital providers. In most cases, companies use a combination of the two (i.e. debt and equity). However, there is a fundamental difference between debt and equity. Debt providers lend capital for a specific term and expect periodic interest mostly independent of the project performance. In the event of default, debt providers can often force the liquidation (selling) of the borrowers' assets to recover their dues. Thus, creditors undertake comparatively lower risk to lose their initial investment. Debt and equity capital are also not equally treated from the income-tax standpoint. Since interest payment to creditors is considered as an expense, the interest paid is often tax-deductible. However, dividends are often regarded as distribution of profit and do not enjoy tax deductibility at the firm's end.

From above discussions, it is clear that equity providers assume more risk than debt providers. As risk and returns are proportionate, equity providers will expect higher returns (in line with the higher risk undertaken) while debt providers will expect relatively modest returns.

Debt

A firm must first determine the type of debt that should be used for financing: secured or unsecured, fixed or variable rate, short term or long term, amortizing (positively or negatively) or interest-only and so on. Numerous contracting aspects such as the priority of claims, maturity term and source of the debt as well as various other clauses, must also be determined. Since hotels are considered relatively riskier compared to other commercial real estate asset classes (such as apartments or offices), stricter underwriting criteria are applied by creditors when it comes to the hotel industry (Elgonemy, 2002; Jang, Tang, & Chen, 2008; Weissenberg, Cohen, & Culliney, 2003). Due to several benefits (such as lower cost, tax deductibility of interest paid, etc.), it has been found that a firm has a tendency to maximize the debt financing despite increased the bankruptcy risk that debt financing poses.

Secured and Unsecured Debt

Secured debt, as the name suggests, offers some security to the creditor in the case of default. In a secured debt contract, the borrower usually offers some fixed asset (such as real estate) as a security to the loan (also known as "collateral"). In case of a loan default, the creditor may have a right to sell the collateral. Unsecured debt, on the other hand, is not tied to a collateral and the creditors only have a promise from the borrowers to repay the loan and pay interest additionally. Thus keeping all else the same, an unsecured loan will be riskier to the creditor than a secured loan. As a result, secured loans tend to have relatively lower cost of capital. Harper (2008) argues:

... the theory behind the requirement for secured lending goes straight to the heart of the theory of capitalism... (A) creditor who has the benefit of a formal valuation will be able to offer more competitive rates to their customers as the loan is effectively less expensive to them, and the savings can be passed to their customer.

Secured debt has several advantages. It has been argued that securing a loan with a collateral lowers lender's risk by either reducing the administration costs (in case of bankruptcy) or by making default a prohibitively expensive alternative for the borrower (Barro, 1976; Benjamin, 1978). Due to its lower cost of capital, secured debt allows a firm to invest in additional assets thus improving the firm value (Ooi, 2000). Besides, since secured creditors already enjoy some risk-protection (offered by the collateral), they tend to sign otherwise less restrictive contracts with the borrowers (Shah & Thakor, 1987).

The conventional wisdom suggests that a firm should borrow as much secured debt as possible since the resulting costs are lower than those of unsecured debt. However, this logic ignores the cost involved with issuing several types of secured debt instruments (such as secured bonds) (Scott, 1977). The associated costs can be significant enough to affect the decision of adopting this type of financing. For instance, the use of secured debt implies a loss of flexibility for the borrowers as the collateral cannot be disposed of freely, without renegotiating the debt (Stulz & Johnson, 1985). The collateral cannot be sold without the creditors' permission. A number of studies show that firms with more tangible assets use more secured debt in particular and overall debt in general (Bevan & Danbolt, 2004; Kale & Shahrur, 2007; Ooi, 2000; Titman & Wessels, 1988; Wald & Long, 2007; Zingales & Rajan, 1995). However, it is also found that smaller firms have a higher propensity to use secured debt.

Mezzanine

Among financiers, creditors supersede equity investors in the priority of claims: both in performing and non-performing loans. If the collateral is foreclosed upon, creditors will be paid off first. However, a project may use multiple creditors. Therefore, a priority of claims must be decided within creditors. In general, creditors who have higher priority are called "senior" lenders whereas those with lower priorities are termed as "junior" lenders. Mezzanine loans are riskier than senior loans and often have lowest priority of claims in the debt stack (Rubock, 2007). According to Linneman (2011): "mezzanine is any type of junior debt, whether holding the second position, the seventh position, or unsecured." Large hotel chains such as Starwood and Marriott are increasingly using mezzanine loans to undertake refurbishment and improvement projects as an incentive for longer management contract or franchise agreements (Weissenberg et al., 2003).

Mezzanine loans are often short-term (e.g. five years) and of interest-only type, (sometimes they may also be of amortizing type). Mezzanine creditors prefer lending to stabilized, existing property owners and avoid offering loans to properties under development or acquisition. Sometimes, mezzanine financing may be secured by the borrower's equity in other entities (Berman, 2005).

Equity

Beyond debt, a project could use funds from the owners. New capital could be raised from existing (or new) owners in addition to using the retained earnings. Such capital is called "equity." Thus, equity capital providers are technically the owners of the project who have to participate in the success (or failure) of their project. In other words, they may benefit from a well-performing project (continually or occasionally); but must also be willing to take a loss if the project does not perform. Such capital could be raised through several mechanisms, as described below.

Common Equity

Common equity is the most common type of equity capital. Common equity providers may expect dividend payments (income return) as well as growth in the value of their capital over time (appreciation return), or both. However, none of these returns is promised. On one hand, the absence of this requirement (to distribute dividends, in particular) offers flexibility to the project managers in terms of using the cash flows generated, while on the other hand, they create cash flow risk for the investors. Unlike debt capital, there is often no maturity period of equity capital. Higher risk in equity investments, therefore, makes the equity capital costly. Besides, equity investors may have substantial say in how the project is managed. This may potentially dilute the control in a project that managers enjoy.

Preferred Equity

Preferred equity providers usually expected a fixed "preferred dividend," although contingent on the availability of cash. Such investors enjoy priority over common equity investors in terms of cash flow distribution. Therefore, preferred equity exhibits some properties of equity and some of debt capital. In particular, the promise of preferred dividend makes it similar to debt. However, since preferred dividends are contingent on the availability of cash, they lead to uncertainty in the dividend stream rendering such investment similar to equity. Preferred dividend tends to be larger than the cost of debt; but lower than the cost of equity.

Many variations of preferred equity may be employed. "Participating" preferred equity holders may have a right to earn some common dividends, usually a percent of dividend distributed to common equity holders, beyond their preferred dividends. In "cumulative option" contracts, the promised preferred dividend which was not distributed due to lack of funds may be deferred such that the preferred-equity holders later receive the unpaid dividends before common shareholders when the cash is available. "Redeemable" preferred equity could be bought back by the project managers even before the project winds up. Otherwise, the invested capital could only be returned when the project is completed. "Convertible" preferred equity may have an option to be exchanged with common equity during certain time at a certain price.

Sweat Equity

The term "sweat equity" colloquially refers to the hard work, expertise and professional time invested by some partners in a project. Usually, the managing partner of a project (or occasionally some external experts) who need not offer hard capital but expect an equity claim in the project are referred to as sweat-equity holders. For example, a consultant may be hired as a "fee developer" of a new hotel development project. The fee-developer, after rendering her services may expect to become an equity-holder in the project, but may offer no capital at the outset. Although there is no formal method to estimate the cost of sweat equity, in successful projects such a stake may generate substantial cash flows even after all other parties have received their fair-share of the project cash flows.

CAPITAL STRUCTURE

As discussed earlier, beyond the respective costs of various types of capital raised, the mix of these types is crucial in determining the WACC. The overall allocation of the capital to these sources is called "capital structure." Roughly, the loan-to-value (LTV) ratio provides useful insights on the capital structure. LTV refers to the ratio of debt capital over the total capital employed in a project. Use of debt capital is also referred to as "leverage." It may appear that since debt is cheaper, higher LTV would essentially imply lower WACC. However, there are several theories surrounding this topic which confound this perception. Excessive leverage minimizes the shareholders' potential to generate cash flows from a project (as a large part of cash flow must mandatorily be distributed to creditors) and increases the bankruptcy risk.

The finance literature has long researched the optimum capital structure. Modigliani & Miller (1958) first claimed that the capital structure did not impact the firm's value (which is a function of WACC given a stream of cash flows) in a frictionless market, that is, a market without any transaction costs or taxes. However, their findings in a market with friction are sensibly different and show that the firm's value is maximized when it is entirely financed by debt, which is seen as the foundations of the trade-off theory (Jang et al., 2008; Modigliani & Miller, 1963). In fact, since Modigliani and Miller, different theories have arisen but none of them has proven to be universal. Myers's (2001) article on the subject delineates capital structure research into three leading theories, one of them being the trade-off theory, the two other ones being the pecking order, and the free cash flow theory. Market timing is another one of those theories that have been widely examined in the finance literature and that is described below (Harrison, Panasian, & Seiler, 2011).

Trade-off Theory

The trade-off theory states that the optimum capital structure is reached when the marginal benefits of holding debt (tax deductions) equal the marginal costs (increase of bankruptcy, financial distress and agency costs) and firm performance is maximized (Jang et al., 2008; Tang & Jang, 2007). In practice firms will set a target LTV ratio at which costs offset the benefits and progressively move towards it. This theory justifies moderate debt ratios by tax-paying firms (Myers, 2001). MacKie-Mason (1990) showed that companies with low marginal tax rates issued more equity than debt. However, no explanation is given on why profit-making companies with high tax rates do not always issue debt. The trade-off theory is therefore currently thought to be able to only explain a portion of the capital structure story (Dalbor & Upneja, 2004).

Pecking Order Theory

The second approach, the pecking order theory, is an alternative to the trade-off theory based on the assumption that there is information asymmetry among investors, which was first proposed by Myers & Majluf (1984). This theory states that firms try to avoid cost to issuing common stock. Since corporate management have more information than external investors, it is believed that new equity will only be issued if stocks are perceived as overvalued, leading to a decline in stock prices. Therefore, firms tend to have a predetermined pecking order when it comes to funding and prefer to use internal financing (retained earnings) before calling on external funds (i.e., firms prefer debt over equity). If they use external financing the pecking order is usually low-risk debt, then convertible securities and equity as last resort, because of the difficulty in its issuance along with the inability to deduct dividend payments, according to Dalbor & Upneja (2004): "In a study of 255 firms from the Fortune 1000 list, in which 71.2% of the respondents indicated that the pecking order theory best describes their approach to capital structure formation. 68% further verified the underpinning of the theory by suggesting that the main reason for an unfavorable market reaction to new equity issues is the negative signaling effect based on asymmetric information. The respondents also indicated that firms maintain financial slack by holding on to cash, marketable securities, and any unused debt capacity in order to avoid having to approach the equity markets. The same sort of thinking also affects dividend policy. In conclusion, the study suggests that companies are much more interested in avoiding the equity market than in using a textbook approach to a balanced capital structure (Block 2011)."

However, Nuri & Archer (2001) established that the amount of debt used in the UK lodging industry was more important than in the UK retail industry. According to them the trade-off theory was more consistent with the lodging and retail industry than the pecking order theory. (Block, 2011)

Market Timing Theory

The market timing theory has become a popular capital structure theory with the publication of Baker & Wurgler (2002). Under this theory, managers of firms will use the unique knowledge they have on the intrinsic value of their securities to choose the time of issuance, in order to maximize the shareholders' wealth. Thus, decision to issue equity depend on market performance (Korajczyk, Lucas, & McDonald, 1992). In a well performing market, if the managers perceive their stocks as overvalued they will issue equity to finance their projects. On the other hand, in a slow market they will prefer debt or wait for the mis-valuation of the stock price to be reduced sufficiently.

A positive relationship between issuance of equity by firms and the business cycle has been found by Choe, Masulis and Nanda (1993), Bayless and Chaplinsky (1996), and Baker and Wurgler (2000). The market timing theory, unlike the pecking order theory, assume that the market is not strongly efficient (i.e. new information in the market does not spread pervasively and quickly among all affected parties), allowing some firm to benefit from the information barriers. While the pecking order theory predicts equity issuances to be rare, Fama & French (2004) have found some evidences that equity issuances have been increasingly frequent and equity used in situations where internal funding or debt could have been preferred. Empirical evidence also supports the link between share price performance and equity issues decisions (Baker & Wurgler, 2002; Zingales & Rajan, 1995).

Free Cash Flow Theory

The free cash flow theory of Jensen (1986) is based on an assumption that the interests of the managers and stockholders are not aligned. Jensen argued that managers have an incentive to use available cash for perquisites. They have a propensity to increase the amount of assets under their control even if it requires undertaking poor projects or reduces the firm's value in general for this purpose. In fact, management might not want to pay dividends to the shareholders and prefer to see the funds kept within the firm (Opler, Pinkowitz, Stulz, & Williamson, 1999). As they have to invest the retained earnings they will choose the most profitable project available that will not always maximize the shareholders' wealth. This behavior depend highly on the availability of free cash flows, which can be tightened through debt financing. Agency problems might be reduced with higher debt levels. Jensen and Meckling (1976) argue that "higher levels of debt also allow to keep the insiders' fraction of equity high and thus improves their incentive to work in the interests of shareholders". The free cash flow theory is designed for mature firms that are prone to overinvest (Myers, 2001).

To summarize, none of these theories can be designed as the only approach managers adopt when it comes to financing decisions. In general, it has been found that the proportion of funds coming from internal or external sources varies over the business cycle (Ronald & Edgar, 2008). Therefore, investors usually have specific preference for the capital structure in a project which may vary with time, place and project type. For example, during healthy market conditions, firms rely more on external funds whereas during bad economic conditions, fewer investment opportunities are available. In particular, hospitality industry is sensitive to systematic risk which makes capital structure decisions important (Andrew

& Schmidgall, 1993). An optimal capital structure is critical to minimizing the cost of capital so as to maximize the shareholder wealth.

CONCLUSION

The value of a firm or a project could be maximized by improving its cash flows through operational interventions. Besides, using investment management skills, for a given set of future cash flows, the value could be further elevated by minimizing the discount rate (WACC). WACC is based on the cost of various sources of capital and how they are mixed. In fact, a manager has little control on the cost of a specific capital as it is primarily determined by the capital market. Besides, operational improvements are often not within the direct purview of investment managers. Therefore, minimizing the WACC is a critical task. In this chapter we illustrate various broad types of capital available to hotel investors. We discuss the characteristics of these types of capital and how they warrant different costs. Finally, we discuss the concept of an "optimal" capital structure that could lead to the minimum WACC.

The following pedagogic case illustrate a realistic hotel investment scenario in which various sources of capital are analyzed in order to minimize the WACC.

CASE STUDY: "JAMES DECIDES ON WACC"

James is a hotel developer. When a development opportunity is recognized, he offers his services as a fee developer. In this business model, he does not invest any cash. Rather, he invests his "sweat equity": the managerial expertise. From the project's cash flows, the claims for various debt and equity investors are paid out first. What he claims is the residual cash flow. Obviously, he likes to minimize the cost of capital.

In this particular project he is working with Key Hotels Group (KHG) that has commissioned James to study the feasibility of developing a new hotel ("NewHotel") in town for them. Total development costs, $200,000,000 will be raised at the beginning of the project and saved in the bank account of NewHotels LLC (NHLLC). The LLC has been able to sign conferences and businesses in advance such that the cash flow for the first year is projected to be well above $20 mi. After a survey of potential creditors and equity investors, James lists the finalists to allocate funding sources:

Personal Equity from KHG

KHG will be the General partner in NHLLC and is willing to invest any amount between $1 mi to $2 mi. Ms. Kirsten Hampton, the CEO of KHG says: "we expect that you will incorporate at least one million dollar in personal equity by KHG in the NewHotel development. We could even go up to two millions. However, we expect at least 18% in annual returns."

Secured Loan

A large nationalized bank which has been doing business with KHG for over a decade agrees to provide a long term, interest-only loan. However, they agree on a maximum loan-to-Cost ratio of 55%. In other words, the remaining 45% of the development cost must come from other sources. They offer an interest rate of 6%. The hotel is offered as a security to the loan.

Unsecured Loan

Assuming that the maximum secured loan and personal equity are exhausted, the project is still running short of funds. Among other options available, James selects a local pension fund to lend some interest- only unsecured debt, up to a maximum of $50 mi at 6.5% annual rate. However, since the debt is not secured by any collateral, the pension fund wants equity investors to take substantial risk. "An additional condition is that our loan amount will be maximum up to triple the equity investment coming from all sources except common equity", says the underwriter at the Pension fund "because we want to have those 'special' equity investors- KHG or preferred equity investors - to have some skin the game".

Preferred Equity

The above three sources are not sufficient. Therefore, James has struck an additional, preferred equity deal with a REIT. The REIT offers equity dollars up to a maximum of $35 mi, and expects a return of 7% annually.

Common Equity

The remaining funding could come from rather expensive, common equity investors who expect 10% return on such investments.

James decides that the first year WACC will be used as the discount rate for the DCF analysis. However, capital allocation to funding sources does not seem simple and is encumbered with numerous conditions. Assume that the creditors and investors will only be paid annually. Only the loan interests are tax deductible. What will be the optimal discount rate? (Hint: Start with equal allocation to all sources. Use solver.)

Model Solution to the Case Study

First we need to list each capital source, the cost associated with it and the conditions imposed, as shown in Figure 1. For simplicity, equal amounts are allocated to each financing source before we optimize the capital structure. The WACC of 9.50% is computed using the SUMPRODUCT function [Cell F11 =SUMPRODUCT(E3:E7;F3:F7)/200000000]. The sum of the allocations [=SUM(F3:F7)] is populated in cell F8.

The next step is to use SOLVER to determine the amount that will be allocated to each financing source (Figure 2). The SOLVER function[3] is available in "Data Menu" under "Data Analysis". The objective is to minimize ["Min"] the WACC [F11] by changing the capital allocations to various sources [F3 to F7].

Next, as shown in Figure 3, all the constraints shown in columns C and D (Figure 1) must be entered. Here are the steps to add the first condition:

Click "Add" and select the first allocation (cell F3) as the reference cell.
In the drop down list, select bigger or equal to ("> =")

Figure 1. List of sources of capital

	A	B	C	D	E	F	
1							
2		Source	Condition 1	Condition 2	Cost	Allocation	
3		Personal Equity	>= 1mi	<=2mi	18%	$	40'000'000
4		Secured Loan	LTC=55%		6%	$	40'000'000
5		Unsecured Loan	<=50 mi	<=3*(Personal+Preferred Equity)	6.50%	$	40'000'000
6		Preferred Equity	<=35 mi		7%	$	40'000'000
7		Common Equity	residual		10%	$	40'000'000
8		TOTAL	fixed			$	200'000'000
9							
10							
11					Before Tax WACC	9.50%	

Figure 2. Setting up the SOLVER objectives

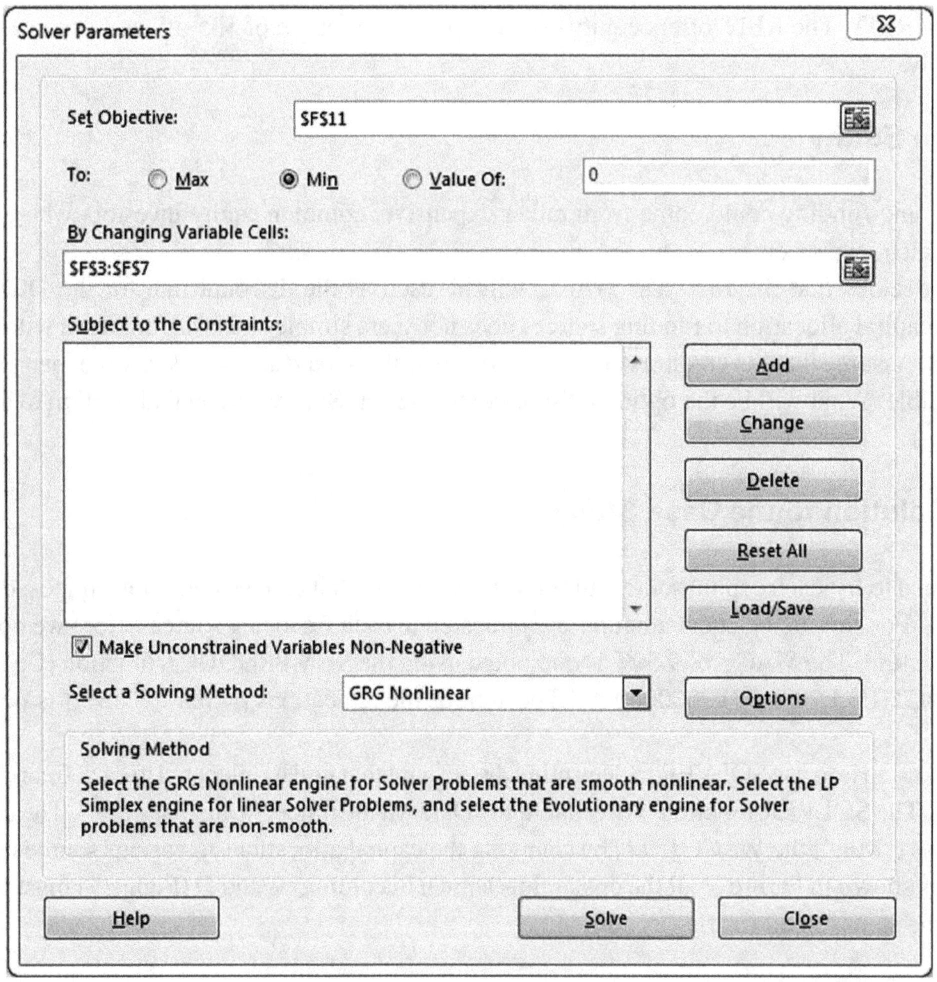

Figure 3. Adding a constraint

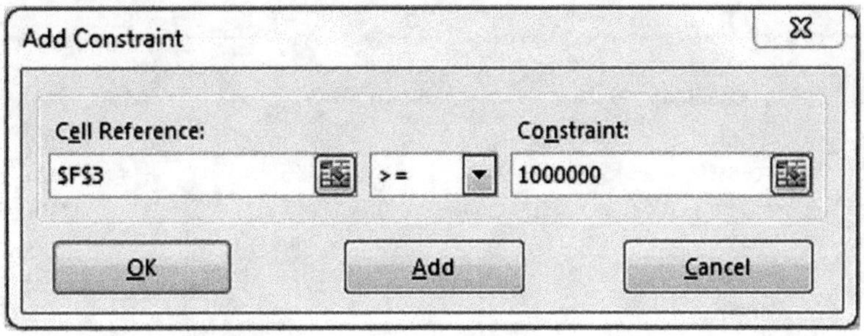

Figure 4. Setting up all the constraints

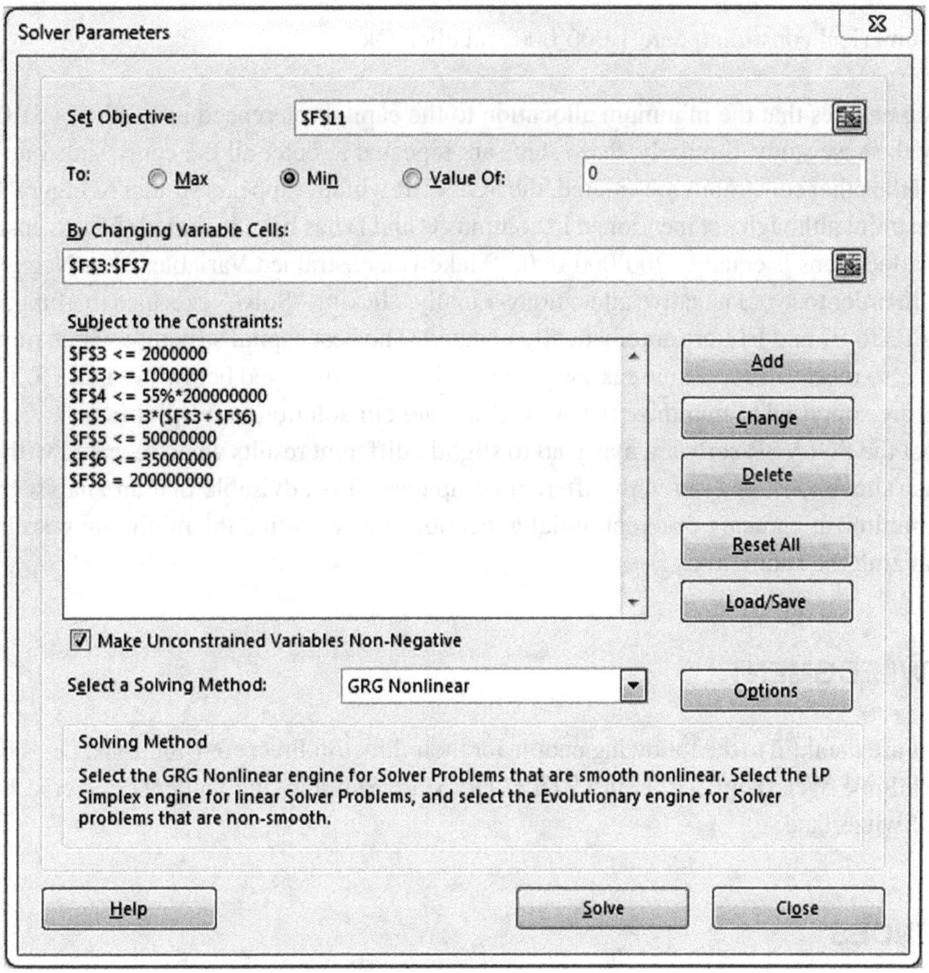

Figure 5. SOLVER-minimized WACC

Source	Condition 1	Condition 2	Cost	Allocation
Personal Equity	>= 1mi	<=2mi	18%	$ 1'000'000
Secured Loan	LTC=55%		6%	$ 110'000'000
Unsecured Loan	<=50 mi	<=3*(Personal+Preferred Equity)	6.50%	$ 50'000'000
Preferred Equity	<=35 mi		7%	$ 35'000'000
Common Equity	residual		10%	$ 4'000'000
TOTAL	fixed			$ 200'000'000
			Before Tax WACC	6.44%

Enter the numerical constraint, here 1'000'000 and click Ok

This step ensures that the minimum allocation to the capital referenced in cell F3 is $1,000,000, as dictated by the case study. Similarly, these steps are repeated to enter all the constraints one-by-one.

Once all the other constraints are entered, the SOLVER window appears similar to Figure 4. Note that the last constraint although not mentioned in columns C and D has been included (F8), to ensure that the sum of all allocations is equal to 200'000'000. "Make Unconstrained Variables Non-Negative" option is checked in order to avoid negative allocations. Finally, clicking "Solve" executes the function. Values in the cells F3 to F7 and F11 are automatically updated. The best capital structure which minimizes the WACC but also meets all conditions, as suggested by Excel is displayed below in Figure 5. The updated allocations are supposed to minimize the WACC and are our solution to the problem.

Note that the SOLVER software may lead to slightly different results when repeated with a different set of initial allocations, or even with different computers. It is advisable that an analyst tries several SOLVER minimization using different initial allocations (so as to find the minimum possible WACC) before finalizing the solution.

ACKNOWLEDGMENT

The authors are thankful to the following people for their direct/indirect contributions: Dr. Ramya Aroul, Dr. Jean-Philippe Weisskopf, Dr. Laura Zizka, Dr. Marc Steirand and students at Ecole hotelière de Lausanne, Switzerland.

REFERENCES

Andrew, P., & Schmidgall, R. (1993). *Financial Management for the Hospitality Industry*. Washington, DC: AH&MA.

Baker, M., & Wurgler, J. (2000). The equity share in new issues and aggregate stock returns. *The Journal of Finance*, *55*(5), 2219–2257. doi:10.1111/0022-1082.00285

Baker, M., & Wurgler, J. (2002). Market timing and capital structure. *The Journal of Finance*, *57*(1), 1–32. doi:10.1111/1540-6261.00414

Barro, R. J. (1976). The Loan Market, Collateral, and Rates of Interest. *Journal of Money, Credit and Banking*, *8*(4), 439–456. doi:10.2307/1991690

Bayless, M., & Chaplinsky, S. (1996). Is there a Window of Opportunity for Seasoned Equity Issuance? *The Journal of Finance*, *51*(1), 253–278. doi:10.1111/j.1540-6261.1996.tb05209.x

Benjamin, D. K. (1978). The use of collateral to enforce debt contracts. *Economic Inquiry*, *16*(3), 333–359. doi:10.1111/j.1465-7295.1978.tb00507.x

Berman, A. R. (2005). "Once a Mortgage, Always a Mortgage" - The Use (and Misuse of) Mezzanine Loans and Preferred Equity Investments. *Stanford Journal of Law. Business & Finance*, *11*, 76.

Bevan, A. A., & Danbolt, J. (2004). Testing for Inconsistencies in the Estimation of UK Capital Structure Determinants. *Applied Financial Economics*, *14*(1), 55–66. doi:10.1080/0960310042000164220

Block, S. (2011). Does the Weighted Average Cost of Capital Describe the Real-World Approach to the Discount Rate? *The Engineering Economist*, *56*(2), 170–180. doi:10.1080/0013791X.2011.573618

Choe, H., Masulis, R. W., & Nanda, V. (1993). Common stock offerings across the business cycle: Theory and evidence. *Journal of Empirical Finance*, *1*(1), 3–31. doi:10.1016/0927-5398(93)90003-A

Dalbor, M. C., & Upneja, A. (2004). The Investment Opportunity Set and the Long-Term Debt Decision of U.S. Lodging Firms. *Journal of Hospitality & Tourism Research (Washington, D.C.)*, *28*(3), 346–355. doi:10.1177/1096348004265024

Elgonemy, A. R. (2002). Debt-financing Alternatives: Refinancing and Restructuring in the Lodging Industry. *The Cornell Hotel and Restaurant Administration Quarterly*, *43*, 7–21.

Fama, E. F., & French, K. R. (2004). The Capital Asset Pricing Model: Theory and Evidence. *The Journal of Economic Perspectives*, *18*(3), 25–40. doi:10.1257/0895330042162430

Harper, D. (2008). *Valuation of Hotels for Investors*. Estates Gazette.

Harrison, D. M., Panasian, C. A., & Seiler, M. J. (2011). Further Evidence on the Capital Structure of REITs. *Real Estate Economics*, *39*(1), 133–166. doi:10.1111/j.1540-6229.2010.00289.x

Jang, S., Tang, C. H., & Chen, M. H. (2008). Financing behaviors of hotel companies. *International Journal of Hospitality Management*, *27*, 478–487. doi:10.1016/j.ijhm.2007.08.010

Jensen, B. M. C. (1986). Agency Costs of Free Cash Flow, Corporate Finance, and Takeovers. *The American Economic Review*, *76*, 323–329.

Kale, J. R., & Shahrur, H. (2007). Corporate capital structure and the characteristics of suppliers and customers. *Journal of Financial Economics*, *83*(2), 321–365. doi:10.1016/j.jfineco.2005.12.007

Korajczyk, R. A., Lucas, D. J., & McDonald, R. L. (1992). Equity issues with time-varying asymmetric information. *Journal of Financial and Quantitative Analysis*, *27*(3), 397–417. doi:10.2307/2331327

Linneman, P. (2011). Real Estate Finance and Investments - Risk and Opportunities. Linneman Associates, Ed.

MacKie-Mason, J. (1990). Do taxes affect corporate financing decisions? *The Journal of Finance*, *45*(5), 1471–1493. doi:10.1111/j.1540-6261.1990.tb03724.x

Microsoft. (n.d.a). *Load the Solver Add-in*. Retrieved July 8, 2016, from https://support.office.com/en-us/article/Load-the-Solver-Add-in-612926fc-d53b-46b4-872c-e24772f078ca

Microsoft. (n.d.b). *Solver Uses Generalized Reduced Gradient Algorithm*. Retrieved May 5, 2016, from https://support.microsoft.com/en-us/kb/82890

Modigliani, F., & Miller, M. (1958). The cost of capital, corporation finance, and the theory of investment. *The American Economic Review*, *48*(6), 261–297.

Modigliani, F., & Miller, M. H. (1963). Corporate income taxes and the cost of capital — A correction. *The American Economic Review*, *53*(3), 433–443.

Myers, S. C. (2001). Capital Structure. *The Journal of Economic Perspectives*, *15*(2), 81–102. doi:10.1257/jep.15.2.81

Myers, S. C., & Majluf, N. (1984). Corporate Financing and Investment Decisions when Firms have Information that Investors do not Have. *Journal of Financial Economics*, *13*(2), 187–221. doi:10.1016/0304-405X(84)90023-0

Nuri, J., & Archer, S. (2001). Target adjustment model against pecking order model of capital structure. In *European Financial Management Association Annual Meeting*.

Ooi, J. T. L. (2000). An empirical investigation on the incidence of secured debt. *Journal of Property Research*, *17*(3), 185–201. doi:10.1080/09599910050119976

Opler, T., Pinkowitz, L., Stulz, R., & Williamson, R. (1999). The determinants and implications of corporate cash holdings. *Journal of Financial Economics*, *52*(1), 3–46. doi:10.1016/S0304-405X(99)00003-3

Ronald, W. M., & Edgar, A. N. (2008). *Introduction to Finance; Markets, Investments, and Financial Management*. Thirteenth.

Rubock, D. B. (2007). *US CMBS and CRE CDO : Moody' s Approach to Rating Commercial Real Estate Mezzanine Loans*. Academic Press.

Scott, J. H. Jr. (1977). Bankruptcy, Secured Debt, and Optimal Capital Structure. *The Journal of Finance*, *32*(1), 1–19. doi:10.1111/j.1540-6261.1977.tb03237.x

Shah, S., & Thakor, A. V. (1987). Optimal capital structure and project financing. *Journal of Economic Theory*, *42*(2), 209–243. doi:10.1016/0022-0531(87)90086-X

Stulz, R., & Johnson, H. (1985). An analysis of secured debt. *Journal of Financial Economics*, *14*(4), 501–521. doi:10.1016/0304-405X(85)90024-8

Tang, C. H., & Jang, S. (2007). Revisit to the determinants of capital structure: A comparison between lodging firms and software firms. *International Journal of Hospitality Management*, *26*(1), 175–187. doi:10.1016/j.ijhm.2005.08.002

Titman, S., & Wessels, R. (1988). The Determinants of Capital Structure Choice. *The Journal of Finance*, *43*(1), 1–19. doi:10.1111/j.1540-6261.1988.tb02585.x

Wald, J. K., & Long, M. S. (2007). The effect of state laws on capital structure. *Journal of Financial Economics*, *83*(2), 297–319. doi:10.1016/j.jfineco.2005.10.008

Weissenberg, A. F., Cohen, J., & Culliney, P. (2003). Mezzanine financing for hospitality property owners. *Journal of Retail & Leisure Property*, *3*(2), 115–121. doi:10.1057/palgrave.rlp.5090168

Zingales, L., & Rajan, R. (1995). What Do We Know about Capital Structure? Some Evidence from International Data. *The Journal of Finance*, *50*(5), 1421–1460. doi:10.1111/j.1540-6261.1995.tb05184.x

ENDNOTES

1 If you don't have the Solver command, please refer to the support website of Microsoft, (n.d.-a)

Chapter 9
Introduction to Tourism Security:
Tourism in the Age of Terrorism

Maximiliano Emanuel Korstanje
University of Palermo, Italy

ABSTRACT

The present essay review explores the problem of terrorism and security in tourism fields. Certainly, plans and policies provided by guide-books are not being followed in disaster-contests simply chaos and disorder are the nature of emergencies. Beyond any protocol, crises and security are not properly defined by scholars. In this essay-review, we will not pay attention to define what tourism security means, lest by the lens of three senior scholars, Sevil Somnez, Abraham Pizam and Peter Tarlow, who have accomplished this task. They have explored not only the roots of terrorism but security over 20 years. Despite the criticism, they deserve recognition for this legacy. Based on substantial point of divergence, these specialists are concerned by the financial dependency of societies respecting to mass-media and its coverage of terrorist attacks.

INTRODUCTION

After 9/11 the problem of security concerned policy makers of tourism and hospitality industries as never before. The importance in discussing terrorism as a main threat for Western countries posed as a necessary point of agenda to strengthen the security of international tourist destination. At some extent, one of the most affected industries was tourism and hospitalities (Ryan 1993; Tarlow, 2006; Bianchi 2007; Reichel, Fuchs & Uriely 2007). The process of globalization triggered the intersection of democracy with others radical thinking in which case resentment is rechanneled by terrorists to accomplish their private-laden goals. Not only the production of terrorism-related studies triplicated after World Trade Center attacks, but also terrorism situated as a buzz-word for tourism scholars. This essay review attempts to discuss the legacy and trajectory of three senior lecturers who have focused their insight in the problem of terrorism

DOI: 10.4018/978-1-5225-1054-3.ch009

and its effects on tourism. For Sevil Somnez, the Luxor Massacre was the cornerstone of her studies. The tragic events of 9/11 were very far from her foci. Rather, Abraham (abe) Pizam alternates diverse disciplinary methods to understand the tourist security. In this token, Rabi Peter Tarlow acknowledges that security is a matter of further complexity than terrorism. The question whether terrorism follows a religious nature is fine for Pizam and Somnez, but not for Tarlow. He goes on the opposite direction. Terrorism only gains strength according to the degree of terror it can generate. This sentiment of panic can be exploited by the imposition of violence and brutality to the extent the "Other" is objectified. At some extent, terrorists are indifferent to human suffering. Religion has nothing to do with terrorism, Tarlow adds. With strengths and weaknesses, each one gives a clear diagnosis about the role of state not only in homeland security but also in the protection of tourists. The experts agree that the success of "international terrorism" to cause political instability in western states is based on two main points. First and foremost, local economies have developed a strong dependency from tourism industry. The second point of entry in this discussion is that this dependency, conjoined to the coverage terrorism received from the media, jeopardizes the image of tourist destination cause serious losses in the periphery. Terrorist not only appeals to kill "innocent travelers" to shock western societies, these acts cause direct damage whenever the international demand declines. Though it is a conceptual chapter, its importance is given by the fact it provides readers with a fresh viewpoint of what has been written in terrorism as well as the leading voices of Pizam, Somnez and Tarlow respecting to how tourism scholars understand this deep-seated issue. It is unfortunate that marketing and management are not doing the correct thing in preventing terrorist attacks because there is some point of misunderstanding respecting to what terrorism is. The fact is that policy makers and practitioners of tourism industry would never focus on the optimization of security issues without understanding conceptually against what they struggle. The current extent literature is based on the belief "terrorists are hatred-filled maniacs who do not tolerate the liberal ideals brought by modern tourists, or the tenets of democracy which are revitalized by the industry of hospitality, in other cases, some voices claimed that terrorism exhibits an incompatibility of religious values between Christendom and Islam. In perspective, as Tarlow explains terrorism has nothing to do with religious life. Rather, oft-used tactics of terrorists remind us that extortion is their primary cultural value (Tarlow, 2014). Many of the attacks planned in Middle East against tourist targets are perpetrated by Muslims against other Muslims. One might speculate that the problem of security is not limited to terrorism alone, but today this seems to be the most urgent issue to address. During cold-war cells as ETA and JAMAS planned attacks in urban cities but selecting important chief police officers, or politicians. Nowadays, all we are the targets… innocent tourists, travelers or journalists became in targets of crueler and virulent groups which appeal to terror to captivate the attention of state. Unlike other times where attacks rattled the periphery, after 9/11 cells have been arrived to the core of most popular cities as NY, Madrid, London, Paris and now Brussels. Since terrorism has mutated to a new shape, there is an urgent necessity to update the literature.

BACKGROUND

The scourge of terrorism is a very hard issue to grasp. Particularly, in this section we will limit our review to the main theories that focused on the roots of terrorism and its impact on tourism. These families of theories focused on different aspects of terrorism:

1. Religion,
2. Democracy, and
3. Production.

Following the legacy of Samuel Hungtinton (1997), some voices have alarmed on the religious incompatibility of Islam and Christendom as the main factor to promote an anti-western sentiment which today is what terrorism feeds back. Even, the attacks to tourist destinations not only evinces these incompatibilities but also the clash of different cultural values. One of the exponents in this direction, Boris Vukonic argues that the problems of religiosity does not rest in how secular tourism is, but in the discrepancies of values which may affect locals. What is in discussion in tourism is the practices of visitors when these acts may aggravate long-simmering resentment (Vukonic, 2002). Henderson (2003) adds that the encounter of tourism in Middle age opens the doors for questions but not for that less opportunities to enhance the understanding with the Other. In her study case Malaysia, shows that the discrepancy is enlarged in the moment the tourist demands are exacerbated over local habits. In this stage, governments play a vital role harmonizing interests of different stakeholders. In this vein, Erik Cohen explores the case of Israel a risky destination where the three Abrahamic faith coexist. He distinguishes micro from macro levels. While the former is negatively affected by terrorism the latter one keeps a constant growth which attracts visitors from all the world. For example, US Jewish segment are more insensitive to terrorism than other tourists. It suggests that religion would be a factor of integration and cohesiveness even in risky places (Cohen, 2014). Limitations of the family of religion was reviewed by Peter Tarlow (2014) who reconsider religion not as the real reason for terrorism, but as a discourse or pretext to vulnerate innocents in order for achieving the self-goals. In effect, the problem of terrorism for other scholars is intended to much deeper inefficiency to control internal violence. Therefore, the second family of studies delves into the position of democracy (and even tourism) as catalysts to more egalitarian societies. Countries where democracy did not take place, have further probabilities to develop "extractive institutions", which monopolize not only the wealth but also political power. Pressed to live in peripheral lands, whole portion of population is subject to prolonged states of famine and starvation (Pizam, 2002). Since financial loans poses as necessary measure to keep these institutions working, states appeal to tourism as a mechanism towards development. Far from achieving the original goals, some sectors of society see in tourism a moral danger which should be eradicated. In contexts which are hostile to democracy, not surprisingly, tourists are hosted, and even killed (Crouch & Ritchie 1999;Kittner, 2007;Wilkinson, 2001;Korstanje & Clayton 2012). Of course, these authorative voices developed a conceptual framework of terrorism as encapsulated in undemocratic nations, never imagining that the effects of their hate will have real consequences in democratic countries as Europe and US. The fact is that many of terrorists who perpetrate the attacks are natives of the societies they attack. Over recent decades, this resulted in a serious criticism of some authors who considered neither the theory of poverty nor democracy explain successfully why tourists are targeted by terrorists. The last family reviews the means of production and capitalism as the main factors behind the rise of terrorism. In his recent book entitled *A Difficult World* Korstanje adds in this sense, that modern terrorism derives from the expansion of capitalism. A midst of XIXth century, the first anarchists were disciplined by the capitalist state to abandon violence. In doing so, more substantial benefits in leisure consumption, and mobility were introduced to calm working force. As a result of this, tourism, unionization and terrorism were historically intertwined. To understand the issue, we have to explore the inconsistencies of the means of production of capitalism. Without terrorism, leisure industries never flourished which means that tourism corresponds with a disciplined form of terrorism (Korstanje, 2015). Interesting coincidences were found in Bianchi and Stephenson (2014) who dissect the fallacies of neo-liberal discourse which promotes tourism and mobilities as right to oppressed citizens or at least as a mechanism of peace-keeping. The same material asymmetries produced by tourism in under-developed countries is rechanneled for terrorists to pose their claims against central nations. It generates a vicious circle which is worsened by current globalization. Far from being solved, the social resentment is aggravated in the host and guest meeting. Whatever the case may, Westerners have expanded capitalism in almost 300 years to worldwide. The primary dilemma of Occident, which is posed by terrorism, appeals to re-consider the contradictions of the meaning of security.

Discussing the Concept of Security

The problem of security rests on a great paradox since while it is imposed, a shared sentiment of vulnerability surfaces. The ideal of security is based on its own reversibility (Cavalletti, 2010; Foucault, 2001, 2006). Each culture, country or any type of human organization develop their own conception of what safe and dangerous. While United States poses terrorism as one of the most terrible threats they have historically faced, Argentina is strongly concerned by local crime instead (Dammert & Malone, 2002; Korstanje, 2009). In earlier studies, Korstanje observed that societies develop different means of production, which are inextricably linked to how the world is experienced, lived and negotiated with others. This suggests not only that fears vary from culture or time, but each culture can be dissected by the fears of the lay-citizens (Korstanje, 2011, 2013a, 2014). In this respect, George, Inbakaran and Poyyamoli (2010) argues that tourism is based on two contrasting tendencies. The curiosity to explore new landscapes, cultures and peoples is contrasted to the sense of being familiar with the visited land. Basically, if the uncertainness trespasses the limited degrees, tourists make the decision of returning home.

As the previous argument given, Peter Tarlow (2011a, 2011b) acknowledges that the hyper-globalized world is particular sensitive to the new risks. In view of that, the international destinations are particularly affected when a terrorist attack is covered by the media. Policy makers have realized that these new times requests outstanding efforts in planning and development. One of the priorities of experts seems to be the rapid location, identification and mitigation of those risks or threats which may jeopardize the reputation of destination. In doing so, we need to expand our current understanding about the types of risks, its roots and consequences over daily life. Most certainly, if the risk is not dully detected, the societal order can collapse. Lorri Pennington gray in her colleagues discovered that sometimes the technology, which originally is posed to protect human beings, produces new unseen risks (Pennington-Gray, London, Cahyanto, & Klages, 2011).

Though risk perception was widely developed by Phycology over more than 40 years and even in tourism fields, the term was originally introduced by Roehl and Fesenmeier, no less true is that 9/11 is the turning point that helps tourism scholars to see in risk perception a fertile ground to understand the issue. Risk management programs would be helpful in order for enhancing the process of communication so that the organic image of destination not to be seriously affected (Henderson, 2008; Korstanje 2009). A serious attempt to classify risks has been done by Fuchs and Reichel (2004, 2010) in two main families. Those man-made risks should be separated from others triggered by nature. Risks associated to crime, terrorism, political instability and thefts are produced by us, while disasters, quakes, flood are determined by external forces. The success of a tourist destination depends on the trust generating levels produced by policy-makers. This is an important idea in the recovery policies simply because exhibits that planning plays a vital role in avoiding the collapse of destinations.

In retrospect, George, Inbakaran and Poyyamoli (2010) propose an interesting model to understand the complex play between curiosity and personal safety. Tourism industry rests on two contrasting drives. On one hand, the self is motivated to escape towards new landscapes (tourist-motivation), but at the same time, there is a "nativist" motivation which appeals to the sense of security the visitor needs to feel safe. Whenever "nativist motivation overweight "tourist one", the self-opt to return to home. The role played by the media is vital to boost or decline destinations. Sometimes, the aims of policy makers diluted in minutes because the risk-management programs lack of an efficient communication policy. Following Michael Hall's insights, the media covers news that shape the opinion public audience keeps respecting to what happens beyond our understanding. The communication policy post and pre disaster context can activate a great variety of responses. The attacks of World Trade Centre not only affected the American tourism industry, but also changed the ways other citizens developed to travel abroad (Hall, 2002, 2010). The process of crises-management faced three important changes from 80s up to date. The first was the economic recession surfaced in North America through 80s. This echoed in Asia with replication of domestic economies from 90s onwards. Finally, the most turbulent stage comes with the advent of 9/11 in 2001 and the last financial crisis in 2008/2009. Based on the satisfaction on intangible services, tourism industry develops a high sensibility to crises and disasters (Hall, 2010).

As discussed, the scourge of terrorism moves into three conceptual directions. At a first glance, tourists as leisure-consumers are sensitive to episodes of extreme violence, and of course terrorism does not serve as good advertising. Though originally, tourists need to be in quest of something news, they are extremely of those aspects that jeopardize their ontological security. Though the sense of security is individual and varies person to person, it is important to discuss new programs to keep efficiency in the communication. Secondly, as Yuan observed (2005), for some reasons people feel that terrorism is a problem of urban context, leaving unaffected rural destinations. This suggests that social trust, which turns stronger in rural zones by the tradition, plays a crucial role in configuring of tourist destinations. In third, terrorism instill in tourists a much deeper sentiment of vulnerability, which is previously determined by the position of the self with others. Park and Reisinger (2010) found interesting evidence that shows those travelers who are accompanied by other relatives feel more secure than alone tourists. Ideology or participant militancy in politics plays an important function in developing attachment or aversion to destinations. Left-Wind tourists would be more prone to Cuba than US and vice-versa (Reichel, Fuchs, & Urely, 2007). This happens simply because certainness or the previous degree of familiarity in regards to the visited space is vital for tourists feel safer. Whenever tourists are not fluency in the local language their specter of exploration undermines in comparison with they are moved in more compatible language destinations (Qi, Gibson, Zhang, 2009).

It is tempting to confirm the concept of security not only depends on conjuncture or environmental factors, but also from individual predispositions or psychological profiles. Risk perceptions enables a selective process in which case what scares some tourists attracts others. Even, we need to discuss further the commonalities and differences between risks and other derived sentiment as fear, worries or even panic. As Svein Larsen noted, we need understand that risks and worries are different concepts. At some extent, destinations do not decline so faster as policy makers preclude. Under some conditions, the fear instilled by terrorism is commoditized to become further attractiveness. Starting from the premise that risks rest on a cognitive dimension, which does not entail emotionality, some places can be perceived as insecure without any emotional involvement. Fieldworkers have found interesting outcomes at seeing a dissociation between risk perception and worries in travelers. While Middle age can be perceived as a dangerous place for holiday making, this does not mean that travelers would opt not to travel there.

Destinations only are avoided when emotionality takes the control of cognitive system (Larsen, 2009). Following this, Maximiliano Korstanje notes that the concept of security is beyond the perception of subjects. It relates to a set of complex variables, enrooted in the level of information that helps reducing the uncertainness. The concept of integral security is linked to the traveler's perception respecting to its possibilities to be harmed. Not only the accessibility, but also the presence of police, the migration regulation would be key factors for the tourists have a good experience (Korstanje, 2012). This is the reason why; we have orchestrated multi-disciplinary efforts to explore the world of safety. Policy makers who struggle to keep the destination image, not only should address objective variables as crime, but also subjective valorizations. As L. Grunewald put it, destinations where the security cannot be granted, face serious disadvantage to compete with other international destinations. Besides, 9/11 changed forever the cosmology of security-led experts worldwide. If in years earlier to this event the safety was inextricably intertwined to local crime and robbery later terrorism marked the pace in the programs of risk-maintenance as the main threats of tourism industry. It is not an exaggeration to say that the security was at the tapestry of tourism studies during some awhile.

This begs a more than interesting question:

- Is the personal safety a right granted by democracy?
- Or a simple asymmetry produced by late-capitalism?

To respond these points, in the next lines we will place the work of three Seniors sociologists, Sevil Somnez, Peter Tarlow and Abe Pizam under the lens of scrutiny.

Over decades, the specialized literature evolved and maturated by the contribution of many scholars, but we have selected only three because they have persisted a career and life in the study of risk perception and terrorism. Not only, needless to say, Somnez, Tarlow and Pizam heavily influenced to expand the current understanding of terrorism and its effects in tourism, but also they were touched by external events that formed their understanding about this deep-seated issue. While Luxor tragedy in Egypt struck Sevil Somnez, 9/11 do the same in Pizam. Undoubtely, Rabbi Peter Tarlow who was originally concerned by the role of religion in terrorism saw in ISIS one of the main dangers of West. By understanding these scholars is a valid attempt to discuss the strengths and limitations of terrorism-related studies.

Sevil Somnez

On 17 November of 1997, in Deir el-bahari, one of the most famous tourist attractions alongside the Nilo's River (Luxor-Egypt), a contingent of 58 international tourists and four policemen were brutally killed by an Islamic group known as "Al-Gama'a al-Islamiyya". The Details of a massacre of this caliber not only shocked the world, but also Sevil Somnez. Luxor's massacre exerted a considerable influence on her text, "Tourism, Terrorism, and Political Instability", published on year later at Annals of Tourism Research. She was ahead of her time in denouncing the fact that "terrorism" will be one of the biggest problems of West in the next decades. Why tourists are targeted by international terrorism? they are vulnerable persons who are not responsible for the politics in Middle East?

She considers terrorism as

Premeditated, politically motivated violence perpetrated against civilians and unarmed military personnel by sub national groups…usually intended to influence an audience, and international terrorism as involving citizens or the territory of more than one country. (Somnez, 1998, p. 417)

To put this in other terms, terrorism, viewed as a form of political expression, was a result of social discontents and structural poverty of under-developed societies. If states are hand-tied to intervene, terrorism may traverse all cultures expanding resentment against West. This is the reason why western government should declare terrorism as a universal threat. Terrorists have showed a thirst for killing tourists and innocent visitors who not only are not familiar with the destinations, but also cause a high impact in the western imaginary. Underpinned in the proposition that terrorists killed to gain attention and legitimacy in their claims, Somnez argues that the media plays a vital role in covering and making public their demands and tactics. Western nation-state should take personal direction of military operations to prevent the use of mass-weapons against us. The high technology as well as the degree of information is employed against West to cause panic and disorganization. The relevance of media in the configuration of terrorism's message is conceptualized by means of three elements:

1. A transmitter of message (who is denominated as terrorist),
2. The recipient or audience of the message (target of attack), and
3. The resulted feedback between terrorists and popular wisdom.

The role of experts is more than important to predict where and under what circumstances the next attack will take shape.

In this vein, Somnez goes on to say

Experts speculate that terrorists will continue to select vulnerable –soft- targets, that their attacks will become more indiscriminate, that terrorism will become institutionalized and spread geographically as a method of armed conflict, and that the public will witness more terrorism than ever before due to the media's improved ability to cover terrorist incidents. (Somnez, 1998, p. 419)

Respecting to the reasons of terrorism Somnez precludes that Marxist/Leninist movements set the pace to "Islamic Fundamentalism". This new wave of violence will cover many under-developed and undemocratic pour countries for what the nations around the world should join "forces" in counter-terrorism strategies, law enforcement and policies. There was unfortunately an inextricably interconnection between turmoil, political instability and terrorism. Frequently, those countries where civil wars and ethnic disputes predominate are a fertile source for the surface of political violence and even terrorists. Many examples validate the idea that prolonged terrorist attack against a certain target not only affect the destination-image but also impinges on tourism industry as a whole. As the previous argument given, Somnez enumerates a set of diverse countries where terrorists acted which oscillate from Egypt toward Northern Ireland or Mexico. Logically, terrorist goals achievements and publicity of their acts are associated to a downright ideological resistance to tourism.

Terrorism against one own citizen may in fact go unquestioned by a media controlled by the hostile government. The reason is simple and obvious and has been demonstrated by numerous incidents: when nationals of other countries become involved, new coverage is guaranteed. This way, terrorist know they will secure media attention while curtailing their government's ability to censor new content. When tourists are kidnapped or killed, the situation is instantaneously dramatized by the media, which also helps the political conflict between terrorists and the establishment reach a global scale. Terrorists achieve the exposure they crave and the media increases its circulations and/or ratings. (Somnez, 1998, pp. 424-425)

For Somnez, tourism represents the most important aspect of capitalism, but it may be considered as a resource for expanding peace to the world. Without tourism, democracy has no opportunity of survival. The failures of development are not flaws coined by the system or the ideology, but aftermaths of the local discontent and resentment. States and scholars must devote considerable efforts and money in order for destination not being in decline. If this happens, the cycle of discontent accelerates the frustration engendering more violence. Somnez said that tourists and local stakeholders who encouraged tourism may find not only linguistic but religious barriers. Some cultural values linked to hedonism, sex, drugs abuse, are restricted by fundamentalist and Islam. Also, the problem with terrorism is associated to short-circuit in the communicative process. From this perspective, Somnez introduces to the discussion a new concept which remained unchecked by scholarship, risk. As a negative and constraint, it does not allow the business and market success. Agents if in danger are prone to modify their attitudes and cancel their reserved travels. Involuntarily, this suggests that people is a rational agent able to maximize their benefits at time their disbenefits are minimized. Of course, as rational consumer, any tourist avoids those zones fraught with political instability or violence. Since tourism ensures stability, she considers that represents a good instrument for peacekeeping. Not only risk is conducive to terrorism, but also to the economic losses of tourist destinations. The specialized literature emphasizes on the idea that Mass-media influence in the coverage of tragedies and terrorist events at the time it gives strategies and policies to absorb its costs.

At a closer look, globalized terrorism endorses to "United States" a reason to be victimized and overvalorized. The tourist-experience is the second element Somnez employs to quantify her discriminatory theory. Since travel is or should be an enjoyable experience any hazard, as terrorism, can be immediately dysfunctional to the apprehension of deep emotions as fear and angst. The publicity of tourism entails peace and safety coupled with democracy or development, both makes from this industry an efficient instrument for betterness. More than a political mechanism, tourism works as a witness evincing the entire world what is happening in countries where democracy has never flourished. In fact, Somnez adheres to the belief that democracy is the best possible government, which establishes the cultural background to prevent terrorism. Its potentiality is often associated to the possibility to establish democracy as primary priority worldwide, even in emergent nations. At some extent, she recognizes (but does not give further explanation about the reasons) that tourism and terrorism are inextricably intertwined:

War is a negative influence on tourism demand. This can be viewed positively or negatively, depending on the country's position -since safer countries not involved or less involved in hostilities- might gain visitors lost by the most dangerous destinations. a second dimension of war-tourism relationship involves the creation of a new kind of tourist, attracted by the conflict (tourism of war). (Somnez, 1998, p. 436)

Respecting to Somnez, two major assumptions should be made. At a first glance, she developed an ethnocentric view of terrorism, which leads to over-valorizes technology and tourism as the best of the feasible worlds. Secondly, she never explores the role of democracy as the precondition for terrorism. Given some conditions, whenever democracy is imposed over other regimes which never had a democratic experience, it results in serious riots and clashes among kinships and classes dormant earlier than the dictatorship plummeted. In fact, Somnez never imagined the effects of 9/11 over American democracy. A critique on Somnez's text reveals an over-emphasis put in Tourism as the horse of democracy and peace. Terrorism, in these terms, would be resulted from resentment and the political instability created by undemocratic governments. In a brilliant research, James Piazza has documented how the radicalization of terrorists not only is present in democracies, but also is previously determined by a weaker system of political parties. (Piazza, 2006, 2007).

Furthermore, terrorism seems not to be associated to religion, or to economic resentment. Rather, it corresponds with a tactic of radical politics where the subject is gradually isolated from the external world, which is seen as morally corrupt, dirty and filthy. The discourse of leaders is more effective in small groups, where the interactions with outgroup is minimal. The lack of permeability, conjoined to strong norms of coaction is two of the key factors proper of terrorism (McCauley & Moskalenko, 2008). Lastly, a lot of other studies have denounced that tourism far from being a creator of prosperity, follows a conflictive nature aimed at creating instability and discrepancies among stakeholders (Eubank & Weinberg, 1994; Sandler, 1995; Li, 2005; Hobsbawm, 2007; Brooks, 2009; Korstanje, 2013[a]). These gaps or caveats should be continued by other senior scholar, Abraham (Abe) Pizam.

Abraham Pizam

Professor and dean of the University of Central Florida, Pizam has situated as leading scholars in terrorism and security issues for tourism fields. However, in sharp contrast to Somnez, he deepened the problem from a broader perspective; even years earlier than 9/11, SARS outbreak, natural disasters and so forth, all these events caused tremor in the industry. He understands that it is vital for policy makers and marketing experts to conduct rich empirical research that helps making valid and efficient plans to mitigate risks. That way, destination may adapt to radical changes looming the next millennium (Pizam & Mansfeld, 1996). The world is subject to a plenty of risks which should be prevented in order for society to achieve tolerable levels of existence. If tourism goes on sustainable tenets, it would inevitably result in the decline of ethnical conflict, and of course, the end of terrorism. In view of that, four are the factors to keep in mind to implement success security programs:

1. Local crime,
2. Terrorism,
3. Warfare and conflicts,
4. Political instability.

The efficiency of security forces to keep the peace in the destination would be of paramount importance in the program of risk-maintenance. It is clear how each risk resonates in culture in different way. Those acts which affects negatively social imaginary in US surely have no further impacts in Europe. This suggests how difficult and complex is risk perception. What is clear for Pizam, seems to be that any state of financial dependency between underdeveloped and developed nations paves the pathway for the rise of terrorism. In fact, it comes from the impossibilities of state to control domestic riots and violence (Mansfield & Pizam 2006). In this point, Somnez and Pizam agree. However, unlike other experts in Pizam violence plays a crucial role as the epicenter of all his works. The faster the violence is encapsulated, further legitimacy the government. This seems to be one of the reason why officials need to understand and limit violence in their respective societies. (Pizam, 1999). If local crime or terrorism is a result from resentment, Pizam adds, we have to question is tourism an agent of peace-keeping and a liberty as many advocates proclaim?

The attacks to World Trade Centre in 2001 shocked Pizam, who in an editorial note, call the attention to the problem of vulnerability of tourism before terrorism. We are clear, Pizam notes, that tourism is not a peace-generator or an industry that expands the democracy and prosperity to underdeveloped world, many scholars precluded. In fact, this violence exerts considerable correlation to process of crises, and economic decline. At this stage, affected peoples balance their frustration by attacking to third persons (retaliation). Unless the state does not intervene at the market to improve the security in the public space, preferably space of mass consumption and leisure, terrorism will have serious probabilities to impose their claims (Pizam, Tarlow & Bloom, 1997; Pizam, 2002). Not only he never imagined terrorism will arrive to US, but also it defied the "precautionary principle", a pragmatist doctrine where almost all tourism literature has been structured. This doctrine held the thesis that if direct intervention of western states is successfully done, possibilities of terrorism to expand are meager. Obviously, 9/11 evinced amply this idea rested on shaky foundations.

As the previous argument given, the spill-over effects affect not only neighboring destinations but also expand a sentiment of panic that involved the whole region. Terrorists are moved to cause damage on the tourist coming from central countries, they think, have been responsible to create the miserable condition where they live. The luxury, proper of international hotels and chains, and ostentation are not recommendable practices for tourist enterprises (Fuchs & Pizam, 2011). A modest hotel for example situated in the Middle age has fewer likelihood to be attacked than luxurious projects. Quite aside from this, additional limitations to achieve standardized protocols for security seems to be the indiscipline of police who are not concerned on tourist motivations. In some nations, there is nothing like programs of protection for foreigner tourists, which aggravates the conditions of hostility against Americans. Any precautionary plan, earlier to be formulated, should follow the needs of visitors, tourists and international travelers, understanding their hopes, fears and motivations. However, no less true is that the over-valorization on tourists respecting to local population engenders sentiments of hate and resentment. Locals feel not only their government do not represent them but strangers are of further worth (Pizam, Tarlow & Bloom, 1997). While tourism may very well revitalize poverty in order for protecting locals (from terrorism), it triggers poverty relief otherwise would be not affordable for citizens. Tourism is important not only for lay-people, but for governments. To what an extent any bad experience prevents the tourist comes back to the destination is one of the main aspects that make Pizam's legacy more than interesting. In a seminal study, Holcomb & Pizam alert that impacts of risk should be evaluated in combination of individual and social models. The former signals to the bad experience as a result of minor problems, for example the baggage lost. The latter ones, rather, alludes to the psychology of tourist, which reacts whenever its physical integrity is at stake. Being victim of robbery, or suffers an act of vandalism are cases very well can be typified under this angle. The outcome of their investigation reveals that tourists can come back even they have been victim of robbery. In view of a process of desensitization, the decision making process is crucial. The person subordinates the events to its previous decision. If we made the decision not to travel Middle East in view of its peligrosity, any effort to make the destination safer is marginal. In contrast, once the destination is selected, negative news rarely affects the original decision (Holcomb & Pizam, 2006).

To cut the long story short, unlike Somnez, Pizam goes beyond the problem of terrorism to reconstruct an all-encompassing model of security. If Chernobyl has showed something, it was the contradictory nature of technology and the human use. To what an extent an efficient plan may work? All our efforts to embellish the city can be counter-productive, captivating the attention of radical groups as terrorists and insurgents. This question Pizam is not responding adequately, is continued by another senior sociologist, Peter Tarlow.

Peter Tarlow

In opposition to Pizam and Somnez, Rabi Peter Tarlow is strongly concerned by a wider set of questions that enhance security. His knowledge in biblical texts, conjoined with sociological literature, leads to explore the connection of terrorism and religion. In Tarlow's view terrorism has nothing to do with religious life (Tarlow, 2012). It is unfortunate that plans backfire because of its misconception of what risks are. The plans should be prepared not only to prevent the potential risky behavior, but also make better the life of peoples (Amorin, Gandara, & Korstanje, 2012). The decline of tourism in some cities is accompanied by the rise of unemployment, poverty and economic stagnation. This represents a vicious circle in view of that resentment arises. Terrorists attack state hosting tourists, it creates a bad image of the country which affects the economy. More stagnation instills further resentment which is conducive to terrorist's discourse.

One of the main problems to study risk is the dispersion of definitions and meanings that create misunderstanding and confusion. In this respective, Tarlow contends that terrorism is not a product of economic crises and resentment. Many pour destinations not only are kindly but also offers a wider hospitality to tourists. The hostility against them is based on the lack of compromise in the other, a type of new instrumentalism which leads us to be insensible to other' sufferings. The other seems to be instrumentalized to achieve the own goals. In view of that, terrorists will attempt to administrate fear in order for their claims to be accepted (Tarlow, 2011). Religion, far from being adjoined to terrorism, Tarlow notes that extortion is the primary value of terrorist cells. These observations usher Tarlow to make a critique not only to Islamic terrorism, but to the political manipulations and tergiversation accomplished by American government. If we have learned a lesson, it was 9/11 has changed our lives as well as the borderlands where the walls of civilization are erected. Unless we understand what is happening, ethnocentrism will engender a biased image of the other. America was enthralled in the right of making direct intervention in the world, violating the current international jurisprudence. Internally, expansion of American imperialism has portrayed radical and frightful image of aliens (Korstanje & Tarlow, 2012). The cultural entertainment industry is forging a dangerous message, where the conception of uphill city, is pushing America to undemocratic form of politics. This not only affects tourism, but also civility as well. The mass media alludes to disasters as outstanding events. In doing so, they introduce a sense of superiority of Americans over the rest of the world (Korstanje & Tarlow, 2013).

It is important not to close this section without reviewing his last great project, *Tourism Security*. In this pungent book, he defines the etymological roots of travels to expand the current understanding we have around tourism. The conception of people traveling for pleasure has gained consensus among scholars, though Tarlow adds, it is very hard to grasp a common-shared meaning of tourism; this "lack of unified vocabulary" leads to confuse the conceptualization of tourist from visitor and vice-versa. Besides a business, tourism has been widely studied over the last decades as a social phenomenon. Whatever the case may be, the logic of travels has been shifted by the advent of technological revolution which made from the environment a safer place. One of the aspects that determine tourism seems to be the social trust posed on the hospitality of hosts. As an act of trust, touring corresponds with a question of safety and networking. This pungent book explores the axiom based on the belief "tourists simply avoid dangerous destinations" creating a serious economic loss for policy makers, tourist community, and involving stakeholders.

Following this backdrop, the tourist trip alternates two contrasting tendencies, the curiosity to experience new sensations with the sense of ontological safety. The biblical and other mythological sources reveal that the sacred space of leisure is constructed in basis of a positive precept that mandates the man to relax, but at the same time, its vulnerability increases. Ancient poets and philosophers emphasized on the fact anything can happen at any time at a banquet, or public game. The rise of new modern risk and challenges for tourism industry is what Tarlow prioritizes as the most significant aspect to debate in his text. Now we are subject to a set of globalized apocalyptic risks which range from natural disasters to terrorist attacks. The attacks to World Trade Centre in 2001 represented a turning point in the security fields of US and the world. In this troublesome context, may tourism have opportunities to survive?

Undoubtedly, the success of policy makers to address successfully the problem of safety seems to be related to how the communication is framed. As prey of our emotions, we often are subject to panic-flight by insignificant issues while others more important are ignored. The pardox of tourism security is that While criminals exploit tourists as a source of profits for their purposes, terrorists are not only beyond the commercial interests, they want to create panic to destabilize the political order of state. The third chapter focuses on the safety inside hotels and lodging industry. To ensure its loyalty to the hotel, the tourist should have a memorable experience. The role of security officer, at the establishment, is of paramount importance to control the front-line security personnel as well as the guest's room. In what is the best chapter, the fourth is fully reserved to risk assessment and crises-led management. Since the industry in United States witnessed a set of diverse airplane hijackings in its history, up to 9/11 where the planes were employed as weapons against civil targets, Tarlow argues that though experts in security claims for the hardening of control in terminal spaces, there is no possibilities to have a destination 100% secure. We need to remind that everything we can do is to control some variables to improve the quality of people's life. Doing the best to have safer airports, bus terminals or tourist destination is not the same to enter in paranoia claiming the state for more security, embracing the zero-risk paradigm or any other populist flag.

In this vein, Cass Sunstein, other senior sociologist concerned by the risk perception and its effects in public opinion, acknowledges that "the problem of precautionary principle" consists in blurring the rational causation among events. The neglect of risk leads us to undermine some probabilities which are not fitted into our cognitive structure, while others are exaggerated (Sunstein, 2005). This happens because emotions obscure the factors to make reasonable decisions. Even Jean Baudrillard (2006), has warned on the problems the "precautionary principle" applied on risk perception. One day, he adds, the world would be as the film "minority report" has portrayed. The criminal acts would be punished before to be committed. The legal jurisprudence and its application, as it has been formulated by roman right, will set the pace to a new more radical forms where the future will replace the present. This raises a philosophical concern; may a person face a conviction by criminal acts which have not been yet committed?

The precautionary doctrine would say yes, but this is not what Tarlow thinks. Rather, even it opens the doors for a future discussion. To the moment, we have no a clear answer to this question but understands on the importance to live beyond the paradigm of risk-zero doctrine (Korstanje, 2013).

FUTURE RESEARCH DIRECTIONS

In a seminal text, Graham Fuller showed how would happen if Islam would never have existed. In fact, history reminds both Christians and Muslims share their Abrahamic roots. There are no cultural values among civilizations that explain terrorism. Fuller adheres to political causes as the main reason why Middle East have inherited the Anti-Western sentiment from Byzantine Empire. In the same direction, the present chapter is intended to reconsider the already existent myopia in tourism-related scholars. Although this review-chapter recognizes not only the importance of tourism by revitalizing the economic and social frustrations of local communities, but how terrorism affects pour nations, it is important to understand to implement efficient programs to tackle the issue which means that marketing experts, officials and tourism-managers should understand what they are struggling. Neither religion nor poverty are factors that coadyuvated in the formation of terrorist minds. Rather, as Tarlow puts it, we have to turn out the attention on hate and extortion. Future approaches should be oriented to discuss not only how democratic tourism is, but also much profound intersections between tourism as a disciplined form of terrorism (Korstanje, 2015).

CONCLUSION

In this chapter-review, we have discussed the contributions of Peter Tarlow, Sevil Somnez and Abraham Pizam in regards to tourism security. From diverse perspectives, these scholars have devoted considerable time and efforts to give a legacy. Many underdeveloped communities adopt tourism as main alternative to revitalize their economies. Although the industry offers many possibilities to enhance the economic progress, no less are the opportunities for terrorism to arrive. It is important to discuss how the theory of development has led many peripheral countries to ask for financial aids, and loans which reinforced a strong dependency between the Centre and its periphery. Whether tourism is adopted as the main option to develop an economy, terrorism has more probabilities to multiply and flourish. This was a widespread belief, after the attacks to WTC in 2001, an event that altered the geo-politic relations of states and the policies of surveillance in the global system. At some stage, Sevil Somnez pointed out that the terrorist's mind is moved to balance the personal frustration and psychological resentment. Tourism paves the ways for the meeting of two contrasting values, democracy and dictatorship. If the tourist is the torchbearer of democracy, terrorists embrace the opposite values. Following Somnez's thesis, tourists are killed by their values, their respect to liberty and individual rights. Rather, Abraham Pizam holds the belief that terrorist is moved by the resentment tourism creates in peripheral economies. As a mechanism of subordination, centre alludes to tourism to create dependency from periphery. It engenders material asymmetries among classes and human groups, which are rechanneled to "jihad".

Last but not least, Peter Tarlow argues convincingly that terrorism alludes to the instrumentalization of the other, who is diminished to an object. Far from being a religious problem, terrorism appeals to exploit the other, as a mean to achieve the personal goals. Neither the poverty, nor the precautionary principle explains why many pour peoples are not following terrorist tactic to extortionate their states. If we have something clear, Tarlow adds, this is that terrorism exploits the weaker one to dissuade the stronger.

REFERENCES

Amorin, E., Gandara, J., & Korsanje, M. E. (2012). Seguridad percibida en la ciudad de Curitiba. Un sondeo exploratorio en residentes locales y turistas. *Revista de análisis turístico*, *14*, 1-9.

Baudrillard, J. (2006). Virtuality and Events: The hell of power. *Baudrillard Studies*, *3*(2), 1–15.

Bianchi, R. (2007). Tourism and The Globalization of Fear: Analyzing the politics of risk and (in)security in global travel. *Tourism and Hospitality Research*, *7*(1), 64–74. doi:10.1057/palgrave.thr.6050028

Bianchi, R., & Stephenson, M. (2014). *Tourism and citizenship: Rights, freedoms and responsibilities in the global order* (Vol. 11). Abingdon: Routledge.

Brooks, R. (2009). Researching democracy and terrorism: How political access affects militant activity. *Security Studies*, *18*(4), 756–788. doi:10.1080/09636410903369027

Cavalletti, A. (2010). *Mitología de la seguridad: la ciudad biopolítica*. Madrid: Adriana Hidalgo.

Cohen, E. H. (2014). Tourism and Terror: a Case Study: Israel 1948-2012. *International Journal of Religious Tourism and Pilgrimage*, *2*(1), 13–26.

Crouch, G. I., & Ritchie, J. B. (1999). Tourism, competitiveness, and societal prosperity. *Journal of Business Research*, *44*(3), 137–152. doi:10.1016/S0148-2963(97)00196-3

Dammert, L., Malone, M. F. T., & Dammert, L. (2002). Inseguridad y temor en la Argentina: El impacto de la confianza en la policía y la corrupción sobre la percepción ciudadana del crimen. *Desarrollo Economico*, *42*(166), 285–301. doi:10.2307/3455944

Eubank, W. L., & Weinberg, L. (1994). Does democracy encourage terrorism? *Terrorism and Political Violence*, *6*(4), 417–435. doi:10.1080/09546559408427271

Foucault, M. (2001). *Defender la Sociedad*. Buenos Aires: Fondo de Cultura Económica.

Foucault, M. (2006). *Seguridad, Territorio, Población*. Buenos Aires: Fondo de Cultura Económica.

Fuchs, G., & Pizam, A. (2011). *The Importance of Safety and Security for Tourism Destinations. Tourism Destination Marketing and Management: Collaborative Strategies, 300.* CABI.

Fuchs, G., & Reichel, A. (2004). Cultural Differences in tourist destination risk perception: An exploratory study. *Tourism*, *52*, 21–37.

Fuchs, G. & Reichel, A. (2010). Health Tourists Visiting a Highly Volatile Destination. *Anatolia: An International Journal of Tourism and Hospitality Research,21*, 205-226.

Fuller, G. (2010). *A World Without Islam*. New York, NY: Back Bay Books.

George, B. P., Inbakaran, R., & Poyyamoli, G. (2010). To Travel or Not to travel: towards understanding the theory of nativistic motivation. *Turizam: znanstveno-stručni časopis, 58*, 395-407.

Grunewald, L. (2012). La Seguridad en el marco de la competitividad de los destinos turísticos. *Revista de Ciencias Sociales*, *21*, 141–158.

Hall, M. (2002). Travel Safety, terrorism and the media the significance circle of the issue attention cycle. *Current Issues in Tourism*, *5*(5), 458–466. doi:10.1080/13683500208667935

Hall, M. (2010). Crisis Events in Tourism: Subjects of crisis in Tourism. *Current Issues in Tourism*, *13*(5), 401–417. doi:10.1080/13683500.2010.491900

Henderson, J. C. (2003). Managing tourism and Islam in peninsular Malaysia. *Tourism Management*, *24*(4), 447–456. doi:10.1016/S0261-5177(02)00106-1

Henderson, J. C. (2008). Managing Crises: UK Civil Aviation, BAA airpots and the August 2006 terrorist Threat. *Tourism and Hospitality Research*, *8*(2), 125–136. doi:10.1057/thr.2008.12

Holcomb, J., & Pizam, A. (2006). *Do incidents of theft at tourist destinations have a negative effect on tourists' decisions to travel to affected destinations? In Tourism, Security & Safety: From Theory to Practice* (pp. 105–124). Oxford, UK: Butterworth-Heinemann.

Huntington, S. P. (1997). *The clash of civilizations and the remaking of world order*. New Dehli, Penguin Books India.

Kittner, C. C. B. (2007). The role of safe havens in Islamist terrorism. *Terrorism and Political Violence*, *19*(3), 307–329. doi:10.1080/09546550701246791

Korstanje, M & Tarlow, P. (2013) Disasters, tourism and mobility: the case of Japan's earthquake. *Pasos: revista de turismo y patrimonio cultural,11*, 17-32

Korstanje, M. E. (2009). Miedos democráticos: nuevas expresiones del temor en Latinoamérica. *Eikasia: revista de filosofía*, *28*, 109-123.

Korstanje, M. E. (2011). La fobología,¿ ciencia o forma de entretenimiento? *Nómadas. Revista Crítica de Ciencias Sociales y Jurídicas*, *31*, 175–189.

Korstanje, M. E. (2012). ¿Se puede ponderar la seguridad turística? Un ensayo conceptual. *Turismo e Sociedade*, *5*(2), 368–390. doi:10.5380/tes.v5i2.26332

Korstanje, M. E. (2013a). Empire and Democracy. A critical reading of Michael Ignatieff. *Nómadas. Revista Crítica de Ciencias Sociales y Jurídicas*, *38*, 69–78.

Korstanje, M. E. (2013b). Preemption and Terrorism. When the Future Governs. *Cultura (Asociación de Docentes de la Universidad de San Martín de Porres)*, *10*, 167–184.

Korstanje, M. E. (2013c). Estar protegido: el rol de las compañías de asistencia al viajero en el sistema turístico. *El Periplo Sustentable: revista de turismo, desarrollo y competitividad*, *25*, 43-76.

Korstanje, M. E. (2014). Towards an Index of Fear: The Role of Capital in Risk' s Construction. *International Journal of Cyber Warfare & Terrorism*, *4*(1), 19–26. doi:10.4018/ijcwt.2014010103

Korstanje, M. E. (2015). *A Difficult World, examining the roots of Capitalism*. New York, NY: Nova Science Publishers.

Korstanje, M. E., & Tarlow, P. (2012). Being lost: Tourism, risk and vulnerability in the post-9/11entertainment industry. *Journal of Tourism and Cultural Change*, *10*(1), 22–33. doi:10.1080/1476 6825.2011.639455

Larsen, S., Brun, W., & Øgaard, T. (2009). What Tourists worry about: Construction of a scale measuring tourist worries. *Tourism Management*, *30*(2), 260–265. doi:10.1016/j.tourman.2008.06.004

Li, Q. (2005). Does democracy promote or reduce transnational terrorist incidents? *The Journal of Conflict Resolution*, *49*(2), 278–297. doi:10.1177/0022002704272830

Mansfeld, Y., & Pizam, A. (Eds.). (2006). *Tourism, security and safety: from theory to practice*. Oxford, UK: Elsevier.

McCauley, C., & Moskalenko, S. (2008). Mechanisms of political radicalization: Pathways toward terrorism. *Terrorism and Political Violence*, *20*(3), 415–433. doi:10.1080/09546550802073367

Park, K., & Reisinger, Y. (2010). Differences in the perceived influence of natural disasters and travel risk on international travels. *Tourism Geographies*, *12*(1), 1–24. doi:10.1080/14616680903493621

Pennington-Gray, L., London, B., Cahyanto, I., & Klages, W. (2011). Expanding the tourism crisis management planning framework to include social media: Lessons from the Deepwater Horizon Oil Spill 2010. *International Journal of Tourism Anthropology*, *1*(3/4), 239–253. doi:10.1504/IJTA.2011.043708

Piazza, J. A. (2006). Rooted in Poverty? Terrorism, Poor Economic Development, and Social Cleavages. *Terrorism and Political Violence*, *18*(1), 159–177. doi:10.1080/095465590944578

Piazza, J. A. (2007). Draining the swamp: Democracy promotion, state failure, and terrorism in 19 Middle Eastern countries. *Studies in Conflict and Terrorism*, *30*(6), 521–539. doi:10.1080/10576100701329576

Pizam, A. (1999). A comprehensive approach to classifying acts of crime and violence at tourism destinations. *Journal of Travel Research*, *38*(1), 5–12. doi:10.1177/004728759903800103

Pizam, A. (2002). Tourism and terrorism. *International Journal of Hospitality Management*, *21*(1), 1–3. doi:10.1016/S0278-4319(01)00037-8

Pizam, A., & Mansfeld, Y. (1996). *Tourism, crime and international security issues*. New York, NY: John Wiley & Sons.

Pizam, A., Tarlow, P. E., & Bloom, J. (1997). Making tourists feel safe: Whose responsibility is it? *Journal of Travel Research*, *36*(1), 23–28. doi:10.1177/004728759703600104

Qi, C. X., Gibson, H. J., & Zhang, J. J. (2009). Perceptions of risk and travel intentions: The case of China and the Beijing Olympic Games. *Journal of Sport & Tourism*, *14*(1), 43–67. doi:10.1080/14775080902847439

Reichel, A., Fuchs, G., & Uriely, N. (2007). Perceived risk and the non-institutionalized tourist role: The case of Israeli student ex-backpackers. *Journal of Travel Research*, *46*(2), 217–226. doi:10.1177/0047287507299580

Roehl, W., & Fesenmaier, D. (1992). Risk Perceptions and Pleasure Travel: An exploratory analysis. *Journal of Travel Research*, *30*(4), 17–29. doi:10.1177/004728759203000403

Ryan, C. (1993). Crime, violence, terrorism and tourism: An accidental or intrinsic relationship? *Tourism Management, 14*(3), 173–183. doi:10.1016/0261-5177(93)90018-G

Sandler, T. (1995). On the relationship between democracy and terrorism. *Terrorism and Political Violence*, *7*(4), 1–9. doi:10.1080/09546559508427315

Sönmez, S. F. (1998). Tourism, terrorism, and political instability. *Annals of Tourism Research*, *25*(2), 416–456. doi:10.1016/S0160-7383(97)00093-5

Sunstein, C. R. (1990). *After the Rights Revolution. Reconceiving the regulatory state*. Cambridge, MA: Harvard University Press.

Sunstein, C. R. (2005). Laws of fear: Beyond the precautionary principle (Vol. 6). Cambridge, MA: Cambridge University Press.

Tarlow, P. (2011b). Tourism Disaster Management in a age of Terrorism. Int. Journal of Tourism Anthropology, 1, 254-272. doi:10.1504/IJTA.2011.043709

Tarlow, P. (2012). Rosh Ha` Shanah Day. *International Journal of Safety and Security in Tourism and Hosp*, *3*, 1–19.

Tarlow, P. (2014). *Tourism Security*. Oxford, UK: Elsevier.

Tarlow, P. E. (2006). Terrorism and tourism. In J. Wilks, D. Pendergast, & P. Leggat (Eds.), *Tourism in turbulent times: Towards safe experiences for visitors* (pp. 79–92). New York, NY: Routledge. doi:10.1016/B978-0-08-044666-0.50014-2

Tarlow, P. E. (2011a). Tourism risk management in an age of terrorism. *Economía Autónoma*, *4*, 18–30.

Vukonic, B. (2002). Religion, tourism and economics: A convenient symbiosis. *Tourism Recreation Research*, *27*(2), 59–64. doi:10.1080/02508281.2002.11081221

Wilkinson, P. (2011). *Terrorism versus democracy: The liberal state response*. London: Routledge.

Yuan, M. (2005). After September 11: determining its Impacts on Rural Canadians travel to U.S. *E-Review of Tourism Research,3*, 103-108.

ADDITIONAL READING

Altheide, D. (2006). *Terrorist and the Politics of Fear*. Oxford: Altamira Press.

Bauman, Z. (1998). On glocalization: Or globalization for some, localization for some others. *Thesis Eleven, 54*, 37-49.

Bauman, Z. (1998a). *Globalization: The human consequences*. New York: Columbia University Press.

Bernstein, R. J. (2013). *The abuse of evil: The corruption of politics and religion since 9/11*. New York, NY: John Wiley & Sons.

Derrida, J. (2000). *Of Hospitality: Anne Dufourmantelle Invites Jacques Derrida to Respond* (R. Bowlby, Trans.). Stanford, CA: Stanford Univ. Press.

Farrington, J. H. (2007). The new narrative of accessibility: Its potential contribution to discourses in (transport) geography. *Journal of Transport Geography*, *15*(5), 319–330. doi:10.1016/j.jtrangeo.2006.11.007

Germann Molz, J. (2014). *Travel Connection: tourism. Technology and togetherness in a mobile World.* New York, NY: Routledge.

Hannam, K., Sheller, M., & Urry, J. (2006). Editorial: Mobilities, immobilities and moorings. *Mobilities*, *1*(1), 1–22. doi:10.1080/17450100500489189

Korstanje, M. E. (2010). Ironman. *Cultura International Journal of Philosophy of Culture and Axiology*, *7*(2), 188–203. doi:10.5840/cultura20107214

Korstanje, M. E. (2013). The archetype of terrorism: How 11/9 changed the world? *Sincronía*, *63*, 8–22.

Lash, S. M., Urry, S. L. J., & Urry, J. (1993). *Economies of signs and space*. London: Sage.

McCauley, C., & Moskalenko, S. (2008). Mechanisms of political radicalization: Pathways toward terrorism. *Terrorism and Political Violence*, *20*(3), 415–433. doi:10.1080/09546550802073367

McCauley, C., & Moskalenko, S. (2011). *Friction: How radicalization happens to them and us*. Oxford: Oxford University Press.

Moskalenko, S., & McCauley, C. (2009). Measuring political mobilization: The distinction between activism and radicalism. *Terrorism and Political Violence*, *21*(2), 239–260. doi:10.1080/09546550902765508

Piazza, J. A. (2007). Draining the swamp: Democracy promotion, state failure, and terrorism in 19 Middle Eastern countries. *Studies in Conflict and Terrorism*, *30*(6), 521–539. doi:10.1080/10576100701329576

Piazza, J. A. (2008). Incubators of terror: Do failed and failing states promote transnational terrorism? *International Studies Quarterly*, *52*(3), 469–488. doi:10.1111/j.1468-2478.2008.00511.x

Possner, R. A. (1983). *The Economics of Justice*. Cambridge: Harvard University Press.

Sheller, M., & Urry, J. (2004). *Tourism mobilities: places to play, places in play*. London: Routledge.

Skoll, G. (2011). Travel Insecurity in North America: the Mexico-US border. *International Journal of safety and Security in Tourism, 1*, 44-63.

Skoll, G. (2016). *Globalization of American Fear Culture: the empire in the Twenty-First Century*. New York, NY: Palgrave-Macmillan. doi:10.1007/978-1-137-57034-5

Skoll, G. R. (2007). Meanings of terrorism. *International Journal for the Semiotics of Law-Revue internationale de Sémiotique juridique, 20,* 107-127.

Skoll, G. R. (2008). Torture and the Fifth Amendment Torture, the Global War on Terror, and Constitutional Values. *Criminal Justice Review*, *33*(1), 29–47. doi:10.1177/0734016808315585

KEY TERMS AND DEFINITIONS

Democracy: It signals to a form of political organization based on the rights for all the people to be involved in decision making process. Democracy varies on time and culture.

Fear: Psychological emotion or disposition whose functionality is given to the survival of organism.

Financial Dependency: It corresponds with a conceptual theory which underlines the intersection of central and peripheral nations. While the former develop centralized means of production to issue loans, the latter are subject to ask for them.

Globalization: It derives from a process of integration culturally and economically that engaged different nations and economies worldwide.

Security: It consists in cognitive perception of protection developed by individuals to adapt at their environments.

Terrorism: It is a form of political violence that exerts extortion in order for their claim to be accepted.

Chapter 10
An Overview of Tourism Supply Chains Management and Optimization Models (TSCM – OM)

Jonnatan F. Avilés-González
Tecnológico de Monterrey, Mexico

Sonia Valeria Avilés-Sacoto
Tecnológico de Monterrey, Mexico

Leopoldo Eduardo Cárdenas-Barrón
Tecnológico de Monterrey, Mexico

ABSTRACT

Around the world tourism industry represents economic benefits to the countries in where the tourist attractions are located. The purpose of this chapter is to search and review recent researches related to the area of tourism under supply chain management and optimization models perspectives. The main aim of this chapter is to identify and discuss how the tourism supply chain is studied when it is subject to different economic, market, and optimization strategies. Considering the period of 2005 to 2016, a systematic review was performed using research studies in the area of tourism supply chain management. The results show that game theory is used as a theoretical base in the majority of the cases, but several novel approaches are also incorporated to the analysis. This review can be used as a complement of the previous works and a valuable information source for the decision makers involved in the tourism area.

INTRODUCTION

Talking about tourism implies an intensive management of several activities including planning, evaluation of tourism services, promotional and discounts tracking, and the different offers to attract new

DOI: 10.4018/978-1-5225-1054-3.ch010

Figure 1. Example of tourism flow routes and stakeholders

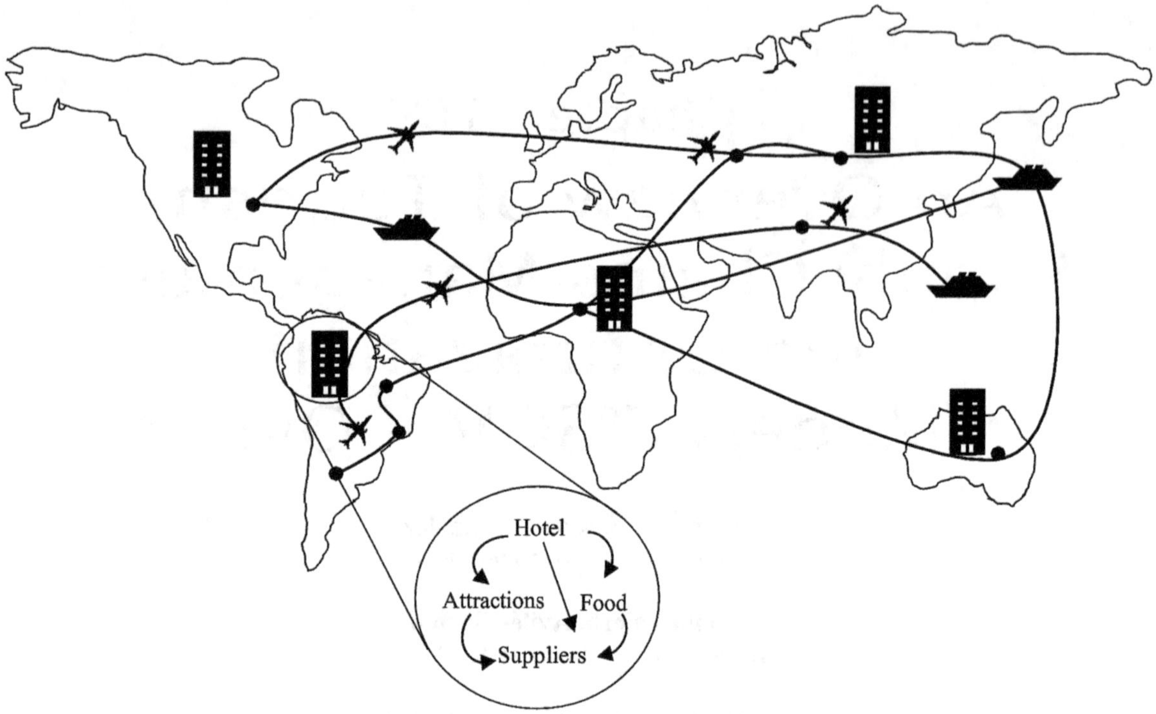

customers to the destiny. This idea leads us to think that it is necessary to develop an analysis of the cost-benefit relation applied to all the stakeholders involved in this field.

From the customer's perspective, tourism is used to satisfy the human desire for fun, recreation, and the undefined motive to seek and explore the unknown and unseen (Tisdell, 2013). It includes activities of people travelling to and staying in places outside their conventional environments. Another concept given by Franklin (2003) associates the tourism with modernity, where the society experiences novel changes and ways of communication, thus the outcome is an extension of the spatial range from the home to outer space.

On the other hand, economically speaking, tourism is a fast-growing economic activity in many countries, benefiting the development of nations due to the employment generation, stimulation of investment, infrastructure development, and foreign exchange earnings. In this direction, some previous studies have reported an increase interest in the tourism industry. Literally, there are thousands of papers related to tourism. In fact, worldwide, tourism is considered as one of the fundamental economic activities, that contributes in the progress of the economy. It is important to remark that the success of the tourism industry relies in the optimal use of all the factors belonging to the productive tourism chain. Some of them are the tourism resources, tourism services, public services, social and cultural activities, among other offerings. Figure 1 shows an example of how the tourism flow is linked and how the stakeholders are related.

However, the competitive environment forces the tourism industry to create a synergy enhancing its components in order to generate value to tourists. It is here, when a problem arises, due to the fact that there is a high number of variables associated with the tourism chain; making hard to optimizing it. For example, in many cases the activities, such as promotional and marketing, are set aside in the evaluation of the supply chain, generating mislead results due to the missing variables during the problem modeling. This particular situation has motivated to the researchers around the world to take the challenge of developing models and algorithmic tools to assess the entire productive chain in the tourism real world problems.

From other point of view, the tourism incorporates information technologies (ITs) in the development of its activities. For instance, the new commercial formats (e-Tourism, online travel agencies, trip-online advisors and others) have changed drastically the tourism industry into a more complex system. Additionally, the globalization and the fast communication among the stakeholders in the tourism supply chain have evolved into a difficult environment where dynamic models play an important role for taking decisions. Therefore, the tourism industry needs to be analyzed from an integrated perspective.

This has led to the authors to respond the need to classify the models from which the tourism can be evaluated; this sort will help to develop an analysis employing both perspectives: the optimization and supply chain models.

The main purpose of this chapter is to identify and describe briefly different approaches applied to the tourism supply chain management, focusing on the taxonomy of different mathematical models. If the reader needs more details about the mathematical expression of the models, please refer to the cited reference. For this, a literature review was done in order to identify a possible connection between the supply chain management and the optimization models. In advance, several settings such as tours, sustainable tourism, hotels and destinations are employed to classify the research works according to application.

BACKGROUND

Drawing on an extensive range of sources, it can be found that some researchers set out three approaches in which the tourism can be evaluated, those are conceptual, empirical and quantitative points of view; see for instance the works of Lafferty and Fossen (2000) and Pearce et al. (2007). It is worthy to mention that the scope of this research is based only on quantitative models. Therefore, this section introduces two main categories of how the tourism can be evaluated.

This section gives a brief overview of the two categories of the quantitative models. The first part deals with the supply chain management, in which the tourism is considered as a set of different stages, parts or factors that are necessary to be linked and evaluated as an entire measurement of the activity. The second part attempts to show the procedure under the category of operations research (OR). Here, all the OR methods and models in the tourism sector are briefly discussed.

CATEGORIES UNDER SUPPLY CHAIN MANAGEMENT APPROACH FOR TOURISM

As it was mentioned above, there is a considerable amount of literature that has been published on tourism; which leads to seek new approaches, tools, and perspectives in order to explore this interesting and challenging area. One of these approaches is the tourism supply chain management (TSCM). Zhang et al. (2009) define the "TSCM as a set of approaches utilized to efficiently manage the operations of the tourism supply chain (TSC) to meet the needs of tourists from the targeted source market(s) and accomplish the business objectives of different enterprises within the TSC." One of the characteristics of the TSCM is that it can be considered as an oligopolistic market; this means that there is a market where a number of players are in competition to reach a specific objective. These players can be independently or jointly working in order to obtain the desired objective. According to Cremer et al. (2016) the objective is to control the supplier resources or the prices in the market. Due to these characteristics the majority of researchers apply game theory, which is a good approach to model reality circumstances, in order to modeling the TSCM; see for example, Sheng (2011). The versatility of the game theory allows to decision makers to measure the interactions among tourism supply chain members and their strategic options, such as price and competition and coordination (Song, Dwyer, Li, & Cao, 2012), which results in a better understanding of the stakeholders behavior in the tourism supply chain.

In a natural way, the tourism is intrinsically linked to marketing, economy, and politics; for this reason a diverse group of conceptual and qualitative strategies are applied during the supply chain modeling. Two common strategies are:

- Pricing, and
- Coopetition (competition-coordination) (Guo, Ling, Dong, & Liang, 2013; Guo, Zheng, Ling, & Yang, 2014).

These concepts are explained below.

Pricing

For many businesses, the pricing decision is a critical strategy, because for the customers the price is perceived as an indicator of service quality that generates satisfaction. It is important to mention that a business can apply a diversity of pricing strategies when it is vending a service or product. The price can be established to maximize profitability for each unit vended or from the market overall. It can be used to protect an existing market from new competitors, to increment market share within a market or to go into a new market. Businesses may benefit from decreasing or increasing prices, depending on the needs and behaviors of the customers in the particular market (Lee & Staelin, 2016; Ordover & Willig, 1981; Song, Yang, & Huang, 2009).

Coopetition

It is a network where the stakeholders are cooperating and competing to create maximum value in a selected market. For example, internet and mobile technologies have made it even more necessary for companies to both cooperate and compete, by enabling relationships through information sharing as well as integrating and streamlining processes (Czakon, Mucha-Kus, & Rogalski, 2014).

A key feature of the TSCM problems is that the problem definition changes according to the elements in the supply chain. An example of this is shown in Figure 1; where the supply chain starts from the customer contacts through internet, agencies and tour operators to the direct suppliers of the service such as attractions, lodging, transports, catering; and then to the direct supplier's suppliers who include maintenance, basic service, infrastructure, food and beverages. Additionally, the decision variables required to model the system depend on the strategy applied. Zhang et al. (2009) proposed a classification of decision variables for TSCM into three categories:

- **Variables of Strategic Decisions:** Tourism taxation policies of the local government, capital investment in the tourism industry, capacity building of tourism facilities, and the entry or deterrence of potential tourism businesses.
- **Variables of Tactical Decisions:** Purchasing and production decisions, pricing strategies, product differentiation, advertising, and inventory policies.
- **Variable of Operational Decisions:** Tour scheduling, route planning, and issuing quotations on tour products.

CATEGORIES UNDER OPTIMIZATION MODELS APPROACH FOR TOURISM

Operations research (OR) is a combination of mathematical problem solving tools to tackle and solve complex decision-making problems. Here, mathematical, statistical and computer methodologies have the main objective of finding the most feasible and approximate results to develop decisions, plans, and policies (Swart, Var & Gearlinh, 1978).

Operations research has been widely used in different areas, and one of them is the tourism. In 1978, the publication of Swart et al. (1978) described an accurate division of the different OR methodologies to deal different problems in tourism. Figure 2 illustrates Swart et al.'s (1978) classification.

Tourist Forecasting Models

These approaches measure the number of tourist arrivals to a touristic place of destination. These models are assorted as it follows:

Figure 2. A simple representation of a tourism supply chain

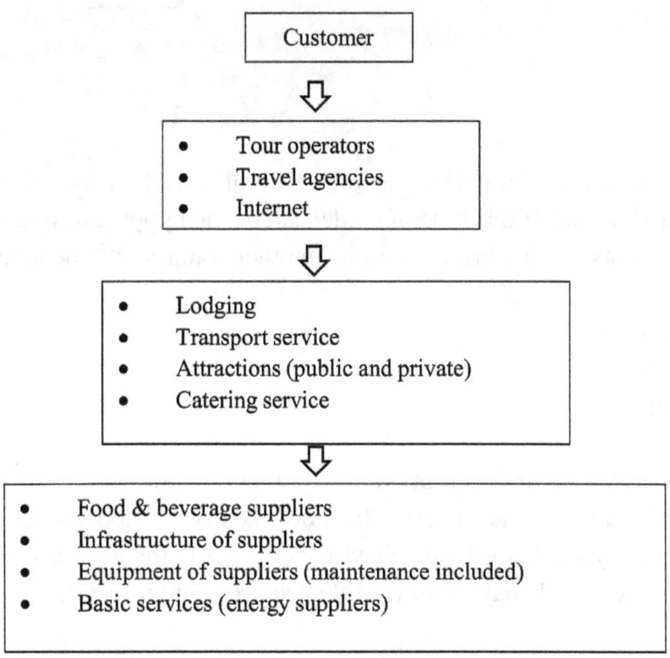

Figure 3. Classification of the operations research models in the tourism area

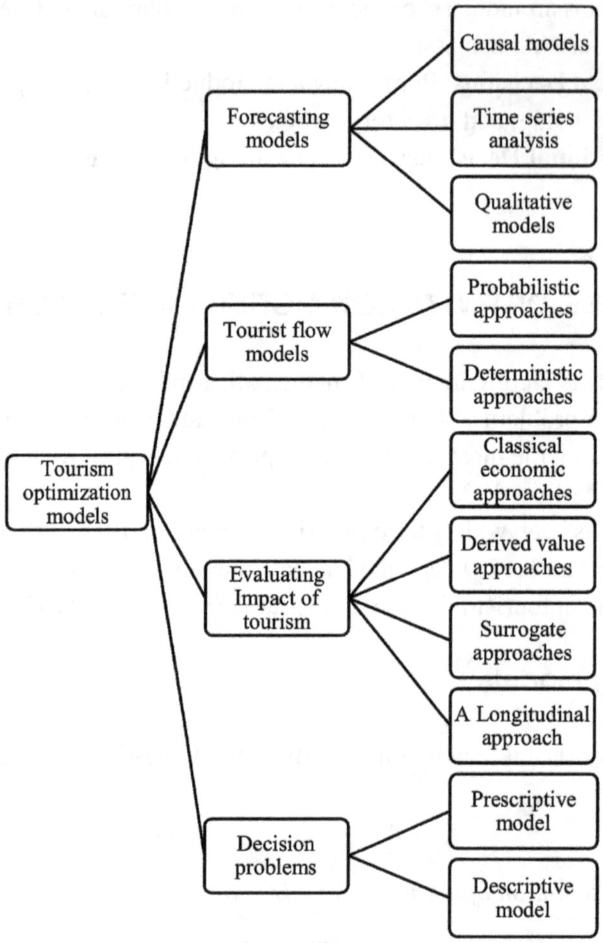

Causal Models

The causal models quantify the relationship between a set of causal, or independent variables, and the variable of interest. An example of this type of model can be the hypothesis of a simple relationship between the number of tourists visiting one country by another country and the independent variables are:

- The per capita income,
- Language similarity,
- Attractiveness, etc.

These can be progressively introduced in order to obtain the most satisfactory outcome. Regression and econometric models fall into this category. In short, the causal model determines what has caused whatever trends, seasonality's, and cyclical behavior may exist in the data, and that an extrapolation of the trend, seasonal, and cyclical behavior will yield an appropriate forecast.

One advantage of this type of models is identifying the variables that are critical in the chain and how they are modifying the behavior of the different participants in the model. The causal models help the researchers to model complex structures based on behaviors, expected results, and importance of events. For more detailed information the reader can refer to Aledo et al. (2010).

Time Series Analysis

Time series analysis quantifies the trend, seasonal, cyclical components of several time series associated with the tourism industry. These models are not concerned with explaining the reason why the forecast is what it is, while in the causal model in which the subjective factor is expressed as in index. Cases like:

- Tourists arriving and departing by air,
- Tourists arriving,
- Foreign currency-income from tourism,
- Residents departing,
- And bed nights in tourist hotels broken can be included in the analysis.

More detailed analysis is focused on the impact of government policies alongside the behavior of the tourism in different zones, or the behavior of tourism demand in different touristic attractions.

Qualitative Models

The previous models require past data, however qualitative forecasting models are designed to obtain and capture the consensus of experts in the field of interest with regard to what the future will bring. So in the qualitative models, the decision of visiting a place comes from a mental cognitive process, which considers and evaluates the characteristics of various possible courses of action in the light of one's needs, goals, and limitations (time, skills, funds, and others). By using this information, the planner or manager can forecast the potential visits. Following this idea the decision making method is a common analysis in this category, usually combines two points of view:

- Technical, and
- Social.

These features can be useful in the research of tourism and supply chain. The Delphi technique also belongs to the qualitative models, in which the discussion and important opinions of a group of experts can help to forecast the visits of a tourism place.

Tourism Flow Models

The tourism flow model predicts tourist flows between the various touristic attractions in a geographical area. The aim of this model is to determine how many will be visiting another location within that area at a subsequent time. Probabilistic (stochastic) and Deterministic approaches belong in this group.

Probabilistic Approaches (Stochastic)

For this approach, different variables can be assessed to determine tourism flow. The study of Algieri and Kanellopoulou (2009) gives the list of those variables, such as tourist arrivals, overnight stays, real revenues, real expenditures and visits per head of the origin's population.

For example, in the study of Swart et al. (1978), exposes the use of the Markov chains, in which the transition probability matrix illustrates the relationships, which can be the economic regions as locations of overnight stopping points. The results of the model yield the average number of overnight stops and their distribution throughout places for visitors' first stopping.

The work of Algieri and Kanellopoulou (2009) explains the core performance of tourism exports. For that, an unobserved component in the form of time-varying trend in the considered tourism demand equation to pick up stochastic un-modeled behaviors of the series and to avoid spurious regressions.

Deterministic Approaches

In this approach the tourism flow model is part of a multi-period planning model for tourism development. The predicting changes in tourist flows are the result of the implementation of new changes (touristic attractiveness), so the consequence, existing tourist flows will be redistributed to reflect this change. This type of models can be evaluated as causal models, as well due to the structure and analysis of the factors involved.

Evaluating the Impact of Tourism

The relevance of this assessment is related to the income generated enhancing the economic sectors due to the tourism impact, which also encompasses the social, cultural, and political impact as well.

Classical Economic Approaches

This approach reflects the contribution to the total net income in foreign currency resulting from tourism activities. By econometric model is easy to estimate the contribution of the visitor industry to economic growth. Examples of this approach involve:

- Evaluating the financial appraisal,
- Economic efficiency analysis based on the prices, and
- The social analysis in terms of the investment consequences.

Other approach is integer-programming model for tourism economical planning.

Derived Value Approaches

Other benefits than economic, such as the social welfare, are evaluated under this approach. Other examples rely on the pursuit of outdoor recreation, as valuable experience; identify patterns and trends in the usage of outdoor recreational facilities, and the use of Delphi panels to evaluate the evolution of tourism.

Surrogate Approaches

The profit or benefit is obtained by other measures in the tourism supply chains. An example of this measure is the touristic attractiveness to evaluate the development of tourist accommodations and attractions in a given area. Diverse researchers applied multi-attribute utility functions.

Additionally, it is possible to determine net foreign exchange earnings due the implementation of development projects that contribute the touristic attractiveness. By these approaches, it is also possible to evaluate the impact of tourism in terms of demand/supply analysis of the tourism sector.

A Longitudinal Approach

In the tourism area it is very common to set plans in the early stages. However, changes in those plans occur frequently due to the consequence of the predictions made. These approaches measure the impact in the change in the semblance of the initial plan.

Decision Models

These types of models evaluate problems where it is necessary to choose a specific alternative among a groups of alternatives. The decision models help to solve the problems encountered by the tourism planner. The different alternatives have different characteristics and resource requirements, but at the same each one has many benefits.

The decision models can be defined as the essence of operations research and is one of the characteristics that distinguishes operations research from other disciplines. These models can be subdivided in prescriptive and descriptive.

Prescriptive Model

The problem is modeled in a way to prescribe a best course of action. Some examples of the application of this type model can include:

Tourism investment planning policies, which can be obtained by considering the budget allocated to that sector, thus, enhancing the regions touristic attractiveness by a given amount. For that it is necessary to consider technological and practical limits, or multiple planning periods and carrying capacities of touristic area. Other studies involves maximize the net income in foreign currency from tourist activities, however, for that it is necessary to consider certain restrictions related to:

- Economic efficiency,
- Accommodation facilities,
- Manpower, and
- Recreational sites, the demand, and preferred seasons.

Other scopes also include additional information such as:

- Amounts of capital investment,
- Maintenance costs,

- Land requirements, and
- Overall state recreation requirements.

Descriptive Model

This type of model describes or evaluates the effect of a decision, which was postulated by the user of the model, considering several alternatives that can be of great value. From a macro-point of view is a descriptive model based on simulation considering constrains as limits of each situation. Examples of these models usually focus on planning models for the tourism sector, routing models for the tour sectors, and others.

Figure 4 displays the most common models applied when the tourism supply chain is designed.

Figure 4. Summary of model to analyze tourism supply chain

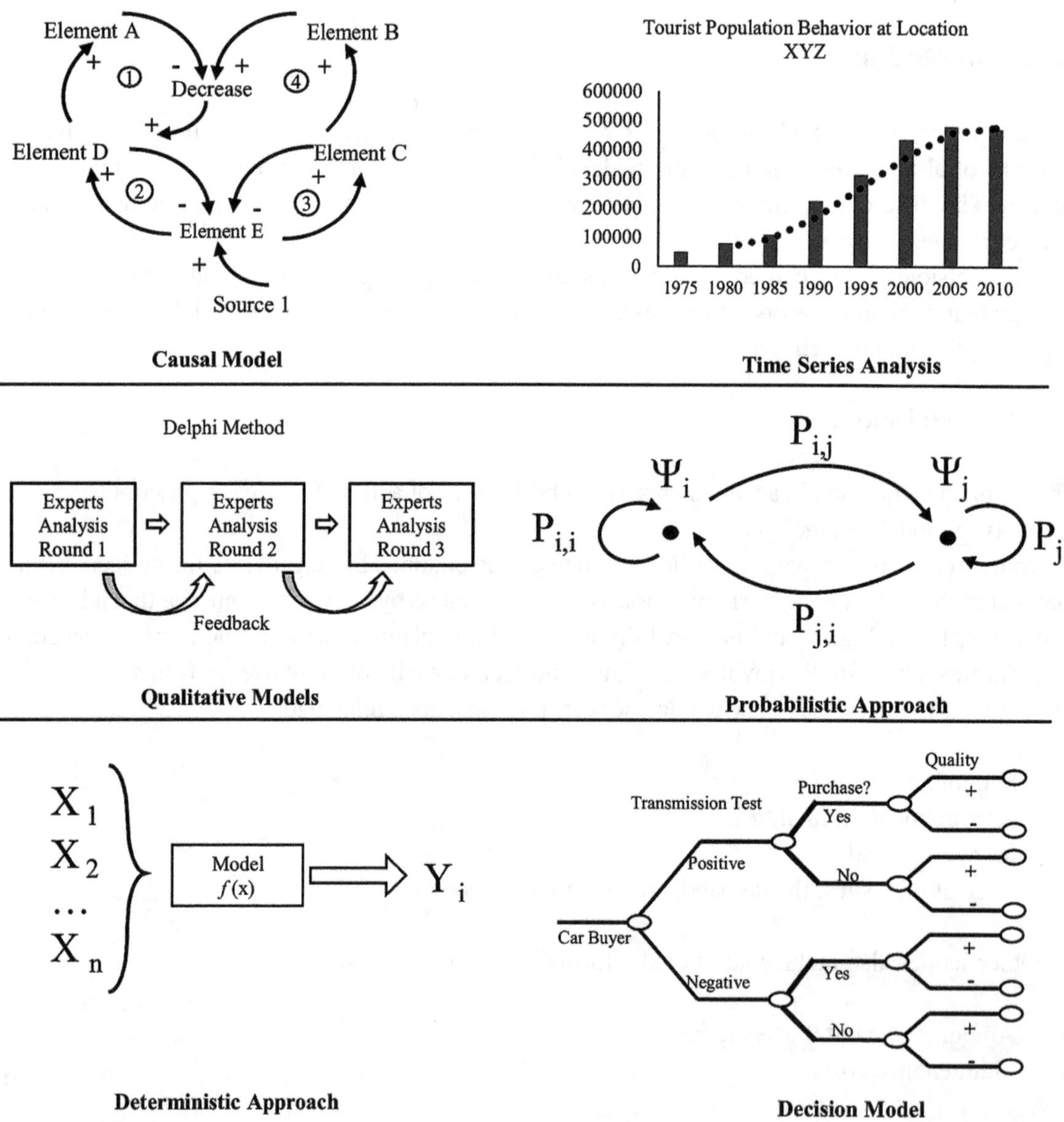

SUPPLY CHAINS MANAGEMENT AND OPTIMIZATION MODELS FOR TOURISM (TSCM – OM)

The research work done by Zhang et al. (2009) includes a review of several studies in the field of tourism supply chain management. In the analysis, they took into account articles from the year 1997 to 2007, in which the authors divided them in terms of methodologies employed. This division encompasses quantitative, conceptual and empirical categories.

Since the interest of this chapter involves both, the supply chain management merged with the optimization models, the study of Zhang et al. (2009) is used as a starting point considering only those articles that belong to the quantitative category. Although in the work of Zhang et al. (2009) thirteen articles were categorized under quantitative analysis, several of them are outside of the scope of the current analysis, because some articles are focused on forecasting and econometric approaches mainly. As a result, only one article actually incorporates an optimization model in its methodology and therefore it is included in this chapter.

For this research work, the articles analyzed were taken only from Science Direct Database. The searching process was done using as key words "supply chain" and "tourism", and filtering the timeline from the year 2006 to 2016, and the results are a set of 711 articles. Using this set, a second round of searching was performed, now considering the key words "optimization" and "mathematical models". As consequence of this filter, only 41 articles are inside of the scope of the analysis. Then, the last set of articles were revised with the aim to determine which belong to the final group that includes both, TSCM and optimization model. Table 1 illustrates the summary of the analysis performed using 15 selected articles sorted chronologically.

In Table 1, the column "Application" refers to the research fields applied in each study. The Table 1 is divided in two sections:

- Supply chain, and
- Optimization considerations.

The section "Supply chain considerations" was designed taking into account the common strategies applied in the majority of articles selected. This section is structured as follows:models that are

1. Applying game theory as fundamental model for supply chain,
2. Using the market behavior with an oligopolistic approach,
3. Considering pricing as strategy to model the behavior of the members in the supply chain,
4. Including coopetition as strategy to model the interaction among elements in the supply chain.

The column "Approach applied" refers to the mathematical model used in order to design the supply chain considering the different strategies applied.

The section "Optimization considerations" has three columns. The column "Flow model" indicates which articles focused on optimize the network using approaches of routing, transporting, and traveling salesman person. The column "Decision making" shows the articles whose objective includes mathematical models merged with decision-making strategies. The column "Approach applied" specifies which mathematical models were applied to find a solution using the decision variables employed in

Table 1. Overview optimization models in tourism supply chain

	Author	Application	Supply Chain Considerations					Optimization Considerations		
			Game Theory	Oligopolistic Market	Pricing	Coopetition	Approach Applied	Flow Model	Decision Making	Approach Applied
1.	Wie (2005)	Cruise lines tours	•	•	•		Dynamic model		•	Non cooperative Nash equilibrium capacity
2.	Garcia and Tugores (2006)	Destinations	•		•	•	Two stage model	•		Partial equilibrium analysis and vertical differentiation model
3.	Candela and Cellini (2006)	Destinations	•	•	•		Dynamic model	•		Open loop Nash equilibrium
4.	Accinelli et al. (2008)	Sustainable tourist	•				Tourism game with central planner		•	Nash equilibrium
5.	Parolo et al. (2009)	Protected areas					Stochastic Genetic Algorithms	•		Allocation of tourist infrastructure
6.	Song et al. (2009)	Destination and tours	•		•	•	Stackelberg approach		•	Tourism utility problem; Nash bargaining
7.	Huang et al. (2010)	Hotels, destinations and tours	•	•	•		Two stage model		•	Non cooperative models
8.	Batabyal and Beladi (2011)	Guided tours					Discrete-time Markov chains		•	**
9.	Tseng et al. (2011)	Destinations			•		Multicriteria		•	Fuzzy and decision Making Trial
10.	Guo et al. (2013)	Hotels and websites	•		•		Stackelberg approach		•	Nash equilibrium
11.	Andergassen et al. (2013)	Destinations	•		•		Two stage model		•	Optimal degree of Tourism sophistication
12.	Huang et al. (2014)	Hotels					Modified two stage model		•	Network DEA
13.	Guo et al. (2014)	Hotels and online travel agencies	•		•	•	Stackelberg approach		•	Benchmark problem
14.	Nilashi et al. (2015)	Hotels					Multi-criteria collaborative filtering		•	Expectation maximization algorithm and adaptive neuro-fuzzy inference system
15.	Hong et al. (2015)	Destinations	•				Two stage model	•		Travel agency supplied tourism packages problem

** Optimization model is not developed totally

each model. A relevant point is that sometimes due to the assumptions and modeling applied a NP hard problem is raised, for example the study done by Parolo et al (2009) faced this situation.

By analyzing the Table 1, it can be recognized that the studies were applied in destinations, hotels, guide tours and others. Destinations refer to the locations (place) where a tourist can arrive and all the services and attractions in the surrounding the area. On the other hand, the majority of the optimization models that were found involve two categories:

1. Flow models, and
2. Decisions making.

The first one is related to routes and network analysis, while the second one refers to an optimization methodology based on specific decision variables.

It was found that the majority of the articles described the TSCM applying game theory concepts. From Table 1, it can be recognized that game theory is combined with other approaches depending on the market strategy. The backup methods are:

- Two-stage model,
- Multi-criteria model,
- Dynamic models,
- Stochastic approach,
- Stakelberg game approach and others.

Two Stage Models

These models analyze the supply chain as a combination of two features, identifying the competitors and the cooperation characteristics among the stakeholders in the chain. One characteristic of these models is that they can apply similar or different features for each component in each stage. Due to the dynamic of the elements in this models, the econometrics approaches are usually applied, however fuzzy concepts can cover the requirement of the two stages models as well. Garcia and Tugores (2006) described a two-stage model between two hotels by measuring the quality service and the competition of the price service. The main objective, according to the authors, is to develop a policy to remain valid the idea of tourism industry towards high quality popularity and environmental issues. One attribute of this study is that the authors consider a duopoly as a market strategy instead of oligopolistic market, which it is frequently used in these situations. The study aims to analyze the role of local consumers and the nature of external effects under optimal conditions. The authors pretend to optimize by considering a partial Nash equilibrium analysis simultaneously with a vertical differentiation model. One relevant assumption made by the authors is considering that there are no externalities affecting the welfares of the last components in the supply chain. The objective function is to maximize the social welfare among the stakeholders.

A similar approach is suggested by Huang et al. (2010). They evaluated a supply chain formed by tour operators, theme parks and hotels – accommodation providers. Looking from the two-stage point of view, the study examined the competition among the tour operators and the hotel-accommodation providers considering the operation cost, sector size and product differentiation in the tourism supply chain (TSC). The authors assume an oligopolistic market under a pricing strategy. The contribution of the authors is the application of different optimization models for each of the members in the supply chain, looking for the benefits interactions among members in the supply chain. The authors study the two-stage model analyzing the price competition among hotels and the behavior of a new competitor in the chain. For example, if a new competitor enters to the market, the decision makers of the other firms in the same sector are forced to reduce their product prices. But the other sector in the same line will get more tourists and profits, benefiting from intensified price competition in its complementing sector.

The supply chain analyzed has three parts,

1. Multiple tours operators,
2. Multiple hotels and accommodation providers, and
3. Destinations.

The objective function proposed by the authors is to reduce operating costs with the aim to attract additional tourists from its competitors in the same sector, and thus improving the general profitability. The two main assumptions that are considered during the modeling are:

1. Providers are assumed to form a hotel chain and offer accommodation at a unified cost level due to their internal competition. In reality, package holiday products may use hotels of different grades,
2. The competition dynamic is examined through sensitivity analysis assuming that changes of operating parameters in the tourism supply chain are small enough that they will not lead to changes during the equilibrium optimization.

In the same line, Andergassen et al. (2013) study a two-stage model focusing on the optimal degree of sophistication of the tourism product, and the optimal institutional arrangement in terms of coordination. The members of the supply chain are studied under a decision tree, considering the destination and the operators (local or foreigner). The authors do not assume an oligopolistic market in their analysis. The two-stage model is based on the optimal development strategy of a tourism destination by identifying and analyzing two key economic factors:

1. The long-term choice of whether to invest in the improving of natural and/or cultural resources or to increment the degree of sophistication of the tourism product, investigating a diversity of complementary services to accommodation that are requested by tourists;
2. The short-term choice of whether or not to implement price strategies among local firms.

Some limitations are clearly defined by the authors:

1. The model proposed is a partial equilibrium one, since the destination is analyzed under isolation form,
2. Accounting explicitly for production and coordination costs, which may reverse the result that price coordination is optimal and might affect in a non-trivial way investment and development strategies in the first stage of the game,
3. Assuming that differentiated tourism related goods and services are produced by more than one firm with some market power, that is, abandoning the assumption of monopolistic competition and framing the model in the context of oligopolistic competition, one could explicitly study the more general price coordination problem where complementarity and substitutability, the core elements of the tourism product, coexist,
4. Tourists being crowding-averse, which is tantamount to assuming that a snob-effect is at work, has to be extended to the alternative assumption of mass tourism, where a band-wagon effect would be at work instead, finally
5. It is assumed that prices and types of investments are strategies without interactions, so an extension would be to transform the multiple-stage problem into a simultaneous equilibrium in which both prices and types of investment are jointly determined.

Hong et al. (2015) propose a two-stage model using game theory whose attribute rely on a network created by several links among tourist destinations and other destinations in the location. The outcome of implementing this strategy is a distinctly hierarchical network of tourism destinations (NTD). The optimization model is developed considering equilibrium analysis but not using an oligopolistic market they propose a set of different assumptions. This research focuses on the relationship between networks of tourism destinations and tourism flows. First, the authors propose a new model to demonstrate networks of tourism destinations using travel agency that supply tourism packages. They demonstrate how to build the network matrices using operating steps applied to tourism markets in China. Next, the authors construct a two-stage game model that assumes that the more links a tourism destination has with other destinations, the greater will be its tourism flows. The study presents that the two variables, road (the length of roads per capita in the destination) and star (the number of star-rated hotels in the destination), are negatively correlated with tourism flows, which is inconsistent with most of the existing literature.

In contrast to the previous articles, the work performed by Huang et al. (2014) proposed a modified two-stage model, just considering the optimization approach by using data envelopment analysis (DEA). DEA is a mathematical tool employed to evaluate the relative efficiency score or performance of one entity related to others of the same nature. The result obtained is a measurement of the level of performance of the entity. It explains how efficiently are the resources used or consumed by one unit, in this case a hotel, compared to others. The objective of this analysis is to measure tourist hotels' productive efficiency, by evaluating the capacity of the service to generate its outcomes (people occupying the hotel and the catering services). The study does not explain clearly the supply chain strategy that are considering instead the authors focuses on the efficiency measure of three frameworks characterizing the properties of value-added chain and resource allocation in a hotel.

Multi-Criteria

Inside this category, the tourism industry is evaluated under formal analysis of every judgment considered in a particular event. For example, in the study done by Tseng, at al. (2011) the authors provided a way to analyze the integration between marketing and supply chain management to help companies fulfill the customers' demands without concerning the demand's uncertainty. For this, fuzzy and decision making trial are used to identify the right destination not only under a pricing strategy, but also incorporating other subjective qualities such as:

- Safety,
- Beauty,
- Strangeness of the destinations.

In the analysis done by the authors, they used the decision-making trial and evaluation laboratory (DEMATEL) research method in uncertainty to evaluate the integration. The different steps followed in this methodology are described herein. First, it is necessary to identify decision goals and obtain the relevant information to evaluate the advantages and disadvantages, and monitoring the results to ensure the goals be achieved. An expert committee should be formed to help this. The second step requires the evaluation criteria and a survey instrument to evaluate the relationships. The next step involves converting and interpreting the linguistic information into fuzzy linguistic scale. For the last step, it is necessary to carry out an analysis involving causal-effect diagram to describe the relationships (goal).

Another approach is proposed by Nilashi et al. (2015) by using collaborative filtering (CF) techniques, which attempt to predict what information will meet a user's needs based on data coming from similar users. The distinguishing approach proposed by the authors is the incorporation of statistical techniques such as the Gaussian mixture model with expectation maximization (EM) algorithm and adaptive neuro-fuzzy inference system (ANFIS), and the principal component analysis (PCA). The aim is to model a map of the inputs and outputs, and find a function for a given input vector in order to predict an output answer. Figure 5 shows a block diagram of the two stage and multicriteria models.

Dynamic Modeling

In tourism, the stock of the resources usually faces a dynamic change over time. Thus, an increased complexity of the analysis appears, where dynamic modeling can tackle this situation. For example, Candela

Figure 5. Block diagram for two of the common approaches applied in tourism supply chain optimization models

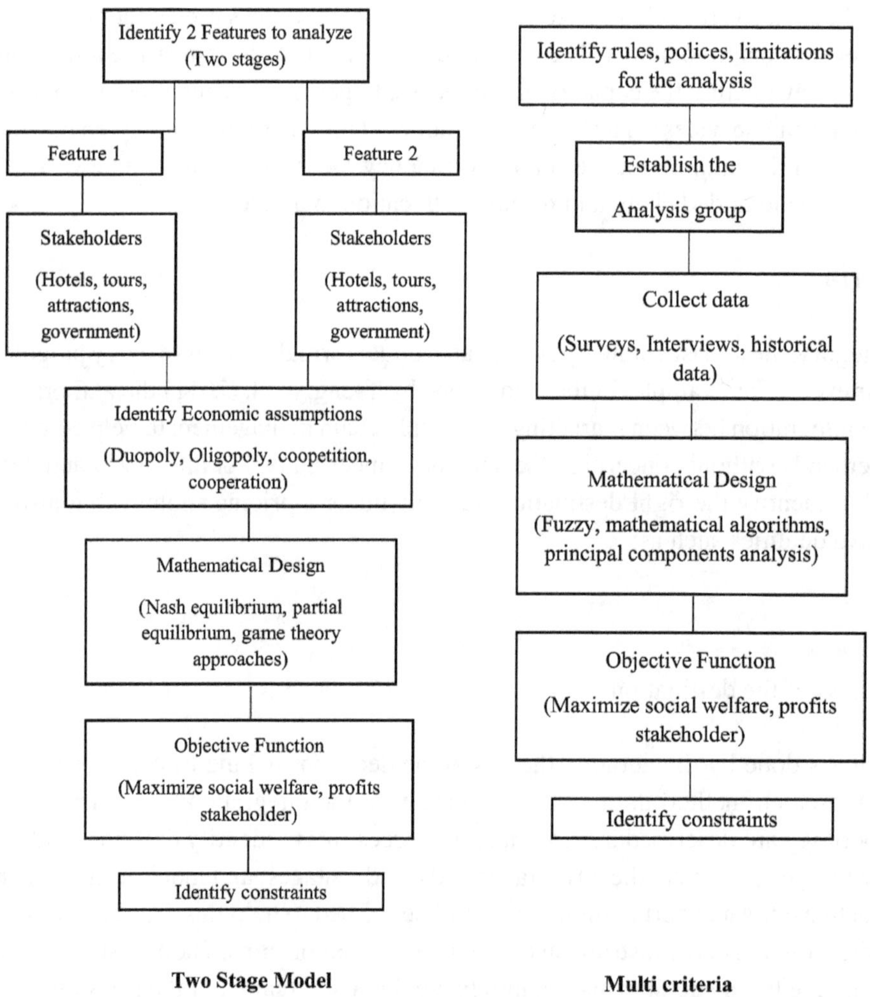

and Cellini (2006) proposed a differential game model of oligopoly, where the number of destinations are large, with diverse goods to study the development strategies of competing tourism destinations. Furthermore, the number of intermediary organizers of the tourism flows is limited and, the markets are highly concentrated. In their study, they identified the optimal path of tourist presence over time, for a given destination. The aim of the study was maximize the present value of the flows of its profits over time. Open-loop Nash equilibrium of the game among destinations was employed to derive an optimization model. The supply chain strategy in this study covered the relationship between the scale of tourism and investment effort and, the product differentiation and the number of available products.

In a similar line, Wie (2005) analyzes the dynamic evolution of capacity investment decisions that lead to maximize the profits of cruise line over a fixed planning horizon. One characteristic of the research was to consider the market as an oligopoly but not in a cooperative way among the parts. The authors used Nash equilibrium capacity in an open-loop information structure. The authors suggested that, this analysis should also be done under a close-loop structure, and considering some factors such as Market Expansion, economic prospects, and competition opportunities. In terms of the price, it also requires an analysis of price differentiation and price elasticity of demand.

Stochastic

Stochastic analysis is a different approach applied in the study of tourism industry. For example, the exploration by Batabyal and Beladi (2011) suggest a modeling of the slack season provision of guided tours to tourists and destinations. The authors considered heterogeneity and tourist demands perspectives. The analysis used the Markov-chains to model the two possible scenarios, where the probability is based on that a particular tourist will continue visiting the rest of the destinations during a specific period of time. Thus, the excursion time of an arriving tourist is exponentially distributed with a fixed mean. However, the authors excluded from the analysis if the stochastic process reflects arrivals in batches, or what is called as batch arrival queuing model. Although, from the optimization point of view is not proposed, the equations can be used to optimize business matters.

Parolo et al. (2009) employed a stochastic genetic algorithm to propose a model to allocate tourist infrastructures (refuges and camping sites). This approach replaces the multi-criteria analysis to obtain a complex interplay in the form of a dynamical simulation where candidate solutions are interactively evaluated. Through a four-step method they solved this problem.

- First they conceptualized a model for the minimization of tourism impacts within protected areas through the use of a spatially explicit genetic algorithm;
- Then they used the model for planning new tourist infrastructures (e.g. Refuges, camping sites) to be added to existent ones;
- After that, they measured the degree of success (performance) of each of the genetically fittest solutions on the basis of the initial criteria, followed by a comparison of the performances of the resulting tourist infrastructures with the performances of existing ones;
- And finally of the least fit genetic solutions (i.e. The worst possible solutions) within the study area, in order to quantitatively and objectively evaluate the advantages provided by the proposed methodology.

Stakelberg Game Approach

The Stackelberg model is an approach used in settings including a leader and follower(s), whose objective is optimize the same set of variables. In the work of Song et al. (2009) is detailed an example of how to apply Stackelberg model to formulate the relationship between a destination and a tour operator, with the aim to maximizes the two players' profits. For this, it is necessary to develop some strategies to increase the demand. Although, the authors pointed out that the primary element that affects the demand is the price, they excluded other important parameters such as seasonality, economic condition and even epidemics that could reflect significant changes in the results. The authors optimize tourism utility problem by the using Nash bargaining approach.

Another approach is the one from Guo et al. (2013) which identifies the optimal pricing strategy for tourism hotels with a third party website cooperation. The game between the hotels and the third party website follows a Stackelberg model in which the hotels play as the leaders and the third party website as the follower. While the game among the hotels follows a Nash game model and in which each hotel decides the commission price to obtain the optimal effort level from the website. Following this idea, and using Stackelberg model Guo et al. (2014) propose a new optimization model based on benchmark modeling considering the players as a single. The aim is to maximize the profit for the whole chain and also determine the cash back issue between the hotel online marketing channel and the online travel agencies.

Other Approaches

Accinelli, et al. (2006) proposed a different approach using game theory. The authors analyze the interdependence between environmental protections of resources, the activities related with the tourism and the behavior of a local population. In the analysis, the players in the supply chain are:

- The central planner (CP),
- Tourists (T), and
- The local inhabitants (L).

The objective is that the central planner (CP) must choose between invest or not on tourism activities, with the minimal social impact. Nash equilibrium methodology is applied to model the needs of the players in the supply chain.

FUTURE RESEARCH DIRECTIONS

Being a specific and brief overview of supply chain management and optimization models for tourism, this research work leaves to interested researchers some future research directions. A relevant future research avenue is to conduct an analysis that identifies which tourism places are losing their attractiveness. For instance, it would be relevant to study whether or not those tourism places are affecting the economy of a country. Since it is increasingly significant for tourism industry knowing and understanding the dynamic and behavior of tourists by an integrative approach of the optimization models and supply chain concepts; it requires an improvement of the tourism industry in order to maximize benefits for all parties involved.

Another future line is the application of time multi-stage DEA analysis in tourism supply chain. This analysis will evaluate the tourism industry over time, measuring the industry activities prior, during and posterior the service offered. This assessment will contribute this industry because it will help to identify how the different elements are contributing more in the global efficiency of the supply chain, and, it will identify which stage is generating a greater impact in the industry.

Since other point of view, tourism supply chain might be modeled as a NP hard problem. In fact, during the literature review just a few authors designed tourism based in a NP hard problem. The advantage of work under this structure is that there exist many models and researches that can be used as a base to initiate tourism NP hard problems models to optimize the supply chain.

Finally, it is also important to study the reliability of the different approaches, such as econometrics, statistical, conceptual and empirical, considering situations when a synergy among these approaches can be used, which outcome will be near to the behavior of the supply chain elements in tourism.

CONCLUSION

Undoubtedly, tourism industry provides huge economic benefits to the countries. This chapter gives an overview of recent articles that consider supply chain management and optimization models applied to the field of tourism. It was identified and discussed some articles from 2005 to 2016 related to the tourism supply chain and optimization models. It was noted that, the game theory and economic features are applied in almost of all the cases, but also some other new approaches are also involved.

The common economic feature in the analysis is modeling the supply chain as an oligopolistic market. This implies that an econometric approach is required to model the tourism supply chain. Several authors proposed the use of equilibrium models; in fact, the most used model is Nash equilibrium considering all its variants.

Something that is usual in the articles is the use of pricing as strategy to define the objective function. In contrast, coopetition strategy is rarely applied due to the complexity in analyzing the interaction and behavior of the elements of the supply chain, when they are cooperating and competing simultaneously. The dynamic of the supply chain is analyzed considering two stage models, stochastic and multi-criteria models, genetic algorithms and novel approaches. Even though, a great number of assumptions are considered by the authors, thus a gap is generated for future research. Since the optimization point of view, the majority of models are studied as decision-making models, focusing on the price as decision variable. Some models considered flow models as transport, integer programing, and linear programming. According to the Table 1, the last researches in tourism supply chain apply optimization methods based on probabilistic events and using expectations maximization algorithm, fuzzy inference, fuzzy decision, among others.

REFERENCES

Accinelli, E., Brida, J. G., & Carrera, E. (2008). A good policy of sustainable tourism. *Revista de Administración, Finanzas y Economía, 2*(2), 150–161.

Aledo, A., García, H., & Ortíz, G. (2010). Análisis de mapas causales de impactos del turismo residencial. *Empiria: Revista De Metodología. Ciencia y Sociedad*, (20), 2061–2086.

Algieri, B., & Kanellopoulou, S. (2009). Determinants of demand for exports of tourism: An unobserved component model. *Tourism and Hospitality Research*, *9*(1), 9–19. doi:10.1057/thr.2008.39

Andergassen, R., Candela, G., & Figini, P. (2013). An economic model for tourism destinations: Product sophistication and price coordination. *Tourism Management*, *37*, 86–98. doi:10.1016/j.tourman.2012.10.013

Batabyal, A. A., & Beladi, H. (2011). An alternate approach to modeling the slack season provision of guided tours to tourists. *Tourism Management*, *32*(5), 1047–1049. doi:10.1016/j.tourman.2010.09.001

Candela, G., & Cellini, R. (2006). Investment in tourism market: A dynamic model of differentiated oligopoly. *Environmental and Resource Economics*, *35*(1), 41–58. doi:10.1007/s10640-006-9005-3

Cremer, H., Marchand, M., & Thisse, J. (1989). The Public Firm as an Instrument for Regulating an Oligopolistic Market. *Oxford Economic Papers*, *41*(2), 283–301.

Czakon, W., Mucha-Kus, K., & Rogalski, M. (2014). Coopetition Research Landscape – A Systematic Literature Review 1997-2010. *Journal of Economics & Management*, *17*, 122–150.

Franklin, A. (2003). *Tourism: An Introduction*. London: SAGE Publications Ltd. doi:10.4135/9781446220108.n2

Garcia, D., & Tugores, M. (2006). Optimal choice of quality in hotel services. *Annals of Tourism Research*, *33*(2), 456–469. doi:10.1016/j.annals.2005.12.010

Guo, X., Ling, L., Dong, Y., & Liang, L. (2013). Cooperation contract in tourism supply chains: The optimal pricing strategy of hotels for cooperative third party strategic websites. *Annals of Tourism Research*, *41*, 20–41. doi:10.1016/j.annals.2012.11.009

Guo, X., Zheng, X., Ling, L., & Yang, C. (2014). Online coopetition between hotels and online travel agencies: From the perspective of cash back after stay. *Tourism Management Perspectives*, *12*, 104–112. doi:10.1016/j.tmp.2014.09.005

Hong, T., Ma, T., & Huan, T. C. T. (2015). Network behavior as driving forces for tourism flows. *Journal of Business Research*, *68*(1), 146–156. doi:10.1016/j.jbusres.2014.04.006

Huang, C. W., Ho, F. N., & Chiu, Y. H. (2014). Measurement of tourist hotels productive efficiency, occupancy, and catering service effectiveness using a modified two-stage DEA model in Taiwan. *Omega*, *48*, 49–59. doi:10.1016/j.omega.2014.02.005

Huang, G. Q., Chen, W., Song, H., & Zhang, X. (2010). Game-theoretic study of the dynamics of tourism supply chains for package holidays under quantity competition. *Tourism Economics*, *16*(1), 197–216. doi:10.5367/000000010790872150

Lafferty, G., & van Fossen, A. (2001). Integrating the tourism industry: Problems and strategies. *Tourism Management*, *22*(1), 11–19. doi:10.1016/S0261-5177(00)00021-2

Lee, E., & Staelin, R. (1997). Vertical Strategic Interaction : Implications for Channel Pricing. *Marketing Science*, *16*(3), 185–207. doi:10.1287/mksc.16.3.185

Nilashi, M., bin Ibrahim, O., Ithnin, N., & Sarmin, N. H. (2015). A multi-criteria collaborative filtering recommender system for the tourism domain using Expectation Maximization (EM) and PCA-ANFIS. *Electronic Commerce Research and Applications*, *14*(6), 542–562. doi:10.1016/j.elerap.2015.08.004

Ordover, J. A., & Willig, R. D. (1981). An Economic Definition of Predation : Pricing and product innovation. *The Yale Law Journal*, *91*(1), 8–53. doi:10.2307/795848

Parolo, G., Ferrarini, A., & Rossi, G. (2009). Optimization of tourism impacts within protected areas by means of genetic algorithms. *Ecological Modelling*, *220*(8), 1138–1147. doi:10.1016/j.ecolmodel.2009.01.012

Pearce, D. G., Tan, R., & Schott, C. (2007). Distribution channels in international markets: A comparative analysis of the distribution of New Zealand tourism in Australia, Great Britain and the USA. *Current Issues in Tourism*, *10*(1), 33–60. doi:10.2167/cit290.0

Sheng, L. (2011). Regional Competition and Sustainable Development: A Game Theory Model for Tourism Destinations. *European Planning Studies*, *19*(4), 669–681. doi:10.1080/09654313.2011.548470

Song, H., Dwyer, L., Li, G., & Cao, Z. (2012). Tourism economics research: A review and assessment. *Annals of Tourism Research*, *39*(3), 1653–1682. doi:10.1016/j.annals.2012.05.023

Song, H., Yang, S., & Huang, G. Q. (2009). Price interactions between theme park and tour operator. *Tourism Economics*, *15*(4), 813–824. doi:10.5367/000000009789955099

Swart, W. W., Var, T., & Gearing, C. E. (1978). Operations research applications to tourism. *Annals of Tourism Research*, *5*(4), 414–428. doi:10.1016/0160-7383(78)90321-3

Tisdell, C. A. (2013). *Handbook of Tourism Economics: Analysis, New Applications, and Case Studies*. Singapore: World Scientific Publishing Company. doi:10.1142/7956

Tseng, M. L., Chiu, A. S., & Vo, M. P. N. (2011). Evaluating the tourists demand to develop Vietnamese tourism performance. *Procedia: Social and Behavioral Sciences*, *25*, 311–326. doi:10.1016/j.sbspro.2011.10.550

Wie, B. W. (2005). A dynamic game model of strategic capacity investment in the cruise line industry. *Tourism Management*, *26*(2), 203–217. doi:10.1016/j.tourman.2003.09.020

Zhang, X., Song, H., & Huang, G. Q. (2009). Tourism supply chain management: A new research agenda. *Tourism Management*, *30*(3), 345–358. doi:10.1016/j.tourman.2008.12.010

ADDITIONAL READING

Acuña, S., & Varela, J. (2011). *Handbook of Optimization Theory: Decision Analysis and Application*. New York: Nova Science Publishers, Inc.

Baggio, R., & Klobas, J. E. (2011). *Quantitative Methods in Tourism: A Handbook*. Bristol, UK: Channel View Publications.

Baur, A., Klein, R., & Steinhardt, C. (2014). Model-based decision support for optimal brochure pricing: Applying advanced analytics in the tour operating industry. *OR-Spektrum*, *36*(3), 557–584. doi:10.1007/s00291-013-0338-3

Brânzei, R., Dimitrov, D., & Tijs, S. (2008). *Models in Cooperative Game Theory*. Berlin: Springer.

Brewer, A., Button, K. J., & Hensher, D. A. (2001). *Handbook of logistics and supply-chain management*. Amsterdam, New York: Pergamon.

Fung, W., & Fung, R. (2014). The development of a supply chain model for tourism crisis management. *ICMIT 2014 - 2014 IEEE International Conference On Management Of Innovation And Technology*, 284-289.

Hillier, F. S., & Lieberman, G. J. (1974). *Operations research*. San Francisco: Holden-Day.

Huang, G., Song, H., & Zhang, X. (2010). A comparative analysis of quantity and price competitions in tourism supply chain networks for package holidays. *Service Industries Journal*, *30*(10), 1593–1606. doi:10.1080/02642060903580631

Jackman, M., & Greenidge, K. (2010). Modelling and forecasting tourist flows to Barbados using structural time series models. *Tourism and Hospitality Research*, *10*(1), 1–13. doi:10.1057/thr.2009.23

Kelly, A. (2003). *Decision Making Using Game Theory: An Introduction for Managers*. Cambridge, UK: Cambridge University Press. doi:10.1017/CBO9780511609992

Mahdavi Mazdeh, M., & Karamouzian, A. (2014). Evaluating strategic issues in supply chain scheduling using game theory. *International Journal of Production Research*, *52*(23), 7100–7113. doi:10.1080/00207543.2014.937880

Nordström, J. (2005). Dynamic and Stochastic Structures in Tourism Demand Modeling. *Empirical Economics*, *30*(2), 379–392. doi:10.1007/s00181-005-0238-8

Petropoulos, C., Nikolopoulos, K., Patelis, A., & Assimakopoulos, V. (2005). A Technical Analysis Approach to Tourism Demand Forecasting. *Applied Economics Letters*, *12*(6), 327–333. doi:10.1080/13504850500065745

Robinson, M., & Jamal, T. (2009). *The SAGE Handbook of Tourism Studies*. Los Angeles: SAGE Publications Ltd.

Rusko, R., Kylänen, M., & Saari, R. (2008). Supply Chain in Tourism Destinations: The Case of Levi Resort in Finnish Lapland. *International Journal of Tourism Research*, *11*(1), 71–87. doi:10.1002/jtr.677

Sequeira, T., & Nunes, P. (2008). Does tourism influence economic growth? A dynamic panel data approach. *Applied Economics*, *40*(18), 2431–2441. doi:10.1080/00036840600949520

Tóth, G., Dávid, L. D., & Vasa, L. (2014). The Role of Transport European Tourism Flows. *Acta Geographica Slovenica-Geografski Zbornik*, *54*(2), 311–320.

Vodeb, K. (2012). Competitiveness of Frontier Regions and Tourism Destination Management. *Managing Global Transitions*, *10*(1), 51–68.

Wierenga, B. (2008). *Handbook of Marketing Decision Models*. New York: Springer. doi:10.1007/978-0-387-78213-3

Zheng, X., Xu, R., Wang, S., & Peng, Y. (2015). Tourism Service Composition Based on Multi-objective Optimization. *Proceedings - 2015 IEEE 12Th International Symposium On Autonomous Decentralized Systems, ISADS 2015*, 117-122. doi:10.1109/ISADS.2015.18

KEY TERMS AND DEFINITIONS

Close Loop: When a player does not know the strategies of the other players.

Data Envelopment Analysis (DEA): A model and method used to evaluate the relative efficiency or performance or an entity among a set of entities called decision making units (DMUs), which performance cannot be evaluated by a single measurement such as profit; instead of that it uses multiple inputs (resources) and multiple outputs (outcomes), by solving linear programming problems for each decision making unit (DMU) according to the observed data.

Decision Making: The process of finding and choosing one alternative among several choice possibilities. Every decision-making process carries out a final result that may or may not prompt action.

Fuzzy Set: Provide a scheme for handling a variety of problems considering several rules not only true or false. It is a good approach to model the entropy in systems.

Game Theory: As a method of applied mathematics, it helps to explain a large collection of economic behaviors, including behaviors of firms, markets, and consumers.

Markov Chains: A random process that undergoes transitions from one state to another on a state space. It must possess a property that is usually characterized as "memorylessness": the probability distribution of the next state depends only on the current state and not on the sequence of events that preceded it. This specific kind of "memorylessness" is called the Markov property.

Multi-Criteria: It implicates problems where a finite set of alternative actions should be assigned into a predefined set of preferentially ordered categories.

Nash Equilibrium: The result of a non-cooperative game involving two or more players, in which it is assumed that each player knows the equilibrium strategies of the other players, and no player has anything to gain by changing only their own strategy. If each player has selected a strategy and no player can benefit by changing strategies while the other players keep theirs unchanged, then the current set of strategy selections and the corresponding payoffs constitutes a Nash equilibrium. The reality of the Nash equilibrium of a game can be tested using experimental economics method. It can be applied in two loops (*Open Loop* and *Close Loop*).

Oligopoly: A market situation where there are only a few competing producers and where each producer must take into account what each other producer does.

Open Loop: When a player knows the strategies of the other players.

Operations Research: The area that deals with the employment of advanced analytical methods and methodologies to help make better decisions.

Optimization Model: Consists on the selection of a best element (with regard to some criteria) from some set of possible alternatives.

Stackelberg Game: A strategic game in economics in which the leader firm moves first and then the follower firms move sequentially. In game theory, the players of this game are a leader and a follower and they compete on quantity.

Stakeholders: The individuals or groups that are likely to influence or be influenced by a proposed action, and classifying them according to their impact on the action and the impact the action will have on them.

Stochastic Process: A collection of random variables, representing the change of some system of random values over time. This is the probabilistic counterpart to a deterministic process (or deterministic system). Instead of describing a process which can only evolve in one way, in a stochastic or random process there is some indeterminacy: even if the initial condition (or starting point) is known, there are several (often infinitely many) directions in which the process may emerge.

Supply Chain Management: The administration of a network of relationships within a company and between interdependent organizations and business units formed by material suppliers, purchasing, production facilities, logistics, marketing, and related systems that facilitate the forward and reverse flow of materials, services, finances and information from the very first producer to final customer with the benefits of adding value, maximizing profitability through efficiencies, and achieving customer satisfaction.

Tourism: The name given to the relationships originated from the interaction of tourists, business suppliers, host governments, and host communities in the process of attracting and accommodating these tourists and other visitors.

Two Stage Model: In the supply chain, these models can be defined as combination of two features, steps, points, or phases, where the subjects under study can be evaluated.

Chapter 11
Benefits and Value of Investments in Information Systems:
The Case of Enterprise Resource Planning (ERP) Systems in the Hospitality Industry

Paula Serdeira Azevedo
University of Algarve, Portugal

Carlos Azevedo
University of Algarve, Portugal

Mário Romão
University of Lisbon, Portugal

ABSTRACT

Organisations constantly have to address the problem of identifying tangible and intangible benefits that can be achieved in result of investments made in Information Systems/Information Technology (IS/IT). This paper has the objective of approaching the problem in the perspective and context of Enterprise Resource Planning (ERP) systems in the Hospitality Industry. It focuses on implementation project strategies, namely in the particular and relevant aspects of their return on investment as well as benefits expected and delivered. When an organisation is confronted with the question of knowing if a certain investment in IS/IT may obtain a considered positive financial revenue, and if that investment means valuable benefits for the business, the answer is often far from being conclusive. Projects are often implemented on-time, on-budget and are technically appropriate, nevertheless the expected benefits are not achieved. This paper seeks to answer these questions in the context of ERP systems, in particular in the sector of hospitality industry, through a case study developed in Portugal. The investigation was developed on a group of hotel units, held by a Portuguese company, for which ERP meant an opportunity to harmonize and improve business processes.

DOI: 10.4018/978-1-5225-1054-3.ch011

INTRODUCTION

Organizations have sometimes serious doubts about a conclusive answer, when asked for the benefits of IS/IT investments, namely: does it have a positive ROI (Return On Investments) or how will the business benefit from those investments?

Although many IS/IT projects (but not limited to this area, we can feel the same problem on other areas such as infrastructures, construction, urban projects, among others) are criticized for not achieving their goals, promises, budget and schedule, many others are often implemented within the expected time frame and budget, and are technically adequate, but nevertheless the expected benefits are not achieved.

One of the objectives of the chapter is to draw attention to this fact, normally absent in project evaluation.

Enterprise Resource Planning (ERP) systems are packages of software:

- Oriented for the organization as a whole,
- Claiming for the benefits of a holistic approach,
- Integration of data and processes,
- Dramatic improvement of valuable information,
- Reporting and finally an alignment with business strategy, dissolving the barriers of technology towards simple and efficient business processes.

In the case of ERP systems, the projects expand through the whole organization, even to customers and suppliers sometimes, and as such they have wider goals than traditional IS/IT projects focused on smaller areas of the organization´s value chain. Normally, ERP projects create expectations resulting from:

1. Discontinued processes,
2. Process improvement, or
3. New processes.

In other words, the context in which investments of this kind are made may be oriented towards existing tasks or innovative ones.

This chapter deals with these matters, in the context of an investigation developed on a group of hotel units, held by a Portuguese corporation, to which ERP meant an opportunity to harmonize and improve business processes.

The Application Portfolio technique used in the investigation (and described later in this chapter) focus on the view the organization has on their strategy towards IS/IT systems and in what degree they are aligned (or are not aligned at all) to business strategy.

The other contribution of this chapter is the identification of directions to evaluate how the organization obtains value from their IS/IT investments, with particular emphasis in the case they are using ERP systems.

The chapter also refers to future directions in similar research projects.

ERP SYSTEMS IN THE HOSPITALITY INDUSTRY

The modifications offered by the impulse of technologies, forces the change in the way business is conducted in the hospitality industry. IS/IT in the hospitality industry must provide the required flexibility and efficiency by facilitating the understanding of customer needs and adapt consequently (Beldona, Cai, & Pearson, 2001).

ERP Systems emerged as a way to automate repetitive processes and provide managers with a global vision and real-time all operations, solving the problems of disintegration and fragmentation of information (Beldona, Cai, & Pearson, 2001). The problem of fragmentation of information is felt, as in other activities, within the hospitality industry (Heart, Pliskin, Schectman, & Reichel, 2001).In this industry, ERP Systems have a particular relevance for the diversity of applications and specific systems for various functions. The predominance of legacy systems, of difficult upgrading, maintenance and incompatibility, makes its updating more complex (Beldona, Cai, & Pearson, 2001).

The international chains, or larger units and groups, have adopted ERP Systems traditionally used in other industries. SAP is market leader, in particular for larger hotels (Panorama Consulting Group, 2009). With regard to the specific processes of the hotel units, primarily front office, for example,

- Reservations,
- Stock and supply,
- F&B (Food & Beverage), or
- Point of sale management (POS – Point of Sales),

specific solutions are adopted, often from suppliers who do not offer integration with the back office and the implemented ERP System (Heart, Pliskin, Schectman, & Reichel, 2001). As a consequence, integration of applications is an increasingly relevant issue.

In a study carried out at the hospitality industry in Israel (Heart, Pliskin, Schectman, & Reichel, 2001) found that 97% of the hotel units were computerized (but not by what could be considered ERP Systems), including those who had less than 50 rooms. The most common front office applications were the reservations and the front desk; the most common back office application was accounting. Unlike other industries, less common applications were sales and marketing, and human resources (Heart, Pliskin, Schectman, & Reichel, 2001).

In the same study, Heart et al. (2001) found the absence of ERP Systems targeted for the hospitality industry in Israel, which forced the hotel units to acquire isolated applications from different suppliers. He also noted the absence of integrated applications that share the same database. The diversity of applications from different suppliers increases the complexity of the updating and maintenance of the applications implemented (Heart, Pliskin, Schectman, & Reichel, 2001).

METHODOLOGY

With the purpose of identifying major critical success factors when implementing ERP Systems, as well as factors that may obstruct organizations from being competitive, a case study was carried out based on data gathered on an important economic Portuguese group. The study also intended to understand how the hospitality industry can solve the problem of the disintegration of information of the various implemented business applications. The research did not seek to confirm or refute hypotheses or the quantitative measurement of the influence of variables in a particular phenomenon. Instead, it attempted to answer proposed questions, interpreting, through the systematic analysis of the collected data, ratings, perceptions, needs and limitations of IS/IT submitted by the people involved in the study.

Therefore, an inductive logic followed, with emphasis on the analysis of qualitative data and using the case study method. Although the dominant approach in research processes in the area of IS has been, until some time ago, the quantitative analysis, the research using the case study method, with qualitative data collection, has become increasingly accepted in the area of IS. This appears to stem from a growing recognition of the potential to help researchers understand the interpretations and meanings that govern activities of organizational stakeholders, as well as how technology is faced and used.

The research presented followed an interpretative approach. In view of the issues to investigate, an approach to understanding and interpreting facts was adopted, by the insertion of researchers in the organizational context of hotel units. This interpretative option permits that a group of people, including managers, directors of IS, or users, in an organizational context, can express, conceptualize and assess the objectives defined by the research.

The research method adopted combines several techniques, such as:

- Semi-structured interviews with the Chairman of the Board of Directors of the hotel group,
- The director of IS/IT's department of the group, as
- Well as the heads of logistics, human resources and financial areas of the IS/IT's department of the hotel group,

This is to find inefficiencies and inconsistencies in the information used and facilitate the analysis of the processes involved. Interviews with key elements of the organization are a way of ensuring an experienced vision of who is inserted in it and could induce a faster knowledge of what is analyzed. A questionnaire survey for 66 key users was also used.

Respondents are users of IS, heads of each department in each hotel unit, with the purpose of understanding the users' satisfaction in relation to the IS/IT implemented in the group, serving, at the same time, the purpose of confirming data consistency. The criteria for choosing key users included:

- Professional experience in the tourism sector;
- Experience on IS/IT of the group;
- Active participation in decisions and implementation of IS;
- Leading roles or responsibilities in their departments;
- Represent the geographical dispersion of the group.

The process was completed through documentary analysis of data collected in the hotel units and by direct observation of the use of installed applications. As a way to collect opinions and suggestions on the information collected, a panel of experts in the field of IS/IT was consulted. This process lasted for the first 3 quarters of 2011.

Figure 1. Top benefits expected from the implementation of an ERP system
Source: (Adapted from Davenport, 2000).

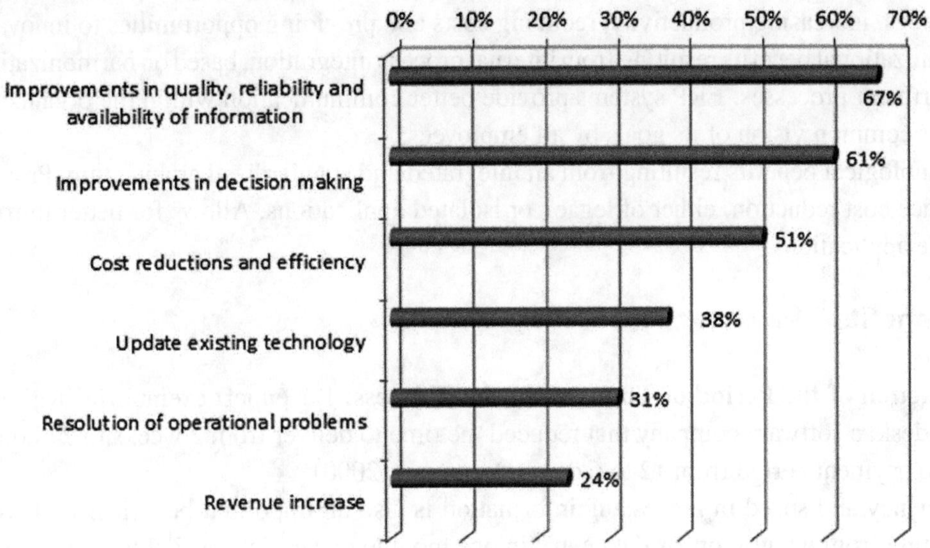

EXPECT BENEFITS FROM ERP SYSTEMS

Top Benefits

The implementation of ERP systems means deep changes in structure and culture of any organization. Although some inherent problems are to be expected, the benefits from this kind of systems are numerous (Mirchandini, 1997; Sumner, 1999; Davenport, 2000; Hayman, 2000; Scheer & Habermann, 2000; Ilfinedo & Nahar, 2006). Whether it comes to financial, operational or management processes, whether it is about the relationship with consumers or suppliers, ERP may contribute significantly to improve business competitive edge (Davenport, 2000).

Davenport made an enquiry to executives of several organizations with the objective of identifying the top benefits expected from the implementation of an ERP system (Davenport, 2000). The results of the study are shown in Figure 1.

The results are not statistically inducible, although they are the outcome of an in-depth study. Therefore, about 2/3 of the managers consider "quality of information available" to be the fundamental issue, and other results suggest that ERP systems are considered an opportunity for an upgrade on existing technology.

To Shang and Seddon, expected benefits from an ERP system may be classified in 5 dimensions (Shang & Seddon, 2002):

1. Operational benefits obtained by automation and rationalization of daily and routine tasks, reducing or avoiding human intervention. They provide cost reduction, reduced time to deliver, generate more efficiency and improve customer service;
2. Management benefits, obtained by centralized data with analytical capabilities, providing better planning and support to decision and management tasks;

3. Strategic benefits resulting from internal and external integration potential of ERP systems. Organizations benefit from an integration of the value chain, including business partners, creating alliances, increasing productivity, reducing costs and providing opportunities to innovation;
4. Organizational benefits resulting from internal process integration, based on harmonization of cross department processes. ERP systems provide better communication within the organization and a better common vision of its goals by all employees;
5. Technological benefits resulting from an integrated and standardized architecture. Provides maintenance cost reduction, either of legacy or isolated applications. Allows for better introduction of future applications.

Other Benefits

1. **Reduction of the Period of Time to Finish a Process:** Davenport exemplifies with the case of Autodesk, a software company that reduced the time to deliver from 2 weeks to 24 hours, as well as the payment period from 12 to 6 days (Davenport, 2000)
2. Efficiency and speed in processing information is also an important benefit from ERP systems, resulting from integration, as data entry in one module makes data available to the whole system (Themistocleous, Irani, O'Keefe, & Paul, 2001). This eliminates or simplifies manual tasks. Davenport underlines IBM´s example, which experienced an increase in process speed, like pricing, stock replacement, among other processes that were performed manually and changed into automated ones (Davenport, 2000).
3. **Quality of Information:** It is an important benefit, given that information integration avoids redundancy and data inconsistency (Themistocleous, Irani, & Love, 2002). Decisions are not accurate when information on the same subject is erratic, coming from different origins (Davenport, 2000).
4. Another important benefit is better data viewing on each process, increasing the knowledge of a certain object. The Monsanto corporation, after implementing its ERP system, had a knowledge base over ground operation of its workers, as well as new decision rules, improving decision making (Davenport, 2000). Also ELF, the petrochemical company has improved its relationship with customers and increased sales after implementing the ERP system with optimized business processes, leading to greater operational margins (Davenport, 2000).
5. **SI Harmonization:** Globalization promotes geographical growth of organizations, forcing systems harmonization and centralized coordination (Davenport, 2000). Microsoft, according to Davenport, had important benefits after harmonizing their financial and supply systems, leading also to better reporting and management. Dow Chemical implemented an ERP system in order to harmonize all processes and information on their subsidiaries scattered over the world, creating a platform for future systems.
6. **Impulse towards E-Business and Supply Chain Management:** E-business is currently the major business support to many companies. ERP systems streamlines the purchase, tracking and product deliver processes (Davenport, 2000). It´s not enough to have a website; all internet based transactions must be linked to the supplier´s transaction systems.

The structure of the value chain must include all players, from consumers to suppliers and distributors, along with a front-end for demand management (Hayman, 2000). This is also important for future growth and even survival of the organizations (Themistocleous, Irani, & Love, 2002). According to Davenport, CISCO managed to reduce yearly costs by US$500 million (Davenport, 2000). Fujitsu reduced their delivery cycle from 20 to 2 days in 90% of the cases, and increased their on-time deliveries by 25%, thanks to the potential of the internet provided by their ERP system (Hayman, 2000).

HOW TO OBTAIN VALUE FROM ERP SYSTEMS

One of the major challenges of management is to guarantee that IS/IT may generate benefits, and so organizations must find ways to measure its value and understand how they deliver sustainable advantages (Peppard, Ward, & Daniel, 2007). According to Ward, what is at stake here is the need to evaluate the contribution of IS/IT to the business, in order to decide where to invest (Ward, Griffiths, & Whitmore, 1990).

Ward refers that only 30% of IS/IT investment projects deliver the expected benefits (Ward, Peppard, & Daniel, 2008). Therefore, IS/IT must be managed like any other investment, contributing also to a strategic balance in the long run (Davenport, 2000). According to Ward there is a difference between investment goals and its benefits. Investment goals must be approved by all stakeholders, and benefits are the advantages the investment brings as a result of the goals achieved (Ward, Peppard, & Daniel, 2008). Incapability to attain benefits reduces the knowledge the organization should have about the value IS/IT may bring to business. This leads to failure to decide correctly about investment or to establish the right priorities to improve results or support new development strategies (Ward, Peppard, & Daniel, 2008).

According to Peppard, there are two types of IS/IT investments:

- The problem-based, oriented to pre-defined objectives or oriented to problem solving;
- And those innovation- based, aiming at new opportunities and the creation of competitive advantages. The difference between the two is related to technological capacity to attain goals, the changes needed to the business and the achieved improvements (Peppard, Ward, & Daniel, 2007).

To measure the value and the success of ERP projects has been a goal perceived by several authors and several methods have been adopted. However, there is consensus that traditional investment analysis should not be used when evaluating ERP systems (Stefanou, 2001).

The scope a project of this nature has, including its strategic and cost aspects, make it a project of fundamental importance (Stefanou, 2001). Some organizations consider it the larger investment in IS/IT ever made. That is why the evaluation of ERP systems shouldn´t be confined to financial criteria, like NPV (Net Present Value), ROI (Return on Investment) or IRR (Internal Rate of Return) (Teltumbde, 2000; Stefanou, 2001). The strategic and benefits dimensions should also be considered, like efficiency, decision making or customer satisfaction (Stefanou, 2001; Themistocleous et al., 2001; Murphy & Simon, 2002; Kumar et al., 2003).

Shang and Seddon define a set of four steps to realize the expected benefits of ERP systems (Shang & Seddon, 2002):

1. Search for information on ERP benefits, problems and reasons to adopt ERP systems: on existing literature as well as on white papers from vendors;

2. Review literature on IS/IT investment evaluation and design a reference model to evaluate the ERP system. The expected benefits may be categorized as follows: operational, managerial, strategic, organizational and technological;
3. Use the reference model to analyse success cases from several vendors;
4. Build a consolidated list of all expected benefits, one that shall be used as a checklist on further implementation steps.

According to Mirchandani of the Gartner Group, benefits can be achieved by the following factors (Mirchandini, 1997):

1. Automate tasks formerly done manually;
2. Improve quality of data, especially in organizations with many departments;
3. Changes and improvement of processes, induced by changing technology;
4. Changes on the organizational structure;
5. Optimizes information, providing better demand forecast;
6. Replacement of legacy systems;
7. Inclusion of new systems providing new services;
8. Investment opportunity in infrastructure;
9. Achievement of intangible benefits, hard to account (Shang & Seddon, Assessing and Managing the Benefits of Enterprise Systems: The Business Manager's Perspective, 2002).

Davenport, using the Mirchandani list, grouped the benefits in three categories: savings due to new approach of business processes, savings due to replacement of legacy systems and generated income (Davenport, 2000). Davenport underlines several examples, like automated reporting, automated data reconciliation or centralized shared services.

RESULTS OF THE CASE STUDY

As previously referred on chapter 3, the results emerge from a comprehensive case study and are not statistically inducible. According to Davenport, the strategic objectives must be known in order to understand how the organization may offer value to its customers and how can it differentiate from the competition (Davenport, 2000).

The hotel group analysed presents a global strategy defined over three lines of action:

- Growth by acquisition of hotel units, integrating it in the group´s organics and image.
- Positioning as an organization offering quality services in the national an international hospitality industry.
- Positioning as an organization with diversified offer: hotels, resorts, golf courses, health clubs and congress centres, allowing to create an association among all those products with a unique style.

In order to implement this strategy, the group is formally a three tier corporation, with one company managing fixed assets, other dealing with touristic activities and other managing shared services.

The implementation of fan ERP system forces a strategic adjustment (Teltumbde, 2000; Baki & Çakar, 2005). Therefore, it must define improve and measure the key-processes in order for the ERP system to adjust to what´s required (Davenport, 2000), and have total commitment from top managers (Summner, 2000; Fitzgerald, 1998).

From the interviews conducted in the case study it is possible to conclude that top managers don´t believe there are competitive advantages emerging from the IS/IT systems running at the time, although they refer the importance of SAP (covering back-office), OPERA (Front-office) and Microsoft CRM (Marketing). Other respondents also consider the group doesn´t obtain any competitive advantages from IS/IT,

This analysis was confirmed by the application portfolio matrix (Table 1), as no high potential applications were identified, meaning that there were no applications that may become strategic in the future of the organization.

CRM software can be considered strategic in what respects to future sustainability, given its nature of a customer oriented software. The foreseen evolution is to be positioned as a key operational application.

For the respondents, IS/IT must be at the same level as the competition and they consider that what differentiates the group and makes it attractive to customers are the following factors:

1. Quality of customer service;
2. Quality of facilities;
3. Brand recognition;
4. Location of facilities (close to places of attraction like beaches, lagoons and protected landscape);
5. Diversification of services offered (entertainment, conference centres, golf courses, health clubs and spa).

In what processes are concerned all interviewed users consider that they require continuous assessment and the announced improvements reflect the expected benefits. Benefits are perceived as resulting from one of the strategic goals: improve customer service.

In particular, some processes and related expected benefits were referred as shown in Table 2.

Table 1. Present application portfolio

Strategic	**High Potential**
Microsoft CRM (Marketing and conference room management)	
Key Operational	**Support**
SAP (MM module - procurement) OPERA (front office) MICROS – POS (sales/invoicing) BELAVITTA (health centers - SPA) CEDIS (health centers) CONCEPT (golf camps) VINGCARD (room access control)	SAP (modules: HR – human resources; FI – financials; BO – reporting) TARGET (cash management)

Table 2. Expected benefits of processes

Processes	Benefits
Sales (booking on-line)	Make the corporate website the major instrument for sales; Effective management of online sales channels
Sales (planning)	Consolidated analysis of several business units
Marketing	Communication; Share of operational data
Financial Management	Cash management (bank reconciliation, loans, leasing)
Human Resources Management	Recruitment; Performance evaluation

FUTURE RESEARCH DIRECTIONS

Given the growing complexity and demanding from the business environment (competition and emerging technologies), the question of IS/IT benefits and value are among the most debated ones in the field

From the investigation described in this chapter several propositions emerged that can be followed, some of them as a natural development of the conclusions achieved, while other may benefit from the work developed but may be structured as independent investigation projects.

One of the possible directions would be to complement the investigation with the impact of ERP systems in other agents of the value chain of the hospitality industry (operators, customers, travel agencies and suppliers).

During the investigation, other hotel units and groups were identified as having the same kind of problems as the hotel group studied. A project that would gather a few independent hotel groups in the same investigation could possibly validate the conclusions of the current one.

Another possible direction would be to investigate the case of a hotel group with a globalisation strategy allowing for some conclusions against a purely "national" context. Confronting strategies could be of value.

On the other side, the current investigation could be complemented with a study on small/medium scale companies, in order to evaluate market proposals oriented to this kind of organizations and confronting eventual problems of scale when it comes to evaluate benefits from IS/IT investments.

Finally, this investigation may be envisaged as a contribution for multiple studies in this field, enlarging the knowledge base of the implementation of IST/IT systems, their benefits and value, even with other investigation methodologies.

CONCLUSION

The case study confirmed that quality and diversified offer is part of the hotel group strategy, namely through complementary services (golf courses and health clubs), as some authors also pointed out (Connoly, Olsen, & Moore, 1998) and (Ribalaygua, 2000). From the interviews conducted in the case study it´s possible to conclude that top managers don´t believe there are competitive advantages emerging from the IS/IT systems and also that they find it secondary compared with aspects like brand recognition, quality and location of facilities. As Ward underlines (Ward, Peppard, & Daniel, 2008), this perspective some organizations have about the value of IS/IT investments – not considering it strategic – reduces the capacity to attain benefits from its usage. Strategically using IS/IT is a way to respond to global business requirements, as state Miguel et al. (2000) and Ribalaygua (2000).

This is a case where IS/IT are considered source of benefits for current processes without any perspective of innovation on new processes. The study confirmed that the reason is the fact that IS/IT are not considered strategic in the organization´s future performance.

REFERENCES

Alshawi, S., Themistocleous, M., & Almadani, R. (2004). Integrating diverse ERP systems: A case study. *The Journal of Enterprise Information Management*, *17*(6), 454–462. doi:10.1108/17410390410566742

Beldona, S., Cai, L., & Pearson, T. (2001). Impact of the Internet on the hospitality value chain – towards an information sharing network. Advances in Hospitality and Tourism Research.*Proceedings of the Sixth Annual Conference and Graduate Students Research in Hospitality and Tourism*, 6, 25-29.

Connoly, D., Olsen, M., & Moore, R. (1998). The Internet as a Distribution Channel. *The Cornell Hotel and Restaurant Administration Quarterly*, *98*(4), 42–54. doi:10.1177/001088049803900408

Davenport, T. (2000). *Mission Critical: Realizing the Promise ok Enterprise Systems*. Boston, MA: Harvard Business School Press.

e-Business W@tch. (2006). *ICT and e-Business in the Tourism Industry: ICT adoption and e-business activity in 2006*. The European Commission, Enterprise & Industry Directorate General.

Heart, T., Pliskin, N., Schechtman, E., & Reichel, A. (2001). Information Technology in the hospitality Industry: The Israeli Scene and Beyond. *Information Technology & Tourism*, *4*(1), 41–64. doi:10.3727/109830501108750886

Ilfinedo, P., & Nahar, N. (2006). Prioritization of *Enterprise Resource Planning* (ERP) Systems Success Measures: Viewpoints of Two Organizationaal Stakeholder Groups.*Proceedings of the 2006 ACM symposium on Applied computing*, (pp. 1554-1560). doi:10.1145/1141277.1141636

Kirk, D., & Pine, R. (1998). Research in Hospitality and Technology. *Hospital Management*, *17*(2), 203–217. doi:10.1016/S0278-4319(98)00016-4

Lee, J., Siau, K. e Hong, S. (2003). Enterprise Integration with ERP and EAI. *Communications of the ACM, 468*(2), 54-60.

Markus, L., Petrie, D., & Axline, S. (2000). Bucking the Trends: What the Future May Hold for ERP Packages. *Information Systems Frontiers*, *2*(2), 181–193. doi:10.1023/A:1026548107263

Martínez, J., Majó, J., & Casadesús, M. (2006). *El uso de las Tecnologias de la Información en el Sector Hotelero*. Academic Press.

Miguel, M., Fernández, E., Olmeda, I., & Segui, M. (2000). Aplicación de las Nuevas Tecnologías de la Informatión al Sector Turístico. *Estudios y Perspectivas en Turísmo*, *9*, 5–23.

Muscatello, J., & Chen, I. (2008). Enterprise Resource Planning (ERP) Implementations: Theory and Practice. *International Journal of Enterprise Information Systems*, *4*(1), 63–83. doi:10.4018/jeis.2008010105

Panorama Consulting Group. (2009). *2009 ERP Report: Hospitality and Entertainment*. Relatório.

Ribalaygua, L. (2000). Estrategia Tecnológica en ele Sector Hotelero. *Estudios y Perspectivas en Turísmo*, *92*, 99–111.

Ross, J., & Vitale, M. (2000). The ERP Revolution: Surviving vs Thiring. *Information Systems Frontiers*, *2*(2), 233–241. doi:10.1023/A:1026500224101

Rus, R. (2009). The use of Information systems in Hotel Market of Cluj-Napoca. *Studia universitatis babes-bolyai, 3*, 51-59.

Themistocleous, M., Irani, Z., O'Keefe, R., & Paul, R. (2001). ERP Problems and Application Integration Issues.*Proceedings of the 34th Hawaii International Conference on System Sciences: An Empirical Survey (HICSS-34)*.

Ward, J., Griffiths, P., & Whitmore, P. (1990). *Strategic planning for Information Systems*. John Wiley & Sons.

Ward, J., Peppard, J., & Daniel, E. (2008). Building Better Business Cases for IT Investments. *MIS Quarterly Executive*, *7*(1), 1–15.

Zornoza, C. (1995). Dirección hotelera y tecnologías de la información: Un Análisis Empírico de la Empresa Hotelera Valenciana. *Estudios turisticos, 125*, 19-52.

Chapter 12
Risk, Terrorism, and Tourism Consumption:
The End of Tourism

Korstanje Maximiliano
University of Palermo, Argentina & University of Leeds, UK

ABSTRACT

The present chapter discusses to what extent the rise of new risks and dangers, modernity has brought, has causing the end of tourism. The answer to the above formulated questions is not easy, and of course, it exhibits a fertile ground to be explored in other approaches. The ceaseless news about violence, cruelty, wars and deaths have serious negative impacts on audience worldwide. Though policy makers have devoted their time and efforts in looking for new alternative segments where death replaces the allegory of beautiness, giving as a result new products as dark-tourism, slum-tourism, disaster-tourism or doom-tourism, other problems arise. The curiosity for tourists in visiting spaces of mass death or extreme suffering contradicts not only its nature as a mechanism of revitalization and relaxes but may very well lead to narcissism which in fact is the end of tourism.

INTRODUCTION

Although tourism seems to be a globalized and growing industry dotted of a great resiliency to face disasters as earthquakes, floods, virus outbreaks and even terrorist blows, no less true is that some indicators show that we are next to the end of tourism. Policy-makers today not only are urged to answers problems generated in other geographical points or by other agents who are external to tourist system, but also are tested to offer innovative solutions to slippery matters. Tourist experts have no real solution to deter terrorism, but should deal with the idea a next attack can suddenly happen anytime and anywhere, which means that the urgency of policy makers to present competitive products is conducive to the attractiveness these places cultivate for terrorist. Security and safety have been posed as the main interests for international demand. This happens because of two main reasons. At a first glance, unlike our grant-parents, we are accustomed to follow our curiosity at time holidays. The quest for novelty or

DOI: 10.4018/978-1-5225-1054-3.ch012

something new has transformed in the primary motive of holidays. Secondly, quite aside from these heritage seekers, tourism has consolidated as one of the most important industries of the world. As a result of this, some cells have acknowledged that the fact of killing tourists is a useful tool to create political instability. Once the credibility of state is seriously affected, their demands would be echoed in an easier way. Sociologically speaking, the anatomy of nation-state was based on welfare paradigm, where officialdom understood that intervention was an efficient instrument to prevent risks, and uncontemplated dangers, leading society towards a zero-risk atmosphere.

From the end of WWII, industrial societies appealed to the conception of risk-zero programs, which proclaimed that rational planning aimed at protecting the citizenry may eradicate risks to make from our neighborhoods a safer place to dwell on. Not only this not happen, but the decline of Soviet Union, the globalization of economies accelerated the complexity of a system which shows serious problems to mitigate the negative effects of risks, even these risks are generated by the introduced technology by West. A type of Run-away modernity in terms of Giddens (1999) paves the ways for new fears and problems that affected the functionability of tourism industry as never before. The experts and planners devoted considerable attention to promotes programs or models that help nuancing the effects of sudden disasters (Pforr & Hosie, 2009; Ritchie, 2004; 2009; Korstanje, 2011b; Blake & Sinclair, 2003; Cohen & Neal, 2010). In view of the limited time and space, this essay review does not focus on the connection of globalization with modernity, but explores to what extent risk perception is enrooted in the process of communication. In the discussion the goals are twofold, on one hand, we review the existent literature in risk perception theory applied in tourism, but at the same time we alert on the exhaustion of tourism. As Gale (2008) noted, we are living in a world where tourism is dying. Since this is an essay, a conceptual piece which does not need further methodologies, we opt to avoid any methodology (qualitative or quantitative) to pose the issue in the tapestry. Rather, the discussion is aimed at stimulating the debate on the possibility tourism is declining as a result of risk inflation and the advance of terrorism.

What Motivates Terrorists?

Undoubtedly the main threat for policy makers in tourism fields is terrorism, like for West. This is the reason why a section respecting to some risk perception theory applied on terrorism is reserved. One decade before 9/11, an event in Egypt shocked the world, *Luxor Massacre*. In this attack, perpetrated in 1977, 58 international tourists and 4 policemen were in cold blood murdered changing the paradigms of geopolitics forever. Nation-states have fomented a much wider process of securitization which consists in the orchestration of different policies in order for First world travellers (US and Europe) to be protected while they are abroad. Unlike other decades, where attacks were perpetrated against politicians, today tourists are targeted so that some separatist cells cause political instability in developed nation-states. As Ambassadors of their cultures, tourists not only are vulnerable for locals, but also appeal to the sense of hospitality where modern nation-state is based. For those states which cannot protect aliens their credibility is undermined. In this vein, Somnez argues that terrorism seeks to destabilize nation state in order to instill terror. That way, their demands are easily addressed by politicians. One of the most important aspects to take in mind is the fact that because tourists are vulnerable agents who are unfamiliar with the visited soil, are target of these cells (Somnez, 1998). Some voices commanded by Abraham Pizam will explore the role played by psychological resentment in the configuration of terrorism. Per these scholars, those peoples who are living in pour conditions or subject to pauperism still is fertile ground for introducing a message of hate against Western travellers. Though political instability is a key factor to understand terrorism it does not suffice. Researchers should pay attention to social cultural backgrounds where terrorists lived, the social imaginary that interprets what media portrays, and finally resentment which resonates in third world to develop a hostile attitude against US and Europe. As Pizam puts it, we thought that tourism is a vehicle towards peace, but our fears materialized after 9/11. In countries where state is not present, gaps are fulfilled by hatred groups that lead locals to terrorism. Starting from the premise that terrorism affects seriously the image of tourist destination, Pizam adds, there is a spill-over syndrome that generates serious losses in neighboring countries. Because hotels and tourism are bulwarks of liberal consumption and mobile capitalism, not surprisingly they have to pay the costs (Pizam, 1999; Pizam 2002; Pizam & Mansfield, 1996; Pizam Tarlow & Bloom, 1997). Surely other study cases as Bali, were mediated and disseminated by journalism which caused a shift of paradigm just later 9/11.

Rather, other senior sociologist as Peter Tarlow claimed for diagnosis takes another contrasting direction. As Rabi and sociologist, Tarlow acknowledges that poverty can be reverse worldwide, as Pizam, but he is adamantly convinced terrorism is enrooted in politics. Nor economy neither religion coadyuvated in the formation of terrorist's mind. Though terrorist embraces religion to recruit new candidates, their interests are based on extortion, speculation and crime. We, media and policy makers, should avoid any ethnocentric discourse at time of treating Islam. It is important not to lose the sight that we have been educated thinking we are rational agents; we are not that. Rather, our behaviour is based on emotional pillars. Terrorism exploits our emotional roots to cause a real damage to nation-states. Killing innocent tourists is the oft-used tactics not only because it questions the legitimacy of states, but also disseminates panic in the population. As this backdrop, policy makers and experts in tourism security should recognize we live in a system of globalization where travels and mobilities are the main pride of the civilization (Tarlow, 2011, 2012, 2014). In this token, efforts of founding parents of tourism to struggle against terrorism need another turn of mind in order for reaching efficient mechanism of resilience for affected communities. This complex cultural background, which was above discussed, indicated that we live in a mobile world, even for risks and terrorist cells. Is globalization responsible of terrorism? to what extent, we hold the thesis that globalization unleashed the all our worst nightmares.

Globalizing Risk

Undoubtedly, the question of risk is discussed along with two contrasting poles. Culturalists will emphasize on the social nature of risk, while probabilists will delve into the algorithms that help preventing the negative effects of risks. In this vein, perception plays a crucial role by configuring the state of alarm of human mind. As Quarantelli put its (1975, 1982, 1990, 2001), not surprisingly psychology found now that attention, commitment and emotions are of paramount importance to understand why some events are over-valorized while others simply ignored. He reminds the paradox given by the earth-quake which hit Los Angeles in 1979 where the audience, convoked to enjoy of a football encounter, ignored the tremor because the game reached its zenith. This reminds that cognitive attention and the signs of dangers do not correspond with reality in some circumstances. In perspective, this global world has engendered a new cosmology where ordinary person feel nobody is safe anywhere and anytime. In fact, the efforts aimed to keep under control the states of emergency, as the case of Fukushima, may cause parallel effects. Sociologists agree that the production of knowledge as well as the shortly-time events are covered and disseminated in the world -by the media- over-stimulates a risk- centered prone that leads state to conduct populist programs to calm the anxiety of population (Sunstein, 2007). Instead of fixing the problem, as Sunstein put it, these programs, which are based on the precautionary principle, backfire.

As noted above, the problem of risks lingers over for more than 40 years within social sciences. While psychology advanced to produce a coherent conceptual framework of the issue, sociology confirmed risks are culturally constructed in order for status quo`s legitimacy does not decline. As L Tierney observed, the great knowledge's dispersion generated caused a misleading interpretation of the issue, which today is very difficult to decipher. In sociology and anthropology, the term was introduced by Ulrich Beck respecting to the disaster of Chernobyl in Ukraine. If the classical economy proclaimed the use of technique as a valid instrument to prevent threats, this event showed the opposite. The same technology constructed to make the life of ordinary people safer, entrapped the society between the wall and blue sea. These new risks were engendered by the human technology (Beck, 2006). Following this reasoning, Beck clarifies that risk and modernity has been inextricably intertwined. The globalization of nations, their economies and boundaries created a more complex system where a subtle disruption may lead us to a catastrophe. The delocalization of the connectivity between reasons and effects has blurred the understanding of risk for modern experts; in a world, where the market destabilizes the social bond, the risk is the only reason for people to have communication (Beck, 2006). Not all sociologists agree with Beck´s development. Niklas Luhmann (2006) adds that it is important to avoid the increasing sentiment of alarmism in social scientist, which is feed by the media and journalism. In doing so, one should distinguish the conception of risk and threat. The former signals only to the "principle of contingency", this means to the possibility for the subject to avoid the consequences of risk. Besides, risks are often resulted from a previous decision making process. In contrast to risks, dangers, which are external fabricated, are imposed to lay-people with less or no possibility to avoid its negative effects. The difference of both is very simple. Those peoples who made the decisions never face the risks these decisions created. An airplane accident is far for being a risk for lay passengers whose actions or ideas have not direct incidence in the misfortune. Since the responsibility is with the owners of air- company. Tourists only are passive agents who face the results of an outside danger. Luhmann acknowledges that not only the risks are generated by the elite, which never face its consequences, but social science alludes to a tactic of blaming the victim when misjudges the difference between risk and threat. Lay people in postmodernity are educated to think they are reliable for their decision, but they are not. The efficacy of capitalized modernity consisted in making believe that "self-management" is a good option towards progress (Luhmann, 2006). Anthony Giddens, another senior sociologist, would criticize Luhmann by two main reasons. Returning the ball to Beck, Giddens says that the dichotomy between risk and threats associated to decision making would be applicable only for Medieval Age. In the modernity, every subject is deciding, even deciding not to make a decision. So, risks are everywhere in our life no matter than our responsibility in their inceptions.

As the previous discussion given, a third alternative can be cited. Cass Sunstein (2006) conceives the inflation of risks as effects of bad decisions, which are the combination of two psychological mechanisms, the neglect of probability, and the heuristic of risk. The theory says that the risk assessment is previously determined by emotional disposition instead of rational evaluation –cognitive activity-. The probability neglect exhibits sometimes people delve into the worst possible situation failing to inquire properly on real causes of dangers. This mechanism conjoins to the heuristic of risk to evaluate the correct diagnosis. It is common to trivialize some serious risk when they do not defy our threshold of control, while others are over-exaggerated. Although the statistics shows the industry of airplanes are safest ways of transportation, an accident causes serious damage to the trust of flying. Inversely, cars are very dangerous but our sense of control is higher than an airplane. Last but not least, Korstanje & Skoll have explored the role of technology in the globalized economy in diverse studies. It is important not to lose the sight risk, like taboo, can develop a protective role for some communities. Basically, as a social construe not only risk varies on society, but also it alludes to the economic order (Skoll & Korstanje, 2012; Korstanje & Skoll, 2013). Zygmunt Bauman (2013b) argues convincingly that the sentiment of fear is fabricated by modernity to intellectualize the uncertainty, but mediating between humans. A closer look shows that postmodernity has advanced by the imposition of a liquid consuming life, which people and goods are equaled to similar conditions of exploitation. Risks, like fears, would serve as mechanism to interact with others, commoditizing the fears to mark those consumers who have made a bad decision. The state of perpetual emergency, Bauman adds, is oriented to curiosity to be accepted. The lay-citizen, in the postmodern world, looks to be accepted by other peers not to be excluded from the circle of mass-consumption. The television show commoditizes disasters as a new form of entertainment. The sense of security of modern citizenry is only affected whenever their decision leads to be out of the consuming world; this represents the main fear of West today (Bauman, 2009, 2013). Since the real dangers are ignored by industrial societies, the disaster is only a question of time (Korstanje, 2009, 2011). In tourism and hospitality fields, risk-directed research has been adopted post 9/11 (the attacks that shocked the world against World Trade Centre and other American targets) (Faulkner, 2001; Ritchie, 2004). Although, experts as Roehl and Fesenmaier (1992) have conducted, years earlier, innovative investigation in the matter of risk and the perception of international destinations, an exploding interest in risk as scientific object of study arose (Korstanje & Olsen, 2011). In next section we will discuss the approaches and limitation of risk-perception in tourism studies.

Risk in Tourism Fields

Over years, specialists discussed the applicability of used methods to study risk. The quantitative-led vs. the qualitative-led approach led the debate to a deadline. While some studies evidence some problems of correlation among variables, others claim for the adoption of new techniques that help overcoming the algorithm inconsistencies. What is more than interesting to debate is to what an extent a correlation gains further explanatory power to understand the meaning of risk. Whether the question women perceive further risk than men, a qualitative view would shed the light this correlation is not given by genre differences, but in the way the woman is educated (to protect their home and family) (Zinn, 2010). Examples like this in the epistemology of risk research are everywhere. Quite aside from this, no less truth is tourism fields have adopted a quantitative-view respecting to risk research (Dolnicar, 2005a, 2005b). The second problem is given by the dichotomy between risk and perception. As Professor Svein Larsen put it, we may feel attraction for a destination even if it is dangerous. This does not happen because we are risk-prone travelers, but the worry and risk work in diverse levels for the mind. A worry activates an internal emotion to avoid certain objects, while risk are only associated to the cognitive structure of the subject (Larsen, 2007, 2009). The importance of image to cause stability within the tourist system was a key factor for scholarship to conduct risk-research post 2001. Virus outbreaks, terrorist attacks, disease and natural disasters were enough motives to understand the world was changing to new states (McCartney, 2008). A serious attempt to classify risks has been done by Fuchs & Reichel (2010, 2004) in two main families. Those man-made risks should be separated from others triggered by nature. Risks associated to crime, terrorism, political instability and thefts are produced by us, while disasters, quakes, flood are determined by external forces. The success of a tourist destination depends on the trust generating levels produced by policy-makers. This is an important idea in the recovery policies simply because exhibits that planning plays a vital role in avoiding the collapse of destinations.

Starting from the premise, the industry faced a set of diverse dangers, the priorities to use risk-management to avoid disasters is of paramount importance so that the destination survives. Risk and disaster-management evaluate the potential steps to follow when the state of normalcy set the pace to emergency (Henderson, 2008). Peter Tarlow (2011), one of the founding fathers of tourism safety, explains that visitors develop adverse sentiment to those countries where their safety cannot be granted. This affects the quality of life of many people and local agents who live from tourism. Combining deep psychological advances with risk-management plans should be the priorities of all states. The divorce of experts and politicians leaves a gap reducing the possibilities for a community to recover after a disaster hit.

Scholars who assert that risk perception is limited or enhanced by the trust travelers have were very popular in the first decade of the twenty-first century. They proposed that those travelers who were accompanied by relatives or friends perceive less risk than other groups. The sentiments of vulnerability are awakened when the travelers go beyond the security of home would be controlled by the social trust associated with accompanying companions (Roehl & Fesenmaier, 1992; Yuan, 2005; Park & Reisinger, 2010). Reichel, Fuchs, and Uriely (2007) found that those who prioritized the political instability as the main threat at time of vacationing, travelers in company of others are more likely to suffer physical wounds than single travelers.

The perception of self-vulnerability is potentiated by the lack of familiarity of unknown visited land. The fluency of language to ask for help or assistance is one of the first reasons for first world tourist to buy all inclusive package tours (Anderson, Juaneda & Sartre, 2009). Similar outcomes were found by Qi, Gibson and Zhang (2009). Per an experiment, participants evidenced serious concerns at time of visiting a destination with different mother-tongue or in context of political instability. Paradoxically, this would be a problem for practitioners or planners who see the cultural differences between hosts and guest as the primary reason for violence and conflict (Sönmez, Apostolopoulos, & Tarlow, 1999; Prideaux, Laws & Faulkner, 2003; Abdel-Azim, 2010). George, Inbakaran, & Poyyamoli propose an interesting model to understand the complex play between curiosity and personal safety. Tourism industry rests on two contrasting drives. On one hand, the self is motivated to escape towards new landscapes (tourist-motivation), but at the same time, there is a "nativist" motivation which appeals to the sense of security the visitor needs to feel safe. Whenever "nativist motivation overweight "tourist one", the self opts to return to home. Ultimately, the role played by the media is vital to boost or decline destinations. Sometimes, the dreams of policy makers diluted in minutes because the risk-management programs lack of an efficient communication policy. Following Michael Hall's insights, the media covers news that shape the opinion public audience keeps respecting to what happens beyond our understanding. The communication policy post and pre disaster context can activate a great variety of responses. The attacks of World Trade Centre not only affected the American tourism industry, but also changed the ways other citizens developed to travel abroad (Hall, 2002). The process of crises-management in tourism and hospitality experienced three important changes from 80s up to date. The first was the economic recession surfaced in North America through 80s. This echoed in Asia with replication of domestic economies from 90s onwards. Finally, the most turbulent stage comes with the advent of 9/11 in 2001 and the last financial crisis in 2008/2009. Based on the satisfaction on intangible services, tourism industry develops a high sensibility to crises and disasters (Hall, 2010).

Geoffrey Skoll and Maximiliano Korstanje set forward a conceptual model to classify risk depending on its effects for public opinion. Public audience goes into panic when the event defies its possibilities to confront the period of crisis. The psychological impact of terrorism in the media and on the audience is based on three points:

1. Control threshold,
2. Probability of repetition, and
3. The status of the victims.

The control threshold refers to the ability of the system to predict the time and place of the attack. The probability of repetition measures the mobilization of resources to prevent traumatic events. The probability of repetition is proportionate to the effects of events that cannot be avoided. Finally, whenever victims are valorized by their high social status, the impact on the social imaginary is greater than cases where low status people are affected (Skoll & Korstanje, 2014a, 2014b).

The Instability in Global Economies

Raoul Bianchi calls the attention to the problem of globalized risk. Tourism would play a pervasive role cementing the hegemonic practice of market to mark the ethnic-groups and other minorities. The globalization of risk, Bianchi adds, not only creates a state of instability that affects global tourism, but also expands disciplinary mechanism of control that objectifies the others. Non-western tourists are discriminated or scrutinized as a dangerous other whose presence is undesired (Bianchi 2007). Similar observations have been done a couple of years ago, by Scott Lash and John Urry. The hyper mobility engendered by the last modernity flourished following the technological advances in transport and information, but no less truth these shifts brought a new way of organizing economies. The modes and developed skills to access to certain types of mobilities confer different status to citizens. The real and symbolic barriers to move are of paramount importance to know the status of a person. In the hyper-mobile world, the cars and mobility have emptied the real geography producing a broader sentiment of omnipotence and control. The trajectory (exchange) of goods and humans (by the commerce or tourism) has created an empty space accelerating the decline of trust and social bond among persons. The history and tradition have been blurred in the hegemony of allegory. Today, the existent tension between experts and lay-people is enlarged. From this viewpoint, the decision before planning the holidays seems to be made according to the abstract information the client may collate. The personal experience and mouth-to mouth recommendations set the pace to new ways of indoctrinating the client's needs. In West, during XIX and XXth centuries, a new type of reflexibility has been established so that traveller may appreciate the visited landscapes, but this reflexibility is not hierarchical but aesthetical. The possibility for travellers not only to experience new sensations and landscapes, but to classify diverse geographical natures was, jointly to tourism, conducive to the assertiveness of cosmopolitism. The schedules of travels reproduced, in this vein, the aesthetic logic of modern society by limiting the presence of otherness to be a simple commodity to be consumed. The ability of travel agents was to create the risk to propose the solution. Risk plays a pervasive role because on one hand it confers further legitimacy to expertise-nets, but on the other, it poses as an instrument of communication. Experts, in this token are an interpreter, for lay-people to keep the trust in their institutions. In effect, travellers perceive as risks they previous beliefs communicated by experts. If times have shifted, the consumers today allude to more selective forms of consumption than other decades. The world has been drawn in geographies of domesticated spots and geographies of danger (wild-life zones) (Lash & Urry, 1993).

As the previous argument given, Bianchi and Stephenson (2014) in a recent seminal study exert a considerable criticism on the cultural mobility as it has been formulated in some circles of tourism industry. At the time first world tourists go through the vast geographies of this world, peripheral migrants are scrutinized, immobilized and monitored by the existent technology. The formation of political poles (cold-war) that leads the industry of travels to draw geography of submission. As this backdrop, tourism served to re-define the potentialities of citizenships as well as enhanced the nationalism in many sense. Not only tourism grew dramatically from 1960 to 1970 positing as one of the leading industries of the world but paved the ways for the accentuations of new material asymmetries inside the societies. Certainly, it is important not exaggerate the role of tourism in local economies nor accepting the promises of globalization as unquestionable truths. What Bianchi and Stephenson like to debate is to what extent tourism and globalization developed contrasting landscapes in the tourist-receiving countries. The material asymmetries prompted by tourism are fertile ground for the upsurge of violence and sentiments against West. The configuration of global elite instilled a new consciousness about culture, mobility and travels. As a maiden of empires, tourism encourages a double-edged mobility which is used as a disciplinary mechanism of control (Bianchi & Stephenson, 2014).

Most likely, one of the troubling aspects of sociology and anthropology to understand risk rests on the dissapationately spirit to evaluate the consequences of crises on daily life of ordinary work-force. Nor interested in the social effects of crises, neither interested in intervening in risk-management plans, social scientists want to be privileged-witnesses of how people reacts to crises. The problem lies in the financial dependency of some peripheral countries respecting to tourism activates the needs to implement urgent plans of contingency before disasters, virus outbreaks in an ever-changing world. However, this poses a serious dilemma for the smooth functioning of industry. At the time, we desensitize to the appalling effects of terrorism or any other danger, further cruelty would be needed for captivating the nation-state's attention. This will lead to a spiral of violence that very well can be absorbed by different mechanisms of resiliency the society developed. For example, tourism offers some new landscape where human suffering, genocide or mass-death is the main attraction. It represents a valid effort for destination not to lose its attractiveness but it is important to remind that this policy has some costs. Not only tourism needs for the "other's pain" to subsist but its original function is distorted. Originally, tourism as any other a rite of passage was aimed at revitalizing the psychological frustrations of daily life in order for the society do not collapse. Now this original function has changed to new roles where relax or rest sets the pace to the quest for novelty. As we will see in the next chapters, this opens the doorsteps to the end of tourism.

Qualitative vs. Quantitative Viewpoints in Risk Perception

Tourism research has advanced over recent years based on an economic-center viewpoint, risk perception is not an exception. To what extent, the combination of both techniques would be fruitful for making efficient policies to prevent terrorism worldwide, but taking into consideration that it seems to be one of the major points to be discussed in this section. In many underdeveloped economies the dependency of governments for tourism paves the ways for a difficult relation. The attacks on tourist destinations not only offer the worst advertising for the affected country but also for neighboring nations as well.

Quantitative-led investigation applied on risk perception is aimed at mitigating the negative cognitive effects on advertising which very well may affect the organic image of destinations. Programs in security and safety encompass key factors that helps policy makers to expand their current understanding how perception works. Further discussion on the quantitative works applied on risk perception is a very hard issue to grasp. Because of limitation in time and space only few of them are here discussed. Henderson understands the importance for risk-management to mitigate external threats over tourism industry, whenever risks can be controlled, disasters are prevented to save lives (Henderson, 2008). At a first look, four main families of theories prevailed:

1. Social Bonds,
2. Nationalities or Cultural Incompatibilities,
3. Residency, and
4. Personality.

The first family of theories signals on the role played by trust in the configuration of a secure destination. Applied-research in this vein evinces how accompanied travelers develop further trust at time of going abroad than others who travel alone. It seems as though the sentiment of vulnerability rises whenever trust declines (Roehl & Fesenmaier, 1992; Yuan, 2005; Park & Reisinger, 2010). Rather, other studies suggest that nationalities conjoined to cultural asymmetries in some causes caused by tourism, are key factors that explain why tourists are attacked. Sacket and Botterill gathered information that validates the idea British and American tourists feel further risk than other nationalities. This happens because the Anglo-alliance caused some concerns in lay-citizens who dangle the possibilities to be killed by actions or decision made by their respective governments. The same was replicated by Dominguez, Burguette and Bernard (2003) through an interesting investigation in Mexico. Ertuna and Erutuna, in this vein, found that correlation between religious affiliation and risk perception is more than important to unveil. Protestant and Catholics are more prone to perceive further risk than Buddhists. Lorri Penninton Gray and colleagues evaluated the effects of 9/11 in New Yorkers and neighbor residents, understanding that the place of residency plays a vital role by enhancing or reducing risk perception (Floyd, Gibson, Pennington-Gray, & Thapa, 2004). Whenever terrorist attack takes hit, survivors associate the environmental background to probabilities the same would be repeated. Therefore, In the same way, rural farmers feel further secure and urban city-dwellers (Yuan, 2005). Last but not least, the theories of personalities triggered by Stanley Plog (1973, 1991) amply showed how some personalities are frightening to new sensations others are more risk-seekers. The legacy of Plog was useful for many colleagues who explored the intersection of curiosity and fear.

As this argument given, Lepp and. Gibson (2008) wrote that tourist travel is subject to two contrasting sentiments: the quest for novelty and the need for safety. To some extent, the cultural incompatibility between tourist originating and receiving countries may reduce the travelers' feelings of safety. Their degree of adaptation to new landscapes is partially determined by their personalities. While some tourists are sensation seekers, others are risk aversive. Kozak, Crotts and Law found the following:

1. Risk attracts some travelers who change their original destination to seek it.
2. Those personalities based on higher degree of tolerance to change do not need to alter their plans in context of adversity.
3. News of disasters or catastrophe not only affect the place of occurrence, but also neighboring countries.
4. Risks are not restricted to specific locales, but extend too much wider regions.
5. While travelers coming from industrial societies are concerned about terrorism, travelers form underdeveloped nations fear virus outbreaks.
6. Personality variations explain why some travellers perceive more risk than others.

Quantitative research in risk studies proliferated after 9/11 conferred to policy makers and tourism practitioners an interesting conceptual background to understand why dissidents, local crime and terrorism vulnerate tourists as well as paved the ways for the configuration of preventive programs to protect tourists while traveling abroad. However, some limitations were found by Korstanje (2011a, 2011b) in the way fieldwork is conducted. Under some situations, the administration of open and close-ended questionnaires left other interesting aspects behind. At the same time, some studies are not completely clear about the dissociation between risk avoiders and risk seekers. Whatever the case may be, risk correlation is not the same than risk meaning. In this position, qualitative research may complement the good steps initiated by quantitative viewpoint. Korstanje adds that, ethodologically, if we conduct investigation prioritizing quantitative techniques, the outcome will not explain the behaviour. We will see only correlations between two or more variables. To understand what is happening in the field, we need to introduce qualitative methods to complement the quantitative one. Quite aside from the size of the sample, the correlation of both variables does not entail explanations of why that correlation occurs. That is, there is no causal connection, no accounting for the mechanisms producing the correlations. For example, considerable evidence suggests that women perceive more risk than men. Following a quantitative reading, one might infer the gender is a variable of risk perception but this exhibits an ecological fallacy. Males are socialized and educated to repress their emotions. Though they feel fear, they avoid voicing or otherwise exhibiting fearful emotions. On the contrary, females are socialized to communicate fear and other emotions. This is the reason why quantitative and qualitative methods should be entwined.

Discussion: Terrorism in Middle East

To respond why terrorists, select carefully some targets while avoid others, Boris Vukonic (2010) explores the importance of tourism in the Middle East. The Muslim World has suffered many changes caused by the advance of modernity and tourism. Although some countries keep a friendly relationship of cooperation and diplomacy with the West, others, like Iran, kept an atmosphere of rivalry and conflict. What remains clear for Vukonic and the extant literature, seems to be that Western travelers are sometimes attacked not only taking advantage of their lack of familiarity about the environs they visit, but also the cultural values they represent. Unless state intervenes in their protection, tourist values may be offensive along with local tradition and customs. In this vein, the specialized literature marked a vast trajectory in studies of host/guest encounters, focusing in the problems as well as opportunities these types of meetings engender. However, Vukonic convincingly explains that the religion/tourism connection may be detached in three primary answers:

1. Religion somehow supports tourism,
2. Tourism somehow influences religion, and
3. Religion & tourism stand in opposition.

It is widely accepted that religion gives support to tourism only when religious norms are not defied or in danger. The world witnessed how the four great religions surfaced by means of pilgrims and holy land explorations. To adhere to religious doctrine, psychical displacement and travels play a crucial role because they pave the way for the construction of infrastructure, shrines, and designated holy places. Under some circumstances, tourism exerts considerable influence on tourism depending on the viewpoints of residents and participants. Some countries at odds may reinforce their mutual understanding by means of tourism, even borrowing religious values from each other. Religion and tourism could be irreconcilable enemies; this view dampens the peace-keeping process prioritizing inter-cultural differences rather than commonalities. Some practices introduced by guests may be incompatible with the religious beliefs of their hosts. In this case, two or more values may cause non-negotiable discrepancies that lead groups to conflicts, and even warfare. Since tourism accelerates the intergroup connection, in such a context, conflict surfaces. Throughout Middle East and beyond, developing countries adopted tourism as a primary industry to adjust some economic asymmetries resulting from problems of governability, corruption, and bad administrations. Although in the short run, tourism brings fresh investments to move host economies, states can develop a dependency on the tourism industry. Terrorists and insurgents may take advantage of this situation to affect the destination image. This way, they not only attack the state but also jeopardize its economy. Pro-tourist policies sometimes contrast with some radicalized beliefs. Therefore, terrorism is for Vukonic a glitch in the process of communication, an anomaly that makes tourism and religion not understand each other. Following Vukonic´s argument is that in an inter-cultural meeting, some groups are favourably encouraged to improve their situations, while others do not.

In 2012, Maximiliano Korstanje & Anthony Clayton wrote a polemic essay admitting tourism and terrorism share more commonalities than one would like to accept. Both scholars argue convincingly that terrorism wakes up some interest and curiosity in affected destinations which not only recover their attractiveness during a considerable time elapsed but even exceeding pre-existent volumes of demand. Thematised as dark or Thana tourism spaces of mass death and suffering as Ground Zero are visited annually by thousands of tourists. By the industry of entertainment, the system recycles the disastrous effects of terrorism into a new commoditized version to be commercially exploited. The problem with this rests on the fact societies never learn the lesson behind cementing the possibilities another new attack takes place in a near future (Korstanje & Clayton, 2012). This triggered a hot-debate in academicians as Shrabani Saha and Ghialy Yap who conducted an extensive quantitative research to correlate terrorism with tourism. They analysed the political instability and its effects on tourism in almost 139 countries during 1999-2009. The original goals of this investigation tilted to measure to what extent tourism is affected by terrorism. Although in some destinations, which kept a financial dependency respecting to tourists-delivering countries, terrorism wreaks havoc in tourism industry, it is not accident to see how in others terrorism produces a real stimulation for demand. To understand this better, authors divide countries who experience a low-political instability as Europe or US from others with higher-levels of risk towards political instability as Middle East (Saha & Yap 2014).

Basically, the problem of terrorism in Middle East is not new. After the US-led invasion, destinations as Syria and Iraq paralyzed their activities by the political instability caused by the installation of semi-democratic order. The action of ISIS and Muslim jihadism posed not only a serious challenge for tourism in these countries, but for western hospitality too. Recently bombing in Paris and Brussels ignited racists discourse against Muslim migration, as well as activated the struggle of extreme right wind parties in both countries. In retrospect, Steiner & al-Hamarmeh (2005) explored the limitations of tourism to revitalize areas or countries whose have ransacked by terrorism. Undoubtedly, everything started with the blow to World Trade Centre on 2001 which prompted the declaration of war by Bush administration, to terrorism which means to Muslim World. It was unfortunate, not only tourism but also development in Middle Age stagnated by action of terrorism. In that occasion, there are losers and winners. Islamic Tourism reacted positively to the negative image constructed in US and Europe around Middle East.

The predicted wide-ranging collapse of the tourism industry in the Arab countries after 9/11did not take place. Different countries were affected differently. In the context of tourism, we should discuss Arab markets rather than one single Arab tourist market. In the last two decades, diverse concepts for development of the tourism industry have been implemented in various Arab countries. The character of tourists' activities (leisure, cultural, pilgrim/religious, medical, shopping, etc.)[16] and the countries of tourists' origin point to four main concepts of tourism development in Arab countries:

1. *European/North American-oriented leisure tourism (Tunisia);*
2. *Arab-oriented leisure tourism (Lebanon);*
3. *multi-ethnic-oriented mixed-character tourism (Egypt, Morocco, Jordan, UAE); and*
4. *multi-ethnic-oriented cultural and pilgrim tourism (Saudi Arabia, Yemen, Syria, Palestine). (Steiner & Al-Hamarneh, 2004 , p. 175)*

The fact is that tourism is sensitive to terrorism, and when some destinations are affected after an attack takes room, other new segments arise instead. The rotation of tourists from one to another destination not only gives policy makers some hope respecting the resiliency of this industry, posed serious worries whenever Al-Baghdadi, former leader of ISIS declared the jihad against all tourist destinations of the world. It represented an urgency for western nations which were unable to protect their own citizens in leisure-spots or train stations. The message that terrorists wish to disseminate is that citizens are no longer safe anytime and in any place. It was unfortunate, while some experts thought terrorism would have controlled in periphery, the first world would remain secure. However, 9/11 showed there was no secure place to hide in the coming years. Because IS found the vulnerability of tourists are combined by the weak lines of surveillance in leisure spots, we understand that tourism is in jeopardy. This message not only threatens the functioning of tourism but also may mark what some specialists dubbed "the end of tourism". Since 9/11, West closed or tightened the borders to aliens. This means that the sense of hospitality that characterized the Western Spirit is dying, as a result of the rise of radical jihadism worldwide. Terrorism posed a paradox on Westerners, since while Syrian refugee are welcomed by some European countries which are moved by humanitarian reasons, IS declared that some infiltrated members are real Horse of Troy in Europe. Police and Intelligence suspected that the attacks on Paris was perpetrated by one infiltrated jihadist who arrived recently to Europe. Real or not, this represents a real problem for the industries of hospitality and tourism.

The Ends of Tourism

In this section, we have exposed an exploratory thesis: the end of tourism is coming-up. Not only risks such as disasters, virus outbreaks or bloody civil wars are unproductive for leisure and tourism, but also the decline of current economy in developed countries after the financial crash in 2008. If our parents started their holidays after a long year of hard work, now holiday-makers ask for loans in the bank system to travel abroad. Some estimation reveals that we are not hyper- mobile as we thought, since only 1% of humankind is legally and financially invested to travel. This begs a more than interesting question, to what extent we can consider that visiting spaces of extreme suffering as Ground-Zero is under the sphere of ethics?

This question is brilliantly addressed by Clare Weeden & Kala Boluk in the book, *Managing Ethical Consumption in Tourism*. Tourism is today one of the most promising industries in the world. However, the current tourist practices start to change towards new trends and forms, which certainly disorientate experts. The classic sun and beach product sets the pace to new forms of tourism where the concept of beautiness has been replaced by other morbid orientations. Dark-Tourism, Slum-tourism, Alcohol-tourism and sex tourism (among others) are gradually being posed as valid segments with a growing demand. This is the reason why Managing Ethical Consumption in tourism, a recent book edited by Clare Weeden and Karla Boluk is a recommended reading. In this text, editors formulate the needs to reconsider the role of ethics in tourism consumption, in a moment where the boundaries between ethics and law are blurred. While some social practices can be legal in some countries as sex with children or adolescent, they remain as unacceptable behaviour from ethics. Of course, despite the numerous studies in tourism consumption, ethics plays a marginal role. At some extent, tourist destinations, as Weeden and Boluk put it, embrace some ethical values whether they can be commoditized or used to gain further advantages in an ever-changing market. The lack of interests for tourism ethics correspond with the belief that responsibility, an optional decision created along with "certification schemes". To put this in other terms, doing a responsible tourism is part of our duties as practitioners, but we are not legally forced. A second more epistemological reasons lie in the fact, studies in tourism ethics are scattered or published in humanitarian journals more associated to philosophy, sociology or anthropology instead of coordinating all produced knowledge in only one magazine or platform.

What is unique about ethics is that more than any of these other fields of inquiry, it gets to the job of illuminating the tension that exists between what is good and bad, right and wrong, and authentic and inauthentic in tourism. It does this by forcing us to examine the human condition first, knowing something of ourselves, as a way to understanding tourism. (Preface Fennell 2014, XVI)

From a broader sense, this fine book is self-oriented not only at instilling a hot but not for that less rational debate respecting to ethics in tourism, but pose as a valid attempt to create an all encompassing model that explains why ethics is so important for practitioners and scholarship. Although sustainability has recently monopolized a great variety of research which centre on failure in developmental programs, or negative effects of tourism in local communities, a profit-oriented spirit is embraced. At some extent, there is a clear disassociation between consuming and ethics, which explains why some tourists are insensitive to "the Other" suffering. Rather than being the result of some had hoc reflections or five-minute papers, this book is the result of years of fieldwork, which to our understanding will become in a seminal text in the next decades. The main thesis is that ethic tourism centers on a conceptual contradiction because the first term (ethic) signals to the moral rules that mark what is good from evilness but at the same time, tourism looks a recreational and selfish nature where the goals is the maximization of pleasure. It is important to note that ethical tourism not only helps reverting the old legacy of colonialism by exotizicing non-white natives, but overcoming the negative aftermaths of what Korstanje dubbed "the postmodern exploitative tactics of consumption", which consists in presenting the "other's suffering" to reinforce the supremacy of West over other cultures and civilization. Following this argument, Korstanje adds, these new trends as dark, thana-tourism or slum tourism are a direct result from the acceptance of "social Darwinism" where the survival of strongest is valorized over other possible alternatives. Visiting places of extreme poverty or suffering not only reminds we are living in the up-hill city, in an exemplary centre where others cannot enter, but serves as an ideological mechanism that makes us proud of our society, accelerating the social ties disorganization. Paradoxically, once we look for "Others", they are not genuinely recognized unless by strengthening our sentiment of superiority (Korstanje, 2015).

A second point of entry in this discussion relates to the financial problems to pay all inclusive packages offered by the market to those holiday-makers who prioritize their security. In a recent re-edited book, entitled *Tourism & Development,* Sharpley and Telfer acknowledged that from the first publication of their book in 2002 many things happen. Despite the promotion of tourism as vehicle towards development, little success reached the peripheral nations. The dilemma lies in the fact despite the abundance of development-related programs adopted by some tourist destinations, under-developed nations failed to enhance their economies. Almost one decade later the first edition, tourism and development have left many open questions. A great variety of radical changes accelerated not only our current system of production but the ways economists see the future trends in financial investment.

What global trends show is that tourism not only is a resilient industry which has faced a lot of risks and problems but also it ensures the transfer of wealth among nations of north-hemisphere. At the time, some peripheral countries introduce tourism as the only instrument to revitalize their economies, a great financial dependency surfaces. This happens because of two main reasons. The first and foremost is the lack of infrastructure and capital necessary to captivate international demands; secondly, the emergent destinations enter in an extreme competence with their neighbors to pose at the top of mind of consumers. The tourism-development doctrine has faced some shifts over the last decade. The problem of wealth inequalities or the asymmetries produced by capitalist system appears to be enlarged by the advent of globalization. Starting from the premise that tourism is enrooted in a hedonistic act of consumption, feeding up the individual ego-enhancement; it is difficult to understand how it should contribute to the well-being of others which is the backbone of development.

What this text ignores is that globalization and of course development are enterprises that face extinction after the financial crash of 2008 in Wall Street. Then, there is incipient evidence not only tourism is dying, but also style of consuming are radically undermined. Whether our grand-parents were accustomed to work hard throughout the year to enjoy their holidays, they did not pay the travels in installments. Their savings were enough for tourism and touring elsewhere. However, this model sets the pace to new more flexible ways where holiday-makers should go into debt in order for them to get the "dream trip". Various factors coadyuvated for the emergence of a new tourist archetype more sensitive for injustices or interested in authentic experiences, but the fact is that, beyond its altruism, the dichotomization of globalized economy remains. Marketing campaigns have devoted considerable efforts and resources to confer to holidays a sacred-nature, centered in what Lash and Urry dubbed "the economy of signs". The costs of transport were increased to levels that lay-people are unable to absorb. The magic of marketing was used to boost consumption in a decaying system where stability, well-fare state and employment set the pace to the rise of uncertainness. As Bauman (2013a) noted, in the postmodern world consumers have converted in commodities to be consumed by capital (Bauman, 2013). In a consuming life fraught of vulnerabilities, following Bauman´s metaphor, debt is created to engage consumers into the process of purchasing. The needs of consuming are psychologically placed from outside of citizens. Their rights for accessing to public services as education, health and security depend on their income or class where they belong. Unlike in other decades wherein rest was a result of labor, the current holiday-makers work hard in the year to pay off their debts with banks. Vacations are not the consequences of work, but the sign of status that mark the difference exploited and exploiters (Korstanje, 2015). The rise of new more dangerous risks conjoined to the decline of post-modern means of production is enough to hint tourism is dying. Terrorism has advanced beyond the contours of Middle East at the doorstep of Anglo-Saxon Empire. One of the effects of

9/11 was not related to the economic loss tourism temporarily suffered, but to the adoption of a radical discourse, where "the Other" who is important for tourism should be subordinated to the cultural values of centre. This allegory paved the ways for the advent of a new radical ethnocentric view in regards of the periphery, which reinforced the protestant view of the manifest destiny, or the supremacy of US over the world. Terrorism has been successful not only in instilling the germen of fear in US soil, but introducing "preemption" as a valid mechanism of defense. As a result of this, US closed its borderlands initiating a new era of anti-hospitality. Tourism as an activity based on hospitableness has little probabilities to prosper in a world where the obsession for security and terrorism persist.

CONCLUSION

The answer to the above formulated questions is not easy, and of course, it exhibits a fertile ground to be explored in other approaches. The ceaseless news about violence, cruelty, wars and deaths have serious negative impacts on audience worldwide. Though policy makers have devoted their time and efforts in looking for new alternative segments where death replaces the allegory of beautiness, giving as a result new products as dark-tourism, slum-tourism, disaster-tourism or doom-tourism, other problems arise. The curiosity for tourists in visiting spaces of mass death or extreme suffering contradicts not only its nature as a mechanism of revitalization and relaxes but may very well lead to narcissism which in fact is the end of tourism. If we pay attention enough to the evolution of dark tourism, we realize that far from being an interest for "others` pain" as specialized literature suggests, it exhibits an act of egoism. George H. Mead, one of the fathers of symbolic interactionism, questioned why paradoxically many people are prone to read or listen of bad news presented by journalism, at the time they show preference by these types of news. What is our fascination for other's suffering? He assertively concludes that the self is configured by its interaction with others. This social dialectic alludes to anticipation and interpretation as two pillars of communication-process. The self feels happiness by other's suffering, because it represents a rite necessary to avoid or think in own pain. Starting from the premise the self is morally obliged to assist the "others" to reinforce its sentiment of superiority, Mead adds, this is the ethical nature of social relationship (Mead, 2009). The same applies for dark tourism.

In medieval times, death was present in almost all institutions, representation of the daily life but paradoxically, pilgrims may not be equaled to dark tourism sites by many reasons. Unlike modern sight-seers, medieval travellers move to sacred sites looking two important aspects to redeem their sins, forgiveness or the mediation of Saints to negotiate with God, a solution to their pains or big troubles. Although venerated, for medieval travellers death was not a problem like modern tourists, but also the beginning for a new better life. In this respect, dark tourism exhibits the opposite dynamic. "Secular tourists" are not interested in the life of others, nor in their heritage, or biography. They want to anticipate and avoid their own death. The specialized bibliography focuses on those modern tourists understand death through the lens of others. Rather, our argument goes in opposite direction. Tourists exorcise death ritualizing other's death to expand their own life expectances.

Zygmunt Bauman has reminded that life has no meaning without death. On his book *Consuming life* and *Liquid Fear.* The capitalist ethos has changed the mind of citizens, who passed being part of the production machinery. As commodities, workers are exploited to congeal the mass-consumption encouraged by capitalism. The big brother is an example how people enter in competence, as commodities, to be selected and bought by others. Participants in this reality show know that only one will win, and the rest will die. Big Brother, for Bauman, emulates the life in capitalist societies which enhance the style of life of few by producing pauperization for the whole. The modern state set the pace to the advent of liberal market to monopolize the sense of security for people. This does not mean that states are unable to keep the security, but also the market is re-channelling the consumption by the imposition of fear. If human disasters as Katrina show the pervasive nature of capitalism which abandoned thousands of pour citizens to death, no less truth is that the "show of disaster" unbinds of responsibilities for the event. The sense of catastrophe, like death, serves to cover the inhuman nature of capitalism (Bauman, 2008, 2013a). This society only has an answer to crisis, when its economic system is at risk. Since the real reason for disaster are ignored by the allegory of death, which persisted in the media and famous TV series where technicians and forensic experts look to solve the crime, the disaster comes sooner or later. What do we really know on the real causes of Auschwitz or 9/11? May a simple museum explain us the complexity of human nature? Bauman will say, absolutely not. Any attempt to sacralise the dying as a spectacle, at the bottom, represents the prelude of its neglect. Last but not least, dark tourism is not different than other issues as FIFA World Cup, reality-shows, Big Brother and so forth. They proclaim ideologically that only one may be the winner. As a result of this, the climate of labor is vulnerated in favor of an eternal competence where workers are challenged to be co-active managers of their own destiny. In shark opposition to the medieval traveler, dark tourism consumers seek to reinforce their life as their other's death. In contrast to what the specialized literature suggests, dark tourism reinforces the modern egocentrism to enjoy in brother's tragedy. Protestantism has brought the idea that life should be symbolized as a great trace where only one will be the winner. Of course, this means that the rest will loose. If tragedy confers to survivors the aura of exemplary civilization, it runs higher costs. Happiness for the other's death is a sign we still remain in contention for a final fight. From the Big Brother to The Hunger Games, the salvation of one by the ruin of whole has been a constant ideology of our modern world. The ethics of dark tourism emulates a new economic form of exploitation that characterizes the capitalist-ethos, which evinces that tourism is at the final stage.

FUTURE RESEARCH DIRECTION

Why risk should be added as a topic in a book dedicated to techniques of optimization of tourist destinations? Well, the importance of risk perception studies adheres to the process of communication that affects their attractiveness. In rare occasions, human beings observe the events as they really are. The psychological cognitive and emotional distortions lead persons to over-valorize some risks, while others are neglected. In recent decades, sociology widely demonstrated that the rise of a plenty of uncertainness, which is dubbed as "inflation of risks". Lay citizens are frightening to unemployment, natural disasters, terrorism, and others. In tourism fields, two events paved the ways for the adoption of risk perception literature, Luxor Massacre and World Trade Centre attacks. Policy makers enabled to fight against terrorism witnessed from now on, tourists were clear targets of terrorist cells. This is the reason why we, in a globalized world where decisions resonate elsewhere. Some indicators as central economies plummeted in seconds, at the same time, that new elaborated and complex terrorist cells (leaded by Islamic State) perpetrated shocking attacks on the metropolis of West as New York, London and recently Paris. Understanding not only how risks are perceived by population but how the process of communication (mass media) deals with them is the first step in order for tourism specialist, officials, and policy makers to plan sustainable tourist destinations in a near future.

REFERENCES

Abdel-Azim, T. S. (2010). The Relationship between the perception of Risk and the Decision Making process of Travel of French Tourists. The case of Egypt. *Tourismos: An International Multidisplinary Journal of Tourism, 2*, 29-47.

Anderson, W., Juaneda, C., & Sastre, F. (2009). Influences of pro-all-inclusive Travel Decisions. *Tourism Review, 64*(2), 4–18. doi:10.1108/16605370910963482

Bauman, Z. (2008). *Liquid fear*. Buenos Aires: Paidos.

Bauman, Z. (2009). Freudian Civilization revisited – or whatever happened to the reality principie? *Anthopological Psychology, 21*, 1–9.

Bauman, Z. (2013a). *Consuming life*. New York, NY: John Wiley & Sons.

Bauman, Z. (2013b). *Liquid fear*. London: John Wiley & Sons.

Beck, U. (2006). *Risk society, towards a new modernity*. Buenos Aires: Paidos.

Bianchi, R. (2007). Tourism and The Globalization of Fear: Analyzing the politics of risk and (in)security in global travel. *Tourism and Hospitality Research, 7*(1), 64–74. doi:10.1057/palgrave.thr.6050028

Bianchi, R., & Stephenson, M. (2014). *Tourism and Citizenship: rights, freedoms and responsibilities in the global Order*. Abingdon: Routledge.

Blake, A., & Sinclair, M. T. (2003). Tourism crisis management: US response to September 11. *Annals of Tourism Research, 30*(4), 813–832. doi:10.1016/S0160-7383(03)00056-2

Cohen, E., & Neal, M. (2010). Coinciding crises and tourism in contemporary Thailand. *Current Issues in Tourism, 13*(5), 455–475. doi:10.1080/13683500.2010.491898

Dolnicar, S. (2005a). Fear Segment in tourism. In *Proceedings of the 14 International Research Conference of the Council for Australian University and Hospitality Education*. CAUTHE.

Dolnicar, S. (2005b). Understanding barriers to leisure travel, tourists fears as marketing basis. *Journal of Vacation Marketing, 11*(3), 197–208. doi:10.1177/1356766705055706

Domínguez, P., Burguette, E., & Bernard, A. (2003). Efectos del 11 de Septiembre en la hotelería Mexicana: reflexión sobre la mono-dependencia turística. *Estudios y Perspectivas en Turismo, 12*(3-4), 335-348.

Ertuna, C., & Ertuna, Z. I. (2009). The Sensitivity of German and British tourists to new shocks. *Tourism Review., 64*(3), 19–27. doi:10.1108/16605370910988809

Faulkner, B. (2001). Towards a framework for tourism disaster management. *Tourism Management, 22*(2), 135–147. doi:10.1016/S0261-5177(00)00048-0

Fennell, D. A. (2014). Foreword. In C. Weeden & K. Boluk (Eds.), *Managing Ethical Consumption in Tourism*. New York, NY: Routledge, xv-xvii. doi:10.1016/B978-0-12-411577-4.09985-8

Floyd, M. F., Gibson, H., Pennington-Gray, L., & Thapa, B. (2004). The effect of risk perceptions on intentions to travel in the aftermath of September 11, 2001. *Journal of Travel & Tourism Marketing*, *15*(2-3), 2–3, 19–38. doi:10.1300/J073v15n02_02

Fuchs, G., & Reichel, A. (2004). Cultural Differences in tourist destination risk perception: An exploratory study. *Tourism*, *52*(1), 21–37.

Fuchs, G. & Reichel, A. (2010). Health Tourists Visiting a Highly Volatile Destination. *Anatolia: An International Journal of Tourism and Hospitality Research, 21*(2), 205-226.

Gale, T. (2008). The end of tourism or endings of tourism? In P. Burns & M. Novelli (Eds.), *Local-Global Connections*. Wallingford: CABI. doi:10.1079/9781845934040.0001

George, B, Inbakaran, R. & Poyyamoli, G. (2010). To Travel or Not to travel: towards understanding the theory of nativistic motivation. *Tourism: An International Interdisciplinary Journal, 58*(4), 395-407.

Giddens, A. (1991). *Modernity and Self-Identity: Self and Society in the Late Modern Age*. Stanford University Press.

Giddens, A. (1999). *The Consequences of Modernity*. Madrid: Alianza Editorial.

Giddens, A. (2000). *Run Away World: how globalization is reshaping our lives*. Madrid: Taurus.

Hall, M. (2002). Travel Safety, terrorism and the media the significance circle of the issue attention cycle. *Current Issues in Tourism*, *5*(5), 458–466. doi:10.1080/13683500208667935

Hall, M. (2010). Crisis Events in Tourism: Subjects of crisis in Tourism. *Current Issues in Tourism*, *13*(5), 401–417. doi:10.1080/13683500.2010.491900

Henderson, J. C. (2008). Managing Crises: UK Civil Aviation, BAA airpots and the August 2006 terrorist Threat. *Tourism and Hospitality Research*, *8*(2), 125–136. doi:10.1057/thr.2008.12

Korstanje, M. (2009). Re-visiting risk perception theory in the context of travel. *E-Review of Tourism Research*, *7*(4), 68–81.

Korstanje, M. (2015). *A Difficult World, examining the roots of Capitalism*. New York, NY: Nova Science publishers.

Korstanje, M. E. (2011a). The fear of traveling: A new perspective for tourism and hospitality. *Anatolia*, *22*(2), 222–233. doi:10.1080/13032917.2011.597935

Korstanje, M. E. (2011b). Why risk why now? Conceptual problems around the risk perception in tourism industry. *RBTur*, *5*(1), 4-22.

Korstanje, M. E., & Olsen, D. H. (2011). The discourse of risk in horror movies post 9/11: Hospitality and hostility in perspective. *International Journal of Tourism Anthropology*, *1*(3), 304–317. doi:10.1504/IJTA.2011.043712

Korstanje, M. E., & Skoll, G. (2013). An Essay on the Social Costs and Benefits of Technology Evolution. *Human & Social Studies. Research and Practice*, *2*(2), 13–39.

Kozak, M., Crotts, J., & Law, R. (2007). The Impact of the pereption of risk on international Travellers. *International Journal of Tourism Research*, *9*(4), 233–242. doi:10.1002/jtr.607

Larsen, S. (2007). Aspects of a Psychology of the Tourist Experience. *Scandinavian Journal of Hospitality and Tourism, 7*(1), 7-18.

Larsen, S., Brun, W., & Øgaard, T. (2009). What Tourists worry about: Construction of a scale measuring tourist worries. *Tourism Management*, *30*(2), 260–265. doi:10.1016/j.tourman.2008.06.004

Lash, S. M., Urry, S. L. J., & Urry, J. (1993). *Economies of signs and space* (Vol. 26). London: Sage.

Lepp, A., & Gibson, H. (2008). Sensation Seeking and Tourism: Tourist role, perception of risk and Destination Choice. *Tourism Management*, *29*(4), 740–750. doi:10.1016/j.tourman.2007.08.002

Luhmann, N. (2006). *Sociología del Riesgo [Sociology of Risk]*. Universidad Iberoamericana.

Mccartney, G. (2008). Does one culture all think the same? An investigation of destination image perceptions from several origins. *Tourism Review, 63*(4), 13-26.

Mead, G. H. (2009). Mind, self, and society: From the standpoint of a social behaviourist (Vol. 1). Chicago: University of Chicago Press.

Park, K., & Reisinger, Y. (2010). Differences in the perceived influence of natural disasters and travel risk on international travels. *Tourism Geographies*, *12*(1), 1–24. doi:10.1080/14616680903493621

Pforr, C., & Hosie, P. (Eds.). (2009). *Crisis management in the tourism industry: beating the odds?* London: Ashgate Publishing, Ltd.

Pizam, A. (1999). A comprehensive approach to classifying acts of crime and violence at tourism destinations. *Journal of Travel Research*, *38*(1), 1, 5–12. doi:10.1177/004728759903800103

Pizam, A. (2002). Tourism and terrorism. *International Journal of Hospitality Management*, *21*(1), 1–3. doi:10.1016/S0278-4319(01)00037-8

Pizam, A., & Mansfeld, Y. (1996). *Tourism, crime and international security issues*. New York, NY: John Wiley & Sons.

Pizam, A., Tarlow, P. E., & Bloom, J. (1997). Making tourists feel safe: Whose responsibility is it? *Journal of Travel Research*, *36*(1), 23–28. doi:10.1177/004728759703600104

Prideaux, B., Laws, E., & Faulkner, B. (2003). Events in Indonesia: Exploring the limits to formal tourism trends forecasting methods in complex crisis situations. *Tourism Management*, *24*(4), 475–487. doi:10.1016/S0261-5177(02)00115-2

Qi, X. C., Gibson, H., & Zhang, J. (2009). Perception of risk and Travel Destinations. The case of China and the Beijing Olympic Games. *Journal of Sport & Tourism*, *14*(1), 43–67. doi:10.1080/14775080902847439

Quarantelli, E. (1975). *Panic Behavior: Some empirical observations.* Disaster Research Center, Preliminary Paper 20.

Quarantelli, E. (1982). *People´s reactions to emergency warnings.* Disaster Research Center, Preliminary Paper 75.

Quarantelli, E. (1990). *The Mass media in disasters in the United States.* Disaster Research Center, Preliminary Paper.

Quarantelli, E. (2001). *The Sociology of Panic.* Disaster Research Center, Preliminary Paper 283.

Reichel, A., Fuchs, G., & Uriely, N. (2007). Perceived risk and the non-instituonalized tourist role: The case of Israeli student ex backpackers. *Journal of Travel Research, 46*(2), 217–226. doi:10.1177/0047287507299580

Ritchie, B. W. (2004). Chaos, crises and disasters: A strategic approach to crisis management in the tourism industry. *Tourism Management, 25*(6), 669–683. doi:10.1016/j.tourman.2003.09.004

Ritchie, B. W. (2009). *Crisis and disaster management for tourism.* Bristol: Channel View Publications.

Roehl, W., & Fesenmaier, D. (1992). Risk Perceptions and Pleasure Travel: An exploratory analysis. *Journal of Travel Research, 30*(4), 17–29. doi:10.1177/004728759203000403

Sackett, H. & Botterill, D. (2006). Perception of international travel risk: an exploratory study of the influence of proximity to terrorist attack. *E-Review of Tourism Research, 4*(2), 44-49.

Saha, S., & Yap, G. (2014). The moderation effects of political instability and terrorism on tourism development: A cross-country panel analysis. *Journal of Travel Research, 53*(4), 509–521. doi:10.1177/0047287513496472

Sharpley, R., & Telfer, D. (2015) Tourism & Development: concepts and Issues (2nd ed.). Bristol: Channel View Publications.

Skoll, G. R., & Korstanje, M. E. (2012). Risks, totems, and fetishes in Marx and Freud. *Sincronía, 2*, 11–27.

Skoll, G. R., & Korstanje, M. E. (2014a). Points of Discussion around 9/11, terrorism and tourism revisited. *e-Review of Tourism Research, 11*(1-2), 1-17.

Skoll, G. R., & Korstanje, M. E. (2014b). Terrorism, homeland safety and event management. *International Journal of Hospitality and Event Management, 1*(1), 95–110. doi:10.1504/IJHEM.2014.062869

Sönmez, S. F. (1998). Tourism, terrorism, and political instability. *Annals of Tourism Research, 25*(2), 416–456. doi:10.1016/S0160-7383(97)00093-5

Sönmez, S. F., Apostolopoulos, Y., & Tarlow, P. (1999). Tourism in crisis: Managing the effects of terrorism. *Journal of Travel Research, 38*(1), 13–18. doi:10.1177/004728759903800104

Steiner, C., & Al-Hamarneh, A. (2005). Islamic tourism: Rethinking the strategies of tourism development in the Arab world after September 11, 2001. *Comparative Studies of South Asia, Africa and the Middle East, 24*(1), 173–182.

Sunstein, C. (2006). *Risk and Reason. Safety, law and environment.* Buenos Aires: Katz.

Tarlow, P. (2011). Tourism disaster management in an age of terrorism. *International Journal of Tourism Anthropology, 1*(3), 254–272. doi:10.1504/IJTA.2011.043709

Tarlow, P. (2012). Rosh Ha` Shanah Day. *International Journal of Safety and Security in Tourism and Hosp., 3*, 1–19.

Tarlow, P. (2014). *Tourism Security*. Oxford, UK: Elsevier.

Tarlow, P. & Korstanje M. (2013). Disasters, tourism, and mobility: The case of Japan Earthquake. *Pasos, revista de Turismo y Patrimonio Cultural, 11*(3), 17-32.

Tierney, K. (1994). *Sociology´s Unique Contributions to the Study of Risk*. Disaster Research Center, Preliminary Paper, 204.

Vukonic, B. (2010). Do we always understand each other? In *Tourism and Muslim World* (pp. 31–45). West Yorkshire: Emerald Group Publishing. doi:10.1108/S2042-1443(2010)0000002006

Weeden, C., & Boluk, K. (2014). *Managing Ethical Consumption in Tourism*. New York, NY: Routledge.

Yuan, M. (2005). After September 11: determining its Impacts on Rural Canadians travel to U.S. *E-Review of Tourism Research, 3*(5), 103-108.

Zinn, J. O. (2010). Biography, Risk and Uncertainty – Is there common ground for biographical Research and Risk Research. *FQS, Forum, Qualitative Social Research, 11*(1), 1-19.

ADDITIONAL READING

Bertaux, D., & Kohli, M. (1984). The life story approach: A continental view. *Annual Review of Sociology, 10*(1), 215–237. doi:10.1146/annurev.so.10.080184.001243

Birtchnell, T., & Buscher, M. (2011). Stranded: An Eruption of Disruption. *Mobilities, 6*(1), 1–9. doi:10.1080/17450101.2011.532648

Boniface, B., & Cooper, C. (2009). *World-wide Destinations: the geography of travel and tourism*. New York, NY: Elsevier.

Britton, S. (1982). The Political Economy of Tourism in the Third World. *Annals of Tourism Research, 9*(3), 331–358. doi:10.1016/0160-7383(82)90018-4

Brunt, P., & Courtney, P. (2000). Host Perception of sociocultural impacts. *Annals of Tourism Research, 26*(3), 493–515. doi:10.1016/S0160-7383(99)00003-1

Burnham, J. J. (2007). Childrens Fears: A Pre 9/11 and Post 9/11 Comparison Using the American Fear Survey Schedule for Children. *Journal of Counseling and Development, 85*(4), 461–466. doi:10.1002/j.1556-6678.2007.tb00614.x

Castaño, J. M. (2005). Psicología Social de los Viajes y el Turismo. Madrid: Thomson Ed.

Caton, K. y Almeida-Santos, C. (2008). "Closing Hermeneutic circle?. Photographic encounters with the other" *Annals of Tourism Research,*. 35.1, 7-26

Chauhan, V. (2007). Safety and Insecurity: Perception of tourists visiting Kashmir, India. *Advances in Hospitality and Leisure, 3*, 3–17. doi:10.1016/S1745-3542(06)03001-3

Cohen, E. (2005). Principales tendencias en el turismo contemporáneo. *Politica y Sociedad, 42*(1), 11–24.

Dahlhamer, J. and Nigg, J. (1994). "An Empirical investigation of rumouring: anticipating disaster under conditions of uncertainty". *Disaster Research Center,* Preliminary Paper 216

Dias Varella, M. (2007). La Dinámica y la percepción pública del riesgo y las respuestas del derecho internacional económico. In G. Sozzo (Ed.), *En el Gobierno de los Riesgos* (pp. 105–123). Santa fe: Universidad Nacional del Litoral.

Dolnicar, S. (2005b). Understanding barriers to leisure travel, tourists fears as marketing basis. *Journal of Vacation Marketing*, *11*(3), 197–208. doi:10.1177/1356766705055706

Domínguez, P, Burguette, E y A. Bernard. (2003). "Efectos del 11 de Septiembre en la hotelería Mexicana: reflexión sobre la mono-dependencia turística". *Estudios y Perspectivas en Turismo,* 12. 3-4, 335-348.

Douglas, M. (1992). *Risk and Blame*. New York, NY: Routledge. doi:10.4324/9780203430866

Fabiani, J. L., & Thyes, J. (1987). *La Societeé Vulnerable.* Paris: Ecole Normale Superieure, Floyd, M. y L. Pennington-Gray. (2004). "Profiling Risk: perception of tourist. *Annals of Tourism Research*, *31*(4), 1051–1054.

Frechtling, D. (1993). "Key Issues in the US Travel industry futures". S. Medlik (editor). Managing Tourism. Oxford: Butterworth-Heinemann.

Graham, J. D., & Clemente, K. (1996). "Hazards in the News: Who Believes that?". *Risk in Perspective. Harvard Center for Risk Analysis*, *4*(4), 1–17.

Gut, P., & Jarrell, S. (2010). Silver Lining on a Dark Cloud: The impact of 9/11 on a Regional tourist Destination. *Journal of Travel Research*, *46*, 147–153.

Korstanje, M., & Skoll, G. (2013). "The borders of tourism and hospitality: the dark side of Empire". *Revista Rosa Dos ventos, 5*.1, 77-82.

Korstanje, M. E. (2011). Reconnecting with poverty: New challenges of disaster management. *International Journal of Disaster Resilience in the Built Environment*, *2*(2), 165–177. doi:10.1108/17595901111149150

Korstanje, M. E. (2011). Swine flu in Buenos Aires: Beyond the principle of resilience. *International Journal of Disaster Resilience in the Built Environment*, *2*(1), 59–73. doi:10.1108/17595901111108371

Korstanje, M. E. (2014). Chile helps Chile: Exploring the effects of earthquake Chile 2010. *International Journal of Disaster Resilience in the Built Environment*, *5*(4), 380–390. doi:10.1108/IJDRBE-01-2012-0002

Korstanje, M. E., & Clayton, A. (2012). "Tourism and terrorism: conflicts and commonalities". *World-wide Hospitality and Tourism Themes, 4*.1, 8-25.

Korstanje, M. E., Tzanelli, R., & Clayton, A. (2014). Brazilian World cup 2014: Terrorism, tourism, and social conflict. *Event Management*, *18*(4), 487–491. doi:10.3727/152599514X14143427352391

Mansfeld, Y., & Pizam, A. (2006). *Tourism, Security and Safety: from theory to practice*. Oxford: Elsevier Butterworth-Heinemann.

Pizam, A. Reichel, A. & Stein, F. (1982). "tourism and crime: is there a relationship?". *Journal of Travel Research.* Vol. 20.3, 7-10.

Plog, S. (1973). Why destination areas rise and fall in popularity. *The Cornell Hotel and Restaurant Administration Quarterly*, *13*(3), 13–16.

Plog, S. (1991). *Leisure Travel; making it a growth market. again!* New York, NY: Ed. Wiley and Sons.

KEY TERMS AND DEFINITIONS

Financial Dependency: Material submission from a periphery respecting to its centre.

Globalization: Cultural and economic project that postulate for the international integration of nations.

Mass-Media: A plenty of technologies and communication processes that transmit information to a much wider audience.

Resentment: A basic emotion determined by hatred and hostility to an external object/person. It remains a combination of disappointment, anger and fear.

Risk Perception: Individual judgment based on the contingency which means the posibibilities to avoid a real danger.

Risk: Potential losing or damage perceived which can be real or not.

Terrorism: Any act aimed to inflict violence and terror into the population.

Tourist Destination: A geographical place conditioned to receive tourists and visitors who are interested in gazing certain attractions.

Chapter 13
Capacity Management in Hotel Industry for Turkey

Deniz Efendioğlu
Marmara University, Turkey

Serol Bulkan
Marmara University, Turkey

ABSTRACT

The hotel industry is capital-intensive with high operating leverage in providing accommodations. Examining final costs play very important role in determining of overcapacity and undercapacity. Turkey's experiencing growth in tourism and has remarkable increases in accommodation. This situation possesses great importance for the capital groups investing in this area. How these groups and local tradesman implement a strategy on capacity management is the main question. Single period inventory model is used at the analysis phase and several interests are examined on this topic. Future Room demand is estimated by ARIMA and in estimating cost of undersupply earnings before taxes is used per room night sold. Combining the derived cost ratio with the future room demand and probability distribution estimated from the ARIMA method, the optimal hotel room capacity can be determined. The purpose of the study is to examine hotel room supply and demand for optimal capacity in Turkey.

INTRODUCTION

The hotel industry is capital-intensive with high operating leverage in providing accommodations. For a specific example, the situation of China's tourism can be given here. Owing to increase in American hotels during 1980-90's and improving their facilities, the significant tourist destination of China in Hong-Kong, Macau etc. suffered huge losses, even bankrupted. When America's attitude towards tourism at 1980-90's is taken into account, the impact of this behavior on tourism firms in China has induced capacity management becoming an important topic.

Turkey's experiencing growth in tourism recently has remarkable increases in accommodation and it is still on. Therefore, it is also one of the important issues for the economic situation of Turkey. This situation possesses great importance for the capital groups investing in this area and leads to more

DOI: 10.4018/978-1-5225-1054-3.ch013

investment on the tourism industry. For both these groups and local tradesman, one important issue is the capacity management. How these groups and local tradesman implement a strategy on capacity management is the main question.

In this work it is tried to make inferences for potential acquirements and what have to be done. All required data for the study "including financial information, occupancy levels, past hotel demand, and future hotel supply" is obtained from Ministry of Tourism & Culture and Turkish Statistical Institute (TUIK). Accommodation costs were procured by regarding on the data obtained from municipal and private tourism facilities. Furthermore, these works can be applied for provinces of Turkey, especially, famous for tourist destinations and facilities.

Hotel capacity is one of the most significant topics in tourism sector. Turkey has enlarged the amount of tourism income by summer tourism and historical tourism. Therefore, the places that the tourists stay in this period become the most important topic. Number of hotel capacity and the new investments must be determined clearly, due to this situation to give the decision that whether new constructions must be made or stopped. Due to tourism sector, Turkey increased economical growth rate and try to have provided the stability of economical system from a different perspective that had never been included larger than before. Clearly determining the amount of tourism expectation in this subject, is going to provide more robust designed facilities and planned urbanism in the future for tourists and local people.

If the data is examined, an increase in number of tourist and income of accommodation are generally observed. However, due to some epidemic and economic reasons, several decreases also have been noticed. Specifically, bird flu in 2006 and the global economic crisis in 2009 are the most known ones which cause really wide drops in the number of tourists. Nevertheless, in the coming period and following years, due to deteriorating relations with Russia, Turkey tourism sector is going to have some wounds, especially for 1-2 years. On the other hand, the need of these two neighbor to each other, considering the economical relations, will have an impact to this deteriorated relations. Summer tourism and with relation to accommodation, have been having problem in hotel capacity situation and new investors in here recently. On the other hand, some prevention are trying to be done for this hapless situation, including trying to attract the attention of neighbor countries and other ones for precaution.

Single period inventory model is used at the analysis phase and some interests are utilized examining several articles on this topic. Examining final costs, particularly fixed part of this cost, play very important role in determining of overcapacity and undercapacity. Future Room demand is estimated by using ARIMA and in estimating cost of undersupply earnings before taxes (EBT) is used per room night sold. Oversupplying is defined as the fixed cost per room night available. Combining the derived cost ratio with the future demand and probability distribution estimated from the ARIMA method, we could thus determine the optimal hotel room capacity. The purpose of the study is to examine hotel room supply and demand for optimal capacity in Turkey's hotel industry during 2002–2013 using a single-period inventory model.

THE SINGLE PERIOD INVENTORY MODEL

In many situations, a decision maker encounter with the problem of determining the value of the quantity ordered of products in inventory which is shown "q". After q has been determined, the demand is observed which is assumed by a random variable D.

Considering the values of d and q, the decision maker incurs a cost of c(d,q). While the calculation are making, it is assumed that the person is risk-neutral and try to find a way to minimize expected cost. Since the decision is made only once, this type of model is called a single-period decision model.

Organizations generally face with inventory problems where the following sequence of events happens:

- The organization decides the order quantity. q will be taken as the number of units ordered.
- A demand of d units occurs, with probability p(d). d must be a nonnegative integer and D is the random variable representing demand.

There are certain products which can be transported for only a very limited period of time, before it is determined that it is not going to sold. Such products are perishable products. For such products model single-period inventory models are generally designed. (Fresh foods, vegetables, fashion goods etc.)

Daily newspaper at a news stand may be given as a significant example of perishable product. Since the demand is a random variable, the owner of the news have to choose a daily order quantity that provides an appropriate trade-off between the potential cost of the wasted expense of ordering more than that can be sold and the potential cost of the lost profit from ordering fewer than that can be sold. This types of problem is called as newsboy problem. In the literature, it is seen as the single period stochastic or probabilistic model.

If D is assumed as an integer-valued discrete random variable for the single period probabilistic model with P(D = d) = p(d).

$$E(q) = \sum_{d} p(d) \cdot c(d,q) \tag{1}$$

E(q) = Decision makers expected cost

It can be clearly seen that E(q) is a convex function of q. If we go through from q* which minimizes E(q) the graph of E(q) is going to take a shape of inverted parabola. Because of possessing one global minimization point, q* become the smallest value of q for which;

$$E(q^* + 1) - E(q^*) \geq 0 \tag{2}$$

If E(q) is a convex function of q, the value of q may be found by obtaining the smallest value of q. Putting 1, 2, ..., n to q is not going to change the basic trend of this array. This approach determines q* by repeatedly computing the effect of adding a marginal unit for the value of q*. Because of this reason, this approach generally called marginal analysis (Winston, 1991).

To see how marginal analysis can be used to solve news vendor problems where demand is a discrete random variable, the following form is used c(d,q);

$$c(d,q) = c_o q + (\textit{terms not involving} \, q), (d \leq q) \tag{3}$$

$$c(d,q) = -c_u q + (terms\ not\ involving q), (d \geq q+1) \tag{4}$$

c_0 is the per-unit cost of being overstocked. If d ≤ q, ordering part was more than was demanded, also it is called as overstocked. If the size of the order is increased from q to q-1, then third equation shows that the cost increases by c_0 which is referred as overstocking cost.

Similarly, if d ≥ q+1, the supply is ordered an amount less than demand, in which it is understocked. If d ≥ q+1 and the size of the order is increased by one unit, the supply is understocked by one less unit. Then 4. equation implies that the cost is reduced by c_u which is the per unit cost of being understocked and it is called as understocking cost (Winston, 1991).

Let E(q) be the expected cost, if an order is placed for q units. If the optimal order quantity is derived via marginal analysis, it is assumed that the goal is to find the value of q* which minimizes E(q).

The smallest value of q must be found for which E (q* + 1) - E (q*) ≥ 0. While we are calculating this expression, two possibilities must be considered;

1. d ≤ q. Ordering q-1 units instead of q units, one more unit is going to be overstocked. This increase cost by c_0. The probability that first possibility will occur is basically P (D ≤ q) where D is the random variable which represents demand.
2. d ≥ q-1. Ordering q-1 units instead of q units, give us the chance of the movement of one less unit. Cost will be decreased by c_u. The probability that second possibility will occur in an equation like P (D ≥ q + 1) = 1 - P (D ≤ q) (Winston, 1991).

In summary, a fraction P (D ≤ q) of time, ordering q+1 units will cost c_o more than ordering q units; and a 1- P (D ≤ q) of the time, ordering q-1 units will cost c_u less than ordering q units. Thus, ordering q-1 units is going to cost;

$$c_o P(D \leq q) - c_u [1 - P(D \leq q)] \tag{5}$$

more than ordering q units. Formally, it is shown that;

$$\begin{aligned} E(q+1) - E(q) &= c_o P(D \leq q) - c_u [1 - P(D \leq q)] \\ &= (c_o + c_u) P(D \leq q) - c_u^+ \end{aligned} \tag{6}$$

Then E (q* + 1) - E (q*) ≥ 0 will hold if;

$$(c_o + c_u) P(D \leq q) - c_u^+ \geq 0 \ or \ P(D \leq q) \geq \frac{c_u}{c_u + c_o} \tag{7}$$

When the news vendor scenario considered with the demand D being continuous random variable having density function f(d). Marginal analysis concepts can be shown that the decision maker's expected cost is minimized by ordering q* units, where q* is the smallest number satisfying

$$P\left(D \leq q^{*}\right) \geq \frac{c_{u}}{c_{u}+c_{o}} \tag{8}$$

q* can be found, since demand is taken a continuous random variable. Due to the fact that, in this case, the optimal order quantity may be determined by finding the value of q* satisfying;

$$P\left(D \leq q^{*}\right) \geq \frac{c_{u}}{c_{u}+c_{o}} \; or \; P\left(D \geq q^{*}\right) \geq \frac{c_{o}}{c_{u}+c_{o}} \tag{9}$$

ARIMA (AUTOREGRESSIVE INTEGRATED MOVING AVERAGE)

The feature of time series can be thought that it provides modeling the relationship between collected data in a certain time and makes predictions. Possessing certain trends, cycles or different components is understandable from time series.

In case of neglecting delay or pause, the predictions, we will get, affected by the values we have achieved in the past. Regression from the data related to variables may be mistaken about the statement, explaining the relationship between them.

ARIMA process is default mechanism, assuming as realization of the time series to be modeled. Also, ARIMA modeling methods created for future prediction and fitting series showing behavior in this way. The goal is simply estimating the future value variable, from the past values and the errors made in the past.

White Noise

It defines the assumption that each element in the series having the average of zero, normally distributed and no autocorrelation is randomly selected. Autoregressive (AR) and Moving Average (MA) models used for correction in violates parts of acceptance of the white noise.

Autoregressive (AR) Models

It defines that the model that the value of one period variable found depends on the value of the previous period AR(p); p delayed autoregressive models can be shown as;

$$y_{t} = \mu + \sum_{i=1}^{p} \gamma_{i} y_{t-i} + \varepsilon_{t} \tag{10}$$

μ = Fixed Variable,
γ_p = t-p lagged variable coefficient.

AR(1) can be written as an equation:

$$y_t = \mu + \gamma y_{t-1} + \varepsilon_t = \mu + \gamma\left(Ly_t\right) + \varepsilon_t \; or \; \left(1 - \gamma L\right) y_t = \mu + \varepsilon_t \tag{11}$$

Moving Average (MA) Model

It is used to describe the likelihood of a relationship between a variable and residuals obtained from the previous period. MA (q); q delayed moving average model can be written as;

$$y_t = \mu + \sum_{i=1}^{1} \theta_i \varepsilon_{t-i} + \varepsilon_t \tag{12}$$

θ_q = t-q lagged coefficient of error term

MA(1)'in can be written as a equation;

$$y_t = \mu + \theta \varepsilon_{t-i} + \varepsilon_t \tag{13}$$

Autoregressive Moving Average (ARMA) Model

It is obtained by combining p number of autoregressive term and q number of moving average term. It can be expressed ARMA (p, q)

$$y_t = \mu + \sum_{i=1}^{q} \theta_i \varepsilon_{t-i} + \sum_{i=1}^{p} \gamma_i y_{t-i} + \varepsilon_t \tag{14}$$

Stationarity

ARMA (p, q) modeling process requires stability (stationary). Stationary processes do not have a certain trend and have the mean and variance that it doesn't change with time.

When the process that brake AR(1); $|\rho| \leq 1$ and the situation that εt assumed as white noise, the formula will change to this form;

$$u_t = \rho u_{t-1} + \varepsilon_t \tag{15}$$

Detrending

A variable can be detrend performing regression on a time trend itself and having the residues.

$$y_t = \mu + \beta t + \varepsilon_t \tag{16}$$

$$\varepsilon_t = y_t - \mu - \beta t - y_t \tag{17}$$

Differencing

If determined y_t possess a fixed structure, one simple solution that can be done is to use differentiated variables. Differentiation of the first order done as follows;

$$\Delta y_t = y_t - y_{t-1} \tag{18}$$

If the difference from the first degree provides obtaining a constant process, y_t variable would have been tied from the first degree and shown as I(1). ARIMA (p,d,q); can be thought as p number autoregressive lagged, q number of lagged moving average and degree of d difference of ARMA.

Seasonality

It is a special case of autocorrelative behaviour, occurs for every certain period (monthly, quarterly, etc.). Before fitting the pattern of a time series, it is necessary to pay attention to seasonality.

Dickey-Fuller Test for Stationary

Let's accept that taken AR(1) model. The model does not have a fixed structure or let's assume that unit root is on the carpet for where $|\rho| = 1$ cases.

$$y_t = \rho y_{t-1} + \varepsilon_t \tag{19}$$

$$y_t - y_{t-1} = \rho y_{t-1} - y_{t-1} + \varepsilon_t \tag{20}$$

$$\Delta y_t = (\rho - 1) y_{t-1} + \varepsilon_t = \gamma y_{t-1} + \varepsilon_t \tag{21}$$

The top model is able to predict and is tested depending on the importance of a coefficient γ. The difference of the variable is retrieved and the manipulated variable is repeated with Dickey-Fuller test whether it is a fixed structure. If null hypothesis is rejected,, $\gamma^* < 0$ and yt possess a fixed structure. The variable is going to be used after that calculation.

Replicated Dickey-Fuller Test

In addition to the above models, withdrawal of μ expression and if the additional delay of dependent variable added, the model is as follows.

$$\Delta y_t = \mu + \sum_{j=1}^{p-1} \varphi_i \Delta y_{t-j} + \gamma^* y_{t-1} + \varepsilon_t \tag{22}$$

Time Series Dickey-Fuller Test

The participation of the time series models turn the formula;

$$\Delta y_t = \mu + \beta t + \sum_{j=1}^{p-1} \varphi_i \Delta y_{t-j} + \gamma^* y_{t-1} + \varepsilon_t \tag{23}$$

Autocorrelation Function (ACF) and Partial Autocorrelation Function (PACF)

ACF, can be represented as the rate of autocorrelation of y_t and y_{t-k} and yt which is dependent variable. ACF (k) gives the correlation of between y_t and y_{t-k} from the raw data.

$$ACF(k) = \rho_k = \frac{Cov(y_t, y_{t-k})}{Var(y_t)} \tag{24}$$

Partial autocorrelation function (PACF) is actually, used to denote the simple correlation between the portions related to the delays, y_t and y_{t-k}.

$$\rho_k^* = Corr\left[y_t - E^*\left(y_t | y_{t-1}, \dots, y_{t-k+1}\right), y_{t-k}\right] \tag{25}$$

$$E^*\left(y_t | y_{t-1}, \dots, y_{t-k+1}\right) \tag{26}$$

The expression above is the minimum mean square error of y_{t-1}, ..., y_{t-k+1} due to y_t origin.

Understanding of ARIMA Model

First of all white noise must be investigated. Box-Pierce and "where k delay samples in the autocorrelation is ρ_k " Ljung Box statistic is used.

According to the null hypothesis, the fact that data are white noise, is to say that series are distributed independently. The values obtained from statistics comes from a certain limit in the chi-square distribution which has p degree of independence.

Second, the Akaike Information Criterion (AIC) and the Bayesian Information Criterion (BIC) are two measure tools for goodness of fit. It refers to the agreement between the model fit and complexity.

$$AIC = -2\ln(L) + 2k \tag{27}$$

$$BIC = -2\ln\left(L\right) + \ln\left(N\right)k \quad (28)$$

In here, L is the value obtained by evaluating, the parameters to be estimated of according to likelihood function. N is the number of observations and k is estimated number of parameters. Low values of AIC and BIC contains more relevant results.

Box-Jenkins Methods for Selection of ARIMA Models

Firstly; outliers, unknown values and structural breaks in the data are determined. Variables without having a fixed structure may have a trend or a unsteady variance. Secondly, the data conversion is done if necessary. In this sense, taking logarithms, differentiation and detrending functions are available. Logarithm is used where increased variability of the data over time; in the destruction of differentiation trend. However, differentiation of no more, causes dependency to be introduced when nothing is appeared.

Thirdly, ACF and PACF is examined.ACF and PACF examples are compared for various theoretical ARMA models. By ACF and PACF features, it is attempted to estimate the appropriate model and p, d, q value is determined. It must be tested in several models with experimental data. Slow differentiation in ACF can be used if a delay is needed.

At the estimate stage, ARMA models are examined and analyzed for the various coefficients. The main aim is to have a good fitted, significant factor and trends and in the subject of steadiness and error capturing stingy structure. In the control state, if things are going well, the residuals must be in the remains of a distribution with white noise. Normality is examined from Q-Q diagrams.

Looking at the ACF and PACF diagrams of residuals, in how amount they remain of the white noise. Finally Ljung-Box statistics do autocorrelation test on the size of the correlations as a group. AIC and BIC is examined by looking at the goodness of fit test.

BACKGROUND

The single-period problem (SPP) is applied for identifying the amount of inventory to purchase given a perishable product or single opportunity. It aims to find the order quantity that maximizes expected profit by probabilistic demand distributions.

A number of extensions to the SPP have been proposed. Among those studies, Urban (2008) combined the effects of reference prices, demand uncertainty to find a solution in single period inventory models. Polatoglu (1991) studied the optimal supplement and pricing decisions in a single-period system for deterministic and probabilistic demand.

Wang and Parlar (1994) suggested a game theory model where each of retailers tries to find out the optimum number for order quantity. The outcomes of the study indicate that an optimal decision may be determined for each retailer by the Nash equilibrium, when they behave independently. The authors also studied for collaboration to see whether inventory reduction can be observed or not.

Anderson et al. (2001) enlarged the single period problem and proposed an inventory model considering probabilistic demand for optimizing inventory capacity in coping inventory management having uncertain demand. The model specifically implements to operations possessing two features. First, the operation ought to link largely seasonal or perishable items that cannot be stored. Second, the demand for the inventory item should be uncertain and possess a probability distribution. In view of such features, hotel room capacity is an appropriate candidate to apply the model.

Tsai and Gu (2009) also proposed a single period inventory model which includes optimization of inventory capacity for uncertain demand. Their study estimates optimal guestroom capacity for Hong-Kong hotel industry during 2010-2013 using a single-period inventory model. In their study, they find out that there is a serious overcapacity which ranges from 30.6 to 40.7% that is above the optimal hotel rooms counts. They, also, consider the existing and future development plans and what can be done for the industry.

Zheng and Gu (2011) used the single-period inventory model, to find optimal monthly room capacity for Shanghai's high-end hotels and to cope with overcapacity, examined refraining from developing new high-end hotels, converting to condo hotels or share properties and taking full advantage of recent World Expo and promoting itself as a top destination of meetings options.

Separating Fixed and Variable Parts of Mixed Costs

Besides considering a decision of using single period inventory models, one of the important steps is estimating c_o or cost of ordering one additional unit and later not possessing the chance of selling the unit. In hotel operations, this type of cost, defined as fixed cost which refers to costs whether or not a room is rented out. Schmidgall (2011) and Jagels and Coltman (2007) give three ways to separate variable/fixed costs:

- The high-low two point method,
- The scatter diagram, and
- The regression method.

High-low two point method uses highest and lowest terms to find the portion of a mixed cost that is variable and the portion that is fixed. This method only uses the high and low activity levels to calculate the variable and fixed costs. Scatter diagram is another method of estimating fixed and variable costs. At different times we can observe the increase of total costs. However, the usage of this method can be crucial to visualize the line through the data points, especially if the data is varied.

First, on the y-axis the total cost and on the x-axis the units (costs) must be drawn and the total cost points must be plotted for each unit points. Secondly we try to draw a straight line through the points. The line must have same distance to these points. Third, we determine variable cost per unit by finding the slope. Fourth, we find where the line crosses y-axis. This is the total fixed cost amount. Linear Regression is modeling a relationship between a scalar dependent variable and one independent variable. Unknown model parameters are estimated from data.

Past Studies on Capacity Management

Most capacity management research has been case-focused or industry-specific, since capacity management is sensitive to context.

Capacity decisions affect how well an enterprise responds to the market today and plans for the future (Getz, 1987; Johnson, 1993; O'Reilly, 1986). Decisions related to hotel room capacity are considered to be strategic in nature because capacity cannot be expanded or reduced overnight.

Bitran and Mondschein (1995) studied optimal strategies for renting hotel rooms when there is a stochastic and dynamic arrival of customers from different market segments.

Bendoly (2013) used a laboratory experiment to investigate how different forms of continuous performance feedback influence agent decisions. More specifically, he considered both a revenue-focused metric as well as a metric framed around pricing-curve adherence as the basis of two feedback mechanisms of interest.

Claveria et. al (2015) aim to apply a new forecasting approach to improve predictions in the hospitality industry. To do so, they developed a multivariate setting that allows the incorporation of the cross-correlations in the evolution of tourist arrivals from visitor markets to a specific destination in neural network models.

Chen and Lin (2013) examined the effects of uncertain demand on hotel capacity using the operation data of international tourist hotels in Taiwan from 1996 to 2008. Moreover, their results indicate that effective management of hotel capacities is a more important issue for managers of medium-sized hotels than for small or large-sized hotels in the Taiwanese hotel industry.

Mei and Zhan (2013) tries to use the variable precision rough set (VPRS) approach to establish a customer room choice model (CRCM) based on the logit model, which satisfies the required precision. In terms of their results, hotels can predict customer demand of each rate class and improve their precision, in order to increase hotel revenue.

PROBLEM DESCRIPTION AND METHODOLOGY

Examining the framework which suggested by Anderson (2001) and Tsai and Gu (2011), this study estimated the optimal capacity for Turkey Hotel Industry for the years 2013-2017.

Annual room nights are used to determine the capacity for hotels. At the website of Ministry of Culture and Tourism, all hotels are examined which include municipality and private documentated plants. The optimal quantity of room nights available for Turkey:

$$EL\left(Q^{*}+1\right)=EL\left(Q^{*}\right) \tag{29}$$

Q*= Economic Order Quantity
EL(Q*)= Expected value of loss for (Q*)
EL(Q*+1)= Expected value of loss for (Q*+1)

We may identify the optimal result from the demand function. Moreover, we can estimate under ordering and over ordering cost to find the ratio of

$$C_U \Big/ \left(C_U+C_O\right) \tag{30}$$

C_U = Cost per unit of demand underestimated
C_O = Cost per unit of demand overestimated

Estimating Future Room Demand

We utilized accommodation quantity from 2002 through 2012 to estimate the model for predicting room demand. In ARIMA we must be careful for trend. If we have a trend, then must take differences or make data transformation until we find that the series do not have a trend. If we suspect that there is an increasing or decreasing relationship between the values then some tests can be applied.

By looking at the coefficient of autocorrelation and partial correlation we create an ARIMA model. We decide the model of ARIMA and find the estimation of the model. There must be no data which pass the confidence interval.

Estimating Cost of Oversupply and Undersupply

C_o is defined as the fixed cost per room available, because fixed costs occur whether a room is rented or not. Cost of oversupply is combination of fixed cost and variable cost. The accommodation cost is found from the website of Ministry of Culture and Tourism.

The tourism expenses consist of a lot of section and one of them is accommodation costs which include annual room nights costs. Therefore, we related the information of the hotel room booked from annual room nights costs by dividing total number of tourists coming to Turkey annually. To extract the total fixed costs from total, regression method is used.

The cost of underordering is defined as the loss of not ordering one additional unit and after the losing the chance of ordering finding that it could have been sold if ordered. In this work, EBT per room night sold is used to represent opportunity loss. Similar to the length of period in estimating C_o, the most recent 12-year Turkey hotel operating statistics from 2002–2013 is utilized, to learn realistic estimates for the cost ratio of optimal capacity

Estimating Optimal Room Capacity

The ratio of single period inventory in this study was therefore the ratio of earning before taxes per room night sold to the sum of earning before taxes per room night sold and fixed cost per room night available. Combining the derived cost ratio with the future demand and probability distribution estimated from the regression model, we could thus determine the optimal hotel room capacity for Turkey for each year from 2013 to 2015 using the following standardization function of the values observed in a normal probability distribution.

$$Z = \frac{Q^* - Y}{\sigma} \tag{31}$$

Z is the standard score, Y is the mean of the sample, and σ is the standard deviation of the probability distribution.

COMPUTATIONAL STUDY AND RESULTS

Future Room Demand

An examination of historical room demand in Turkey from 1992 to 2013 (Table 2) shows that the demand growth is approximately linear, but also we may observe some drops in 2006, because of bird flu and the global economic crisis in 2009-2010.

We look for whether there is a trend or not, and at first we see an increasing relationship, so we must make data transformation to obtain not a trend one. After first transformation and take differences we obtain an autocorrelated graph. From PACF and ACF graphs, AR(1) and MA(1) models determined and because of taking differences, the model become ARIMA(1,1,1). When Ljung-Box and t tests for parameter estimates are investigated, significant results are found from the statistics. Therefore ARIMA(1,1,1) will be suitable to forecast the result of our data.

For future demand, ARIMA(1,1,1) model have been utilized and the results shown below. 180.890.060, 192.845.518, 205.634.577, 219.273.024, 233.821.077 and standard errors of the model is 9.360.750

Cost of Oversupply and Undersupply

From Table 3, we can see the tourism income and one of its subhead is accommodation costs. The values shown below and they referred as total costs. To find total fixed costs we will use regression method. From the diagram we find that the equation of total fixed costs and variable cost as,

$$Y = 355.000.000 + 71.812.175 \cdot X \tag{32}$$

In equation 3.1 total fixed cost is 355.000.000 and variable cost is 71.812.172. The values are shown in Table 3. Income before taxes are shown in Table 4

Estimating Optimal Room Capacity

The ratio of $C_u/(C_u + C_o)$ was the ratio of EBT per room night sold. Associating the cost ratio with future room demand and probability distribution found from regression model, we may determine the optimal room capacity for Turkey for each year from 2014 to 2018 using the standardization function of the values taken in a normal distribution:

$$Z = \frac{Q_{frd}{}^{*} - Y_{fd}}{\sigma_{er}} \tag{33}$$

Table 1. ARIMA specialties

	AR(p)	MA(q)	ARMA(p,q)
ACF	No tail	Cut after q lags	No tail
PACF	Cut after p lags	No tail	No tail

Table 2. Distribution of number of arrivals, nights spent, average length of stay, and occupancy rates of establishments by years

Years	Nights Spent			Occupancy Rate (%)			Rooms Available For Sale		
	Foreigner	Citizen	Total	Foreigner	Citizen	Total	Foreigner	Citizen	Total
1992	21,667,842	21,402,256	43,070,098	16.72	21.35	33.23	129,626,949	100,261 705	129,626,949
1993	22,730,132	20,963,255	43,693,387	20.26	18.66	38.88	112,201,367	112,321 032	112,376,964
1994	21,429,565	23,884,655	45,314,220	15.53	17.32	32.82	137,948,061	137,909 884	138,071,313
1995	26,261,954	31,545,831	57,807,785	19.88	19.15	35.09	132,071,225	164,742 557	164,742,557
1996	36,055,785	31,667,969	67,723,754	20.32	17.91	38.17	177,448,290	176,813 671	177,448,290
1997	46,773,913	33,540,021	80,313,934	24.26	17.39	41.58	192,825,588	192,841 626	193,155,209
1998	37,617,417	33,935,349	71,552,766	19.51	17.59	37.08	192,762,010	192,957 630	192,964,211
1999	25,377,854	32,693,041	58,070,895	14.09	18.12	32.18	180,154,367	180,406 651	180,479,430
2000	34,504,508	31,963,448	66,467,956	15.15	14.04	29.19	227,717,216	227,626 612	227,717,216
2001	43,607,631	27,380,820	70,988,451	23.24	14.59	37.83	187,675,680	187,675 680	187,675,680
2002	53,213,533	28,231,253	81,444,786	25.74	13.65	39.39	206,760,724	206,760 724	206,760,724
2003	51,118,310	28,814,909	79,933,219	24.96	14.07	39.02	204,832,786	204,832 786	204,832,786
2004	62,071,774	31,230,128	93,301,902	28.66	14.42	43.08	216,579,497	216,579 497	216,579,497
2005	71,141,390	31,865,789	103,007,179	31.81	14.25	46.06	223,630,823	223,644 731	223,644,731
2006	56,895,351	34,157,426	91,052,777	26.30	15.19	40.50	216,369,553	224,808 911	224,808,911
2007	74,191,840	41,775,470	115,967,310	26.42	14.87	41.28	280,867,205	280,898 332	280,903,431
2008	73,590,885	40,326,871	113,917,756	27.19	14.60	41.25	270,629,613	276,164 794	276,164,794
2009	79,670,940	44,718,277	124,389,217	26.91	15.10	42.01	296,071,571	296,128 712	296,128,712
2010	100,139,752	49,047,190	149,186,942	28.17	13.80	41.96	355,521,327	355,440 448	355,521,327
2011	101,551,112	54,683,603	156,234,715	28.72	15.46	44.17	353,629,964	353,646 174	353,702,320
2012	113,330,412	58,988,300	172,318,712	30.42	15.83	46.25	372,603,838	372,587 707	372,603,838
2013	109,948,630	59,763,570	169,712,200	24.10	15.87	37.20	456,199,832	376,654 017	456,199,832

Source: (Turkish Statistical Institute, 2014).

Table 3. Tourism expenses by type of expenditure, 2003-2013

Years	Tourism Expenditure	Accommodation	Rooms Available for Sale	Expense per Available Room	Fixed Costs per Room
2003	2,424,826,972 TL	396,464,752	204,832 786	1,935.553 TL	1,733 TL
2004	2,954,458,549 TL	566,575,427	216,579 497	2,616.016 TL	1,639 TL
2005	3,394,601,594 TL	593,200,958	223,644 731	2,652.425 TL	1,587 TL
2006	3,270,947,582 TL	534,701,954	224,808 911	2,378.473 TL	1,579 TL
2007	4,043,283,388 TL	644,124,515	280,903 431	2,293.046 TL	1,264 TL
2008	4,266,196,865 TL	748,843,494	276,164 794	2,711.582 TL	1,285 TL
2009	5,090,440,380 TL	897,980,884	296,128 712	3,032.401 TL	1,199 TL
2010	5,874,519,966 TL	1,125,761,798	355,521 327	3,166.510 TL	,999 TL
2011	5,531,485,595 TL	1,080,018,525	353,702 320	3,053.467 TL	1,004 TL
2012	4,593,389,355 TL	1,051,839,000	372,603 838	2,822.942 TL	953 TL
2013	5,253,564,809 TL	1,008,835,928	456,199 832	2,211.390 TL	778 TL
				Average	1,275 TL

Table 4. Tourism income by type of expenditure, 2003-2013

Years	Tourism Income	Accommodation	Rooms Available for Sale	Income per Available Room Night
2003	13,854,868,337 TL	2,106,349,000	204,832,786	9,255 TL
2004	17,076,609,158 TL	2,514,220,000	216,579,497	10,448 TL
2005	20,322,110,498 TL	2,689,794,119	223,644,731	10,824 TL
2006	18,593,949,511 TL	2,348,095,473	224,808,911	9,400 TL
2007	20,942,499,689 TL	2,599,240,212	280,903,431	8,328 TL
2008	25,415,067,156 TL	2,850,710,161	276,164,794	9,290 TL
2009	25,064,481,054 TL	2,534,187,303	296,128,712	7,702 TL
2010	24,930,995,818 TL	2,530,705,087	355,521,327	6,406 TL
2011	28,115,692,635 TL	3,082,349,745	353,702,320	7,843 TL
2012	29,007,002,755 TL	3,053,646,682	372,603,838	7,376 TL
2013	32,308,991,189 TL	3,544,119,767	456,199,832	6,992 TL
			Average	8,533 TL

$Q_{frd}*$ = Quantity of forecasted annual room nights demanded,
Y_{fd} = Mean of forecasted annual room nights demanded,
σ_{er} =Standard deviation of probability distribution.

When considering 11-year average C_u and C_o, we find that the probability that demand will be less than optimal room capacity ought to be, is 8.533 TL/(8.533+1.275) TL = 0,8700.

Determining the annual optimal room nights demanded from Turkey during 2014-2018 to be 191.433.073, 203.388.531, 216.177.590, 229.816.037, 244.364.090. The optimal room capacities during 2014-2018 are 524.474, 557.229, 592.267, 629.633, and 669.490.

When generally looking at the data in Table 2, with the number obtained from the model, regression method in somewhere do not meet the results obtained from data. However, in general the trend is caught by the model, when R^2 statistics and residuals are examined. Due to this statistics, the general model explained enough from the regression line.

From ARIMA, the Ljung-Box statistics is stayed below limit, and the parameter estimates of coefficient is significant due to t test of them. From Table 3 and Table 4 the regression model is created. This result combined with the forecast of Table 1 and pass to the standardization of the functions.

FUTURE RESEARCH DIRECTIONS

Single Period Inventory Model gives a lot of subjects for authors to examine related topics. Hospitality is one of them and tourism sector considers significant topics due to their subjects. A lot of study can be made on the forecast and regression part of tourist expenses and incomes. On the forecast side neural networks, support vector designs and various methods can be used to find and compare with the result of ARIMA. Forecasting techniques can be more functional due to the subject they deal with.

Looking for literature review also can be an answer to understand which methods are suitable for the subject experts are investigating. In determining a model from expenses and incomes of hotel rooms, maybe another approximation made due to the headline of this subject. In here regression methods used for explaining the model. More specific subjects under forecasting techniques may be suitable with the subjects investigated. Except from tourism sector, the business of housing or entertaining can be investigated to see the potential of their income. The subjects where inventory models can be applied can be examined under this situation.

CONCLUSION

The purpose of the study is to examine hotel room supply and demand for optimal capacity in Turkey's hotel industry during 2002–2013 using a single-period inventory model. This study examines a single period inventory model to explain optimal room capacity for Turkey during 2014-2018. Underordering and overordering costs are main part of the single period inventory model formula. Part of fixed and variable costs of tourism expenses are derived by a regression model. Fixed part of the equation is total fixed cost and variable part of equation is variable cost. Fixed part of equation is defined as overordering cost. To find underordering costs EBT is calculated. Moreover, future room demand is calculated by ARIMA. In this step, the test of chi-square and Ljung-Box is stay below the limit, but better results can be found if there will be more study on these data. The statistics observed from regression part can have much better results for determining the capacity. It can be a future research direction to see whether another topic can have a good answer, when it is compared with regression part.

Extraction of tourism expenses and income is the weak part of this study. Because of the method, we consider fixed cost is stable for every year. Therefore, the numbers we find in this study cannot show the real value exactly. However, in estimating part, the study is compensated In other studies, serious overcapacity and undercapacity problems can be seen. In Turkey when we look for the number in near future we cannot see a problem like this.

In 2013, Turkey possessed fourth place at occupancy rate in Europe and also we observed that there was nearly 27% increase in the number of rooms. For capital groups and local tradesman, there is no need for acceleration in the number of hotel rooms for 1-2 years, however, after 2 years following the data, the observation have to be made clearly to monitor whether there is a possibility of undercapacity or still overcapacity is observed. The cost of undercapacity can have an impact of more than overcapacity and the observations must go on every year to forecast and understand the way the number of tourists go. Due to this situation precautions must be taken and the relations that Turkey have, must examined clearly and in detail.

The worst situation in here is, in 2015 because of airplane crisis between Russia and Turkey, a lot of economical relations have broken down except from primal requirements. Russia government do not advice their citizens to go Turkey for their summer holiday and Russian tourists play very important role on the blast of Turkey's hotel capacity number till this year. On the other hand, the solution for this issue is still being tried to be solved by Turkey government and due to providing a common reconciliation, the issue is not going to effect the amount of tourists coming to Turkey from Russia. Another subject that have to be considered is terrorist attacks. The bomb attacks of Kurdish Dependence Movement (which is called in Turkey as PKK) and ISIS (Islamic State of Iraq and Syria) alarmed countries that have tourists in Turkey and advice them not to travel in Turkey especially for some significant places. Specifically, the bomb attacks of ISIS in Istanbul can have negative impacts from the tourist coming from Germany. However, this type of issue generally happened for a short period of time in the history of Turkey, but their effect gave much more destruction. Considering the stability of economy and powerful potential of young generation, these attacks will be parried much more easily than the past. Although, developing relations with Iran is still going on, the number in hotel capacity may have a negative trend in 2016.

Nowadays, we also observe price inflation. Considering money in construction industry, the steps should be taken cautiously. The investments must be pushed sectors which needs more than constructing for hotel industry. Level of optimal capacity shows that the situation is well, but some higher rates also can be viewed in other countries. From here, it is understood that property of tourism activities have to be improved. Another way is moving the other areas such as Package Tours, Health, Sports, Education and Culture, Souvenirs and etc.

REFERENCES

Anderson, D. R., Sweeney, D. J., & Williams, T. A. (2001). *Quantitative methods for business*. Cincinnati, OH: South Western College Publishing.

Belediye Belgeli Tesislerde Konaklama İstatistikleri. (2013). Retrieved December 15, 2015, from http://www.tuik.gov.tr/PreTablo.do?alt_id=1072

Bendoly, E. (2013). Real-time feedback and booking behavior in the hospitality industry: Moderating the balance between imperfect judgment and imperfect prescription. *Journal of Operations Management*, *31*(1), 62–71. doi:10.1016/j.jom.2012.06.003

Bitran, G. R., & Mondschein, S. V. (1995). An application of yield management to the hotel industry considering multiple day stays. *Operations Research*, *43*(3), 427–443. doi:10.1287/opre.43.3.427

Chen, C. M., & Lin, Y. C. (2013). The influence of uncertain demand on hotel capacity. *International Journal of Hospitality Management*, *34*, 462–465. doi:10.1016/j.ijhm.2012.11.008

Claveria, O., Monte, E., & Torra, S. (2015). A new forecasting approach for the hospitality industry. *International Journal of Contemporary Hospitality Management*, *27*(7), 1520–1538. doi:10.1108/IJCHM-06-2014-0286

Getz, D. (1987). Capacity to absorb tourism: Concepts and implications for strategic planning. *Annals of Tourism Research*, *10*(2), 239–261. doi:10.1016/0160-7383(83)90028-2

Gu, Z. (2003). Analysis of Las Vegas Strip casino hotel capacity: An inventory model for optimization. *Tourism Management*, *24*(3), 309–314. doi:10.1016/S0261-5177(02)00073-0

Jagels, M. G., & Coltman, M. M. (2007). *Hospitality management accounting* (9th ed.). Hoboken, NJ: John Wiley & Sons.

Johnson, P. (1993). The notion of capacity in tourism: A review of the issues. In C. P. Cooper & A. Lockwood (Eds.), *Proceedings in Tourism, Recreation, and Hospitality Management* (Vol. 5, pp. 297–308). New York, NY: John Wiley & Sons.

Mei, H., & Zhan, Z. (2013). An analysis of customer room choice model and revenue management practices in the hotel industry. *International Journal of Hospitality Management*, *33*, 178–183. doi:10.1016/j.ijhm.2012.08.001

OReilly, A. (1986). Tourism carrying capacity: Concepts and issues. *Tourism Management*, *7*(4), 254–258. doi:10.1016/0261-5177(86)90035-X

Polatoglu, L. K. (1991). Optimal order quantity and pricing decisions in single-inventory inventory systems. *International Journal of Production Economics*, *23*(1–3), 175–185. doi:10.1016/0925-5273(91)90060-7

Schmidgall, R. (2011). *Hospitality industry managerial accounting* (7th ed.). Lansing, MI: The Educational Institute of the American Hotel & Lodging Association.

Tsai, H., & Gu, Z. (2007). Optimizing room capacity and profitability for Hong Kong hotels. *Journal of Travel & Tourism Marketing*, *29*(1), 57–68. doi:10.1080/10548408.2012.638563

Tsai, H., & Gu, Z. (2012). Institutional ownership and firm performance: Empirical evidence from US-based publicly traded restaurant firms. *Journal of Hospitality & Tourism Research (Washington, D.C.)*, *31*(1), 19–38. doi:10.1177/1096348006296056

Urban, T. L. (2008). Coordinating pricing and inventory decisions under reference price effects. *International Journal of Manufacturing Technology and Management*, *13*(1), 78–94. doi:10.1504/IJMTM.2008.015975

Wang, Q., & Parlar, M. (1994). A three-person game theory model arising in stochastic inventory control theory. *European Journal of Operational Research*, *76*(1), 83–97. doi:10.1016/0377-2217(94)90008-6

Winston, W. L. (1991). *Operations Research Applications and Algorithms*. Duxbury Press.

Yılı İşletme Belgeli Konaklama Tesislerine Giriş ve Geceleme Sayıları. (2013). Retrieved December 15, 2015, from http://www.tuik.gov.tr/PreTablo.do?alt_id=1072

Zheng, T., & Gu, Z. (2011). Overcapacity in Shanghais high-end hotel sector: Analysis based on an inventory model. *Journal of Convention & Event Tourism*, *12*(4), 253–270. doi:10.1080/15470148.2011.621585

KEY TERMS AND DEFINITIONS

ARIMA: In statistics and econometrics, and in particular in time series analysis, an autoregressive integrated moving average (ARIMA) model is a generalization of autoregressive (ARMA) model. These models are fitted to time series data either to better understand the data or to predict future points in the series (forecasting).

Autocorrelation Function (ACF) and Partial Autocorrelation Function (PACF): ACF, can be represented as the rate of autocorrelation of y_t and y_{t-k} and y_t which is dependent variable. ACF (k) gives the correlation of between y_t and y_{t-k} from the raw data. Partial autocorrelation function (PACF) is actually, used to denote the simple correlation between the portions related to the delays, y_t and y_{t-k}.

Autoregressive (AR) Models: The model that the value of one period variable found depends on the value of the previous period AR(p).

Autoregressive Moving Average (ARMA) Model: It is obtained by combining p number of autoregressive term and q number of moving average term.

Capacity Management: Kind of a management that measuring the production capability of the business.

Detrending: A variable can be detrend performing regression on a time trend itself and having the residues.

Differencing: If determined y_t possess a fixed structure, one simple solution that can be done is to use differentiated variables.

Moving Average (MA) Model: It is used to describe the likelihood of a relationship between a variable and residuals obtained from the previous period.

Single Period Inventory Model: The single-period problem (SPP) is applied for identifying the amount of inventory to purchase given a perishable product or single opportunity. It aims to find the order quantity that maximizes expected profit by probabilistic demand distributions.

Stationarity: ARMA (p, q) modeling process requires stability (stationary). Stationary processes do not have a certain trend and have the mean and variance that it doesn't change with time.

Chapter 14
Shaping and Re-Shaping Tourism Areas: A Network Approach

Vincenzo Asero
University of Catania, Italy

Simona Gozzo
University of Catania, Italy

Venera Tomaselli
University of Catania, Italy

ABSTRACT

Defining the boundaries of tourism destinations has been long recognised as a problem in tourism research. The authors aim to define the spatial configuration of tourism areas including different destinations within a same region. Tourist mobility is employed as a methodological criterion to reveal the network relationships among destinations and explain how tourism areas are being shaped and reshaped. The study combines Network Analysis methods and multinomial logistic regression models, in an approach to processing the data of a sampling survey, carried out in Sicily. The results show that the network structures among destinations affect the shape and dimension of tourism areas. Useful evidence for the spatial planning of tourism regions and destination management strategies are derived.

INTRODUCTION

A tourism destination is a specific geographical area within which tourists enjoy different types of holiday. According to Pike (2008), the destination is the spatial unit of analysis in any modelling of a tourism system. Many scholars highlight the fact that destination is a problematic concept (e.g., Framke, 2002; Saraniemi & Kylänen, 2011), one which can be used in different ways and across a varying range of spatial scale (e.g., Beaumont & Dredge, 2010; Saarinen, 2014). Others recognise that destinations are complex dynamic systems (Baggio, 2008, 2014; Baggio, Scott, & Cooper, 2010). Bieger et al. (2009)

DOI: 10.4018/978-1-5225-1054-3.ch014

noticed that the size of a destination influences the Destination Management Organization (DMO) functions, activities and budgeting. More recently, Pearce (2014) observed that the conceptualization of destinations has practical implications for destination management, since viewing a destination as an open or a closed system, instead of as an entity complete in itself, affects DMO actions. The geographical dimension of tourism destinations is often associated with blurred boundaries, since they can encompass either a single location or a network of different places (Dredge, 1999). In this regard, Richie and Crouch (2003) distinguished different levels and types of destinations, where the biggest are countries and macro-regions, and the smallest are cities and single places (e.g., theme parks). Wall (1997) suggested a classification in terms of points, lines and areas based on spatial characteristics. Following Porter (1998), a destination can be considered as a tourism cluster, since it encompasses a geographical concentration of interconnected firms and institutions. By contrast, in the perspective of Massey (1994), destinations are networks of social relations instead of areas with boundaries. The spatial dimension of a destination can therefore be defined according to cooperation among different locations or strategic partnerships among firms and public entities, instead of administrative conditions.

Buhalis (2000) noted that a destination is also a subjective concept, perceived by tourists on the basis of different elements such as their travel itineraries. Similarly, Jenkins et al. (2011) observed that tourists are more likely to define a destination in relation to their journey, specifically in terms of attractions, services, travel time and entry and departure points. Klepers and Rozite (2010) found that tourists, while travelling, do not notice administrative boundaries.

There is a debate in tourism literature over whether the geographic boundaries of a destination are functional or administrative, fixed or fluid (Pearce & Schänzel, 2013). Recently, Beritelli, Bieger, and Laesser (2014) proposed a dynamic viewpoint, affirming that a destination is a space of 'variable geometry', since it is the playground of different supply networks activated by tourist movements. The analysis of the spatial distribution of tourists, therefore, could be an evaluation criterion of the interconnections among various destinations within an area, defining a tourism system (Carlsen, 1999; Cross, Borgatti, & Parker, 2002). As Burnes and Novelli (2008) observed, understanding tourist mobility among multiple localities could explain how destinations are being shaped and re-shaped. As a consequence, the mapping and modelling of tourists' movements are suitable for understanding the dimension of a tourism system.

Defining the boundaries of tourism destinations and regions is recognised as an important issue in tourism research, since space affects the way an economic system works (Capello, 2009). The spatial configuration of tourism areas, indeed, is crucial for policy makers, firms and destination managers to plan and implement coordinated actions of DMO. In this view, destinations and tourism regions require the evaluation and revision of the boundaries of tourism areas they contain, in order to increase market competition, and to meet globalizing tourism markets. There is, therefore, a need for methods and approaches useful to delimit and understand the structure of network areas, which includes identifying the 'central' or 'peripheral' role of destinations.

Following the concept of 'variable geometry' (Beritelli et al., 2014), this study proposes a holistic approach to define the spatial configuration of tourism areas, employing a demand and supply perspective. Tourist mobility has been considered as a methodological criterion to reveal the network relationships among destinations. With this aim in mind, Network Analysis (NA) methods are combined with regression models to assess the shaping and re-shaping of tourism areas including different destinations within a same region. This procedure shows its effectiveness in drawing up the boundaries of destination areas, overcoming one limitation of NA methods that have been employed in tourism research (Ahmed, 2012). Supply data, furthermore, are used in order to analyse the endowment of accommodation facilities in these areas.

On a practical level, the study provides insights into multi-destination areas for government planners and tourism service providers, in the spatial planning of tourism, and destination management.

TOURISM DESTINATIONS IN A NETWORK PERSPECTIVE

A single destination or connected destinations define a tourism network with geographically marked boundaries, and within this context the corresponding tourism components are actors of the network (Ahmed, 2012). The concept of network is very common in tourism, since a destination comprises different interacting elements (McIntosh, Goeldner, & Ritchie, 1995), which would otherwise be in competition with each other. Thus, a network can be seen as a way of creating competitive advantages in the tourism market (Pavlovich, 2003), and more effective tourism management (Beaumont & Dredge, 2010). As such, Dollinger (1990) underlined the importance of collective strategies in the creation of a network, while Crouch and Ritchie (1999) highlighted the role of cooperative actions for tourism development.

Within a geographic region, each destination has structural characteristics resulting from the links with other destinations in the same area (Shih, 2006). Therefore, a network is a key concept for understanding the links that exist either within a destination or among different destinations that cooperate and interact with each other. Similarly, Woodside and Dubelaar (2002) conceptualised a tourism system as a complex and dynamic pattern of tourists' actions, describing the interconnections among places, activities and experiences. From a spatial perspective, this implies that the dimension of a tourism network may be related to single settings or include hierarchically-ordered centres, where a main location plays the role of the node in the organization of services (Asero, D'Agata, & Tomaselli, 2013). Thus, understanding the network features of destinations is essential for the management and planning of tourism (Hall, 2008).

Tourism networks have been analysed from different perspectives, and the majority of studies focus on their structural characteristics, applying NA methods, which confirm the effectiveness of this approach. In so doing, for example, some scholars examined the structure and cohesiveness of different organizations within Waitomo, in New Zeland (Pavlovich, 2003) and within four tourism destinations in Australia (Scott, Cooper, & Baggio, 2008b). Shih (2006) focused on the network characteristics of drive tourism destinations in Naintou, Taiwan. Other studies concern multi-destination trips (Hwang & Fesenmaier, 2003; Hwang, Gretzel, & Fesenmaier, 2006; Jeng & Fesenmaier, 1998; Kim & Fesenmaier, 1990; Tideswell & Faulkner, 1999). Zach and Gretzel (2012) observed that tourists create dynamic relationships through their choices, which can be conceptualised as 'tourist-activated networks'. Stienmetz and Fesenmaier (2013a) argued that a network paradigm is useful for assessing the economic value of tourism attractions. Asero et al. (2016) re-constructed tourists' travel patterns among destinations, employing NA measurements to explore tourism networks in Sicily.

If, on the one hand, the multifaceted connections in tourism can be of various spatial levels, forms and types of interest (Scott, 2000); on the other, some aspects of destination networks, such as geographical limits, still need to be explicitly explored (Pearce, 2014).

The starting point of this study is the perspective of tourists and their mobility, since destinations can be viewed "as mobile constructions produced as a result of multiple uneven mobilities" (Dredge & Jamal, 2013, p. 559). Mobility could thus be useful in defining the spatial configuration of a tourism destination system (e.g., Burnes & Novelli, 2008; Sheller, 2009).

The authors propose an experimental and innovative procedure, focusing on tourism in Sicily, the largest island in the Mediterranean Sea and the largest Italian region. Visited by more than 4,000,000 tourists per year, it includes the small surrounding islands of Ustica and Pantelleria and the archipelagos of the Aeolian to the north, Aegadian to the west and Pelagie to the south-west. Seaside and natural resources, cultural heritage, food and wine are the finest tourist products on offer.

METHODS AND RESEARCH DATA

In order to define different territorial areas of tourism, procedures of clustering and measuring derived from NA methods are employed.

A network consists of nodes representing objects and ties referring to relationships among objects (Borgatti & Halgin, 2011). Initially developed for the systematic study of social structures (Wasserman & Faust, 1994), the set of NA methods have been employed in tourism research that has underlined their methodological validity (Bhat & Milne, 2008; Hwang et al., 2006; Pavlovich, 2003; Scott, Baggio, & Cooper, 2008a; Scott et al., 2008b; Shih, 2006; Ziakas & Costa, 2010). Since destinations can be viewed as nodes of a network and the tourist routes as links among destinations, the NA approach is suitable to reveal patterns of mobility among the destinations (Stienmetz & Fesenmaier, 2013b).

In this study, NA measures and methods are proposed to examine the tourists' spatial movement and provide evidence about interconnectedness among tourism destinations within Sicily, following methodologies used in the most recent literature using the NA approach (Baggio, 2008; Baggio & Cooper, 2010; Baggio et al., 2010; Baggio & Sainaghi, 2011; Baggio, Scott, & Cooper, 2011). The connections among the destinations are investigated by analysing tourists' travel patterns and the tourism areas are, in succession, spatially defined on the basis of their similarities.

The first NA procedure employed is CONvergence of iterated CORrelation (CONCOR), the most common block-modelling method (Nunkesser & Sawitzki, 2005; Schwartz, 1977), based on the concept of structural equivalence (Breiger, Boorman, & Arabie, 1975; Lorrain & White, 1971). Here, the destinations are defined as structurally equivalent when they are linked to the same places. Formally, two nodes i and j are defined as equivalent if, for all nodes $k = 1, 2, \ldots, g$ $(k \neq i, j)$ and all the ties $r = 1, 2, \ldots, R$, the node i has a tie to k, if and only if the other node j also has a tie to k and i has a tie from k, if and only if j also has a tie from k, in an adjacency matrix with identical rows and identical columns. According to the above conditions, two destinations i and j are equivalent, $i \equiv j$, if they maintain identical routes to and from identical destinations, on all R routes (Wasserman & Faust, 1994). Euclidean distance (Burt, 1976) and Pearson product-moment correlation coefficient are the most commonly used measures of equivalence among pairs of nodes, since both are part of more comprehensive positional analysis procedures.

CONCOR is usually applied to identify patterns of nodes on the basis of how they are related to each other in a positional analysis (Wasserman & Faust, 1994). In this study, the procedure is used since it is specifically suitable for network data, in order to identify clusters of almost equivalent nodes. According to this principle, destinations within similar routes belong to the same group/block. They are, then, closer to being equivalent to each other rather than those in different groups. Compared to other agglomerative methods, CONCOR is more flexible and encompasses more information, grouping localities both with respect to the frequency of the visits and the presence of similar paths among destinations. Coherently, the main result of this procedure is a block-model that consists of two outputs: a partition of localities into subgroups (defined as blocks) and the identification of presence/absence of ties for each pair of places visited within or between the blocks. Consequently, CONCOR shows groups of localities that are similar with respect to their positions in a network. It allows building tourism routes among destinations belonging to the same block or to different blocks. This procedure is useful for defining homogeneous geographical areas, including localities with similar tourist attractiveness.

In this study, the blocks defined through CONCOR are employed to classify tourism areas as 'central' or 'peripheral' according to frequency of visits. The areas are then mapped in the geographical space. The effectiveness of CONCOR is proved by an experimental procedure combining NA methods with causal models. In order to estimate if and how much the predictors significantly affect the building of each geographical tourism area, multinomial logistic regression models are specified.

Multinomial logistic regression models are based on the probability of membership of each $n-1$ category of the dependent variable compared to a baseline or reference category (Greene, 2012). The probability related to each category of the Y-dependent variable explained by X_i-independent variables is computed as follows (Hosmer & Lemeshow, 2000):

$$\log it\left(p\right) = \beta_0 + \beta_1 X_1 + \beta_2 X_2 + \beta_3 X_3 + \ldots + \beta_n X_n = XB \tag{1}$$

where p is the probability of geographical areas as Y-dependent variable.

Formally:

$$P(Y \mid X) = \frac{e^{\alpha+\beta x}}{1+e^{\alpha+\beta x}} \tag{2}$$

The logarithmic transformation is:

$$\ln\left[\frac{P(Y \mid X)}{1-P(Y \mid X)}\right] = \alpha + \beta_x \tag{3}$$

which is equal to the $\log it\left[P\left(Y \mid X\right)\right]$ (Agresti, 1990).

Employing geographical subdivision in areas as a dependent variable, in a first model the predictors are the network properties of the destinations within each tourist travel pattern. Destinations are more or less connected with each other on the basis of the number of links: the more the destinations in a route are chosen by tourists, the more connected they are. The ego-networks, then, are nets built on all the destinations to which an ego is connected. Ego-network measures are computed for each locality (ego or node).

Taking into account the research goals, the ego-network measures chosen are:

- *Size*, defined by the number of nodes of the ego-network, plus ego itself. This is an indicator of integration in a network.
- *WeakComp* is the number of weak components in the ego-network, which are connected to each another only by ego. A weak component is the largest number of connected nodes, disregarding the direction of the ties. The count of components can be normalised by size (*nWeakComp*).

- *2-StepReach* is the number of pairs of nodes with an intermediary within the net. For the whole network, it computes the percentage of all nodes within two links of ego, which are two directed steps away from ego and not only the ego's one-step neighbourhood.
- *ReachEff* normalises the *2-StepReach* measure, dividing it by size. This index measures the number of non-redundant, secondary contacts for each primary contact. The amount of this measurement is an indicator of the capability of ego to reach a wider network through each primary link.
- *nBrokerage* is the number of pairs of nodes not directly connected to each other within the neighbourhood of ego, normalised by the number of pairs. The higher this number, the more ties go through ego. Since ego is connected to every other node, within its network ego can act as a 'broker' among pairs of other nodes, even if they are not connected directly to one another, because it falls on the paths between the others.
- *Effective Size (EffSize)* is the difference between the number of nodes linked to ego and the average number of links which each node has with other nodes. The index measures the overall impact of ego within the net, avoiding redundant ties among other nodes. It refers to the concept of structural holes as very important aspects of positional advantage/disadvantage of nodes, resulting from how they are embedded in neighbourhoods, and implies how and why the ways that a node is connected affect its constraints and opportunities.

In order to examine the connections and network structure among tourist destinations, the data are settled in a single squared, weighted and oriented destination by destination matrix (Lue, Crompton, & Stewart, 1996; Shih, 2006). The matrix is organised by overlapping of *n*-1 origin by destination matrices between the *n* stages (or combined multiple stages of the routes) drawn by *n* destinations reached by tourists.

In the second logistic regression model, geographical subdivision in areas is the dependent variable but the model is specified using as independent variables trip-related attributes of tourists visiting the destinations in order to assess the differences among the areas.

The above methods are employed to process the research data collected by a sampling survey of incoming tourism in Sicily. The analysis is based on the answers of 3,182 self-organised tourists, interviewed departing from the main ports and airports of Sicily after their holiday. The face-to-face interviews were carried out in spring, summer and early autumn, when more than 80% of tourists visit the island, as accounted for by official data on tourist flows in Sicily.

Through the data collected, the destinations shaping the networks are identified by rebuilding travel paths of the tourists. The destinations are those indicated by tourists after their visit unlike in other studies, where the places investigated are identified from a list (Liu et al., 2012; Scott et al., 2008b; Shih, 2006; Stienmetz & Fesenmaier, 2013a).

Finally, considering that the areas are defined as 'central' or 'peripheral' on the basis of demand data, the endowment of accommodation facilities has been analysed to check if supply data also reveal these attributes in a long time perspective. As a consequence, it is expected that in a 'central' area the supply of tourist accommodation is much larger than in a 'peripheral' one.

RESULTS

Tourism Areas from a Tourist Perspective

A first step is to identify the most and least attractive routes for tourism in Sicily. With this aim in mind, a matrix is built describing each displacement from and towards each destination, which allows the application of CONCOR. This procedure selects 'blocks' of destinations, applying the mathematical concept of structural equivalence, distinguishing important routes as defined by tourists on the basis of the destinations visited. The block-model tests hypotheses about the presence of particular ties among destinations, since these ties define sub-groups of localities with structural equivalent positions within a net, as shown in Figure 1.

Figure 1. Main destinations in tourism routes

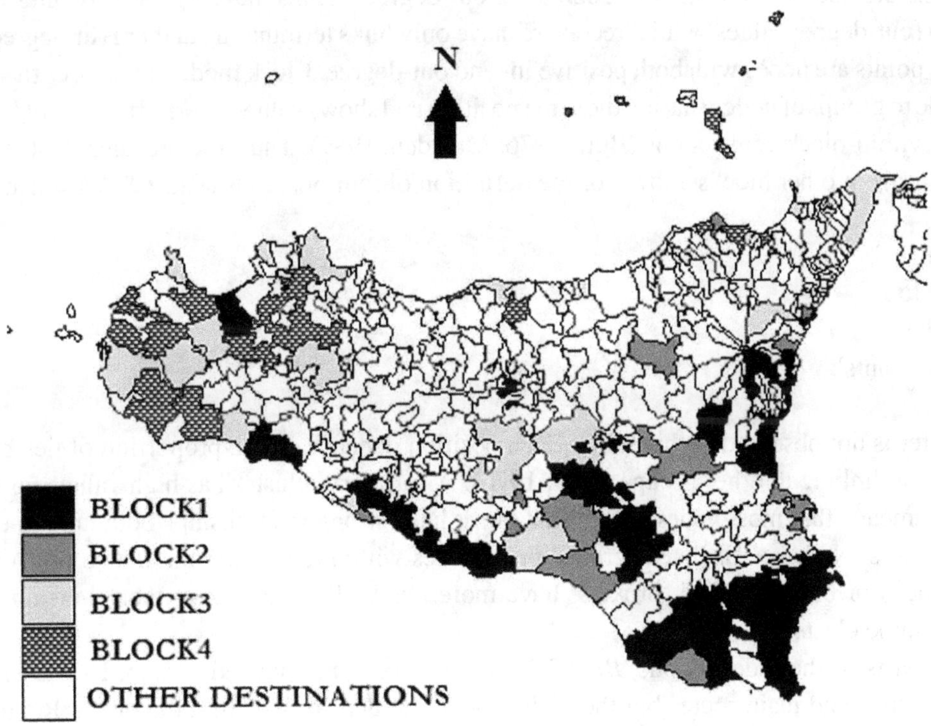

Table 1. Matrix of density

	BLOCK 1	BLOCK 2	BLOCK 3	BLOCK 4
BLOCK 1	**0,973**	0,060	0,304	0,264
BLOCK 2	0,144	0,011	0,022	0,060
BLOCK 3	**0,397**	0,039	**0,399**	0,284
BLOCK 4	0,200	0,047	0,371	0,152

The partition obtained is neither theoretical nor geographical, but defines the most similar areas with respect to paths among places. Four groups of destinations are identified, depending on the presence or absence of direct ties between localities. The density matrix in Table 1, which is built by the density scores for each block (Prell, 2012), shows patterns of intra- and/or inter-block relations. The higher the density, the more important the pattern is. Data show an average network density of 0,23. Comparing each density value between blocks with this cut-off, the first and third block are central groups, while the second and fourth are peripheral, since they cluster less-visited destinations.

The block model procedure is similar to that applied for ties from and towards single nodes – here, routes and destinations, respectively:

- To identify 'isolates',
- 'Transmitters',
- 'Receivers' and
- 'Ordinary' nodes (Wasserman & Faust, 1994).

'Isolates' are nodes with neither in-degree nor out-degree. 'Transmitters' have only links originating from them (out-degree values) while 'receivers' have only links terminating at them (in-degree). Finally, 'ordinary' points are nodes with both positive in- and out-degree. Block model transposes the same ratio from a node to groups of nodes sharing the same position and shows values referred to inter-block relations as well as within-block connections (Burt, 1976; Marsden, 1989). Thus, the prevalence of ties within a block and to/from other blocks allows for the definition of four positions (Faust & Wasserman, 1992):

- 'Primary',
- 'Isolate',
- 'Brokers' and
- 'Sycophants'.

The latter is not observed in this study. Each position depends on the proportion of ties both within the group and linking to other groups (Burt, 1976). A 'primary' cluster has high values for both measures. This means that most nodes within this block have strong relationships both among themselves and with others. Nodes of 'Isolated' groups present ties within the cluster but not so much with other groups. Nodes of blocks named 'brokers' have more ties with other groups than links among nodes within the same cluster.

On the basis of these definitions, *BLOCK1* is a 'central' but 'isolated' group. Here there are both central localities and main routes but the highest value of density is within the block. Routes connect mainly to destinations of the same block. *BLOCK2* is a 'broker' and it includes most of the peripheral localities that are, compared to destinations in other blocks, less connected. *BLOCK3* is a 'primary' group. Its highest density values, in fact, are found for destinations belonging both to the same block and to the first block. Finally, *BLOCK4* is a 'broker' but it includes central localities more than *BLOCK2* (as shown below by image matrix). This block clusters destinations belonging to the main inter-group routes - those also crossing places of *BLOCK1* and *BLOCK3* - rather than to the intra-group routes.

The image matrix in Table 2 helps to distinguish 'central' and 'peripheral' groups. It is computed by average density as cut-off, so the procedure assigns 1 when the density of direct ties within two blocks exceeds the value or 0 otherwise.

Table 2. Block model image matrix for four blocks of destinations

	BLOCK 1	BLOCK 2	BLOCK 3	BLOCK 4
BLOCK 1	1	0	1	1
BLOCK 2	0	0	0	0
BLOCK 3	1	0	1	1
BLOCK 4	0	0	1	0

As a practical implication, destinations belonging to the first and third group reproduce the most selected routes. Here the main tourism destinations are steps of routes that include both destinations within the same block or belonging to different blocks. The destinations within the second block are 'peripheral' as they are the ones less visited by tourists. Finally, the fourth block includes localities chosen by tourists as travel destinations rather than steps of their routes.

Tourism Areas from a Spatial Perspective

The CONCOR procedure shows that structurally equivalent destinations are patchily distributed throughout the whole island (Figure 1). Applying a spatial perspective, structural equivalence can be used as a tool to define macro-areas with common traits. With this aim, the destinations are again grouped but in geo-referenced partitions, which geographically define three spatial areas represented in Figure 2.

Comparing the blocks obtained by CONCOR (Figure 1) and spatial partition (Figure 2), some empirical findings arise. The south-east of Sicily includes 'central' destinations and 'closed' paths (*BLOCK1*)

Figure 2. Visited destinations grouped in spatial partitions

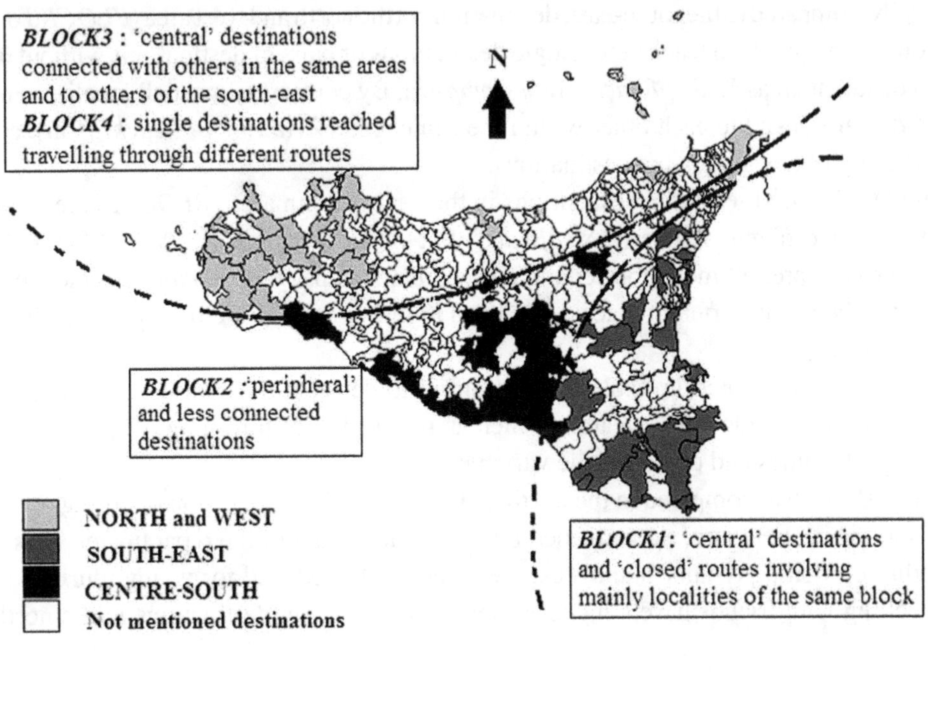

Table 3. Categories of dependent variable

Sicilian Areas	No. of Destinations	%
North and West	35	47.9%
South-East	22	30.1%
Center-South	16	21.9%
Total	73	100.0%

linking places localised in the same area. In the northern and western area, there are 'central' destinations which are both linked with others in the same area and to others of the south-east (*BLOCK3),* as well as unique destinations reached by travelling through different routes (*BLOCK4*), *e.g.* the small Sicilian islands. Compared to the south-east, indeed, locations of this area are more often chosen as single destinations rather than as steps of travel itineraries. Finally, most 'peripheral' destinations (*BLOCK2*) are located in the centre-south of Sicily.

To test the hypothesis on differences in tourism within each area, a multinomial logistic regression model is estimated to check the effectiveness of CONCOR in defining homogeneous tourism areas. Next, the three above-mentioned spatial areas are introduced in the model as a dependent variable, see Table 3.

Two different types of predictors are used:

- Firstly, the ego-network measures of each destination, described in Table 4,
- And secondly, the attributes of tourists travelling in each area.

In the two models, the estimated parameters for each variable X_i are analysed if the critical value of the Wald test, calculated as the ratio between the coefficients β and *SD*, was significant at 0.05 level.

The findings of the first model in Table 5 confirm the validity of the assessments of the geographical mapping of the blocks obtained by CONCOR.

Specifically, compared to the south-east, destinations in the north and west area (*BLOCK3* or *BLOCK4*) are more connecting points to reach other single destinations or pairs of destinations without direct links, probably in different areas (*2-StepReach*, *WeakComp*-in). By contrast, almost all localities in the south-east (*BLOCK1*) are linked to each other within the same area (*WeakComp*, *ReachEff*-in), as shown by the higher density of ties among the destinations.

Compared to the south-east, the destinations in the centre-south area (*BLOCK2*) are more probably crossing points of different routes (*2-StepReach*) but more 'peripheral' *(ReachEff-in)*. Meanwhile, south-eastern routes present more non-redundant secondary steps (*WeakComp*-in), since more destinations can be reached from different places. Finally, a higher presence of closed paths within the area is confirmed (*ReachEff*-in).

A second model, presented in Table 6, is specified to detect if these three geographical areas are different in relation to some tourist features, such as motivation of travelling, sources of information, accommodation facilities and demographic variables

The results show that, compared to the north and west, a higher proportion of young people visit the south-east and the centre-south, as confirmed by the negative sign of the β parameter. This means that the probability of visiting the north and west area tends to be reduced for young tourists. In addition, tourists travelling with friends have a higher probability of visiting destinations in the north and west

Table 4. Ego-network measures

Measures	The Meaning Here
Size	Number of destinations linked to ego, in the same routes.
WeakComp	If this index has a high value, pairs of destinations are connected through non-direct links. This implies that, within the network, ego plays the role of broker among the pairs.
2-StepReach	Presence of destinations connected with ego through another destination, assuming the heterogeneity of ties within the ego-network. Higher indexes mean better links within the network.
ReachEff	This can be considered a measure of the extension of tourism routes departing from a specific location.
nBrokerage	The localities are main tourism destinations or 'landmarks' within tourism routes.
EffSize	How connected is an ego-destination to all the other destinations within a net.

Note: Size, WeakComp, 2-StepReach and *ReachEff* are also computed as *in* and *out* indices. Referring to the direction of routes and, respectively, identifying only destinations towards which tourists go (in) or from which tourists move (out).

Table 5. Multinomial logistic regression of ego-network structures

Reference Category		β	SE	Wald	df	P-Value	Exp(β)
			South-East				
North and West	Intercept	-40,441	22,131	3,339	1	0,068	
	WeakComp	-0,385	0,196	3,858	1	0,049	0,681
	2-StepReach	0,417	0,203	4,226	1	0,040	1,517
	WeakComp-in	0,234	0,097	5,852	1	0,016	1,264
	ReachEff-in	-0,285	0,116	6,064	1	0,014	0,752
Centre-South	Intercept	-50,592	25,031	4,085	1	0,043	
	WeakComp	-0,395	0,193	4,199	1	0,040	0,673
	2-StepReach	0,473	0,225	4,434	1	0,035	1,605
	WeakComp-in	0,291	0,103	7,986	1	0,005	1,338
	ReachEff-in	-0,351	0,122	8,277	1	0,004	0,704
			North and West				
South-East	*WeakComp*	0,385	0,196	3,858	1	0,049	1,469
	2-StepReach	-0,417	0,203	4,226	1	0,040	0,659
	WeakComp-in	-0,234	0,097	5,852	1	0,016	0,791
	ReachEff-in	0,285	0,116	6,064	1	0,014	1,329
			Centre-South				
South-East	Intercept	50,592	25,031	4,085	1	0,043	
	WeakComp	0,395	0,193	4,199	1	0,040	1,485
	2-StepReach	-0,473	0,225	4,434	1	0,035	0,623
	WeakComp-in	-0,291	0,103	7,986	1	0,005	0,748
	ReachEff-in	0,351	0,122	8,277	1	0,004	1,420

Model chi-square = 37,12; $p<0.04$, –2 log likelihood = 115,684, Pseudo R^2 (Nagelkerke) = 0.45.

Table 6. Multinomial logistic regression of tourist features

Reference Category		β	SE	Wald	df	P-Value	Exp(β)
South-East							
North and West	18-23 years old	-,990	,406	5,953	1	,015	,372
Center-South	Non-hotel stay	,492	,217	5,125	1	,024	1,636
	Italian tourists	-,296	,145	4,193	1	,041	,744
	Information from travel agencies	,806	,338	5,700	1	,017	2,240
North and West							
Center-South	18-23 years old	,867	,404	4,590	1	,032	2,379
	Travel with friends	-,684	,305	5,027	1	,025	,504
	Non-hotel stay	,521	,216	5,821	1	,016	1,684
	Info from relatives, friends, experiences	-,632	,290	4,752	1	,029	,531
South-East	18-23 years old	,990	,406	5,953	1	,015	2,690
Center-South							
North and West	18-23 years old	-,867	,404	4,590	1	,032	,420
	Travel with friends	,684	,305	5,027	1	,025	1,982
	Non-hotel stay	-,521	,216	5,821	1	,016	,594
	Info from relatives, friends, experiences	,632	,290	4,752	1	,029	1,882
South-East	Non-hotel stay	-,492	,217	5,125	1	,024	,611
	Italian tourists	,296	,145	4,193	1	,041	1,345
	Information from travel agencies	-,806	,338	5,700	1	,017	,446

Model chi-square = 52,84; $p<0.001$, –2 log likelihood = 96,87, Pseudo R^2 (Nagelkerke) = 0.59.

rather than in the centre-south. Tourists visiting destinations in the centre-south of the island do not choose hotels as their type of accommodation. The coefficient β, indeed, is negative compared with the south-east and the north and west areas. Italian tourists, on the other hand, are more probably localised in the south-east than in the centre-south area. The tourists visiting the north and western area are more likely to use friends, relatives and previous experiences as sources of information than travellers in the centre-south area, who prefer more travel agencies than south-east visitors. This latter result could be related to a 'central' or 'peripheral' attribute, characterising the destinations within these two areas. As a consequence, tourists visiting 'peripheral' destinations need information that is more likely available from formal sources. By contrast, information about well-known 'central' destinations can also be acquired from informal sources.

Tourism Areas from a Supply-Side Perspective

From the supply side, data collected over a period of ten years show that the distribution of the endowment of accommodation establishments reproduces the 'central' or 'peripheral' attributes of the areas defined through NA. Indeed, the more 'central' the area, the larger and wider is the supply, and vice versa. Nevertheless, some dissimilarity is detectable over time, and in relation to the different categories of tourist accommodation establishments.

Specifically, hotels are mostly localised in the north and west area, defined as the most 'central', and decrease in the south-east area, while the smallest amount is in the 'peripheral' area, see Table 7. The endowment of accommodation establishments increases over time, due to the development of tourism.

In the north and west area and in the south-east area, the distribution of the other collective establishments (such as campsite, holiday dwellings, rural tourism facilities, B&B, etc.) is almost identical as shown in Table 8.

Over the decade, the number of B&Bs increased, due to the local policies addressed mainly at recovering the architectural heritage, in order to promote tourism in historical cities. The greatest proportion of B&Bs over other types of accommodation establishment, indeed, is localised in the 'central' south-east area, where many towns exhibit a plethora of late Baroque art and architecture of high quality.

Table 7. Hotels and similar establishments in the destinations of the 3 areas defined by NA

	PA, TP, ME North and West		AG, CL, EN Centre-South		CT, RG, SR South-East		Total
2004	453	61.6%	99	13.5%	183	24.9%	735
2005	488	62.1%	104	13.2%	194	24.7%	786
2006	506	60.7%	108	12.9%	220	26.4%	834
2007	527	60.5%	113	13.0%	231	26.5%	871
2008	537	59.9%	113	12.6%	247	27.5%	897
2009	546	58.5%	125	13.4%	262	28.1%	933
2010	583	59.4%	125	12.7%	274	27.9%	982
2011	565	57.9%	129	13.2%	282	28.9%	976
2012	550	57.4%	129	13.5%	279	29.1%	958
2013	577	58.1%	129	13.0%	287	28.9%	993

Source: (ISTAT - Italian Institute of Statistics, 2015).

Table 8. Other establishments in the destinations of the 3 areas defined by NA

	PA, TP, ME North and West		AG, CL, EN Centre-South		CT, RG, SR South-East		Total
2004	485	49.2%	105	10.6%	396	40.2%	986
2005	594	48.4%	142	11.6%	492	40.1%	1228
2006	701	45.8%	184	12.0%	644	42.1%	1529
2007	759	44.9%	244	14.4%	689	40.7%	1692
2008	793	41.5%	294	15.4%	823	43.1%	1910
2009	823	40.9%	287	14.3%	903	44.9%	2013
2010	1070	45.5%	311	13.2%	972	41.3%	2353
2011	1047	44.1%	334	14.1%	991	41.8%	2372
2012	1129	44.4%	362	14.2%	1053	41.4%	2544
2013	1277	42.6%	401	13.4%	1321	44.0%	2999

Source: (ISTAT - Italian Institute of Statistics, 2015).

Data on accommodation capacity confirm the definition of the three areas in terms of 'central' and 'peripheral'. However, in the 'peripheral' centre-south area, the accommodation capacity of B&Bs is greater than in the other areas, which could well be due to the lower number of hotels and similar establishments.

SOLUTIONS AND RECOMMENDATIONS

The delimitation of tourism areas can be useful for planning tourism facilities, managing tourism routes and defining marketing strategies. In this regard, the study shows that the combination between NA methods and logistic regression models is an effective and innovative procedure of data analysis suitable to investigate tourism networks through tourists' mobility among destinations and depict shaping and re-shaping of tourism areas.

In addition, it solves a problem that has long been discussed in tourism literature with respect to difficulties in defining the boundaries of destination areas (Hall, 2008). Therefore, the procedures applied here could be useful in other contexts, since they guarantee consistency and reliability of results, and hence provide a contribution to make to the field of tourism research.

FUTURE RESEARCH DIRECTIONS

This chapter shows the benefit of using NA to investigate the structure of tourism networks. Specifically, the case study demonstrates how NA can provide insight to help understand the interconnections among destinations, which can then be used to define tourism areas geographically (Cooper & Hall, 2013).

The managerial applications of such findings emerge when one considers that networks coordinate complimentary assets at the level of destination, which includes marketing alliance, promotion strategies and tourism products (Tremblay, 1998). Therefore, by defining tourism areas, it may be possible to recognize the level of interdependency that exists among different local stakeholders and tourism businesses, and create the potential for promoting a cooperative basis on which integrative marketing and management strategies can be developed. This becomes important since tourism actors operate according to different business logics. In fact, individuals agree on the challenges of the destinations, while group-related perceptions of competitive market advantage and competitors are different (Beritelli, 2011). Consequently, financing projects, adopting initiatives and assigning functions to local organizations imply fostering collaboration among stakeholders and balancing the relationships between cooperation/competition and individual/common benefits, in order to achieve market success for both the destination and their individual businesses (Marzano & Scott, 2009; Wang & Krakover, 2008).

In this regard, this study suggests the need for further research that aims to measure the strength and weakness of the relationships among the supply actors within a tourism network, and to investigate how linkages between private and public organizations can inform strategic leverages for the planning and management of tourism areas. The stronger these relationships are, the better private enterprises and public institutions respond to the dynamic change in the competitive market. Therefore, additional research is also recommended in order to understand the role of these linkages in performance and the evolution of networks. Finally, quantitative model-based approaches can be addressed to estimate if, and how much, the supply factors affect tourist choices and behaviours and, successively, the shape of tourism areas.

CONCLUSION

In order to increase market competition, destinations and tourism regions require evaluating, and in many case revising, the boundaries of tourism areas (Bieger et al., 2009). A tourism area can be seen as a supply network, demand-caused, which aims to produce goods and services for tourists, where actors and organizations move at different speeds in responding to market development (Fine, 1999). Consequently, defining the spatial configuration of tourism areas is crucial for policy makers, firms and destination managers, to plan and implement coordinated actions of DMO. For this reason, it is important to study a tourism area from a multi-destination perspective, since each destination within a specific geographic region has structural characteristics resulting from the links with other destinations in the same area (Shih, 2006). Therefore, there is a need for both methods useful for delimiting boundaries and criteria for understanding the structure of network areas, a process which includes identifying the 'central' or 'peripheral' role of destinations along tourism routes.

While it is clear, on the one hand, that structural analysis of a tourism network may inform many important tourism issues such as policy development and destination governance and management, on the other, studies of tourism networks should be further explored, developed and utilized (Ahmed, 2012). Following the concept of 'variable geometry', used to deconstruct the notion of a destination (Beritelli et al., 2013), two perspectives have then informed the present study: the viewpoint of tourists in choosing their routes and the networks they build by bundling destinations.

Tourist mobility has been considered as a methodological criterion to reveal linkages among destinations. A procedure combining NA methods with multinomial regression models has been experimented to process tourist movement data and define the spatial configuration of tourism destination networks. Through measuring and clustering NA methods, tourism macro areas have been classified as 'central' or 'peripheral' assuming the destinations as 'nodes' of a network and the tourist routes as 'links' among destinations. Multinomial regression models, then, have been specified to estimate if and how much the NA measures significantly affect as predictors the shaping and re-shaping of tourism areas mapped in a geographical space.

Different but complementary results have been achieved. Firstly, tourism locations play either a 'central' or a 'peripheral' role, based on the visiting frequency of the tourists. Furthermore, they are chosen by tourists either as unique destinations or as steps of travel itineraries. Secondly, applying a spatial perspective, the network structures among destinations define tourism macro-areas, mapped in a geographical space. Then, the spatial configuration obtained has been validated through the logistic regression model. The significance of the ego-network measures shows that they affect the shape of the three homogeneous tourism areas. In addition, the areas identified are different when considered in the light of certain characteristics of tourists. Finally, the endowment of accommodation facilities in these areas has been analysed to check if data also reveal that in a 'central' area the tourist accommodation supply is much larger than in a 'peripheral' one.

On a practical level, the study contributes by providing insights into multi-destinations areas for government planners and tourism service providers in the spatial planning of tourism and destination management.

REFERENCES

Agresti, A. (1990). *Categorical data analysis*. New York, NY: John Wiley & Sons.

Ahmed, E. (2012). Network analysis. In L. Dwyer, A. Gill, & N. Seetaram (Eds.), *Handbook of research methods in tourism: Quantitative and qualitative approaches* (pp. 472–494). Cheltenham, UK: Edward Elgar. doi:10.4337/9781781001295.00033

Asero, V., D'Agata, R., & Tomaselli, V. (2013). Tourism mobility among destinations: An application of network analysis. In A. M. Oliveri & S. De Cantis (Eds.), *Analysing local tourism. A statistical perspective* (pp. 309–324). Maidenhead, UK: McGraw-Hill Education.

Asero, V., Gozzo, S., & Tomaselli, V. (2016). Building tourism networks through tourist mobility. *Journal of Travel Research*, 55(6), 751–763. doi:10.1177/0047287515569777

Baggio, R. (2008). Symptoms of complexity in a tourism system. *Tourism Analysis*, *13*(1), 1–20. doi:10.3727/108354208784548797

Baggio, R. (2014). Complex tourism systems: A visibility graph approach. *Kybernetes*, *43*(3/4), 445–461. doi:10.1108/K-12-2013-0266

Baggio, R., & Cooper, C. (2010). Knowledge transfer in a tourism destination: The effects of a network structure.The*Service Industries Journal*, *30*(10), 1–15. doi:10.1080/02642060903580649

Baggio, R., & Sainaghi, R. (2011). Complex and chaotic tourism systems: Towards a quantitative approach. *International Journal of Contemporary Hospitality Management*, *23*(6), 840–861. doi:10.1108/09596111111153501

Baggio, R., Scott, N., & Cooper, C. (2010). Network science - A review focused on tourism. *Annals of Tourism Research*, *37*(3), 802–827. doi:10.1016/j.annals.2010.02.008

Baggio, R., Scott, N., & Cooper, C. (2011). Design of tourism governance networks. In E. Laws, H. Richins, J. Agrusa, & N. Scott (Eds.), *Tourism destination governance: Practice, theory and issues* (pp. 150–171). Wallingford, UK: CAB International. doi:10.1079/9781845937942.0159

Beaumont, N., & Dredge, D. (2010). Local tourism governance: A comparison of three network approaches. *Journal of Sustainable Tourism*, *18*(1), 7–28. doi:10.1080/09669580903215139

Beritelli, P. (2011). Do actors really agree on strategic issues? Applying consensus analysis of stakeholder perceptions in tourist destination communities. *Tourism Analysis*, *16*(4), 219–241. doi:10.3727/108354211X13110944386924

Beritelli, P., Bieger, T., & Laesser, C. (2014). The new frontiers of destination management: Applying variable geometry as a function-based approach. *Journal of Travel Research*, *53*(4), 403–417. doi:10.1177/0047287513506298

Bhat, S. S., & Milne, S. (2008). Network effects on cooperation in destination website development. *Tourism Management*, *29*(6), 1131–1140. doi:10.1016/j.tourman.2008.02.010

Bieger, T., Beritelli, P., & Laesser, C. (2009). Size matters! Increasing DMO effectiveness and extending tourism destination boundaries. *Tourism Review*, *57*(3), 309–327.

Borgatti, S. P., & Halgin, D. S. (2011). On network theory. *Organization Science*, *22*(5), 1168–1181. doi:10.1287/orsc.1100.0641

Breiger, R. L., Boorman, S. A., & Arabie, P. (1975). An algorithm for clustering relational data with applications to social network analysis and comparison with multidimensional scaling. *Journal of Mathematical Psychology*, *12*(3), 328–383. doi:10.1016/0022-2496(75)90028-0

Buhalis, D. (2000). Marketing the competitive destination of the future. *Tourism Management*, *21*(1), 97–116. doi:10.1016/S0261-5177(99)00095-3

Burns, P., & Novelli, M. (Eds.). (2008). *Tourism and mobilities: Local-global connections*. Wallingford, UK: CAB International. doi:10.1079/9781845934040.0000

Burt, R. S. (1976). Positions in networks. *Social Forces*, *55*(1), 93–122. doi:10.1093/sf/55.1.93

Capello, R. (2009). Space, growth and development. In R. Capello & P. Nijkamp (Eds.), *Handbook of regional growth and development theories* (pp. 33–52). Cheltenham, UK: Edward Elgar. doi:10.4337/9781848445987.00008

Carlsen, J. (1999). A systems approach to island tourism destination management. *Systems Research and Behavioral Science*, *16*(4), 321–327. doi:10.1002/(SICI)1099-1743(199907/08)16:4<321::AID-SRES255>3.0.CO;2-5

Cooper, C., & Hall, M. (2013). *Contemporary tourism: An international approach* (2nd ed.). Oxford, UK: Goodfellow Publishers.

Cross, R. L., Borgatti, S. P., & Parker, A. (2002). Making invisible work visible: Using social network analysis to support human networks. *California Management Review*, *44*(2), 25–46. doi:10.2307/41166121

Crouch, G. I., & Ritchie, J. R. B. (1999). Tourism, competitiveness, and societal prosperity. *Journal of Business Research*, *44*(3), 137–152. doi:10.1016/S0148-2963(97)00196-3

Dollinger, M. J. (1990). The evolution of collective strategies in fragmented industries. *Academy of Management Review*, *15*(2), 266–285.

Dredge, D. (1999). Destination place planning and design. *Annals of Tourism Research*, *26*(4), 772–791. doi:10.1016/S0160-7383(99)00007-9

Dredge, D., & Jamal, T. (2013). Mobilities on the Gold Coast, Australia: Implications for destination governance and sustainable tourism. *Journal of Sustainable Tourism*, *21*(4), 557–579. doi:10.1080/09669582.2013.776064

Faust, K., & Wasserman, S. (1992). Blockmodels: Interpretation and evaluation. *Social Networks*, *14*(1-2), 5–61. doi:10.1016/0378-8733(92)90013-W

Fine, C. H. (1999). *Clockspeed: Winning industry control in the age of temporary advantage*. New York, NY: Basic Books.

Framke, W. (2002). The destination as a concept: A discussion of the business-related perspective versus the socio-cultural approach in tourism theory. *Scandinavian Journal of Hospitality and Tourism*, *2*(2), 92–108. doi:10.1080/15022250216287

Greene, W. H. (2012). *Econometric analysis* (7th ed.). Boston, MA: Pearson Education.

Hall, C. M. (2008). *Tourism planning: Policies, processes and relationships* (2nd ed.). Harlow, UK: Pearson Prentice Hall.

Hosmer, D. W., & Lemeshow, S. (2000). *Applied logistic regression*. New York, NY: John Wiley & Sons. doi:10.1002/0471722146

Hwang, Y.-H., & Fesenmaier, D. R. (2003). Multidestination pleasure travel patterns: Empirical evidence from the American Travel Survey. *Journal of Travel Research*, *42*(2), 166–171. doi:10.1177/0047287503253936

Hwang, Y.-H., Gretzel, U., & Fesenmaier, D. R. (2006). Multicity trip patterns. Tourist to the United States. *Annals of Tourism Research*, *33*(4), 1057–1078. doi:10.1016/j.annals.2006.04.004

Jeng, J., & Fesenmaier, D. R. (1998). Destination compatibility in multidestination pleasure travel. *Tourism Analysis*, *3*(3), 77–87.

Jenkins, J., Dredge, D., & Taplin, J. (2011). Destination planning and policy: Process and practice. In Y. Wang & A. Pizam (Eds.), *Destination marketing and management: Theories and applications* (pp. 21–38). Wallingford, UK: CAB International. doi:10.1079/9781845937621.0021

Kim, S., & Fesenmaier, D. R. (1990). Evaluating spatial structure effects in recreational travel. *Leisure Sciences*, *12*(4), 367–381. doi:10.1080/01490409009513115

Klepers, A., & Rozite, M. (2010). Administrative border as a challenge in spatial aspects of tourism competitiveness. In A. Clarke (Ed.), *Constructing central Europe: Tourism competitiveness* (pp. 23–33). Veszprém, HU: University Press, University of Pannonia.

Liu, F., Zhang, J., Zhang, J., Chen, D., Liu, Z., & Lu, S. (2012). Roles and functions of tourism destinations in tourism region of South Anhui: A tourist flow network perspective. *Chinese Geographical Science*, *22*(6), 755–764. doi:10.1007/s11769-012-0557-6

Lorrain, F., & White, H. C. (1971). Structural equivalence of individuals in social networks. *The Journal of Mathematical Sociology*, *1*(1), 49–80. doi:10.1080/0022250X.1971.9989788

Lue, C. C., Crompton, J. L., & Stewart, W. P. (1996). Evidence of cumulative attraction in multidestination recreational trip decision. *Journal of Travel Research*, *35*(1), 41–49. doi:10.1177/004728759603500107

Marsden, P. V. (1989). Methods for the characterization of role structures in network analysis. In L. C. Freeman, D. R. White, & A. K. Romney (Eds.), *Research methods in social network analysis* (pp. 489–530). Fairfax, VA: George Mason University Press.

Marzano, G., & Scott, N. (2009). Power in destination branding. *Annals of Tourism Research*, *36*(2), 247–267. doi:10.1016/j.annals.2009.01.004

Massey, D. (1994). *Space, place and gender*. Cambridge, UK: Polity Press.

McIntosh, R. W., Goeldner, C. R., & Ritchie, J. R. B. (1995). *Tourism, principles, practices, philosophies* (2nd ed.). New York, NY: John Wiley & Sons.

Nunkesser, M., & Sawitzki, D. (2005). Blockmodels. In U. Brandes & T. Erlebach (Eds.), *Network analysis: Methodological foundation* (pp. 253–292). Berlin: Springer-Verlag. doi:10.1007/978-3-540-31955-9_10

Pavlovich, K. (2003). The evolution and transformation of a tourism destination network: The Waitomo Caves, New Zealand. *Tourism Management*, *24*(2), 203–216. doi:10.1016/S0261-5177(02)00056-0

Pearce, D. G. (2014). Toward an integrative conceptual framework of destinations. *Journal of Travel Research*, *53*(2), 141–153. doi:10.1177/0047287513491334

Pearce, D. G., & Schänzel, H. A. (2013). Destinations: Tourists' perspectives from New Zealand. *International Journal of Tourism Research*. doi:10.1002/jtr.1955

Pike, S. (2008). *Destination marketing. An integrated marketing communication approach*. Oxford, UK: Elsevier Butterworth-Heinemann.

Porter, M. (1998). Clusters and the new economics of competition. *Harvard Business Review*, *76*(6), 77–90. PMID:10187248

Prell, C. (2012). *Social Network Analysis: History, theory and methodology*. London: Sage.

Ritchie, J. R. B., & Crouch, G. I. (2003). *The competitive destination. A sustainable tourism perspective*. Wallingford, UK: CAB International. doi:10.1079/9780851996646.0000

Saarinen, J. (2014). Transforming destinations: A discoursive approach to tourist destinations and development. In A. Viken & B. Granås (Eds.), *Tourism destination development: Turns and tactics* (pp. 47–62). Farnham, UK: Ashgate Publishing Limited.

Saraniemi, S., & Kylänen, M. (2011). Problematizing the concept of tourism destination: An analysis of different theoretical approaches. *Journal of Travel Research*, *50*(2), 133–143. doi:10.1177/0047287510362775

Schwartz, J. E. (1977). An Examination of *CONCOR* and related methods for blocking sociometric data. In D. R. Heise (Ed.), *Sociological Methodology* (pp. 255–282). San Francisco, CA: Jossey-Bass. doi:10.2307/270759

Scott, J. (2000). *Social network analysis. A handbook* (2nd ed.). London: Sage Publications.

Scott, N., Baggio, R., & Cooper, C. (2008a). *Network analysis and tourism: From theory to practice*. Clevedon, UK: Channel View.

Scott, N., Cooper, C., & Baggio, R. (2008b). Destination networks: Four Australian cases. *Annals of Tourism Research*, *35*(1), 169–188. doi:10.1016/j.annals.2007.07.004

Sheller, M. (2009). The new Caribbean complexity: Mobility systems, tourism and spatial rescaling. *Singapore Journal of Tropical Geography*, *30*(2), 189–203. doi:10.1111/j.1467-9493.2009.00365.x

Shih, H. Y. (2006). Network characteristics of drive tourism destinations: An application of network analysis in tourism. *Tourism Management*, *27*(5), 1029–1039. doi:10.1016/j.tourman.2005.08.002

Stienmetz, J. L., & Fesenmaier, D. R. (2013a). Assessing the economic value of tourist attraction using a network approach. In *Proceedings of the third international conference on the measurement and economic analysis of regional tourism.*

Stienmetz, J. L., & Fesenmaier, D. R. (2013b). Traveling the network: A proposal for destination performance metrics. *International Journal of Tourism Sciences, 13*(2), 57–75. doi:10.1080/15980634.2013.11434673

Tideswell, C., & Faulkner, W. (1999). Multidestination travel patterns of international visitors to Queensland. *Journal of Travel Research, 37*(4), 364–374. doi:10.1177/004728759903700406

Tremblay, P. (1998). The economic organization of tourism. *Annals of Tourism Research, 24*(4), 837–859. doi:10.1016/S0160-7383(98)00028-0

Wall, G. (1997). Tourism attractions: Points, lines and areas. *Annals of Tourism Research, 24*(1), 240–243. doi:10.1016/S0160-7383(96)00039-4

Wang, Y., & Krakover, S. (2008). Destination marketing: Competition, cooperation or coopetition? *International Journal of Contemporary Hospitality Management, 20*(2), 126–141. doi:10.1108/09596110810852122

Wasserman, S., & Faust, K. (1994). *Social network analysis: Methods and applications.* Cambridge, UK: Cambridge University Press. doi:10.1017/CBO9780511815478

Woodside, A. G., & Dubelaar, C. (2002). A general theory of tourism consumption systems: A conceptual framework and an experimental exploration. *Journal of Travel Research, 41*(2), 120–132. doi:10.1177/004728702237412

Zach, F., & Gretzel, U. (2012). Tourist-activated networks: Implications for dynamic bundling and en route recommendations. *Journal of Information Technology and Tourism, 13*(3), 239–257.

Ziakas, V., & Costa, C. A. (2010). Explicating inter-organizational linkages of a host communitys events network. *International Journal of Event and Festival Management, 1*(2), 132–147. doi:10.1108/17852951011056919

KEY TERMS AND DEFINITIONS

Accommodation Establishments: Units providing short-term or short-stay accommodation services. Here, they are classified into two groups, including hotels and similar establishments and other collective establishments.

CONCOR: A block-modelling method to group nodes (here, destinations) into blocks based on structural equivalence.

Density Matrix: A matrix showing intra- and/or inter-block relations measured by density scores. Density is based on the number of ties, as a proportion of the maximum possible number of ties among nodes in a network.

Ego-Network: The concept indicates the amount of all the nodes (here, destinations) to which an ego/node is directly connected and includes all of the ties among nodes in a network.

Network Analysis: Methodological approach to analyse the relational patterns among entities.

Structural Equivalence: Mathematical property of two nodes having the same ties with themselves, each other, and to and from all other nodes in the network.

Tourist Mobility: Spatial movement of tourists defining routes among destinations.

Chapter 15
On Interdisciplinary Intersection of Unconventional Algorithms and Big Data Processing in Real World Problems:
A Real World Example Based on Ho Chi Minh City Traffic

Ivan Zelinka
VSB Technical University of Ostrava, Czech Republic

Marek Běhálek
VSB Technical University of Ostrava, Czech Republic

Martin Kruliš
KSI, Czech Republic & Charles University in Prague, Czech Republic

Tung Minh Luu
Ton Duc Thang University, Vietnam

Jaroslav Pokorný
Charles University in Prague, Czech Republic

ABSTRACT

Optimization algorithms are a powerful tool for solving many problems of engineering applications from different fields of real life. They are usually used where the solution of a given problem analytically is unsuitable or unrealistic. If implemented in a suitable manner, there is no need for frequent user intervention into the actions of the equipment in which they are used. The majority of the problems of real life applications can be defined as optimization problems, for example, finding the optimum trajectory of a robot, optimal data flows in various processes like city traffic optimization or modelling and optimization of the seasonal variances of supply, traffic and facilities occupation in tourism among the others. The structure of this chapter is such that on the beginning are introduced bio-inspired algorithms, then parallelization of algorithms and parallel hardware and at the end, open research on Ho Chi Minh City traffic optimization real world example is introduced. In Conclusion are discussed possibilities of mutual combinations of introduced methods.

DOI: 10.4018/978-1-5225-1054-3.ch015

INTRODUCTION

Heuristic algorithms are a powerful tool for solving many problems of engineering applications from different fields of real life. They are usually used where the solution of a given problem analytically is unsuitable or unrealistic. If implemented in a suitable manner, there is no need for frequent user intervention into the actions of the equipment in which they are used.

The majority of the problems of real life applications can be defined as optimization problems, for example:

- Finding the optimum trajectory of a robot,
- The optimum thickness of the wall of a pressure tank,
- The optimum setting of the regulator's parameters,
- Optimal data flows in various processes like city traffic optimization or modelling and optimization of the seasonal variances of supply,
- Traffic and facilities occupation in tourism, amongst the others.

In other words, the solved problem can be transformed into a mathematical problem defined by a suitable prescription, whose optimization leads to finding the arguments of the objective function, which is the goal – the problem solution.

The structure of this chapter is such that on the beginning are introduced bio-inspired (i.e. evolutionary) algorithms, then parallelization of such algorithms and parallel hardware. At the end, we sketch a problem of our research on Ho Chi Minh City traffic optimization as an illustrative real world example. In Conclusion are discussed possibilities of mutual combinations of introduced methods.

BIO-INSPIRED ALGORITHMS: A LIGHT OVERVIEW

Countless examples illustrating problem of bio-inspired algorithms (Back, Fogel & Michalewicz, 1997) and its use on economic as well as industrial problems can be found. The solution of such problems usually requires working with the arguments of optimized functions, where the definition ranges of these arguments may be of a heterogeneous character, such as, for example, the range of integers, real or complex numbers, etc. Moreover, it may happen (depending on the case) that for certain subintervals from the permitted interval of values, the corresponding argument of the optimized function may assume values of various types (integers, real, complex). Besides this, various penalizations and restrictions can play a role within optimization, not only for given arguments, but also for the functional value of the optimized function. In many cases, the analytical solution of such an optimization problem is possible, nevertheless, considerably complicated and tedious. Evolutionary algorithms solve problems in such an elegant manner that they became very popular and are used in many engineering fields.

From the point of view of the most general classification, the evolutionary algorithms belong to heuristic algorithms. Heuristic algorithms are either deterministic or stochastic. The algorithms of the second group differ in that their certain steps use random operations, which means that the results of the solutions obtained with their use may differ in the individual runs of the program. It is therefore meaningful to run the program several times and select the best solution obtained.

Stochastic heuristic methods are sometimes called metaheuristics, because they only provide a general framework and the algorithms of the operation itself must be chosen (for example, by the operation of crossbreeding and mutation in genetic algorithms, operation of neighborhood in simulated annealing, "tabu search", etc.) in dependence on the problem investigated. Because these methods are frequently inspired by natural processes, they are also called evolutionary algorithms. Depending on their strategy, they can be divided in two classes:

- Methods based on point-based strategy such as, for example:
    - Simulated annealing, (Kirkpatrick, Gelatt & Vecchi, 1983; Cerny, 1985; Telfar, 1996),
    - Hill-climbing algorithm (Russell, Stuart & Norvig, 2003) and
    - Tabu-search (Glover & Laguna, 1997).
        - These algorithms are based on the neighborhood operation of the current solution, in which we are looking for a better solution, for example by mutation of the original-previous solution.
- Population-Based Strategy.
    - Genetic algorithms (Davis, 1996; Goldberg, 1989; Michalewicz, 1996; Chu, 1997) are based on population strategy as well as another algorithms like differential evolution, particle swarm, SOMA (Davendra, Zelinka, 2016; Zelinka, 2004), etc.

These methods differ from classic gradient methods by admitting (with a certain probability) a worse solution into the next iteration; in this manner, they try to avoid local minima. For more details, see for example books by Glover and Laguna, (1997), Goldberg, (1989), Michalewicz, (1996), and Reeves (1993).

As a representative example of evolutionary algorithms e.g.:

- **Genetic Algorithm (GA):** This algorithm is one of the first successfully applied EAs methods (Holland, 1975; Goldberg, 1989). In GAs the main principles of EAs are applied in their purest form. The individuals are encoded as binary strings and represent possible solutions to the optimization problem under study.
- **Evolutionary Strategies (ES):** This algorithm also belongs to the first successful stochastic algorithms in history. It was proposed at the beginning of the sixties by Rechenberg (1971) and Schwefel (1974). It is based on the principles of natural selection similarly as the genetic algorithms. More modern and swarm based algorithm, with very good performance in combinatorial problems is Ant Colony Optimization.
- **Ant Colony Optimization (ACO) (Dorigo & Suttzle, 2004):** This is an algorithm whose action simulates the behavior of ants in a colony and is one of the first swarm based algorithms as well as Particle Swarm (PS) or SOMA (Self-Organizing Migrating Algorithm).
- **SOMA:** A stochastic optimization algorithm that is modeled on the social behavior of cooperating individuals (Zelinka, 2004). It has been proven that the algorithm has the ability to converge towards the global optimum (Davendra & Zelinka, 2016). SOMA works on a population of candidate solutions in loops called migration loops. It is vector oriented algorithm based on central individual called Leader that influences behavior of the whole population. SOMA is in fact swarm based algorithm, too, however it can be also catalogued as the Memetic Algorithms.

- **Memetic Algorithms (MA):** This term represents a broad class of metaheuristic algorithms (Onwubolu & Babu, 2004; Hart, Krasnogor & Smith, 2005; Goh, Ong & Tan, 2009; Schonberger, 2005). The key characteristics of these algorithms are
    - The use of various approximation algorithms,
    - Local search techniques,
    - Special recombination operators, etc.

These metaheuristic algorithms can be basically characterized as competitive-cooperative strategies featuring attributes of synergy.

- **Differential Evolution (DE):** Differential Evolution (Price, 1999) is a population-based optimization method that works on real-number coded individuals. Differential Evolution is robust, fast, and effective with global optimization ability. It does not require that the objective function is differentiable, and it works with noisy, epistatic and time-dependent objective functions.
- Along with those well-known algorithms also exist another (sometimes similar in principle) algorithms like Firefly (Yang, 2009), CoCoo algorithm (Yang & Deb, 2009), Bat algorithm (Yang, 2010) amongst the others.

These algorithms can be divided according to the principles of their action, complexity of the algorithm, etc. Of course, this classification is not the only possibility, nevertheless, because it fits the current state rather well, it can be considered as one of the possible views on the classical and modern optimization methods. There are slight differences in opinions on their classification. One can encounter statements that, for example, simulated annealing does not belong to evolutionary techniques, which is true to some extent. On the other side, other "evolutionists'' state that simulated annealing does belong to evolutionary techniques, at least as their direct predecessor. It is true that if simulated annealing is taken into account with elitism, then one could consider this alternative as an evolutionary algorithm.

The evolutionary algorithms can be essentially used for the solution of very heterogeneous problems. Of course, for the solution of the optimization problems, there are many more algorithms than were indicated here. Because their description would exceed the framework of this text, we can only refer to the corresponding literature, where the algorithms indicated above are described in more details. Beside "standard" EAs a class of evolutionary algorithms used for symbolic regression (SR) (Zelinka, 2011; Zelinka, 2005; Koza, 1999; O'Neill & Brabazon, 2006) also exist. They are used for the model synthesis, comparing to with EAs that are used only to estimate parameters of model defined by user.

Despite fact that above mentioned algorithms are very powerful, there are still problems that cannot be solved in real time with acceptable results and new approach is needed. In this case the simplest way is to parallelize used algorithms. Parallelization then increase its performance significantly. In the next sections are mentioned some basics facts about parallelization, its impact on algorithm performance and computability of some problems.

ON ALGORITHM PERFORMANCE AND PARALLELIZATION

The efficiency of algorithms and their implementations has been one of the key issues of the computer science since its very foundation. It has become even more important in the last decade as the size and complexity of the data being processed grows rapidly and the phenomenon Big Data is getting a more pronounced shape. Even though there are many general ways how to increase the performance of a software application – such as

- Designing better algorithms and data structures,
- Optimizing the implementation for particular hardware, or
- Simply using a faster processor.

– there is a distinct trend both in the hardware design and in application development that leans towards parallel processing. Unlike alternative approaches, parallel processing offers more scalable solutions for many problems, especially when the data or the work itself can be divided into independent parts. On the other hand, parallel programming is much more tedious and error prone since it defies the natural thinking process of the programmer which is typically sequential.

Parallel programming (McCool, 2012) is based on a simple idea to utilize multiple processing units and divide the workload among them, thus reducing the required time or increasing the processing throughput. If we have an ideally parallelizable algorithm which requires time $T_1(X)$ to process data X sequentially (i.e., on a single processor), the time required to process the same data on P processing units is $T_p(X)= T_1(X)/P$. We also define the speedup S_p of the algorithm on P processors as the ratio of its serial and parallel time i.e. as $S_p=T_1/T_p$ which is equal to P in the ideal case.

There are multiple approaches how to divide work among the processing units. The following three are perhaps the most profound ones:

- **Task Parallelism:** The algorithm is divided into separate sub-problems (tasks) which can be processed concurrently. An example of such algorithm could be a computer game, which needs to perform various independent task like rendering the graphic, simulate physics of the objects, or perform computations related to the artificial intelligence of the opponents.
- **Data Parallelism:** The same algorithm is executed independently on multiple data elements. A typical representative of such operations is 3D scene rendering where the same mathematical transformations are performed over many vertices in space.
- **Pipeline Parallelism:** This paradigm is in fact a combination of task parallelism and data parallelism. A pipeline is formed by multiple stages which can operate concurrently and the results of one stage are passed on to a subsequent stage. If the pipeline is given multiple data items to process, the work of the stages can overlap and thus achieve a parallel speedup. Video decoding is a good example of pipeline processing since it decodes a sequence of frames, while each frame has to pass multiple stages (inverse quantization, inverse transform, color space conversion, etc.).

Unfortunately, designing a parallel algorithm does not always fit the described parallel paradigms ideally. A typical problem is that parallel tasks need to share some data or exchange intermediate results. In such case, a synchronization primitive has to be used to ensure mutual exclusion and correct processing order. Furthermore, it is virtually impossible to achieve ideal parallelization. Even algorithms which are designed to run completely in parallel have some sequential parts like the initialization of the computing units or the result aggregation.

We can express the overhead of parallelization by dividing the algorithm into a parallel part and a sequential part, where the parallel part is being processed truly concurrently utilizing all available processors and the sequential part represents the overhead and the parts of the algorithm which cannot be parallelized. In this case the execution time of the parallel algorithm is $T_p = T_{par}/P + T_{seq}$. If we can determine the relative size of the parallel and sequential parts, we can establish a theoretical speedup using the Amdahl's law (Amdahl, 1997).

Amdahl's law defines the maximal theoretical speed up to S_{max} which is determined from the parallel execution time on an unlimited number of processors. In other words, it computes the scalability limits of a parallel solution.

$$S_{max} = \lim_{p \to \infty} S_p = \lim_{p \to \infty} \frac{T_1}{\frac{T_{par}}{p} + T_{seq}} = \frac{T_1}{T_{seq}} = \frac{T_{par} + T_{seq}}{T_{seq}}$$

Hence, the maximal theoretical speedup is the inverse value of the relative portion of the sequential part. It means that the sequential part will take almost the entire processing time when the number of processing units tends to infinity. For instance, if the sequential part takes only 5% of the total processing time on one processor, the algorithm will never be able to run more than 20 times faster than its sequential version no matter how many processors we use. Therefore, one of the main objectives of parallel programming is also to minimize the sequential parts of the algorithms in favor of the parallel parts to ensure scalability to massively parallel platforms and future hardware.

Modern Parallel Hardware

Parallel processing is not a novel concept, but it received much more attention since it has been widely employed in commodity processors (CPUs). In the past, the CPU vendors relied mostly on increasing the operating frequencies of the processors to increase their processing speed. This approach has reached an impasse at the dawn of the 21st century. The main reason is that the power consumption (and thus the thermal output) of a processor is cubically proportional to its operating frequency and the CPU cooling systems could no longer effectively match the required heat dissipation of the chips. The processor designers have abandoned the frequency pursuit and introduced multicore CPUs which allow processing of multiple independent threads of code concurrently, which makes them quite suitable for task parallelism. Furthermore, CPUs have limited data-parallel capabilities as they introduce vector instructions (like SSE), which allows the code to perform the same operation on a short data vector (e.g., four numbers) concurrently.

In addition to multicore CPUs, which are currently present virtually in every desktop, server, or handheld device, various massively parallel platforms have emerged. Typical representatives are graphical processing units (GPUs), which were originally designed for specialized mathematical computations performed in 3D scene rendering. As each new generation of GPUs embraced new functionality, they have evolved in generic-purpose parallel units capable of processing non-graphical tasks. The nature of GPU architectures makes them particularly efficient in data parallelism (Sanders & Kandrot, 2010).

Finally, the rapid development in communication technologies allowed significant advances in distributed computing. Distributed computing represents another level of parallel computing where multiple host systems are used. It introduces new issues such as communication latency or reliability, but it also allows to scale the parallelism much further. Large corporations often employ farms of servers to process much larger data in much shorter time than any single system could. For instance, Google uses hundreds of thousands interconnected computers (Lämmel, 2008) to index the whole contents of the World Wide Web for the purposes of their search engine.

Parallelism and concurrent processing have become virtually omnipresent in the world of computer science. The programmers and algorithm designers must embrace this particular paradigm in order to create faster and more efficient software, especially when the processing time or the processing throughput are important issues.

Parallelization

Various meta-heuristics have been introduced in the last years. There are articles, see for example (Alba, Luque & Nesmachnow, 2013; Pedemonte, Nesmachnow & Cancela, 2011) that try to categorize the most frequently used approaches, how to extend these algorithms to take advantage from the parallel/distributed computing architecture. Let's discuss an example of meta-heuristic called DSOMA, that is discrete version of the SOMA (Davendra & Zelinka, 2016) algorithm. DSOMA (Davendra, 2013) is a population based meta-heuristic, that iteratively searches for better solutions. Iterations in DSOMA are called migrations. The DSOMA population composes of individuals. In each migration, selected pairs of individuals are used to produce new trial individuals. If a better solution is found among the trial individuals, then the successful trial individual replaces the original individual in the population.

From the algorithm point of view, we can define different levels, where we can apply parallel computing.

- **Fitness Function:** Usually the most time consuming function is the computation of the fitness function. It is frequently used in every step of DSOMA. First option how to speed up the computation is to speed up this function. On the other hand, used fitness function depends on the solved problem and some fitness functions may be easier to parallelize then others.
- **Constructing Trial Individuals:** Trial individuals in DSOMA are computed based on two individuals from the population. After the trial individuals are generated, they are repaired and their fitness values are computed. These last two operations are more time consuming then generating the trial individuals, moreover for a given set of trial individuals they are independent and thus suitable to be performed in parallel.
- **Computing a Migration:** A common strategy is to select pairs of individuals and construct trial individuals by combining the current best individual in the population with the remaining individuals. Computing the trial individuals for a given pair does not depend on other pairs and it can be performed in parallel.

To improve the obtained results, DSOMA (similarly to other evolutionary algorithms) utilizes some local search algorithm. In (Davendra, 2013), the *2-Opt* local search is utilized. Because an application of such local search algorithm on a single individual can be even more time consuming then the whole DSOMA computations, its parallelization can be crucial, if we want to speed up the computation.

Moreover, similarly to other population based evolution algorithms, we can add some new layers to improve the overall parallel behavior. For example, if we want to use a computer (cluster) with the distributed memory architecture (they represent a majority between current supercomputers) then none of the previously described levels may be appropriate. The reason is the communication bottleneck, which can outweigh the speed up achieved by the parallel execution.

The most popular and simplest model in the distributed environment is the so called island model (Alba, Luque & Nesmachnow, 2013; Araujo & Merelo, 2011). In this model, the population is partitioned in a small set of sub-populations (islands, colonies). These islands execute the original algorithm and then some (sparse) individuals exchange algorithm is applied to exchange information between such islands. It was shown, that sparse information exchange between such islands brings diversity into the population and thus prevents convergence in local optima. Moreover, the overall parallel behavior is affected by the used communication topology or exchange data rate (Ruciński, Izzo & Biscani, 2010; Kushida, 2013).

In practical applications, some hybrid approaches that combine more than one level of parallelization are used. In general, the higher level of parallelization is coarse-grain implementation (for example the mentioned set of islands) and then each of these islands integrates other parallel model (or models).

Hardware and Software Options for Parallelization

Different levels of parallelization for DSOMA were mentioned (Davendra, 2013). But the *appropriate* level (or levels) of parallelization depends also on the target hardware and used software technologies.

The following paragraphs summarize the most distinguishing options that are state-of-the art in the world of high performance computing (HPC). The first such choice, that greatly affects the solution especially from a programmer's perspective, is a target device that will be used for computing. There are two main options:

- **General Purpose Processors:** Current (super)computers used for HPC contains up to hundreds of thousands of computational cores. Even current mainstream desktop processors contain between 4-8 cores. Considering the usage of general purpose processors, the parallelization usually implements a model - Multiple Program - Multiple Data (MPMD). In this model, each core runs independently of others and can perform a different program with its own data.
- **Many-Core (or Massively Multi-Core) Coprocessors:** The most common devices from this category are graphics processing units (GPUs) that allows general purpose programming. However, recently Intel has also introduced their Xeon Phi (http://www.intel.com/content/www/us/en/processors/xeon/xeon-phi-detail.html), a coprocessor for HPC. These coprocessors usually contain many (tens or hundreds) lightweight cores and frequently implement Single Instruction - Multiple Data (SIMD) model. It describes devices with multiple processing elements that perform the same operation on multiple data simultaneously.

Considering the many-core coprocessors, usually the computation is divided between general purpose processor and many-core coprocessor. Such SIMD coprocessors successfully exploit data level parallelism, but not concurrency. A common approach is to use the coprocessor when appropriate and for remaining computations use the general purpose processors. Moreover, most applications do not use the coprocessor and CPU at the same time (McCool, Reinders & Robison, 2012; Kirk & Hwu, 2012). Some effort has been made to exploit the full computation power of CPUs and coprocessors at the same time for evolutionary algorithms (Van, Taillard, Melab & Talbi, 2012), but this idea is not explored in this chapter any further.

Current computers used in HPC frequently combine both type of devices. For example a computer named Titan (from the *Top 500* list (http://www.top500.org/system/177975)) is a cluster composed from

18 688 nodes where each node contains a traditional general purpose processor with 16 cores and the NVIDIA Tesla K20 GPU accelerator.

Considering general purpose processors, there are two main memory architectures.

- **Shared Memory Architecture:** In this architecture, all processors share the same memory and this memory is used to exchange data and for synchronization. This architecture is frequently used for smaller HPC computers with tenths or hundreds computing cores.
- **Distributed Memory Architecture:** It is the main model in today's supercomputers. In this architecture, a computer (usually a cluster) composes from interconnected nodes. Each of these nodes usually contains multiple computational cores with the shared memory. Nodes can exchange some kind of messages, but they cannot directly access a memory in a different node. Thus the memory is in fact distributed among the nodes.

Considering HPC, the most widely used programming languages are C, C++ or Fortran. In this chapter, we will focus on C++, as our original sequential implementation of DSOMA is also in C++.

Concurrent programming with threads is the simplest way as to how to create a parallel application in the shared memory environment. Nearly every programming language supports this programming paradigm. Also C++ contains libraries for concurrent programming with threads (http://www.cplusplus.com/reference/thread/thread/). Moreover, there are technologies that simplify the development of such applications. Like an example we can list Open Multi-Processing (OpenMP, http://openmp.org/), Cilk and Cilk++ (www.cilkplus.org). For the distributed memory model, the de facto standard is Message Passing Interface (MPI, www.mpi-forum.org).

The technologies for programming on GPUs have quickly evolved in last ten years. Currently, there are various approaches and technologies. Earlier, general purpose computing on graphics processing units was primarily based on modification of graphics pipeline, e.g. in Open Graphics Library (OpenGL) where programmable shaders are used for time consuming Basic Linear Algebra Subprograms (BLAS). Nowadays, Open Computing Language (OpenCL) is the currently dominant open general-purpose GPU computing language. It is an open standard defined by the Khronos Group (www.khronos.org). The dominant proprietary framework is NVIDIA's Compute Unified Device Architecture (CUDA) that was launched in 2006 as an SDK and API that allows using the C programming language to code algorithms executable on GPUs. Microsoft (www.microsoft.com) also introduced its own API called DirectCompute that supports general-purpose computing on GPUs, this time with full support of DirectX.

In the following subsections, technologies that are later used for parallelization of DSOMA are introduced in more detail.

OpenMP

OpenMP is maybe the most widely used technology considering shared memory architecture. Moreover, it is also frequently used in combination with MPI for distributed memory systems. OpenMP is an API that supports multi-platform shared memory programming in C, C++, and FORTRAN. Its support is implemented on most processor architectures and operating systems. It consists of a set of compiler directives, library routines, and environment variables that influence run-time behavior.

The simplest example of the OpenMP usage is the automatic parallelization of for cycles. Algorithm 1 demonstrates a simple for cycle. The pragma directive defines, that this cycle should be parallelized by

Algorithm 1. A simple for in OPENMP

```
const int N = 100000;
int a[N];
#pragma omp parallel for
for (int i = 0; i < N; i++) {
    a[i] = 2 * i;
}
```

Algorithm 2. Compilation and execution of a source code with OPENMP pragmas in g++

```
g++ -fopenm source.cpp -o result
OMP_NUM_THREADS=4 ./result
```

Algorithm 3. A simple MPI example

```
MPI_Init (&argc, &argv);
int myrank ;
MPI_Comm_rank(MPI_COMM_WORLD, &myrank);
if(myrank == 0){
    int data[10];
    MPI_Send(data, 10, MPI_BYTE, 1, 1, MPI_COMM_WORLD);
}
if(myrank == 1){
    int data [10];
    MPI_Recv(data, MPI_BYTE, 10,
    MPI_ANY_SOURCE, MPI_ANY_TAG, MPI_COM_WORLD);
}
MPI_Finalize ();
```

OpenMP. OpenMP is supported by a wide range of compilers. For example, its support is built in widely used GNU Compiler Collection (GCC, https://gcc.gnu.org/onlinedocs/libgomp/Enabling-OpenMP.html). This directive is ignored by a compiler, if there is no OpenMP support or if it is switched off.

Algorithm 2 shows how to enable OpenMP with GCC g++ compiler (parameter -fopenmp). Using the environment variable OMP_NUM_THREADS, it is possible to define the number of threads that are used during the execution.

Even if OpenMP automatically parallelize annotated for cycles, other issues like a concurrent modification of shared data, must be solved by a programmer. OpenMP provides additional constructs controlling data sharing or synchronization.

Message Passing Interface

MPI is a language-independent communications protocol used to program parallel computers or clusters with distributed memory architecture. It defines a message-passing interface, together with protocol and semantic specifications for how its features must behave in any implementation. The MPI applications compose from independent processes where the MPI messages and constructs are the only communication mechanisms. It is used to exchange data and also to solve synchronization issues.

The primary functionality is a point-to-point and collective communication. Algorithm 3 shows a simple example of a point-to-point communication in MPI. Similarly to OpenMP, a user specifies the number of processes when the application starts. Each process is uniquely identified by its rank (a number from 0 to the number of processes). In the following example, two processes are expected. Process ranked as 0 sends some data and the process 1 is waiting until the data arrives.

Even though MPI is relatively simple to use, it represents a quite low-level interface. Hence development of applications directly in C++ with MPI can be laborious and time-consuming. Furthermore, the complexity of creating parallel applications lies also in other supportive activities like debugging and profiling. Even an experienced programmer of sequential applications can spend a lot of time learning a new set of complex tools. Therefore, for many non-experts in the area of parallel computing, it can be difficult to create, debug, and analyze their distributed applications.

GPU Computing with CUDA

Compute Unified Device Architecture (CUDA) (Cheng, Grossman & McKercher, 2014; Wilt, 2013) was introduced by NVIDIA as a general parallel programming and computing platform in 2006. Although utilization of CUDA was very limited at the beginning and the first GPU chips based on G80 Tesla Architecture were too expensive, it became very popular after a few years. NVIDIA developed their own hardware architecture that enables solving known problems in shorter time due to massive parallelism. The GPUs were well-suited to address real problems that could be expressed as data-parallel computations. Except for the gaming community that still plays a very important role for hardware producers and vendors, NVIDIA decided to focus on research and computational areas as well. Nowadays, CUDA is the most popular and supported architecture running on GPUs (Cook, 2012;Kindratenko, 2014; Suh & Kim, 2013), and its environment allows a heterogeneous programming approach.

Simply stated, a part of program pipeline can be processed by GPU, whereas another part by CPU. Moreover, current design patterns for parallel programming (McCool, Reinders & Robison, 2012; Mattson, Sanders & Massingill, 2004; Kirk & Hwu, 2012; White, 2012) strongly suggest to distinguish between CPU and GPU parts to achieve the best application performance. A good program structuring itself does not ensure expected results. Memory arrangement and data alignment belong to the most important tasks that the programmer has to solve. CUDA ready graphic cards have basically five types of memory. The biggest is a global memory, which serves as a communication point between GPU and CPU and a primary storage if the data can not be given somewhere else. Since global memory is too slow, a shared memory (shared by all threads in one thread-block) is often employed to achieve better performance. Registers represent the fastest memory. Constant and texture memory are used in such cases where the read-only data structures can be used. The most important functions performed on GPUs are called kernels. Every kernel needs its runtime configuration that consists at least of CUDA grid and block settings. Both grid and blocks can have up to three dimensions. The grid consists of blocks and every block encapsulates a set of threads. The computation is performed on several streaming multiprocessors independently (Maxwell Multiprocessor SMM on Maxwell architecture). The maximum number of blocks, threads and SMMs depends on GPU specification. We refer to (Cheng, Grossman & McKercher, 2014; Sanders & Kandrot, 2010; Wilt, 2013; Pacheco, 2011) for more details on CUDA programming since it is out of scope of this chapter.

Algorithm 5. A simple kernel in CUDA

```
__global void foo(const unsigned int N, int* a){
unsigned int tOffset = blockIdx.x * blockDim.x + threadIdx.x;
while (tOffset < N){
  a[tOffset] = 2 * tOffset;
  tOffset = blockIdx.x * blockDim.x;
  }
}
```

Internal architecture of GPUs is suitable for vector and matrix algebra operations. That leads to the wide usage of GPUs in the area of information retrieval, data mining, image processing, data compression, etc. Nowadays, programmers usually choose between OpenCL which is supported by all hardware producers, and CUDA which is supported by NVIDIA only. An important benefit of OpenCL is its platform independence; however, CUDA still sets the trends in GPU programming.

CUDA kernels are usually relatively complex and particular implementation can require suitable data arrangement and indexing. Algorithm 5 represents an illustrative kernel implementation.

ON HYBRIDIZATION AND PARALLELIZATION OF METAHEURISTIC ALGORITHMS

In the previous section we have mentioned a light introduction to the bio-inspired algorithms, parallel hardware architectures and also how can be bio-inspired algorithms parallelized in order to get better performance in problem solution, as for example big city traffic is (see next section). Usual naive question is: why? Let's take a look on following Tables 1-3 that shall give a clear answer.

In the Table 1 are mentioned basic different classes of complexity in which n is number of input variables (so called size of instance) for the function that describe problem. One can see that in the set of exponential functions (i.e. problems of exponential complexity) we get a tremendous number of possible results (amongst which is hidden the right one). If we have computer that is able to recalculate one possible result during 1 microsecond (even faster does not play significant role) then time needed to calculate all possible combinations and then select one take an incredible time, see Table 2.

Even if that hypothetical computer is speeded up by arbitrary speed, then in the case of the exponential class of functions does not help, see Table 3. All this shows that any, even hypothetical, computer cannot solve hard problem of exponential class. Does not matter what classical algorithm is used.

An acceptable solution of this problem is to use bio-inspired (evolutionary, heuristic, etc.) algorithms in its parallelized version. They are parallel "in natura" so there is no need to develop complicated way how to make them parallel. Some of them, like Dorigo and Suttzle (2004), and Davendra (2013) are a priori suitable for combinatorial problems like travelling salesman, which is our case example (Ho Chi Minh City traffic) in the last section.

For comparison, the number of protons in the visible universe has 79 digits. Number of microseconds since the "big bang" has 24 digits.

As an example of well written books on that topic can be mentioned (Alba, 2005; Alba, Luque & Nesmachnow, 2013). Based on contemporary results can be stated that parallelized bio-inspired algorithms are reasonable way how to solve hard problems, as for example ACO has been used to success-

Table 1. Complexity classes in general showing the number of possible combinations for different classes of complexity with instance size n.

n Function	10	50	100	300	1000
Polynomial					
$5n$	50	250	500	1500	5000
$n \log_2 n$	33	282	665	2469	9966
n^2	100	2 500	10 000	90 000	1 million (7 cifer)
n^3	1000	125 000	1 million (7 cifer)	27 millions (8 cifer)	1 bilion (10 cifer)
Exponential					
2”	1024	16 digits number číslo	31 digits number number o	91 digits number	302 digits number
$n!$	3,6 millions (7 cifer)	65 digits number	161 digits number	623 ciferné číslo	Giant number
n^n	10 bilions (11 cipher)	85 digits number	201 digits number	744 digits number	Giant number

Table 2. Estimating the duration of f(n) operations if one takes about 1 microsecond

n Function	10	20	50	100	300
Polynomial					
n^2	1/10000 s	1/2500 s	1/400 s	1/100 s	9/100 s
n^5	1/10 s	3.2s	5.2s	2.8 h	28.1 days
Exponential					
2^n	1/1000 s	1 s	35.7 yrs	400 trilion of centuries století	75 digit of centuries
n^n	2.8 days	3.3 trilion yrs	70 digit of centuries	185 digit of centuries	728 digit of centuries

Table 3. Acceleration calculation using n-times faster computers

Function	The maximum Size of the Input Is Manageable within a Reasonable Time		
	Contemporary Computers	**100x Faster Computers**	**1000x Faster Computers**
n	N_1	$100 N_1$	$1000 N_1$
n^2	N_2	$10 N_2$	$31{,}6 N_2$
2^n	N_3	$N_3 + 6{,}64$	$N_3 + 9{,}97$
$n!$	N_4	$N_4 + 1$	$N_4 + 2$

Note that speed up for 2^n and $n!$ is only addition of small constants to the original speed. Thus no influence on computer power is visible for that problems.

fully solve travelling salesman problem for 10000 cities (i.e. 10^{35650} of possible results, see (Zelinka, Celikovský, Richter & Chen, 2010)). Our example of real world problem, that can be regarded as a constrained version of the travelling salesman problem (TSP) is the optimal route finding in real time traffic in Ho Chi Minh City.

Ho Chi Minh City Traffic: A Real Problem Application

As a typical example of problem that represent theoretically hard problems for optimization (remember TSP) and real life problem that is joined with robust, huge and big data based traffic that is seasonally influenced by touristic activities. Such kind of problem can be easily modified to data flow in tourism, management of any big data flow process etc. Suggested algorithms for solution can be as already mentioned ACO, SOMA, DE and PSO amongst the other. And parallelized, if possible of course.

Let's take a brief overview of problem here. "Traffic jam" is always one of the hottest topic to discuss with Ho Chi Minh City (HMC) problems as well as in another big world cities. Especially with the very well developing country like as Vietnam which has attracted lots of foreign investors in recent times. It is also one of disadvantage factors to promote the development of our country. Traffic jam regularly happens in HMC. The transportation system is based mostly on roads. No rails, no subway. Therefore, the number of major transports such as cars, motorcycles, lorry, buses, trucks and other vehicles…are more and more crowded on streets. Let's try to imagine what will happen if the streets are obstructed by full of transports and people while we want to come back home quickly on the bad weather or after a tired working day. At that time, the vehicles move very slowly in HCM City that has about 3300 kilometers of roads and streets

Beside classical personal and industrial traffic, is in HMC public transportation done by different companies and competing and/or cooperating. Our problem is based on fact that company, under our investigation, collect people from different districts (see Figures 1-3) and transport them into one of two collecting centers (East and West) with coach buses. Passengers then continue to its location out of HMC. Thus that problem is kind of TSP dynamically changing in the time (optimized trajectories can change due to traffic jam, accidents, number of waiting passengers, that call company transportation service unpredictably, etc.). Let's describe problem briefly and schematically:

- The HCM city have 24 districts.
- 182 main roads is there.
- The 2 big collecting stations: West station & East station, are there.
- The population is about 14 milion in HMC.
- HMC area is 2.095,6 km^2.

Moreover, beside geographical data, technical and demographical data are

- Each district they have least one bus stop to pick up customer from it to BIG station (W. station & E. station). Now 30th bus stop is there.
- Every day, the customer by the ticket and waiting for transfer to the collecting stations and get on Bus to the province. The customer is about 50.000 persons per day.
- The number of customer at the bus stop is a priori known.

Figure 1. Complex networks of HMC traffic: a map of main road in HCM City

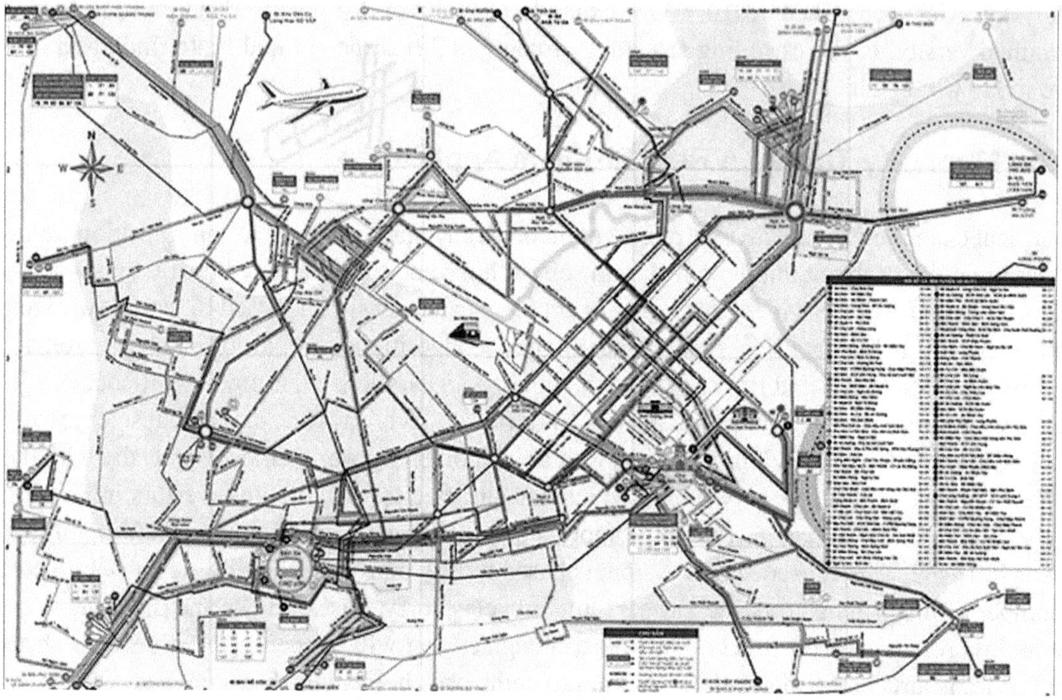

Figure 2. Complex networks of HMC traffic: a map of main road in HCM City with district segments

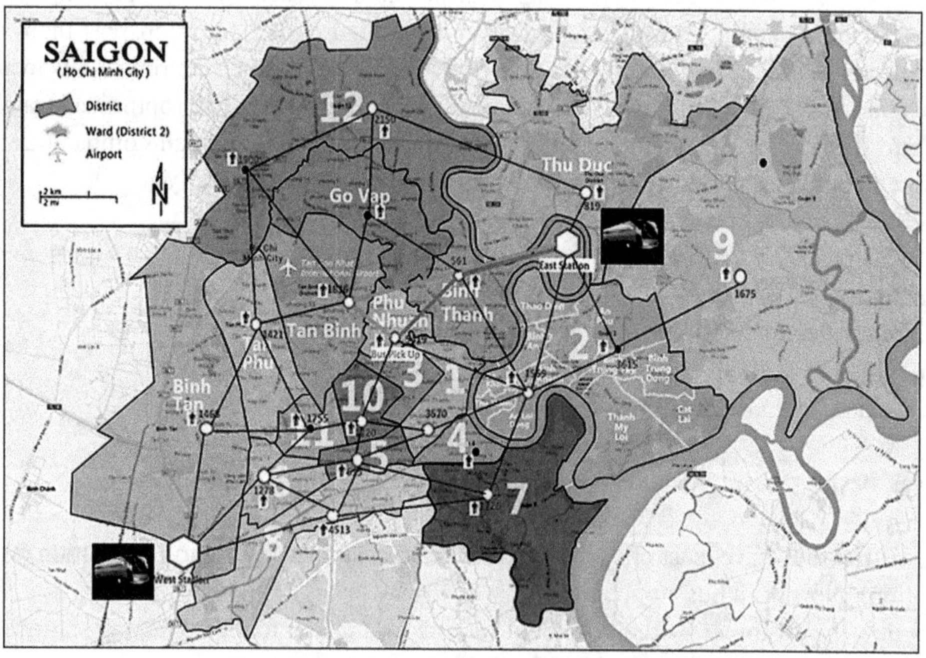

Figure 3. Complex networks of HMC traffic: a schematic map of main road in HCM City with sample data description. This is graph of positon pickup bus of FUTA Bus Lines.

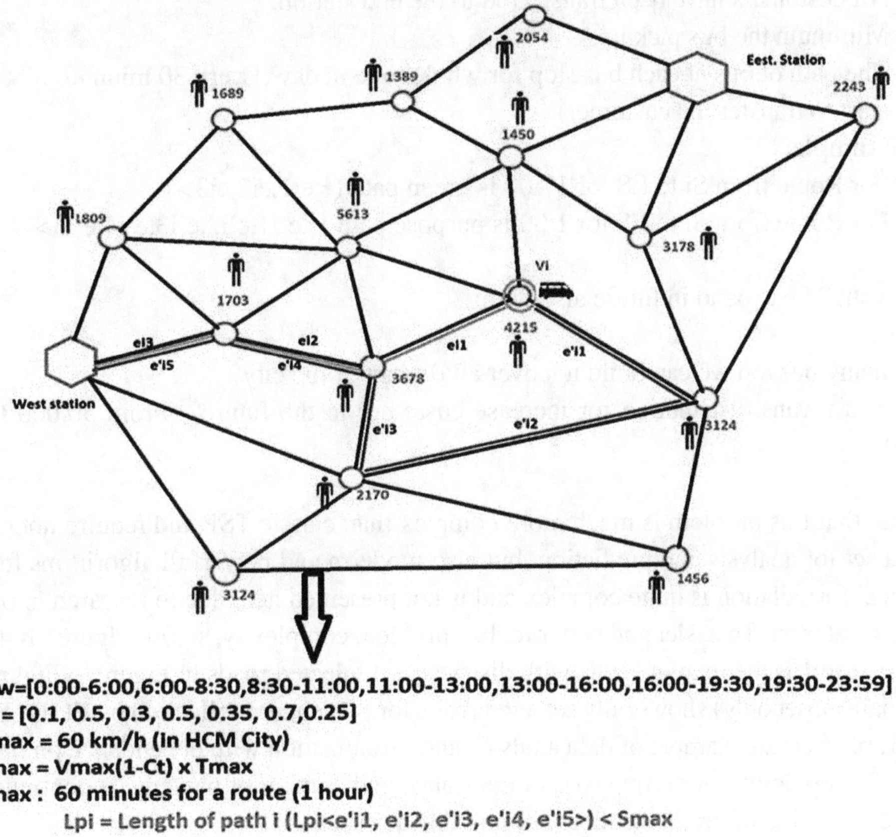

- For passenger transportation the car with 16 seats (more 200 vans) is used.
- The problem is: We must transfer all customer to the BIG station with minimum the van and some cost.

To transform all this into a mathematical problem definition, it is important to realize that optimization task is maximize the profit based on customer transfer from bus stop to the end station. Also minimization of used trajectory is required. Thus we are approaching multiobjective optimization (Deb, 2001). Parameters of problem shall be as:

- **Input:** Graph of stations.
  - **Types:** Zone station – bus stop (sz) and end station (se).
  - **Customer:** 50,000 persons / per day. The distribution of customer for bus stop.
  - **End Station:** 2 Nodes: East Station & West Station.
  - **Number of Bus Pick Up:** Car with 16 seats. (more 200 vans)
- **Condition:**
  - Transfer all customer to end station.
  - Limit of times for transfer is 60 minutes for any bus stop go to end station.

  - Follow the crowded time of traffic jam hours (11:30 to 12:30, 17:00 to 18:30).
- **Goal:**
  - All customers have been transferred to the end station.
  - Minimum the bus pick up.
  - The path of bus at each bus stop for whole time in day (Every 30 minutes, a bus pick up will start to transfer the customer)
- **For Example:**
  - For Route from Si to ES for 17:00 is green path (<ei1,ei2,ei3>).
  - For Route from Si to ES for 1:00 is purpose path (<e'i1,e'i2,e'i3,e'i4,e'i5>).

Problem shall be expand in future also with:

- How many bus top we can build to cover all district of the city?
- How many vans distribution for increase customer in the future? (From 50.000 to 100.000 or more).

It is clear that this problem is much more complex than classic TSP and require not only good historical data set for analysis and prediction, but also moderm and powerfull algorithms for its solution. Mathematical formulation is quite complex and is not presented here due to research in process and its industrial use later on. To easier and better realize problem complexity, see on Figures 1–4 and Tables 4 and 5. On Figure 1 is the map of HMC with districts and selected roads in Figure 2. Figure 3 (based on Table 4; small subset only) shows only selected roads for passengers collection by FUTA Bus Lines. On Figure 4 (Table 5) is an example of data analysis and visualization with dependance on time. The problem change its topology, connecting roads importance and number of passengers continuously in time.

To solve such kind of problems there are possibilities such as:

1. Solve by classical approach.
2. Use bio-inspired - evolutionary algorithms on single core computers.
3. Use bio-inspired - evolutionary algorithms on parallel hardware architecture.

Figure 4. Simple data analysis of HMC traffic

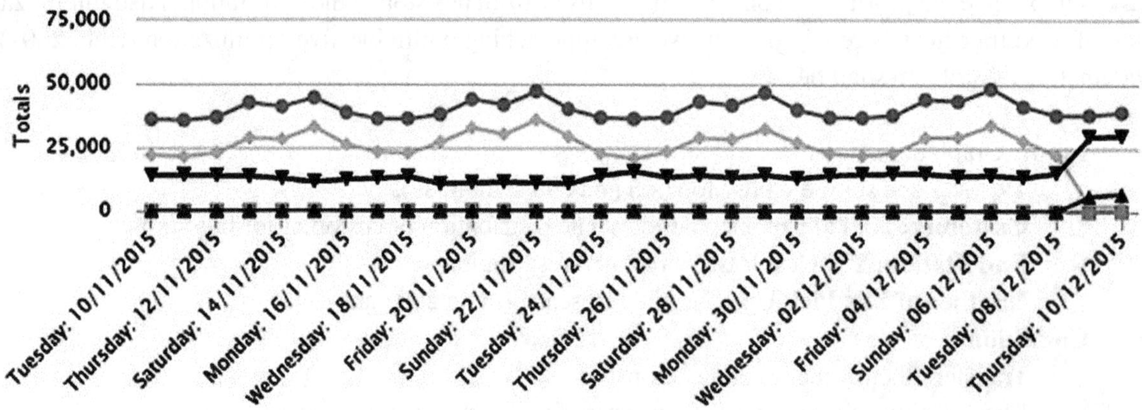

Table 4. Monitoring each segment follow Crw

Segment ID	Cell_x	Cell_y	Frame	Speed	Street_id
12300	25	29	25	19.459	32585800
95092	13	16	25	24.05	193514953
23116	25	18	25	18.5	32666179
9268	22	19	25	16.65	32583181
85458	16	21	25	13.15408603	186347710
113961	13	23	25	7.169773603	220818784
85803	16	20	25	6	187895195
12610	23	30	25	26.8625	32586067
123787	19	16	25	36	237480053
101472	21	13	25	13.63982856	199161611
122079	15	16	25	13	235283467
3893	26	7	25	40.33333333	32578837
59667	18	22	25	27.33333333	136505780

Table 5. Transit person 2012-2016: a small example from a larger dataset

STT	Postion - Bus Stop	January					February				
		2012	2013	2014	2015	2016	2012	2013	2014	2015	2016
1	Q1	5,059	4,939	5,430	4,626	6,339	3,986	5,555	5,879	5,666	3,803
2	Q2	4,473	3,605	9,309	4,311	7,986	3,086	3,737	5,135	7,514	5,205
3	Q3	4,311	4,628	4,353	4,326	5,244	4,177	5,726	6,048	4,641	2,867
4	Q4	-	2,720	2,995	3,141	4,027	-	3,294	6,363	3,183	2,463
5	Q5	9,018	3,557	7,557	13,020	17,362	10,404	9,461	12,262	18,179	10,679
6	Q6	-	4,357	6,055	14,569	18,648	-	8,606	12,738	15,982	8,875
7	Q7	27,142	29,813	72,661	61,655	65,918	21,326	72,551	60,521	63,796	37,388
8	Q8	-	65,807	53,787	59,387	61,969	16,604	69,485	68,598	59,624	29,779
9	Q9	7,377	5,824	22,942	19,759	21,164	5,296	15,357	18,009	22,920	12,301
10	Q10	-	11,554	15,068	18,878	18,816	-	13,875	23,692	20,500	10,099
11	Q11	10,161	17,684	27,100	39,321	36,187	10,147	16,230	34,291	39,984	19,654

4. Use swarm algorithms like ACO, PSO and the others to solve such problem in a "swarm way", i.e. synchronize swarm dynamics with real data in order to mimic city dynamics and then use it to predict – explore the best route to the collection stations.
5. Analyze collected data as demonstrated in (Svaton, Martinovic, Slaninova and Bures, 2014; Svaton, Martinovic, Slaninova and Snasel, 2014) in order to explore "strategies" of traffic swarm and use it with already mentioned methods.

Of course there is a lot of another possibilities, but those above are the main and promissing one.

CONCLUSION

In this chapter we have discussed possibilities on mutual combination of parallelization of bio-inspired algorithms and its use on real world example based on Ho Chi Minh City trafffic under specific condition. The aim of chapter was to give a light introduction into problems of parallelization, bio-inspired algorithms and TSP kind of problems. The proposed topic reported here is still under our investigation with use of approaches as mentioned in previous section. We believe, based on our various succesfull algorithms adaptations to parallel architectures, that such kind of problems are solvable with acceptable results and in real time.

ACKNOWLEDGMENT

The following grant are acknowledged for the financial support provided for this research: Grant Agency of the Czech Republic - GACR P103/13/08195S, VSB-TU internal grant SGS 2016/175 and by The Ministry of Education, Youth and Sports from the National Programme of Sustainability (NPU II) project "IT4Innovations excellence in science - LQ1602".

REFERENCES

Alba, E. (2005). *Parallel Metaheuristics: A New Class of Algorithms*. Wiley-Interscience. doi:10.1002/0471739383

Alba, E., Luque, G., & Nesmachnow, S. (2013). Parallel metaheuristics: Recent advances and new trends. *International Transactions in Operational Research*, *20*(1), 1–48. doi:10.1111/j.1475-3995.2012.00862.x

Amdahl, G. M. (1967). Validity of the single processor approach to achieving large scale computing capabilities. *Proceedings of the April 18-20, 1967, spring joint computer conference*. ACM. doi:10.1145/1465482.1465560

Araujo, L., & Merelo, J. (2011). Diversity through multiculturality: Assessing migrant choice policies in an island model. Evolutionary Computation. *IEEE Transactions on*, *15*(4), 456–469. doi:10.1109/TEVC.2010.2064322

Back, T., Fogel, B., & Michalewicz, Z. (1997). *Handbook of Evolutionary Computation*. London: Institute of Physics. doi:10.1887/0750308958

Cerny, V. (1985). Thermodynamical approach to the traveling salesman problem: An efficient simulation algorithm. *Journal of Optimization Theory and Applications*, *45*(1), 41–51. doi:10.1007/BF00940812

Cheng, J., Grossman, M. & McKercher, T. (2014). *Professional CUDA C Programming*. Wrox.

Chu, P. (1997). *A Genetic Algorithm Approach for Combinatorial Optimization Problems* (Ph.D. Thesis). The Management School Imperial College of Science, Technology and Medicine, London, UK

Cook, S. (2012). *CUDA Programming: A Developer's Guide to Parallel Computing with GPUs (Applications of Gpu Computing)*. Morgan Kaufmann.

Davendra, D., Zelinka, I., Bialic-Davendra, M., Senkerik, R., & Jasek, R. (2013). Discrete self-organizing migrating algorithm for flow-shop scheduling with no-wait makespan. *Mathematical and Computer Modelling, 57*(1-2), 100–110. doi:10.1016/j.mcm.2011.05.029

Davendra, D. D., & Zelinka, I. (2016). *Self-Organizing Migrating Algorithm - Methodology and Implementation*. Springer. doi:10.1007/978-3-319-28161-2

Davis, L. (1996). *Handbook of Genetic Algorithms*. Berlin: Van Nostrand Reinhold.

Deb, K. (2001). *Multi-objective optimization using evolutionary algorithms*. John Wiley & Sons.

Dorigo, M., & Suttzle, T. (2004). *Ant Colony Optimization*. MIT Press.

Glover, F., & Laguna, M. (1997). Tabu Search. Springer.

Goh, C., Ong, Y., & Tan, K. (2009). *Multi-Objective Memetic Algorithms*. Springer-Verlag. doi:10.1007/978-3-540-88051-6

Goldberg, D. (1989). *Genetic Algorithms in Search, Optimization, and Machine Learning*. Addison-Wesley Publishing Company Inc.

Hart, W., Krasnogor, N., & Smith, J. (2005). *Recent Advances in Memetic Algorithms*. SpringerVerlag. doi:10.1007/3-540-32363-5

Holland, J. (1975). *Adaptation in natural and artificial systems*. Ann Arbor, MI: Univ. of Michigan Press.

Kindratenko, V. (Ed.). (2014). *Numerical Computations with GPUs*. Springer. doi:10.1007/978-3-319-06548-9

Kirk, D. B., Mei, W., & Hwu, W. (2012). Programming Massively Parallel Processors: A Hands-on Approach (2nd ed.). Morgan Kaufmann.

Kirkpatrick, S., Gelatt, C. Jr, & Vecchi, M. (1983). Optimization by Simulated Annealing. *Science, 220*(4598), 671–680. doi:10.1126/science.220.4598.671 PMID:17813860

Koza, J. R., Andre, D., Bennett, F. H., & Keane, M. A. (1999). *Genetic Programming III: Darwinian Invention and Problem Solving* (1st ed.). San Francisco, CA: Morgan Kaufmann Publishers Inc.

Kushida, J. I., Hara, A., Takahama, T., & Kido, A. (2013). Island-based differential evolution with varying subpopulation size. In *Computational Intelligence & Applications (IWCIA), 2013 IEEE Sixth International Workshop on*, (pp. 119–124). IEEE doi:10.1109/IWCIA.2013.6624798

Lämmel, R. (2008). Googles MapReduce programming model—Revisited. *Science of Computer Programming, 70*(1), 1–30. doi:10.1016/j.scico.2007.07.001

Mattson, T. G., Sanders, B. A., & Massingill, B. L. (2004). *Patterns for Parallel Programming*. Addison-Wesley Professional.

McCool, M., Reinders, J., & Robison, A. (2012). *Structured Parallel Programming: Patterns for Efficient Computation*. Morgan Kaufmann.

McCool, M., Robison, A. D., & Reinders, J. (2012). *Structured Parallel Programming*. Morgan Kaufmann.

Michalewicz, Z. (1996). *Genetic Algorithms + Data Structures = Evolution Programs*. Berlin: Springer-Verlag. doi:10.1007/978-3-662-03315-9

Michalewicz, Z., & Fogel, D. B. (2000). *How to Solve It: Modern Heuristics*. Berlin: Springer-Verlag. doi:10.1007/978-3-662-04131-4

O'Neill, M., & Brabazon, A. (2006), Grammatical differential evolution. In *Proceedings of the 2006 International Conference on Artificial Intelligence*. Las Vegas, NV: CSREA Press.

Onwubolu, G., & Babu, B. (2004). *New Optimization Techniques in Engineering*. New York: Springer-Verlag. doi:10.1007/978-3-540-39930-8

Pacheco, P. (2011). *An Introduction to Parallel Programming*. Morgan Kaufmann.

Pedemonte, M., Nesmachnow, S., & Cancela, H. (2011). A survey on parallel ant colony optimization. *Applied Soft Computing*, *11*(8), 5181–5197. doi:10.1016/j.asoc.2011.05.042

Price, K. (1999). An introduction to differential evolution. In D. Corne, M. Dorigo, & F. Glover (Eds.), *New Ideas in Optimisation* (pp. 79–108). McGraw Hill, International.

Rechenberg, I. (1971). *Evolutionsstrategie - Optimierung technischer Systeme nach Prinzipien der biologischen Evolution* (PhD thesis). Fromman-Holzboog.

Reeves, C. (1993). *Modern Heuristic Techniques for Combinatorial Problems*. Oxford, UK: Blackwell Scientific Publications.

Ruciński, M., Izzo, D., & Biscani, F. (2010). On the impact of the migration topology on the island model. *Parallel Computing*, *36*(10), 555–571. doi:10.1016/j.parco.2010.04.002

Russell, S. J., & Norvig, P. (2003). *Artificial Intelligence: A Modern Approach* (2nd ed.). Upper Saddle River, NJ: Prentice Hall.

Sanders, J., & Kandrot, E. (2010). *CUDA by Example: An Introduction to General-Purpose GPU Programming*. Addison-Wesley Professional.

Sanders, J., & Kandrot, E. (2010). *CUDA by Example: An Introduction to General-Purpose GPU Programming, Portable Documents*. Addison-Wesley Professional.

Schonberger, J. (2005). *Operational Freight Carrier Planning, Basic Concepts, Optimization Models and Advanced Memetic Algorithms*. Springer-Verlag.

Schwefel, H. (1974). *Numerische Optimierung von Computer-Modellen* (PhD thesis). Birkhuser.

Suh, J. W., & Kim, Y. (2013). *Accelerating MATLAB with GPU Computing: A Primer with Examples*. Morgan Kaufmann.

Svaton, M., Slaninova, & Bures. (2014). Improving strategy in robot soccer game by sequence extraction. In *18th International Conference in Knowledge Based and Intelligent Information and Engineering Systems, KES 2014*.

Svaton, V., Martinovic, J., Slaninova, K., & Snasel, V. (2014). Improving rule selection from robot soccer strategy with substrategies. In *Computer Information Systems and Industrial Management - 13th IFIP TC8 International Conference*. doi:10.1007/978-3-662-45237-0_9

Telfar, G. (1996). *Acceleration Techniques for Simulated Annealing* (MSc Thesis). Victoria University of Wellington, New Zealand.

Van, T., Taillard, E., Melab, N., & Talbi, E.-G. (2012). Parallelization strategies for hybrid metaheuristics using a single GPU and multi-core resources. In *Parallel Problem Solving from Nature-PPSN XII* (pp. 368–377). Springer.

White, T. (2012). Hadoop: The Definitive Guide (3rd ed.). Yahoo Press.

Wilt, N. (2013). *CUDA Handbook: A Comprehensive Guide to GPU Programming*. Addison-Wesley Professional.

Yang, X. S. (2009). Firefly algorithms for multimodal optimization. Stochastic Algorithms: Foundations and Applications, SAGA 2009. *Lecture Notes in Computer Science*, *5792*, 169178.

Yang, X.-S., & Deb, S. (2009). Cuckoo search via Lvy flights.*World Congress on Nature and Biologically Inspired Computing (NaBIC 2009)*. IEEE Publications.

Yang. (2010). A New Metaheuristic Bat-Inspired Algorithm. In *Nature Inspired Cooperative Strategies for Optimization* (NISCO 2010). Springer.

Zelinka, I. (2004). SOMA - Self Organizing Migrating Algorithm. In B. B. Onwubolu (Ed.), *New Optimization Techniques in Engineering* (pp. 167–218). New York: Springer-Verlag. doi:10.1007/978-3-540-39930-8_7

Zelinka, I., Celikovský, S., Richter, H., & Chen, G. (2010). *Evolutionary Algorithms and Chaotic Systems*. Springer. doi:10.1007/978-3-642-10707-8

Zelinka, I., Davendra, D., Senkerik, R., Jasek, R., & Oplatkova, Z. (2011). *Analytical Programming - a Novel Approach for Evolutionary Synthesis of Symbolic Structures*. InTech. doi:10.5772/16166

Zelinka, I., Oplatkova, Z., & Nolle, L. (2005). Analytic programming - symbolic regression by means of arbitrary evolutionary algorithms. International Journal of Simulation Systems. *Science and Technology*, *6*(9), 44–56.

Chapter 16

Cooperative Optimization of Tourism Networks: An Application of a Game Theory Model

Vincenzo Asero
University of Catania, Italy

Sebastiano Patti
University of Catania, Italy

Stefania Skonieczny
Le Mude Ltd., UK

ABSTRACT

A tourism network is determined by close collaboration among various stakeholders working interactively on common problems or issues through formal and informal approaches. As many cases highlight, informal relationships contribute to the formation of formal partnerships. In this study a game theoretical model is applied to explain the decisional process of entrepreneurs about forming partnerships in formal tourism networks. The model shows that, in terms of Nash equilibrium, the cooperative optimization of a tourism network should be achieved when entrepreneurs have the same business goals and a common tourism vision The chapter presents an empirical solution that arises from the case of Business Networks in Italy, which represents an innovation in Italian Law.

INTRODUCTION

A destination comprises different interacting elements: tourists, enterprises, institutions and the host community (McIntosh, Goeldner, & Ritchie, 1995). According to Wang (2011), a destination includes both physical features and management and services. Pavlovich (2003) stated that the relationship among these various components is a critical factor in achieving competitive advantages in the tourism market. Merinero-Rodríguez and Pulido-Fernández (2016) emphasized that relationships are the core element for understanding a tourism phenomenon, proving useful in the planning and management of destinations, although different relational approaches have been developed in tourism research.

DOI: 10.4018/978-1-5225-1054-3.ch016

Closely associated with tourism is the concept of network, which describes the various relationships and transactions that characterise a tourism system. As Bickerdyke (1996) observed, tourism is the economic sector with the highest number of inter-organizational networks. A tourism network is therefore based on the cooperative behaviour of organisations, enterprises and individuals, which would otherwise be in competition with each other. Following Wood and Gray (1991), cooperation in tourism can be defined as a form of voluntary joint action in which autonomous stakeholders interact, sharing rules, norms and structures. In this regard, Edwards et al. (2000) stated that collaborations and partnerships are not a homogeneous form of governance, but consist of a set of different institutions, with diverse focuses, operational scales, history and duration, and funding.

Dollinger (1990) highlighted the role of collective strategies for the creation of a network, while Crouch and Ritchie (1999) affirmed that collective actions have a great impact on tourism development. Similarly, Jamal and Jamrozy (2006) claimed that a destination can be considered as a system in which different stakeholders take part in tourism development and planning, although no one individual stakeholder can entirely control the decisional processes. Consequently, understanding the network characteristics of destinations has practical implications for the competitiveness, management and planning of tourism (Hall, 2008).

The growing interest in tourism networks is largely due to the effects of globalization, which demand the creation of partnerships, alliances and joint projects to improve efficiency and competitiveness. This requires the active involvement of different actors, from both public and private sectors. In this regard, many scholars emphasize that collaborative networks have increased in importance, becoming indispensable to the promotion of coordination between tourism policies and related actions. This has provided a rich vein for theoretical approaches, and empirical studies, that aim to understand and investigate the collaborative forms of governance in tourism.

On the basis of these arguments, in this study a game theory approach is applied as a model to explain the decisional process of entrepreneurs about forming partnerships in formal tourism networks. The game theory model shows that, in terms of Nash equilibrium, a cooperative optimization can be achieved when participants of a network act in the same sector cooperate to reach common goals (Bruwer, 2003; Hall, Sharples, Cambourne, & Macionis, 2000; Meyer-Cech, 2005; Novelli, Schmitz, & Spencer, 2006). As many cases highlight, that needs of collaborative strategies among organizations involved in a network (Vernon, Essex, Pinder, & Curry, 2005). In dealing with this issue, the chapter presents an empirical solution that arises from the case of Business Networks in Italy.

BACKGROUND

Tourism Network Relationships

The network paradigm is certainly not a recent concept in economics. It refers to forms of collective actions linking firms or complementary industries, as preconditions for innovation and market competitiveness (Porter, 1998). Collaborative networks stem from agreements among enterprises that intend to share economic benefits (Bucklin & Sengupta, 1989; Goldhar & Lei, 1991; Ohmae, 1989; Thorelli, 1986). Tremblay (1998) stated that networks play a role in exploiting external economies of scale and scope, and in lowering transaction costs. Furthermore, some literature distinguishes between vertical and horizontal alliances. The former are those between producers and their suppliers or distributors, the latter aim to protect core competencies (Magun, 1996).

Tourism cooperation has been studied using different theories and a great diversity of analyses, such as territorial levels, static and dynamic aspects, including impact on destination developments, evolution and transformation of partnerships (Czernek, 2013). In many countries, the tourism network has become very active in recent years. The term refers to a wide range of cooperative behaviour between organisations linked through economic and social relationships (Hall et al., 2000), which provides increased economic and social benefits for the local community involved (Novelli et al., 2006). Cooperation among local actors represents a way to be more competitive within the market (Bruwer, 2003). In addition, networks allow small actors with few resources to be part of the decision-making process.

As many experiences show, networks provide a wide range of positive externalities, stimulating cooperation and partnerships, which act as drivers of tourism development (Briedenhann & Wickens, 2004; Gibson, Lynch, & Morrison, 2005). Network externalities arise if the benefits enjoyed by anyone depend on the number of actors having a vision of tourism in common (Crouch & Ritchie, 1999). This implies that the more experience in tourism actors within a network have, the more are the benefits (Telfer, 2001). As Vernon et al. (2005) highlighted, collaborative initiatives among local authorities, government agencies, businesses and host communities constitute the principal resource of contemporary tourism.

Tourism networks are increasingly important in shaping tourism planning and development (Dredge, 2006). They may assist in fostering greater cooperation by drawing together public and private sector interests, and maximizing the multiplier effect in the local economy (Clarke, 2005). The formation of a tourism network should not be seen as a spontaneous and simple process, but as a very complex one, determined by close collaboration among various stakeholders (Novelli et al., 2006). It is therefore important to analyse interaction patterns among network members, and to understand the structure of relationships within a tourism network.

Sheehan and Ritchie (2005) provided empirical evidence about the great diversity of stakeholders of a tourism destination, which can be differentiated according to their potential to threaten and cooperate in management strategies. The authors list different private and public stakeholders, based on their levels of salience, arguing that various connected stakeholders in a tourism network can influence the destination management organization (DMO). Therefore, a tourism network is able to build a system of alliances involving environment, institutions, symbols and routines that facilitate the activities of entrepreneurs by giving them access to resources not accessible through individual action alone (Brunori & Rossi, 2000).

As globalization has increased pressure on firms to be competitive, networks offer SMEs the opportunity to operate in an international business environment, mitigating their size disadvantage (Bieger, 2004). By networking and sharing knowledge, small firms are able to internalise assets and competencies that large firms typically internalise through economies of scale (Tayler & McRae-Williams, 2005). Networks provide the firms involved in them with access to knowledge transfer, resources and markets (Inkpen & Tsang, 2005), operating as strategic tools for exchanging and sharing information (Gulati, 1998). Arndt and Sternberg (2000) revealed in their study that small firms are most likely to cooperate with others in their vicinity, highlighting the relevance of spatial proximity and networking. From this perspective, Novelli et al. (2006) emphasised that the creation of tourism business networks can be viewed as innovative opportunities to support tourism SMEs and contribute to local development. Similarly, Tinsley and Lynch (2001) highlighted the importance of networks involving small tourism businesses for destination development.

Network may include firms embedded in a formal structure of business connections or an informal structure of personal relations bounded in a geographical, institutional, or social space (Ireland, Hitt, & Vaidyanath, 2002). In this regard, Vernon et al (2005) affirmed that tourism cooperation involves various stakeholders working interactively on common problems or issues through a formal approach. According to Strobl and Peters (2013), in destination networks informal relationships are important in forming alliances since they are based on trustworthiness. However, the authors underline that informal relationships contribute to the formation of formal partnerships. The latter are related to contractual relationships, which are perceived as safer due to their legal character.

Successful cooperation in tourism is determined by some crucial factors, such as the perception that benefits will affect all partnership members, stakeholder interdependence, the coordination of skills, and the sharing of a strategic plan (Jamal & Getz, 1995). Similarly, Waddock (1989) observed that collaboration between organizations depends on recognition of interdependence, on agreements, and on a belief that benefits will derive from cooperation. According to Reid et al. (2008), effective tourism partnerships require different conditions:

- Decisions should be taken to promote the goals of cooperation.

In addition, based on pre-determined, objective criteria, partners should negotiate their way considering their differences, and results achieved should be evaluated over a long period. Effective cooperation also needs efficient management structures, flexibility and explicit guidelines.

Among determinants of cooperative interactions in tourism, Wang and Fesenmaier (2007) listed

- Competition,
- Economic conditions,
- Technology and problems, which lead potential partners towards shared solutions.

The same authors identify some organizational factors, including leadership, convener attitudes, competence, communication and human and financial resources. Proposing a classification of exogenous and endogenous cooperation determinants, Czernek (2013) emphasised that motives of cooperation are also important and can be similar for private and public organizations (e.g. management of a tourism destination), or can be different between them - e.g. organizing local services (public sector) and generating profits (private sector).

Although the literature has identified a large number of different reasons that can determine cooperation in a tourist region, there are also some country-specific factors that should be considered. They concern, for example, legal determinants and trade agreements arising from foreign and domestic regulations, which can stimulate cooperation through collaborative tourist projects. In these cases, as many experiences show, tourism networks aim at fostering tourism development initiatives established by local actors.

The above discussion suggests that the structure of the network and the manner in which the linkages are formed and maintained can be critical. Therefore, while the heterogeneity of relationships and their specificity provide several opportunities for a network, at the same time these aspects could also pose some problems for management. Constraints that affect networking include low levels of collaboration, lack of leadership, and lack of trust among stakeholders. In particular, in the relationship perspective and according to the network approach, two issues appear critical: how to mobilize the various different counterparts, and how to develop cooperative behaviour and coordination mechanisms in interaction

(Håkansson & Snehota, 1995). As a consequence, opportunistic behaviour and power imbalances with regard to stakeholder functions and roles can destroy the whole atmosphere in terms of relationships, producing negative effects on network structure and goal performance. It is therefore important to produce strategies that recognize the contribution of all tourism actors within a network to the enhancement and optimisation of cooperation.

EVALUATING COLLABORATIVE PARTNERSHIPS: A SEQUENTIAL GAME

Destination development depends on the connection among different actors, as well as on their ability to exert influence in the destination network. Among these actors, entrepreneurs are the main drivers of innovation, and play a key role in destination networks (Koh & Hatten, 2002; Russell & Faulkner, 2004). Given this, the theoretical model proposed in this study concerns the application of a game theory approach to analyse the decisional process of entrepreneurs about forming partnerships in a formal tourism network.

The game theory is the formal study of interacting decision makers, which represents a rational decisional process where several players, with their preferences, information and available strategic actions, make choices that potentially affect the interests of the other players. Game theoretical concepts apply whenever the actions of several agents are interdependent. These agents may be individuals, firms or groups. They are rational hence they play trying to maximize their own utility. It is assumed that the players' rationality is common knowledge, which means that players know their individual payoff and strategies as well as the other players' payoff and strategies. Payoff is the outcome of each player and results from different available options. This is a number, which indicates the utility that each player receives if a particular combination of strategies is chosen. It reflects the desirability of an outcome to players, and therefore represents the preference of players over their action profiles, where an action profile is the list of the possible actions of each player. Payoff can be shown in a table with a cell for each strategy combination. A strategy could be either one of the given possible actions available to each player, which can be chosen simultaneously, as in a strategic form game, or a complete plan of choices, where players take turns choosing actions, as in a sequential game. Strategies and outcomes, relate to each other mathematically, affect the consequences of the strategic decision by all players. A Nash equilibrium is a kind of rational expectations equilibrium, in actions and beliefs. In other words, it is a pair of mixed strategies that maximizes the expected utility of player's choice, given the strategy of the other players (Varian, 1992).

In the following section, a sequential game is used to investigate entrepreneur decision process about the establishment of, and adhesion to, an inter-firm network. The institution of a tourism network can take place due to initiatives promoted by private or public stakeholders, each with different preferences, strategies and contrasting interests. The goal of entrepreneurs is to maximize utility, which means to obtain several competitive advantages in terms of transaction costs, reputation, value, increase of market power, resource transfer, knowledge creation, lower costs of production and profits. Public institutions intend to ameliorate the territorial context.

The game involves the following agents:

- Tourism market oriented entrepreneurs (E^T) and
- Non-tourism market oriented entrepreneurs (E^{nT}).

Specifically, E^T takes into consideration tourist content of production and E^{nT} does not consider it. Different productions imply different management and market strategies and, therefore, these could affect competitive advantages of firms. Although the creation of a network, as well as the coexistence of different agents, presents some problems, it is quite easy to imagine that an organizing and management committee will control the process of selection of the enterprises that enter the network.

Thus, some firms set up the network and some others enter later in the game. The absence of a tourism inter-firm network represents the status quo.

There are three steps in the game.

The First Step: This is the step where the tourism network does not exist yet, and the stakeholders are deciding whether to set up one (a network). The initiative could be organized by both private and public agents together or separately. Their main utilitarian goal is maximising the competitive advantages of a tourism destination, which means tourism development and increased profits.

The Second Step: Other entrepreneurs, who could be tourism and non tourism market oriented, have to decide whether or not to enter the tourism network. They are considered rival with respect to the firms that are already part of the network. The new entry firms prefer to enter the network, and so exploit the advantages in terms of reputation and quality that have already been developed by E^T. The tourism entrepreneurs (E^T) already part of the network consider that entry by the rival (either E^{nT} or E^T) may represent a high cost in terms of their own reputation, (for instance, the rival quality could be lower than the average quality of the firms that established the network). For this reason, they may consider the strategies 'no enter' or 'no cooperate' as the best choices.

The Third Step: Each entrepreneur compares the own utility when the network does not exist with the advantages represented at the end of the second step. The adhesion to the network is confirmed if, and only if, the utility has increased, otherwise entrepreneurs prefer to exit. The model considers tourism (E^T) and non-tourism enterprises (E^{nT}), which can choose between two actions: 'Enter' or 'No Enter'. The payoffs depend on the decisions taken by each firm considering the opponent's decision and are summarized in the following Matrix 1, where:

E^T means tourism enterprises;

E^{nT} means non tourism enterprises;

$A_N^{\ T}$ is the advantage to enter the network for tourism enterprises;

$A_N^{\ nT}$ is the advantage to enter the network for non-tourism enterprises;

$A_0^{\ T}$ represents the advantage to do not enter the network for tourism enterprises, which can prefer to exit in case they are already part of a network;

$A_0^{\ nT}$ represents the advantage to do not enter the network for non-tourism enterprises;

$\varepsilon > 0$ indicates an externality and it can be positive or negative.

As far as the payoff is concerned, it has been assumed that the advantage to enter the network for non-tourism enterprises is greater than that related to the choice 'no enter' ($A_N^{\ nT} > A_0^{\ nT}$) and the advantage to enter the network for tourism enterprises may be less than that related to 'no enter' ($A_N^{\ T} > A_0^{\ T}$). Notations $A_N^{\ nT} + \varepsilon$ and $A_N^{\ T} - \varepsilon$ show the payoff (advantages) both firms obtain when they choose 'enter' the network. The externality (ε) is due to the rival player decision, in game theory terms, to join the network. In other words, non-tourism enterprises consider enter the network as positive thanks to the adhesion of tourism enterprises, while tourism enterprises find this entrance as negative since it can compromise the success of the network due to different quality standards and low reputation of non-tourism enterprises.

Table 1. Matrix 1: the interaction among firms within the game

		E^T	
		Enter	**No Enter**
E^{nT}	**Enter**	$A_N^{nT} + \varepsilon$ $A_N^T - \varepsilon$	A_N^{nT} A_0^T (or exit)
	No Enter	A_0^{nT} A_N^T	A_0^{nT} A_0^T

The game shows two Nash equilibria:

- One in 'Enter - Enter', and
- Another one in 'Enter - No Enter'.

They represent an increase in payoff with respect to the status quo situation for all individuals. In the first case, 'Enter' the network is a dominant strategy for non-tourism enterprises (E^{nT}), where dominant strategy is considered as better than other strategies, no matter what other players might do. Then, non-tourism enterprises (E^{nT}) benefit from an increase in payoff with respect to the status quo situation, since to enter the network yields a major advantage that does not enter ($A_N^{nT} > A_0^{nT}$). As for tourism enterprises (E^T), it needs to discern if

1. The strategy 'enter' is dominant for them; or
2. No dominant strategy exists for them.

In the case (1) 'enter' is a dominant strategy and be positive if entering the network ($A_N^T - \varepsilon$) represents a value greater than or equal to that of not adhere to it (A_0^T)

$$A_N^T - \varepsilon \geq A_0^T \quad (1)$$

Otherwise, in the case (2) 'no enter' the network is a value bigger than 'enter' for tourism enterprises. Therefore, do not enter the network is preferable, since it represents an advantage more than join the network. This occurs when the E^T warns the entry of rival as a threat.

$$A_0^T > A_N^T - \varepsilon. \quad (2)$$

Further consideration can be proposed about the Nash equilibrium 'Enter-Enter'. Although, non-tourism market oriented firms find the Nash equilibrium situation preferable with respect to the status quo, tourism market oriented firms could find the Nash equilibrium worse than the status quo. Hence, non-tourism firms would like to enter the network to obtain advantages such as reputation, quality, visibility, and, as a consequence, greater profits. While tourism firms suspect that the entrance of non-tourism firms can threaten the standards of the network.

Two different interpretations (1A) and (1B) arise from this condition. In (1A), if non-tourism enterprises are more than tourism ones ($E^{nT} > E^{T}$), thus the number of non-tourism entrepreneurs benefiting by entrance to the network is larger than the number in the status quo. Then, game's equilibrium occurs in the following sequence:

- Tourist firms enhance the network and only the non-tourism ones want to enter the network.
- In (1B), on the contrary, if non-tourism enterprises are less than tourism ones ($E^{nT} < E^{T}$), then tourism enterprises prefer the status quo situation where no network does exist, because the presence of non-tourism firms can affect in a negative manner the network management.
- At the same time, non-tourism firms prefer the Nash-equilibrium ('Enter - Enter'), since the establishment of a network is preferable for them compared to the status quo.

In case of the Nash equilibrium 'Enter – No Enter', given the building of the network, while non-tourism firms decide to enter the network, the tourism firms do not enter.

This condition determines two possible situations (2A) and (2B). In (2A), if non-tourism enterprises are more than tourism ones ($E^{nT} > E^{T}$), the game's equilibrium occurs in the following sequence: tourist firms enhance the network and a number of non-tourism enterprises equal to $E^{nT}/E > ½$ adheres to the network. In (2B), if non-tourism enterprises are less than tourism ones ($E^{nT} < E^{T}$), the perfect equilibrium is represented by the status quo situation.

Another interpretation of the game is represented in Table 2, where players can choose between two actions:

- 'Enter' or 'No Enter', and
- 'Cooperate' or 'No Cooperate'.

In the game presented in Matrix 2, it is possible to observe a Nash equilibrium in the situation 'Enter-Cooperate'. Specifically, when non-tourism enterprises (E^{nT}) enter, if the incumbent chooses to cooperate, both obtain advantages (1, 1). In the case 'No Enter-No Cooperate', when non-tourism enterprises stay out and tourism ones (E^{T}) choose not to cooperate, even though the latter obtains a payoff larger than before (0, 2), it is not a perfect sub-game Nash equilibrium. This would show that there is not efficiency, according to the reasons of the network.

The model shows that there is efficiency in terms of Nash equilibrium when enterprises that decide to join to the network cooperate. More precisely, when players decide to cooperate there will be positive externalities originating from the cooperation between tourism and non-tourism market oriented firms. Furthermore, the application of the above model suggests some important considerations, which

Table 2. Matrix 2: the interaction among firms within the game

		E^{T}	
		Cooperate	**No Cooperate**
E^{nT}	**Enter**	1 1	-1 -1
	No Enter	2 0	2 0

confirm results of other studies. First, an inter-firm tourism network, to be successful, requires that entrepreneurs involved have a common tourism goal. This could confirm what Crouch and Ritchie (1999) argued, affirming that the competitive advantage of a destination is more related to a common vision of tourism among stakeholders than to the endowment of tourist resources. This implies the cooperation between tourist and non-tourist market oriented firms, from which will be derived positive externalities for the network. Secondly, it is important that all members assure and maintain the standards (e.g., quality and products) that characterize the reputation of the tourism network. This goal can be achieved by establishing formal rules and contractual relationships, binding all the members of a tourism business network. In this regard, it could be useful to consider the study of Strobl and Peters (2013), which emphasized that in a tourism network, formal relationships are associated with liability and safety due to their legal credibility. Therefore, the respect of formal rules should encourage cooperative tourism actions, dissuading entrance within the network of entrepreneurs who could have different tourism visions and goals (Figure 1).

Finally, it needs to consider the case of zero-sum game in which firm's advantage loss is equal to the other firm's advantage gain (Tucker & Luce, 1959). In this case, networking determines a change in the attitude of entrepreneurs in the long-run moving from a situation where one against the other tactics leading to a zero-sum game to a cooperative optimization.

AN EMPIRICAL SOLUTION: BUSINESS NETWORKS IN ITALY

Although different factors can determine cooperative tourism networks, some country-specific aspects should be considered. An empirical solution arises from the Italian experience, where law and domestic regulations foster tourism partnerships and collaborative actions among actors. More specifically, this

Figure 1. Factors inducing formal tourism network

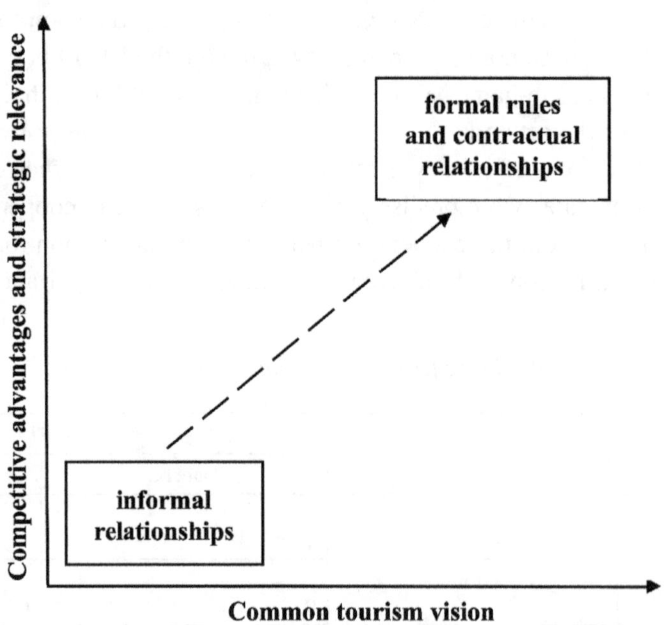

is the case of Business Networks, which represents an innovation in law, involving in the process civil, business and fiscal branches of Italian Law.

The common and inspiring keywords behind the creation of Business Networks are:

- Improvement,
- Competitiveness,
- Effectiveness in global markets, and
- Innovation.

There are three main types of networks that have been created since the beginning.

1. The first kind of business network is an enterprises aggregation focused on the exchange of information and know-how and/or the development of a product or service (statistics on flow of tourism, customer preferences, booking systems, and the like) among the members of the group.
2. The second type - *Contratto di Rete d'Imprese* - is a model of partnership based on a binding contract among enterprises – called *Retisti* - which take actions on certain shared goals. The companies involved maintain their individualities and brands, working together as if the network was a limited company under a common name, guided by an external manager or Management Body and having in common a bank fund (Figure 2).

This kind of network can be vertical or horizontal. It is vertical when companies belonging to different business sectors are all taking part together in a Business Network. For example, in the tourism sector, a vertical network would involve hotels, B&Bs, tour operators, web developers, taxi/shuttle companies, tourist guides, and others. By contrast, the network is horizontal, if it is composed of companies of the same sector. In this case, it would be possible to find a network of hotels and/or hospitality structures and another of tourist guides only. This could also be a super specialized network, if the possibility to join is restricted to a segment of a sector, for example a network of "mountain charm hotels".

Figure 2. Model of network Contratto di Rete d'Imprese

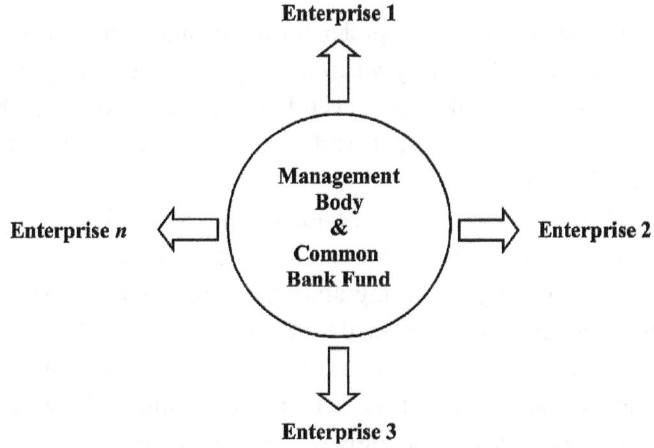

3. The third kind of network - *Rete Soggetto* - has the same structure as the previous one, and also the same options, as the Networks can be vertical or horizontal. However, under Italian fiscal law, the *Rete Soggetto* is considered as a new company itself and not as an aggregation of companies. This means, for example, that to each *Rete Soggetto* is assigned a VAT number, and not just a UTR number as in the previous case.

Since the second typology of formal business networks - *Contratto di Rete d'Imprese* - is the most common of the three, it is also the one that will be further analysed here. To do so, it is important to mention Law 2009, n. 99, because this law has introduced several changes to the effectiveness of business networks.

On 1st July 2014, in Italy, there were 7870 Reti Contratto. The legislator has drawn the contract to enhance the innovative capacity and competitiveness of firms in the local and global markets. This implies the possibility of reaching goals and taking advantage of different means that are otherwise precluded to a single company, such as access to specific financial aid measures, simplified administrative procedures, public call for bids, and cost sharing for strategic activities (e.g. marketing actions, branding).

Retisti also have the possibility to:

- Share customer data, general and specific information,
- Develop new business ideas/opportunities,
- Share suppliers, workers, and consultants,
- Better pricing,
- Sign conventions (for example, with insurance companies or suppliers such as telephone companies),
- Make joint purchases,
- Request export/internationalization support, and
- Carry out lobbying.

In the tourism sector, companies also have the possibility to offer an all-inclusive and modular package.

It is interesting to analyse the reasons why Italian entrepreneurs are willing to bind themselves to a formal business contract while they could, most probably, obtain the same results through an informal network with a simple handshake.

The reasons that encourage enterprises to prefer formal business networks are several. First of all, there is an Italian cultural need for formality when dealing with the Public Administration and other official bodies and entities, such as banks. This need has been well expressed by the Romans, indeed the famous Latin proverb says: Scripta manent, verba volant, suggesting that only written documents are always conclusive in public matters.

Another reason behind formal contracts is the importance given by the legislator only to formal business networks. For example, facilitation and tax breaks are only provided for formal aggregations, and Italian Labour Law only allows job sharing among companies of a formal network. In addition, nowadays, many invitations to tender are designed for Contratti di Rete. By the way, when an invitation to tender is not specifically designed for formal Business Networks, the administrative process always includes Networks among the possible tenderers. In this case, Business Networks would take part to the tender following the same regulation as cooperatives and temporary registered aggregations.

At the same time, regions and other local authorities publish announcements reserved to formal networks, to obtain advantages such as free exposition stands at international events and fairs. Usually this kind of aid - and free consultancy by local governmental internationalization departments - focuses on business networks composed of at least 30 companies.

As suggested before, lobbying is also a very interesting point for formal networks. In Italy, this is even more evident because of the typical structure of the entrepreneurial sector, composed of a majority of micro and small companies. The formality of the aggregation gives companies more power when sitting at the same table with the planning administration, and when trying to obtain specific improvements, or when fighting competition from bigger companies or of similar business categories. At the same time, a written and binding contract ensures the same quality standards and level of offer, by binding each company not only to the contract but also to a common vision, mission and disciplinary framework. Furthermore, if the network grows in number of members, each one of the new companies has to be accepted by the old ones. The members are thus also able to control the standards, vision, mission, available instruments and level of development of the newcomers before accepting them in the network.

From a legal perspective, the regulations laid down by the Italian legislator provide a scheme with the essential content of a contract, but companies are still free to define specific clauses. The contract has to have, by law:

- A specific form,
- A common purpose,
- A programme, and
- A management body (external to the group of companies, so as to be *super partes*).

The number of members is theoretically unlimited, and only the contract can set the criteria for new admissions and those to exit the contract.

Last but not least, companies which are part of a network contract are given access to specific financial aid measures, simplified administrative procedures, and simplified procedures for access to finance,

- Bank packages focused on business network contracts,
- European union funds,
- Italian funds, and
- Allowances.

In the 2014 Economic and Financial Document (DEF), 200 million Euros were allocated to a fund created for the registered Business Network subsidized regime (the Business Network Fund). The allowance consisted in a tax-deferral of the share of profit for the year designated by the Common Fund.

All the above mentioned social, legal and fiscal aspects made the creation of formal, more than informal, networks very interesting to Italian companies, compared to past practices. The enterprises' convenience is even more evident in the tourism sector, where networks can work as actors to develop the area in which they are established.

FUTURE RESEARCH DIRECTIONS

This chapter has discussed the importance of networking to the enhancement of the competitive advantage of destinations, and emphasized that successful tourism networks require a high level of cohesion among different stakeholders. A tourism network grows if a critical mass of organizations and firms cooperate to create a homogeneous set of competencies, strategies, products, and branding. However, as the chapter has indicated, networks are not always naturally established and may hence need to be fostered.

As for tourism SMEs, it has been shown that networking challenges include appropriate integration of firms, and entrepreneurial management. Competitive advantage based on networking sustains the idea of the importance of cooperation among firms. Their relationships enhances tourism cooperative optimization as well as for exchanging information and sharing knowledge. The implications are quite evident if we consider the case of Business Networks in Italy.

This study suggests the need for further research into other possible interactions of specific conditions that determine tourism network formation, such as the role of perceived market opportunities, or the impact of entrepreneurial policies and strategies. Further research into systemic factors not included in this research is also recommended, such as the role of financial capital that may help to develop a successful network. Additional research into the composition, use, and developmental processes of tourism networks in different countries is advisable, in order to identify the specific assets that determine cooperative optimization at regional and local level.

CONCLUSION

Cooperative actions and relationships among different stakeholders of tourism networks have become important factors in tourism, proving to be useful in the planning and management of destinations. As many scholars have highlighted, collaborations within a tourism network avoid the costs of resolving adversarial conflicts among stakeholders, they give stakeholders greater power in the decision-making processes, and improve the coordination of tourism policies and related actions.

In a business scenario, then, the formation of inter-firm networks and alliances can be seen as a great opportunity for collaborations, successful business operations and tourism management. In addition, alliances

- Promote the quality of services,
- Enhance the visibility of partners involved, and
- Allow cross-marketing activities.

Inter-firm tourism networks represent models of cooperative relationship between businesses (Vellas & Becherel, 1999), providing them with new opportunities to operate locally and in a global market (Novelli et al., 2006). They create value and increase innovation (Go & Pine, 1995).

When tourism entrepreneurs intend to promote an inter-firm network there could be some problems concerning the differences between those ensuring high standards - for instance, quality and reputation - and those that do not ensure the same standard. The model applied in this study shows that there could be different Nash equilibria in relation to the decision of entrepreneurs to adhere or not to a tourism network. However, when a network is created, the best situation is when tourist and non-tourist market oriented firms cooperate, originating positive externalities.

This chapter has shown that the cooperative optimization of a tourism network should be achieved when entrepreneurs have the same business goals and a common tourism vision, and cooperate assuring homogeneous standards in terms of quality. Nevertheless, a collaborative perspective between the tourism actors involved (Vernon et al., 2005) is necessary, based on trust and ethics. Hence, cooperative activities add value to the relationships within a network. In the short run, strong relationships can reinforce entrepreneurial cooperation, encouraging further business development in the long run.

If the institution and management of tourism networks needs partnerships among entrepreneurs and local stakeholders, it is important to consider that formal and informal relationships between these actors have important implications for a destination's development (Dredge, 2006). However, as Strobl and Peters (2013) showed, informal relationships are used in preparation of contractual relationships, which are perceived as safer due to their legal character. On this latter topic, this study presented the case of Business Networks in Italy, which represents an innovation in law, showing that Italian entrepreneurs are willing to bind themselves to formal business networks.

REFERENCES

Arndt, O., & Sternberg, R. (2000). Do manufacturing firms profit from intraregional innovation linkages? An empirical based answer. *European Planning Studies*, *8*(4), 465–485. doi:10.1080/713666423

Bickerdyke, I. (1996). Australia: The evolving structure and strategies of business networks. In OECD (Ed.), *Networks of enterprises and local development: Competing and co-operating in local productive systems* (pp. 203–216). Paris: OECD.

Bieger, T. (2004). SMEs and cooperation. In P. Keller & T. Bieger (Eds.), *The future of small and medium sized enterprises in tourism* (pp. 141–150). St. Gallen, Switzerland: AIEST.

Briedenhann, J., & Wickens, E. (2004). Tourism routes as a tool for the economic development of rural areas - vibrant hope or impossible dream? *Tourism Management*, *25*(1), 71–79. doi:10.1016/S0261-5177(03)00063-3

Brunori, G., & Rossi, A. (2000). Collective action: Some insights from wine routes in Tuscany. *Sociologia Ruralis*, *40*(4), 409–423. doi:10.1111/1467-9523.00157

Bruwer, J. (2003). South African wine routes: Some perspectives on the wine tourism industrys structural dimensions and wine tourism product. *Tourism Management*, *24*(4), 423–435. doi:10.1016/S0261-5177(02)00105-X

Bucklin, L. P., & Sengupta, S. (1989). Organizing successful co-marketing alliances. *Journal of Marketing*, *57*(2), 32–46. doi:10.2307/1252025

Clarke, J. (2005). Effective marketing for rural tourism. In D. Hall, I. Kirkpatrick, & M. Mitchell (Eds.), *Rural tourism and sustainable business* (pp. 87–102). Clevedon, UK: Channel View.

Crouch, G. I., & Ritchie, J. R. B. (1999). Tourism, competitiveness, and societal prosperity. *Journal of Business Research*, *44*(3), 137–152. doi:10.1016/S0148-2963(97)00196-3

Czernek, K. (2013). Determinants of cooperation in a tourist region. *Annals of Tourism Research, 40*(1), 83–104. doi:10.1016/j.annals.2012.09.003

Dollinger, M. J. (1990). The evolution of collective strategies in fragmented industries. *Academy of Management Review, 15*(2), 266–285.

Dredge, D. (2006). Policy networks and the local organisation of tourism. *Tourism Management, 27*(2), 269–280. doi:10.1016/j.tourman.2004.10.003

Edwards, B., Goodwin, M., Pemberton, S., & Woods, M. (2000). *Partnership working in rural regeneration: Governance and empowerment?* Bristol, UK: The Policy Press and the Joseph Rowntree Foundation.

Gibson, L., Lynch, P. A., & Morrison, A. (2005). The local destination tourism network: Development issues. *Tourism and Hospitality Planning & Development, 2*(2), 87–99. doi:10.1080/14790530500171708

Go, F., & Pine, R. (1995). *Globalization strategy in the hotel industry*. New York, NY: Routledge.

Goldhar, J. D., & Lei, D. (1991). The shape of twenty-first century global manufacturing. *The Journal of Business Strategy, 12*(2), 37–41. doi:10.1108/eb039401 PMID:10111375

Gulati, R. (1998). Alliances and networks. *Strategic Management Journal, 19*(4), 293–317. doi:10.1002/(SICI)1097-0266(199804)19:4<293::AID-SMJ982>3.0.CO;2-M

Håkansson, H., & Snehota, I. (Eds.). (1995). *Developing relationships in business networks*. London: Routledge.

Hall, C. M. (2008). *Tourism planning: Policies, processes and relationship*. Harlow, UK: Pearson-Prentice Hall.

Hall, C. M., Sharples, L., Cambourne, B., & Macionis, N. (Eds.). (2000). *Wine tourism around the world: Development, management and markets*. Oxford, UK: Butterworth-Heinemann.

Inkpen, A. C., & Tsang, E. W. K. (2005). Social capital, networks, and knowledge transfer. *Academy of Management Review, 30*(1), 146–165. doi:10.5465/AMR.2005.15281445

Ireland, R. D., Hitt, M. A., & Vaidyanath, D. (2002). Alliance management as a source of competitive advantage. *Journal of Management, 28*(3), 413–446. doi:10.1177/014920630202800308

Jamal, T., & Jamrozy, U. (2006). Collaborative networks and partnerships for integrated destination management. In D. Buhalis & C. Costa (Eds.), *Tourism management dynamics: Trends, management and tools* (pp. 164–172). Oxford, UK: Elsevier Butterworth-Heinemann. doi:10.1016/B978-0-7506-6378-6.50028-7

Jamal, T. B., & Getz, D. (1995). Collaboration theory and community tourism planning. *Annals of Tourism Research, 22*(1), 186–204. doi:10.1016/0160-7383(94)00067-3

Koh, K. Y., & Hatten, T. S. (2002). The tourism entrepreneur. The overlooked player in tourism development studies. *International Journal of Hospitality & Tourism Administration, 3*(1), 21–48. doi:10.1300/J149v03n01_02

Magun, S. (1996). *The development of strategic alliances in Canadian industries: A micro-analysis.* Working paper No. 13. Industry Canada.

McIntosh, R. W., Goeldner, C. R., & Ritchie, J. R. B. (1995). *Tourism, principles, practices, philosophies.* New York, NY: Wiley & Sons.

Merinero-Rodríguez, R., & Pulido-Fernández, J. I. (2016). Analysing relationships in tourism: A review. *Tourism Management, 54*, 122–135. doi:10.1016/j.tourman.2015.10.010

Meyer-Cech, K. (2005). Regional cooperation in rural theme trails. In D. Hall, I. Kirkpatrick, & M. Mitchell (Eds.), *Rural tourism and sustainable business* (pp. 137–148). Clevedon, UK: Channel View.

Novelli, M., Schmitz, B., & Spencer, T. (2006). Networks, clusters and innovation in tourism: A UK experience. *Tourism Management, 27*(6), 1141–1152. doi:10.1016/j.tourman.2005.11.011

Ohmae, K. (1989). The global logic of strategic alliances. *Harvard Business Review, 67*(2), 143–154.

Pavlovich, K. (2003). The evolution and transformation of a tourism destination network: The Waitomo Caves, New Zealand. *Tourism Management, 24*(2), 203–216. doi:10.1016/S0261-5177(02)00056-0

Porter, M. (1998), Clusters and the new economics of competition. *Harvard Business Review, 76*(6), 77-90.

Reid, L. J., Smith, S. L. J., & McCloskey, R. (2008). The effectiveness of regional marketing alliances: A case study of the Atlantic Canada Tourism Partnership 2000–2006. *Tourism Management, 29*(3), 581–593. doi:10.1016/j.tourman.2007.05.005

Russell, R., & Faulkner, B. (2004). Entrepreneurship, chaos and the tourism area lifecycle. *Annals of Tourism Research, 31*(3), 556–579. doi:10.1016/j.annals.2004.01.008

Sheehan, L. R., & Ritchie, J. R. B. (2005). Destination stakeholders. Exploring identity and salience. *Annals of Tourism Research, 32*(3), 711–734. doi:10.1016/j.annals.2004.10.013

Strobl, A., & Peters, M. (2013). Entrepreneurial reputation in destination networks. *Annals of Tourism Research, 40*, 59–82. doi:10.1016/j.annals.2012.08.005

Tayler, P., & McRae-Williams, P. (2005). *Internal versus external economies of scale.* Working Paper, Centre for Regional Innovation & Competitiveness.

Telfer, D. J. (2001). Strategic alliances along the Niagara Wine Route. *Tourism Management, 22*(1), 21–30. doi:10.1016/S0261-5177(00)00033-9

Thorelli, H. B. (1986). Networks: Between markets and hierarchies. *Strategic Management Journal, 7*(1), 37–51. doi:10.1002/smj.4250070105

Tinsley, R., & Lynch, P. (2001). Small tourism business networks and destination development. *International Journal of Hospitality Management, 20*(4), 367–378. doi:10.1016/S0278-4319(01)00024-X

Tremblay, P. (1998). The economic organization of tourism. *Annals of Tourism Research, 25*(4), 837–859. doi:10.1016/S0160-7383(98)00028-0

Tucker, A. W., & Luce, R. D. (Eds.). (1959). *Contributions to the theory of games* (Vol. 4). Princeton: Princeton University Press.

Varian, H. R. (1992). *Microeconomic Analysis* (3rd ed.). New York, NY: Norton & C.

Vellas, F., & Becherel, L. (1999). *The international marketing of travel and tourism: A strategic approach*. London: Macmillan. doi:10.1007/978-1-349-27486-4

Vernon, J., Essex, S., Pinder, D., & Curry, K. (2005). Collaborative policymaking: Local sustainable projects. *Annals of Tourism Research*, *32*(2), 325–345. doi:10.1016/j.annals.2004.06.005

Waddock, S. A. (1989). Understanding social partnerships: An evolutionary model of partnership organizations. *Administration & Society*, *21*(1), 78–100. doi:10.1177/009539978902100105

Wang, Y. (2011). Destination marketing and management: Scope, definition and structures. In Y. Wang & A. Pizam (Eds.), *Destination marketing and management: Theories and applications* (pp. 1–20). Wallingford, UK: CAB International. doi:10.1079/9781845937621.0001

Wang, Y., & Fesenmaier, D. R. (2007). Collaborative destination marketing: A case study of Elkhart county, Indiana. *Tourism Management*, *28*(3), 863–875. doi:10.1016/j.tourman.2006.02.007

Wood, D. J., & Gray, B. (1991). Toward a comprehensive theory of collaboration. *The Journal of Applied Behavioral Science*, *27*(2), 139–162. doi:10.1177/0021886391272001

KEY TERMS AND DEFINITIONS

Business Network: A form of voluntary alliance among entrepreneurs who want to reach same goals with common efforts.

***Contratto di Rete d'Imprese*:** A model of partnership based on a binding contract among enterprises taking actions on certain shared goals, which represents an innovation in Italian Law.

Externality: The cost or benefit that affects a party who did not choose to incur that cost or benefit. An externality could be positive or negative.

Game Theory: A model of optimality taking into account benefits, costs and the interaction among agents. It attempts to observe the relationship between players in a model predicting their optimal decision.

Nash Equilibrium: A fundamental concept in the game theory and it is used to predict the outcomes of a strategic interaction in the social sciences.

***Retisti*:** Companies involved in the type of business network called *Contratto di Rete d'Imprese.*

Tourism Network: A form of partnership involving different interacting elements, which include tourists, enterprises, institutions and the host community.

Chapter 17
Optimisation of Wine and Spirit Inventory Assets in Fine Dining Restaurants

J. E. (Joe) Barth
University of Guelph, Canada

ABSTRACT

Research on profit optimisation in the travel sector of hospitality industry has been dominated by the development of effective revenue management techniques to help managers in situations where demand is variable, variable costs are low, assets are fixed and perishable. These techniques have been extended to the restaurant sector, with recognition that variable costs are a larger proportion of total costs and not fixed. In fine dining restaurants substantive wine and spirit inventories present operators with the challenge of how to optimise the sales per dollar of inventory. Theoretical foundations of yield management, retail inventory optimisation and menu engineering are reviewed with direct application to the development of the WINSPID model of wine list and wine inventory optimisation. Data from a successful fine dining restaurant are used to illustrate how the model can be used to improve the sales efficiency of the wine list and inventory. Opportunities to extend the model to spirit inventories are proposed.

INTRODUCTION

Archeological evidence shows that wine taken with food has been part of cultural history for more than 8,000 years. There are many motivations for the consumption of alcoholic beverages, broadly categorized as

- Hedonistic (social, religious, celebratory, etc.), and
- Utilitarian (lubrication, dependency, refreshment, health, etc.) (Charters, 2006).

Wine, beer and spirits were consumed as a safe alternative to water, which was often contaminated with e. coli, typhus, cholera and other disease causing pathogens. Beer was served to children in English public schools in the mid-19th century for health reasons (Barr, 1995). Alcoholic beverages were

DOI: 10.4018/978-1-5225-1054-3.ch017

provided to labourers, served to troops and sailors on campaigns as a healthy alternative to other drinks and to improve morale. Wine, beer and spirits have been sold in inns and restaurants for profit as long as there have been travellers. Regardless of the motivations to consume or serve alcoholic beverages, most people drink because they enjoy it (Heath, 2000).

Wine plays a fundamental role in the enjoyment of the restaurant dining experience (Yuksel & Yuksel, 2002), consequently many types of restaurants offer wine selections to their customers. Berenguer et al. (2009) found that the length of the wine list differentiated top-quality fine dining establishments from those focussed simply on providing excellent food. The style of the wine menu was also a significant factor in the categorization. In many instances, owners of restaurants express their own passion for wine by having wine lists that go well beyond what their customers require. National and regional awards recognise restaurants with exemplary wine menus, judged both on style and selection.

During the global recession of 2009, many fine dining restaurants in North America were challenged due to steep reductions in demand. In particular, wine sales were affected as customers who were buying premium selections switched to lower priced options. In an industry-commissioned study of 300 restaurant industry professionals, Bell (2009) reported that 60 percent of restaurant managers observed a marked decrease in wine sales after the recession and 61 percent felt that customers purchased less expensive wines than in 2008. The study goes on to report that customers were more likely to purchase wine by the glass than previously and 75 percent of the respondents were making changes to their wine lists. Fine dining restaurants often hold large inventories of wine, and the recession resulted in wine lists with many selections for which there is little or no demand. Restaurant managers were motivated to optimise their investment in wine inventory assets by revisiting the wine list, inventory turnover and pricing, however there was little or no guidance available to help them.

Much of the work on the optimisation of asset utility in the hospitality industry has been dominated by Yield or Revenue Management techniques applied to perishable fixed assets such as hotel rooms, airline seats, rental cars, etc. These assets are characteristic of travel industries, but other parts of the hospitality industry such as restaurants have adopted some of these strategies to better utilise fixed seating capacity and revenue per seat hour (Kimes, 1999). Despite these extensions to the optimisation of fixed assets, restaurant inventory assets have been largely ignored by researchers. Restaurants have significant amounts of variable assets in the form of inventory, and the management of these assets is equally important to the overall optimisation of return on assets in the business.

It is important not to confuse inventory optimisation with the management of inventory assets. Inventory optimisation deals with finding the optimum number and quantity of items in stock, and not with the management of the material flows in and out of the stockroom. In the hospitality field, substantial work has been done in the control of inventory (Green & Weaver 2008), however very little has been done regarding inventory turnover and the productivity of capital invested in inventory. Wine and spirit inventories in particular can be large in comparison to the sales they generate for fine dining restaurants, and thus pose an important optimisation problem.

Optimising inventories of wine is more complicated than simply reducing the investment because wine lists serve both as a menu and a statement of a fine dining restaurant's quality and commitment to gastronomy at the highest levels (Gil et al., 2009). A good analogy is a hotel which has a swimming pool. Potential guests will choose a hotel with a pool (even though they will seldom use this amenity) because it is indicative of a higher class of property.

Academic research to assist fine dining managers with the construction of wine lists based on pairing with food (for example Harrington, 2007), wine menu characteristics (Yang & Lynn, 2009) and as a differentiation strategy (Berenguer et al., 2008) is available. The WINSPID (Barth, 2011) model was developed to assist restaurant managers with the task of jointly optimising the wine list, pricing and the financial return on the capital invested in wine inventory. Wine Sales per Inventory Dollar (with apologies to other currencies) is an easily calculated yield statistic, which informs managers of the efficacy of their wine list, inventory and pricing decisions.

In this chapter we will explore several optimisation methods used in travel industries, retail management and menu engineering and show how the WINSPID model is related to these concepts. After describing the model development in detail, data from a restaurant will be used to demonstrate how WINSPID can be used to optimise wine lists and wine inventories in fine dining restaurants.

BACKGROUND

Optimisation of Capacity Utilisation of Fixed Assets

Room nights, seats on an aircraft, car rentals, etc. are well known examples of perishable fixed assets. Perishable assets refer to units of capacity that if unsold are lost revenue opportunities that can never be recovered. Fixed assets refer to units of capacity that do not change in the short to medium term. In circumstances when demand is uncertain and variable, capacity is fixed and perishable, marginal costs are low and advance selling is possible, a special set of challenges and opportunities are created. These special circumstances exist in the travel industry, as well as some other types of businesses such as

- Broadcast advertising sales,
- Banquets,
- Conferences, and
- Fresh food brokerage.

These special challenges and opportunities have lead to the development of optimising techniques which provide enhanced profit for businesses with these operating characteristics. Yield Management refers to profit optimisation obtained from the pricing and capacity utilization of perishable fixed assets. The term Revenue Management is often used to describe similar methods, and is perhaps a more accurate description in that it is primarily based on managing revenue, rather than gross margin or operating profit. See Chiang, Chen and Xu (2007) for a detailed review of the development of these methods.

Given a downward sloping demand curve (more people will buy if the price is reduced) a trade-off develops between the desire to obtain the highest average price and sell all available units of capacity (seats, rooms, tee-off times, etc.). Market segmentation based on time or price sensitivity enables different prices to be charged for the same product depending on the customers' segmentation. The manager has two problems. The first is to know which combination of prices and volume (occupancy) will optimize revenue. The second is to know how many units of capacity (rooms, seats, cars, etc.) need to be held back for customers who will pay higher prices due to time sensitivity. Not having seats available at high prices due to over-selling at lower prices is suboptimal.

Yield management provides solutions to this problem using capacity management and pricing tools. Yield management can be accomplished using computer software programs, but also with heuristic or "rule of thumb" approaches. Regardless of the type of yield management methodology used, available

Table 1. Examples of yield statistic expansions

Description	Price Efficiency ×	Volume Efficiency =	Yield Statistic
Return on Assets	Net Income / ~~Total Sales~~ Profitability	~~Total Sales~~ / Total Assets Asset Turnover	Net Income / Total Assets ROI
Revenue per available room	Revenue / ~~# of rooms sold~~ ADR	~~# of rooms sold~~ / # of rooms available Occupancy	Revenue / # of rooms available RevPar
Revenue per seat hour (Kimes, 1999)	Revenue / ~~Seat hours Sold~~ $ per seat hour	~~Seats hours Sold~~ / Seat hours Available Capacity Utilization	Revenue / Seat Hours Available RevPash
Gross Margin return on Inventory	Gross Margin / ~~CGS~~ Margin Ratio	~~CGS~~ / Inventory Inventory Turnover Ratio	Gross Margin / Inventory GMROI

ADR = Average Daily Rate,
Gross Margin = Sales – Variable cost,
CGS = Cost of Goods sold,
Strike through items: cancel when the multiplication is completed.

capacity is allocated to different consumer market segments. Prices are primarily based on demand level of market segments at various times and to a much lesser degree on the attributes of the product being sold.

Every yield management strategy consists of charging different prices for the same (or similar) product or service according to demand and capacity available. The results obtained from these different prices and corresponding changes in capacity utilization are tracked with a yield statistic (for example REVPAR in hotels). Yield statistics are constructed with a rate efficiency ratio (average price) multiplied by a volume efficiency ratio (such as capacity utilization). Table 1 provides examples of Yield statistic expansions that have been developed for various industries. These yield statistics are quick and easy ways to determine how well a manager's price and capacity decisions are working.

Yield statistics provide the manager with a simple, immediate indication of the efficacy of combined of price and volume efficiency decisions. If the value of the yield statistic increases, yield is being optimised, if it decreases, the decisions are sub-optimal.

When variable costs are small (such as airline seats or hotel rooms) yield statistics such as RevPar inform managers about the efficacy of their price and capacity decisions reliably. However, when variable costs are a significant proportion of sales (such as the relatively lower margins obtained in restaurants, or in the case of some budget hotels), yield statistics can be misleading.

A simple example is provided to illustrate how variable costs affect the reliability of yield statistics with regards to management decision making about price and capacity. The following operating details apply to the analysis in Table 2:

A Budget Hotel has 100 rooms. Management spends $14 per room night servicing rooms that have been occupied, and the hotel has fixed costs of $4,000 per day on a year-round basis.

The hotel experiences three different situations defined by the average rate and occupancy. The manager calculates RevPar to help him/her decide which situation is the most profitable for the hotel.

Table 2. Budget hotel RevPar summary

Daily Data	Situation 1	Situation 2	Situation 3
Total Revenue	**$6,000**	**6,000**	**$6,300**
Number of rooms Sold	100	75	100
ADR (Average Daily Rate)	$60	$80	$63
Cap. Util. (Occupancy)	100%	75%	100%
REVPAR (ADR X Occ.)	$60	$60	$63
Proforma Profit and Loss			
Variable Costs ($14/room sold)	1,400	1,050	1,400
Gross Operating Income	4,600	4,950	4,900
Fixed Costs	4,000	4,000	4,000
Net Operating Income	600	950	900

In situation 1, the hotel enjoys 100% occupancy at a rate of $60 per night. Revenue is $6,000, it costs $1400 to clean 100 rooms, $4,000 is allocated to fixed costs leaving a net operating income of $600 for the day. The RevPar yield statistic (room revenue divided by the number of rooms available) is $60 per available room.

In situation 2, management obtains an average room rate of $80 per night, however only 75 rooms have been sold. Revenue is $6,000 (unchanged from situation 1) however the cost to service occupied rooms is only $1,050 (75 rooms X $14 each), and after the fixed costs of $4,000 have been applied, the net operating income is $950. The RevPar statistic for situation 2 is also $60 per available room.

In situation 3, management has achieved an average room rate of $63 per night and has sold all 100 rooms. In this case, the total revenue is $6,300, $1,400 is spent on servicing occupied rooms and after the $4,000 fixed costs have been applied the net operating income is $900 for the day. The RevPar statistic for situation 3 is $63 per available room.

If the manager relies solely on the RevPar yield statistic for guidance, s/he should be indifferent between situations 1 and 2. RevPar ($60) is the same for both. However, savings obtained from cleaning 25 less rooms ($14 X 25) result in $350 in increased operating income.

Similarly, a manager who relies on RevPar to determine the best situation would choose Situation 3 which has RevPar = $63. While the proforma operating income shows the profit is higher than situation 1, it is less than that of Situation 2 with RevPar = $60.

When variable costs take on higher proportions of revenue, revenue-based yield statistics become less useful as a tool to inform managers of the effectiveness of their price and capacity decisions.

Yield statistics can be a useful method for managers to know if their price and capacity decisions are optimising profits. However, as variable costs become larger (nearly 25% in the example), statistics can mislead managers, and must be used with care and understanding. New yield statistics, such as GOP-PAR (gross operating profit per available room), have been developed to use gross operating profit as the numerator, and avoid confounding due to variable cost. Yield statistics can be particularly misleading when used in restaurants, where prime costs (variable labour and product costs) are 65% or more.

The research on yield management is clear that yield statistics are useful way to assess the joint optimisation of volume and pricing decisions.

INVENTORY OPTIMISATION IN THE MANUFACTURING SECTOR

Prior to the industrial revolution, products were manufactured on a one-off basis, with parts being made as they were needed, each custom made and fitted into place. The basis of mass production was the standardisation of parts so that the manufacture of parts was separate from the assembly process. Henry Ford and others quickly realized that inventories of parts were the buffer that enabled efficient production despite time delays in production, supply and demand. However, stocks of parts waiting for assembly were worth as much or more than the investment in plant and equipment and needed to be managed as part of an optimal low-cost production strategy. The so-called Dupont expansion of Return on Assets (Table 1), attributed to industrialist Pierre Dupont illustrates the relationship between margins on sale and the turnover of assets. By reducing assets to a minimum, the returns on investment could be improved, for any level of profit margin.

It became clear to early industrialists that optimal inventory levels were a trade-off between the holding cost of the inventory, order costs, lead times and demand. The EOQ Model (Harris, 1913) was developed to solve this optimisation problem, and became the basis for more sophisticated inventory models that included stochastic demand, safety stock, variable lead times and other complications involved in the management of inventory (Wilson, 1934). In the 1980's, US automobile manufacturers were in crisis: poor quality and investment in large amounts of work-in-process parts inventories reduced the return on investment to shareholders to unacceptable levels, and major auto company share prices fell. The problem was solved with Deming's quality improvement methods and the adoption of just-in-time production techniques developed by Taiichi Ohno in Japan. Just-in-time methods optimised the returns on assets by minimized inventory throughout the entire supply chain. Today, inventory minimisation is central to the field of logistics, supply chain management and planning system architectures such as MRP (Materials Resource Planning) and MRPII (Manufacturing Resource Planning).

INVENTORY OPTIMISATION IN RETAIL

Inventory has great impact in many service sector industries. Pure services, such as airlines or car rentals define their capacity in terms of units of "perishable" inventory (for example rooms that are not rented are equivalent to tangible inventory items whose utility has expired). Much of the study of Yield Management discussed earlier deals with the joint optimization of perishable inventory utilization and pricing.

Many other service industries (such as retail and restaurants) have goods inventories as well as units of capacity to deal with. They cannot rely on revenue-based yield statistics alone because variable costs are high, and as shown previously, these statistics can be misleading.

The retail industry does not manufacture products and its capacity to supply demand is based on stock-in-trade inventory. Wine inventories are similar to the retail model, since wines are mostly sold in the package in which they are purchased with little additional input other than sales advice and service. Managing retail inventory involves the efficient conversion of the cost of goods on hand into gross margins while providing both a sufficient number of items for selection and a sufficient quantity to meet variable customer demand (Sellers-Rubio & Mas-Ruiz, 2009).

A good example is how furniture retail managers deal with inventory. Customers are attracted to furniture stores by a wide selection of items, colours, styles and prices to choose from. The inventory is a major investment; however it is not perishable and retains its value for a long period of time. As inventory declines, the store becomes less attractive to consumers due to reduced choice and more specifically a reduction in the number of popular items available. Consequently, managers replenish depleted inventory with new items as soon as floor space becomes available. Margins on furniture are substantial, and discounting is used to stimulate the sale of slow moving items at the expense of margins earned. Ultimately, every item is discounted until it is sold.

In order to increase stock turnover, furniture store managers discount items that have not been sold within a period of time after they were acquired. Inventory turnover is defined as sales divided by average inventory. Price discounts reduce the gross margin ratio (defined as sales minus cost of goods sold divided by sales). Using the retail definition of gross margin ratio makes sense for retailers, because every dollar discounted reduces the margin on the sale of that item by the same amount. Discounting may have the secondary effect of increasing sales, because shoppers who save money on a discounted item may be motivated to buy an additional item, or a more expensive item.

Retail managers use a yield statistic called GMROI (gross margin return on inventory) to jointly optimise gross margins and stock turnover. GMROI is similar in construction to other yield statistics (Table 1):

The product of a measure of price efficiency (margin on sale) and a measure of volume efficiency (inventory turnover).

If the retail manager's pricing and product line decisions are more effective, GMROI will increase. If these decisions are less effective, GMROI will decrease. The value of the statistic is easily calculated by busy managers:

Divide the gross margin by the average value of inventory.

This simple calculation can be applied to a single item, group of items, or the entire inventory.

The GMROI statistic used to optimise the trade-off between inventory turnover and gross margin can be readily adapted to manage wine and spirit inventories. Restaurant managers tend not to discount their wine and spirits but rather stimulate sales by promotional activity such as suggestive selling by service personnel, menu tags, sampling, etc. Making Wine Sales the numerator instead of gross margin (as in GMROI) is a better fit with how restaurant managers think about pricing wine. Wine Sales divided by cost of goods sold is referred to as the mark-up ratio. A mark-up ratio of 2 means the wine is sold for twice the cost of purchasing it.

OPTIMISATION OF RESTAURANT MENUS

Restaurants have a mixed manufacturing and service function. Restaurants prepare meals from basic ingredients and serve them to customers. In the restaurant business, inventory is both units of capacity (seat hours, Kimes, 1999) and stocks of basic ingredients, partially prepared items and finished menu items ready for service. Among the items in inventory, wine is purchased ready to use primarily in the same container in which it will be sold, and spirits in quantities to be dispensed or mixed with other ingredients.

Menu engineering (Kasavana & Smith, 1982) is a well known method that optimises menus by gross margin (price efficiency) and popularity (item turnover). Although conceived prior to the popularity and development of yield statistics, the analysis was done using a graphical approach, identifying items that were

- Stars (high margins and high popularity);
- Dogs (low margins and low popularity);
- Plough horses (low margins and high popularity), and
- Puzzles (high margins and low popularity).

Managers could over time, remove dogs from the menu, re-price plough horses, and promote puzzles and thereby optimise the menu. The menu engineering analysis was intuitive, immensely helpful and broadly adopted around the world. Despite its popularity, it does not provide a simple yield statistic to determine if the pricing and promotional activities are having the desired effect.

WINSPID: CONSTRUCTING A MODEL FOR WINE LISTS AND INVENTORIES

The lessons learned from yield management, manufacturing, retail and restaurant methods have been used to develop a simple system for optimising wine lists. The same methodology can easily be extended to spirits menus and other high-value inventories.

The GMROI statistic could be used by restaurant managers by using gross margin as the numerator (sales less cost of goods sold) and average inventory value as the denominator. However, unlike the furniture retail manager who goes through a cycle of price setting followed by discounting, the restaurant manager tends to set prices, and optimise sales by promotional activity. Thus, a statistic with wine sales in the numerator is better suited to the restaurant manager. Furthermore, inventory turnover in the restaurant industry is defined as the cost of goods sold divided by average inventory. Given these differences, a new statistic that fits restaurant industry norms has been adapted from GMROI. The Wine Sales Per Inventory Dollar (WINSPID) statistic expansion is provided in Table 3.

HOW TO USE WINSPID

Restaurant managers are able to manage wine lists by changing three things. They can change the number of selections on the wine list; change the number of bottles held in inventory, or change the prices (and thus the mark-up). Changing any one of these may have an effect on one or more of the others.

Table 3. The WINSPID statistic expansion

	Statistic		Price Efficiency		Volume Efficiency
WINSPID	$\frac{\text{Wine Sales}}{\text{Avg. Wine Inventory}}$	=	$\frac{\text{Wine Sales}}{\cancel{\text{Cost of Wine Sold}}}$	×	$\frac{\cancel{\text{Cost of Wine Sold}}}{\text{Avg. Wine Inventory}}$
		=	Mark-up ratio	×	Inventory Turnover

For example, decreasing the number of selections on the wine list will affect the value of the inventory; changing prices will affect mark-up, sales volumes, and thus inventory. The WINSPID statistic enables managers to evaluate the net result of any changes made to the wine list, the inventory or mark-up ratio.

The use of the WINSPID yield statistic and its expansion into price efficiency (mark-up ratio) and volume efficiency (inventory turnover) is best illustrated with an example, followed by discussion.

MERCADO CENTRAL

The Mercado Central (Santiago, 2011) is restaurant located in Argentina. It has a 46 item wine list with a good variety of prices and types of wines. Prices appear in parenthesis next to the name of each wine selection. Management has recorded sales, inventory and cost information for a 1-month period (Table 4). There is substantial variation in the cost of goods sold and inventory values.

Sales are not necessarily divisible by price, because selective quantity and promotional discounts are a regular part of the restaurant's internal marketing program. For example Familia Gascon Malbec (item #27) was extensively sampled by winery representatives during the month, resulting in low margins and high inventory turnover. Other wines such as the Salentein Malbec (item #44) are sold at substantial margins.

Overall, the Mercado Central wine list yields $13,124 in sales per month, has an average mark-up ratio of 1.63, an inventory turnover ratio of 0.36 turns per month and WINSPID of $0.58 per dollar of inventory held. The wine list includes 10 selections that did not sell at all, and although actual cost of goods sold for these items was $0.00, $1.00 was recorded to avoid a division by zero when the inventory turnover ratios were computed. The inventory value of the wine selections that did not sell is $4,855. If the inventory value of $22,584 were reduced by this amount, a substantive improvement in WINSPID ($0.74) would be achieved.

Table 4. Mercado Central wine list data

#		Sales and Costs			Ratio Computation		
	Wine Item	Sales	CGS	Inv.	ITO	Markup	WINSPID
1	Alamos Cabernet ($65)	$474	$289	$965	0.30	1.64	0.49
2	Alamos Chardonnay ($65)	$102	$64	$1,736	0.04	1.59	0.06
3	Alamos Extra Brut ($70)	$174	$105	$648	0.16	1.65	0.27
4	Alamos Reserve (?)	$0	$1	$60	0.02	0.00	0.00
5	Alma Mora Cabernet ($40)	$45	$36	$109	0.33	1.24	0.41
6	Alma Mora Malbec ($40)	$23	$18	$109	0.17	1.24	0.21
7	Alma Mora Viognier ($40)	$23	$18	$81	0.22	1.24	0.28
8	Angelica Zapata Cabernet ($170)	$0	$1	$91	0.01	0.00	0.00
9	Angelica Zapata Malbec ($170)	$0	$1	$273	0.00	0.00	0.00
10	Callia Alta Syrah ($35)	$553	$234	$231	1.01	2.36	2.39

continued on following page

Table 4. Continued

#		Sales and Costs			Ratio Computation		
	Wine Item	Sales	CGS	Inv.	ITO	Markup	WINSPID
11	Callia Alta Syrah-Cabernet ($35)	$161	$58	$435	0.13	2.76	0.37
12	Callia Alta Syrah-Malbec ($35)	$547	$234	$523	0.45	2.34	1.05
13	Callia Chardonnay ($35)	$0	$1	$66	0.02	0.00	0.00
14	DV Catena Cabernet 2003 ($90)	$143	$100	$450	0.22	1.43	0.32
15	Dv Catena Cab-Mal 2006 ($90)	$180	$86	$1,156	0.07	2.10	0.16
16	Escorihuela Gascon Cab ($70)	$70	$33	$198	0.17	2.12	0.35
17	Escorihuela Gascon Malbec /$70)	$0	$1	$1,485	0.00	0.00	0.00
18	Fin del Mundo extra Brut ($90)	$72	$43	$258	0.17	1.67	0.28
19	Finca La Linda Cabernet 750cc ($60)	$52	$33	$623	0.05	1.57	0.08
20	Finca La Linda Cabernet 375cc ($40)	$94	$62	$124	0.50	1.52	0.76
21	Finca La Linda Chard. 750cc ($55)	$196	$128	$444	0.29	1.53	0.44
22	Finca La Linda Malbec 375cc ($40)	$559	$372	$248	1.50	1.50	2.25
23	Finca La Linda Malbec 750cc ($60)	$670	$356	$578	0.62	1.88	1.16
24	Finca La Linda Syrah 750cc ($60)	$44	$30	$287	0.10	1.48	0.15
25	Flia Gascon Cabernet ($45)	$722	$443	$839	0.53	1.63	0.86
26	Flia Gascon Chardonnay ($45)	$391	$233	$571	0.41	1.68	0.68
27	Flia Gascon Malbec ($45)	$1,088	$932	$606	1.54	1.17	1.80
28	Las Perdices 375cc ($30)	$193	$145	$485	0.30	1.33	0.40
29	Las Perdices Cabernet ($48)	$830	$552	$288	1.92	1.50	2.88
30	Las Perdices Malbec ($48)	$1,767	$1,176	$834	1.41	1.50	2.12
31	Luigi Bosca Res. Cabernet ($100)	$0	$1	$380	0.00	0.00	0.00
32	Luigi Bosca Res. Malbec ($100)	$240	$120	$260	0.46	2.00	0.92
33	Luigi Bosca Res. Sauv. Blanc ($100)	$0	$1	$160	0.01	0.00	0.00
34	Newen Cabernet ($40)	$221	$168	$756	0.22	1.32	0.29
35	Newen Malbec ($45)	$237	$168	$576	0.29	1.41	0.41
36	Newen Sauvignon Blanc ($45)	$243	$168	$216	0.78	1.45	1.12
37	Postales Cabernet ($35)	$374	$215	$517	0.42	1.74	0.72
38	Postales Chardonnay ($35)	$476	$258	$193	1.34	1.84	2.46
39	Postales Malbec ($40)	$1,292	$678	$662	1.02	1.90	1.95
40	Postales x 375cc Cabernet ($25)	$139	$90	$60	1.50	1.54	2.30
41	Postales x 375cc Malbec ($25)	$216	$110	$1,069	0.10	1.96	0.20
42	Saint Felicien Chardonnay ($85)	$0	$1	$1,418	0.00	0.00	0.00
43	Saint Felicien Sauvignon Blanc ($85)	$0	$1	$516	0.00	0.00	0.00
44	Salentein Malbec ($110)	$183	$74	$167	0.44	2.47	1.10
45	Salentein Pinot Noir Reserve ($90)	$0	$1	$407	0.00	0.00	0.00
46	St. Felicien Cab/Merl ($80)	$331	$215	$430	0.50	1.54	0.77
Total/Overall		$13,124	$8,056	$22,584	$0.36	$1.63	$0.58

Figure 1. Mercado Central wine sales vs. inventory value

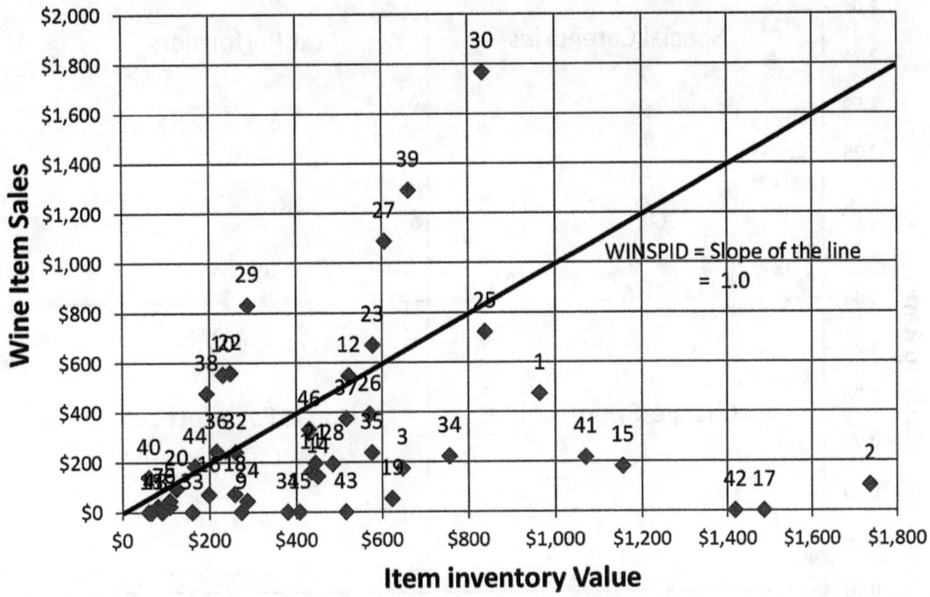

A scatter plot of wine sales for each selection and its corresponding inventory value is provided in Figure 1. The slope of the diagonal line (sales divided by inventory) is WINSPID. In Figure 1, WINSPID is equal to $1 wine sales per dollar of inventory. The wines above the line (slightly less than half) have WINSPID greater than $1.00, and are more efficient selections of the wine menu. Item with WINSPID less than 1 (below the line) play a less productive role, but may have value in enhancing the customer's perception of the wine list and restaurant quality.

With the addition of horizontal and vertical lines corresponding to target values, a scatter plot of Mark-up and Inventory turnover (Figure 2) closely resembles a menu engineering approach. Four (4) categories of wine are revealed by the analysis.

Wines in the "House Favourites" category have lower margins and high turnover rates. They represent exceptional value to customers. Only 1 wine is found in the "House Favourites" category. The Familia Gascon Malbec (#27), as mentioned before was heavily sampled, and thus the mark-up ratio was low. Depending on the success of the sampling, this wine may become a Star performer in the month to follow. Opportunities to improve markup ratio by obtaining quantity discounts or modest price increases can be used to keep these selections near the top of the category.

The "Special Categories" are wine selections (Table 5) with good mark-up ratios, but low turnover ratios. Wines in this quadrant need some attention, hence the name, "Special Categories". There are two possibilities which jointly or independently cause a wine selection to be in this category. Sales of most of these wines are low in comparison to others we have dealt with. Collectively these 19 wines account for 40% of wine sales and 51% of the inventory. They make a positive contribution to the breadth of the wine list offerings, and have good mark-ups. Many of these wines are relatively expensive and sell only 1 or two bottles per month, but serve to communicate the quality of the restaurant's offerings. In all cases, the wines in this category have excessively large inventories of wine on hand. By simply reducing the stock on hand, the inventory turnover ratio can be improved, moving these selections into the Star Performer Category.

Figure 2. Mercado Central mark-up vs. inventory turnover

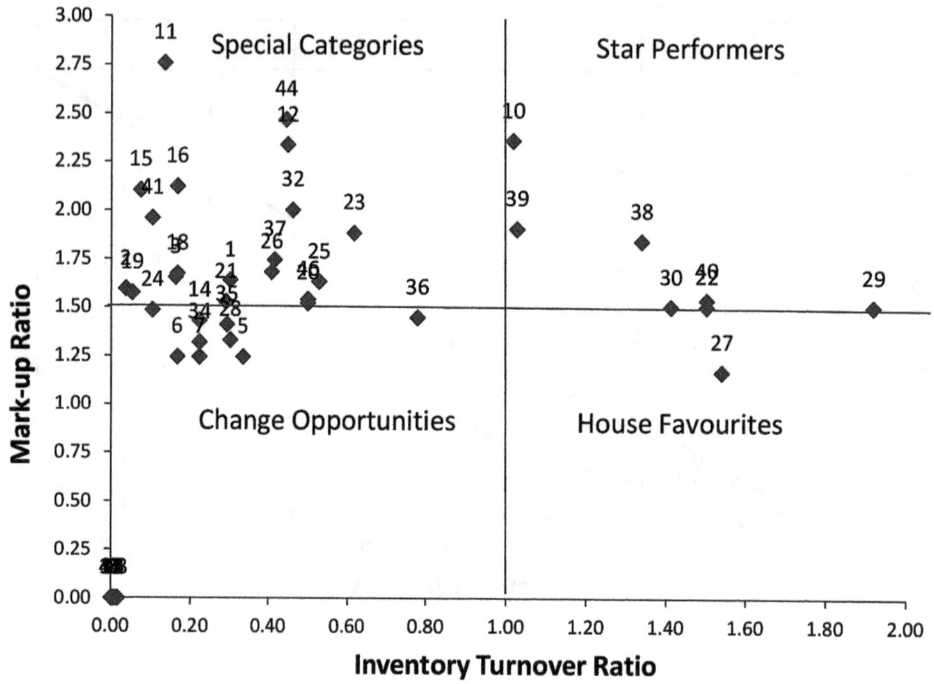

Table 5. Special categories

ID #	Wine Item	Sales	CGS	Inv.	ITO	Markup	WINSPID
11	Callia Alta Syrah-Cabernet ($35)	$161	$58	$435	0.13	2.76	0.37
44	Salentain Malbec ($110)	$183	$74	$167	0.44	2.47	1.10
12	Callia Alta Syrah-Malbec ($35)	$547	$234	$523	0.45	2.34	1.05
16	Escorihuela Gascon Cab ($70)	$70	$33	$198	0.17	2.12	0.35
15	Dv Catena Cab-Mal 2006 ($90)	$180	$86	$1,156	0.07	2.10	0.16
32	Luigi Bosca Res. Malbec ($100)	$240	$120	$260	0.46	2.00	0.92
41	Postales x 375cc Malbec ($25)	$216	$110	$1,069	0.10	1.96	0.20
23	Finca La Linda Malb. 750cc ($60)	$670	$356	$578	0.62	1.88	1.16
37	Postales Cabernet ($35)	$374	$215	$517	0.42	1.74	0.72
26	Flia Gascon Chardonay ($45)	$391	$233	$571	0.41	1.68	0.68
18	Fin del Mundo extra Brut ($90)	$72	$43	$258	0.17	1.67	0.28
3	Alamos Extra Brut ($70)	$174	$105	$648	0.16	1.65	0.27
1	Alamos Cabernet ($65)	$474	$289	$965	0.30	1.64	0.49
25	Flia Gascon Cabernet ($45)	$722	$443	$839	0.53	1.63	0.86
2	Alamos Chardonay ($65)	$102	$64	$1,736	0.04	1.59	0.06
19	Finca La Linda Cab. 750cc ($60)	$52	$33	$623	0.05	1.57	0.08
46	St. Felicien Cab/Merl ($80)	$331	$215	$430	0.50	1.54	0.77
21	Finca La Linda Chard. 750cc ($55)	$196	$128	$444	0.29	1.53	0.44
20	Finca La Linda Cab. 375cc ($40)	$94	$62	$124	0.50	1.52	0.76
Total/Overall		$5,249	$2,901	$11,538	0.25	1.81	0.45

A significant number of wines on the list have no sales at all during the month. These appear in the Change candidates category, along with wines that have both low mark-ups and low inventory turnover rates. Selections in this category contribute less than 9% to the restaurant's wine sales, yet represent 35% of the total wine inventory. Those with higher mark-up ratios and large inventories may be re-positioned by simply reducing the inventory. By evaluating this category over a period of several months, selections that remain in this quadrant should be considered for elimination from the wine list and the wine inventory without adversely affecting the restaurant's wine sales or wine program.

Elimination of selections from the wine list may provide opportunities to add new selections to the list with better WINSPID performance.

Managing Supplier Selection and Evaluation

In the Mercado Central restaurant, a majority of the wine selections have low (less than 1.00) WINSPID values (Table 6). As part of a program to improve the financial performance of the wine menu, managers can make changes to the wine list. For example, an obvious change is to eliminate the ten (10) selections with no sales at all.

Another possibility is to adopt new wines that will outperform existing wines on the list. For example, the Alma Mora selections on the wine list have low WINSPID values. The winery representative may

Table 6. Change candidates

ID#	Wine Item	Sales	CGS	Inv.	ITO	Markup	WINSPID
24	Finca La Linda Syrah 750cc($60)	$44	$30	$287	0.10	1.48	0.15
36	Newen Sauvignon Blanc ($45)	$243	$168	$216	0.78	1.45	1.12
14	DV Catena Cabernet 2003 ($90)	$143	$100	$450	0.22	1.43	0.32
35	Newen Malbec ($45)	$237	$168	$576	0.29	1.41	0.41
28	Las Perdices 375cc ($30)	$193	$145	$485	0.30	1.33	0.40
34	Newen Cabernet ($40)	$221	$168	$756	0.22	1.32	0.29
5	Alma Mora Cabernet ($40)	$45	$36	$109	0.33	1.24	0.41
7	Alma Mora Viognier (40)	$23	$18	$81	0.22	1.24	0.28
6	Alma Mora Malbec ($40)	$23	$18	$109	0.17	1.24	0.21
4	Alamos Reserve (40)	$0	$1	$60	0.02	0.00	0.00
13	Callia Chardonay (35)	$0	$1	$66	0.02	0.00	0.00
8	Angelica Zapata Cabernet ($170)	$0	$1	$91	0.01	0.00	0.00
33	Luigi Bosca Res. Sauv. Blanc ($100)	$0	$1	$160	0.01	0.00	0.00
9	Angelica Zapata Malbec ($170)	$0	$1	$273	0.00	0.00	0.00
31	Luigi Bosca Reserva Cabernet ($100)	$0	$1	$380	0.00	0.00	0.00
45	Salentain Pinot Noir Reserve ($90)	$0	$1	$407	0.00	0.00	0.00
43	Saint Felicien Sauvignon Blanc ($85)	$0	$1	$516	0.00	0.00	0.00
42	Saint Felicien Chardonay ($85)	$0	$1	$1,418	0.00	0.00	0.00
17	Escorihuela Gascon Malbec /$70)	$0	$1	$1,485	0.00	0.00	0.00
Total/Overall		$1,172	$862	$7,923	0.11	1.36	0.15

suggest other Alma Mora products that would perform better. WINSPID can be used to evaluate any one, or a group of wine selections. If the new additions to the wine list have WINSPID greater than 1.0, they are increasing the performance of the wine menu. If their performance is lower, perhaps another winery should be given the opportunity to try their wines in this venue.

Wine merchants can be instructed how their selections will be evaluated (WINSPID) and how they can help improve their product's WINSPID scores. Wine merchants can improve WINSPID score for their products by arranging delivery of smaller orders more frequently (enabling a reduction in inventory), by providing discounts (which reduce inventory value), by sampling and new product introduction promotions.

FUTURE RESEARCH

Wine and spirit inventories are part of the current assets of a restaurant and in the case of the fine dining segment, often represent a large investment that turns over slowly. Under pressure from competition as well as general changes in the economy, restaurants cannot afford to ignore the efficiency with which profits are obtained from the investment in assets of all types. Food items are also significant investments that need to be managed better. While inventory control, optimal order quantity and analytical methods are common, models that jointly optimise the restaurant menu, items in stock, turnover and pricing have not been developed. The need for optimised menus increases as restaurant chains proliferate and franchisors are seeking ever more highly profitable restaurant concepts.

CONCLUSION

In this chapter, we have shown how yield management in travel industries, inventory management in retail and restaurant menu engineering have formed the basis of a model that can be used to improve the financial performance of wine menus. The joint optimisation of the volume and price efficiency of variable assets like wine inventory adds a layer of complexity to the manager's problem. Not only is it necessary to manage price and volume as the travel managers do, but it is also necessary to manage stocks of wine, which ultimately are linked to the wine list and the consumer's perception of restaurant quality. The WINSPID statistic and methodology presented here enable managers to do that. With no changes in typical information reporting or special software, the simplicity of the WINSPID statistic makes it easy for managers to evaluate changes to their margins, promotional activity, purchasing and inventory turnover. While the example provided illustrates its use in wine lists, it could easily be extended to premium spirits (as seen in whiskey bars), or other high-value items carried in restaurant inventories.

REFERENCES

Barth, J. E. (2011). A model for wine list and wine inventory yield management. *International Journal of Hospitality Management*, *30*(3), 701–707. doi:10.1016/j.ijhm.2010.12.004

Barr, A. (1995). Drink: An informal social history. London, UK: Bantham.

Bell, E. (2009). *Glass Half Empty? On the State of Restaurant Wine Sales in the Recession*. Retrieved from http://www.starchefs.com/features/editors_dish/wine-survey/html/index.shtml

Berenguer, G., Gil, I., & Ruiz, M. E. (2009). Do upscale restaurant owners use wine lists as a differentiation strategy? *International Journal of Hospitality Management*, *28*(1), 86–95. doi:10.1016/j.ijhm.2008.04.003

Charters, S. (2006). *Wine and Society: The social and cultural context of a drink*. Burlington, MA: Elsevier Butterworth-Heinemann.

Chiang, W., Chen, J., & Xu, X. (2007). An overview of research in revenue management: Current issues and future research. *Int. J. Revenue Management*, *1*(1), 97–128. doi:10.1504/IJRM.2007.011196

Green, Y. N., & Weaver, P. A. (2008). Approaches, Techniques, and Information Technology Systems in the Restaurants and Foodservice Industry: A Qualitative Study in Sales Forecasting. *International Journal of Hospitality & Tourism Administration*, *9*(2), 164–191. doi:10.1080/15256480801909071

Harrington, R. J. (2007). *Food and Wine Pairing: A Sensory Experience*. Wiley and Sons, Publishers.

Harris, F.W. (1913). How many parts to make at once. *Factory: The Magazine of Management, 10*(2), 135-136, 152.

Heath, D. B. (2000). *Drinking Occasions: Comparative Perspectives on Alcohol and Culture*. Ann Arbor, MI: Taylor and Francis.

Kasavana, M. L., & Smith, D. I. (1982). *Menu Engineering: A Practical Guide to Menu Analysis*. Okemos, MI: Hospitality Publications.

Kimes, S. (1999). Implementing Restaurant Revenue Management: A Five-step Approach. *The Cornell Hotel and Restaurant Administration Quarterly*, *40*(3), 16–21. doi:10.1177/001088049904000315

Santiago, R. (2011). *Yield Management aplicado a la medición de la performance de la carta de vinos del restaurante Mercado Central* (Unpublished undergraduate thesis). Universidad de San Andrés, Buenos Aires, Argentina.

Sellers-Rubio, R., & Más-Ruiz, F. J. (2009). Technical efficiency in the retail food industry: The influence of inventory investment, wage levels, and age of the firm. *European Journal of Marketing*, *43*(5/6), 652–669. doi:10.1108/03090560910946981

Wilson, R. H. (1934). A Scientific Routine for Stock Control. *Harvard Business Review*, *13*, 116–128.

Yang, S., & Lynn, M. (2009). Wine List Characteristics Associated with Greater Wine Sales. *Cornell Hospitality Reports, 9*(11). Retrieved from http://www.hotelschool.cornell.edu/research/chr/pubs/reports/abstract-15089.html

Yuksel, A., & Yuksel, F. (2002). Measurement of tourist satisfaction with restaurant services: A segment-based approach. *Journal of Vacation Marketing.*, *9*(1), 52–68. doi:10.1177/135676670200900104

KEY TERMS AND DEFINITIONS

GMROI: Gross margin return on inventory; a yield statistic used by the retail industry.

Gross Margin: Selling price minus the cost of acquisition.

Gross Margin Ratio: The ratio of gross margin to sales.

Inventory Turnover Ratio: The cost of goods sold divided by the inventory value.

Perishable Assets: A class of fixed assets whose utility is obtained by renting them to customers. If the asset is not rented for a time, the revenue potential is lost, and thus these assets are called perishable assets.

Price Efficiency Ratio: The average price of a product that is sold for different prices.

REVPAR: Revenue per available room; a yield statistic used by the hotel industry.

Volume Efficiency Ratio: The proportion of times an asset is sold.

Yield Statistic: The product of price efficiency and volume efficiency ratios.

Chapter 18
Young Tourists' Perceptions of Hotel Disintermediation: Evidence from Italy

Giacomo Del Chiappa
University of Sassari, Italy

Mariella Pinna
University of Sassari, Italy

Marcello Atzeni
University of Cagliari, Italy

ABSTRACT

Generation Y has been considered to be a sizeable new market. This study, based on a sample of 1131 Italian travellers from Gen Y, investigates their views for and against disintermediation, and analyses how their choices are influenced by user generated content (UGC), rather than by information provided by high street travel agencies. The factor analysis uncovers three dimensions: "Benefits of Travel Agency", "Benefits of Online Reservation", and "Online Trust & Search Behaviour". Further, a series of statistical tests indicate that demographics such as age and education have a significant influence on the respondents' perceptions. Our findings suggest that hotel managers and travel agencies should monitor Gen Y perceptions of the benefits and constraints of using the Internet, UGC and travel agencies for hotel booking. Further, accommodation providers should use online channels to create affective commitment in their young customers. Limitations of the study are discussed and suggestions for further research are given.

INTRODUCTION

In the last decades, the Internet has been experiencing significant growth worldwide. According to Internet World Stats (2016), more than 3.6 billion people in the world are active Internet users, with an increase of 10% since 2015. More specifically, Internet penetration varies considerably among countries, but an

DOI: 10.4018/978-1-5225-1054-3.ch018

average of 46% has been observed worldwide. The United States (88%) and Europe (83%), respectively, show the highest Internet penetration ratios, whereas South Asia (27%) and Africa (29%) record the most negative peaks (We are Social, 2016). The development of the Internet has influenced the daily lives of individuals (Wan, 2002). More than 70% of the global population searches online for products or services to buy, 53% actually purchases on the web (We are Social, 2016), and 58% uses the Internet and peer-to-peer applications to seek advice about brands and products (Communify, 2015). Specifically, 92% of consumers perceive User Generated Content to have a higher degree of trustworthiness and credibility than traditional advertising (Communify, 2015).

Tourism has become the foremost industry in terms of online market share, and travel planning and booking are two of the most popular online activities (e.g., Guo, Ling, Dong, & Liang, 2013). The Internet and User Generated Content have completely reshaped the way in which people plan, buy and consume tourism services (Buhalis & Law, 2008) and select tourism destinations (ITB, 2016). The Internet and User Generated Content have resulted in a gain of hotel efficiency: hospitality companies are now able to reach their customers directly and offer them marketing promotions and sales (Berne, Garcia-Gonzalez, & Mugica, 2012), thus challenging and changing the traditional role of tourism intermediaries (Del Chiappa, 2013).

In the last two decades, there has been a growing academic interest in studies aimed at analyzing the impact of the Internet, Information and Communication Technologies and peer-to-peer applications on consumers, firms and the marketplace (Manjit & Pavlou, 2014). In recent years, the Internet has become one of the most popular and influential information sources in tourism, especially due to the recent and greater use of wireless devices (MacKay & Vogt, 2012), with tourists using it to gather information, comparing alternatives and making purchases and reservations (Money & Crotts, 2003).

The most active group of individuals involved in online behaviours is Gen Y (e.g., Bolton et al., 2013). Gen Y can be considered to be a sizeable new market. In the US alone, this group consists of 82 million people, spending nearly 200 billion dollars annually (e.g., Djamasbi, Siegel, & Tullis, 2010). In 2012, as reported by the UNWTO, youth travel generated US$182 billion, and the total number of international trips by Gen Y is estimated to increase to 300 million trips a year by 2020 (IPK International, 2013). Based on these figures, it is easy to explain why Gen Y is widely recognised as the future market of the travel and tourism industry (Benckendorff, Moscardo, & Pendergast, 2010).

There are several reasons why it is useful to further investigate the behavior of Gen Y. First, in the last decade youth and student travel has experienced a significant growth as the international student population is increasing (Richards & Wilson, 2003). Gen Y will become dominant players in the prime age work-force, thus their prospective purchasing power cannot be ignored. Second, youth travelers are heavy users of the Internet and social media where they frequently upload and share comments, reviews, photos and videos, thus affecting the online reputation of service providers and influencing tourists' choices (Jin, Lin & Hung, 2014). Third, Gen Y is expected to be the next largest cohort, after the baby boomers, that will have a significant influence and impact on the tourism sector, especially given the fact that they seem to be characterized by a higher frequency of travelling and a prospective higher willingness to spend more on international travel (Hritz, Sidman & D'Abundo, 2014). Finally, Gen Y tend to travel more frequently and for longer periods (Richards & Wilson, 2003), they are more digital savvy, and more involved in travel planning (Xiang, Magnini & Fesenmaier, 2015).

Despite the relevant role that Gen Y seems to play in the health of the global tourism travel sector, little research has examined, in depth, the views of Gen Y regarding hotel disintermediation, while taking into consideration the relative power of User Generated Content in influencing tourists' choices with

Table 1. Arguments for and against disintermediation: examples

Arguments for	Arguments against
…it ensures great flexibility and a variety of consumer choices	…travel agencies are professional travel advisers who provide valuable services and advice
…the personnel of travel agencies are poorly trained and barely competent	…the human contact with the personnel reduces uncertainty and insecurity, and helps travellers to save time
…travel agencies are biased in favour of principals who offer override commissions and in-house partners	…travel agencies can better manage trust and relationship building
…travel agencies have less knowledge than experienced travellers	…travel agencies offer better guarantees, including in terms of privacy issues

Source: our elaboration from: (Buhalis 1998; Del Chiappa, 2013; Zehre & Moschl, 2008).

respect to information delivered by travel agencies. This is particularly evident in the context of Italy, where, to our best knowledge, no academic paper has been published regarding this topic. This has occurred despite the importance that the Italian tourism market represents for many countries worldwide, especially in Europe. According to Banca d'Italia (2013), 59.197.000 Italians spent their holidays abroad in 2012, thus generating 256.367.000 overnight stays and 20,512 million Euros.

This study was therefore carried out with a sample of 1131 Italian travellers from Gen Y to analyse their views for and against disintermediation, and to investigate the degree to which their choices are influenced by User Generated Content, rather than by the information provided by high street travel agencies.

BACKGROUND

The disintermediation hypothesis, that is, the idea that the elimination of the intermediary or middlemen by using the Internet (Buhalis, 1998), has been heavily debated by researchers (Zehre & Moschl, 2008). Customers increasingly prefer the information variety provided by the Internet, rather than the travel suggestions of location-based travel providers (Gretzel, Fesenmaier, & O'Leary, 2006). As a matter of fact, traditional travel agencies have been facing the need to modify, at least partially, their business model; luckily, Information and Communication Technologies and peer-to-peer applications are providing travel agencies with a wide range of marketing opportunities (e.g., Huang, 2011).

There are several arguments in favour of and against disintermediation (see Table 1), and they can be identified adopting both a supply-side and a demand-side perspective.

Adopting the consumers' perspective, the Internet and Information and Communication Technologies allow them to save time and costs when searching for information, to stimulate and expand electronic word of mouth (e-WOM), to buy tailor-made products, and to access a wider selection of travel service providers anytime and anywhere (Anckar, 2003). However, consumers are often required to spend a lot of time evaluating the huge number of options available online (i.e., information overload); hence, they may prefer to rely on professional travel agents (Anckar, 2003). Further, traditional travel agents assume the responsibility for all travel arrangements, thus reducing consumers' uncertainty and insecurity (e.g. Law, Leung, & Wong, 2004). Consumers may also be worried about the safety and security of online booking, thus preferring to book with a street travel agent (e.g., Chen, 2001). In addition, travellers may prefer a travel agent, given the possibility of experiencing a personal touch and receiving tailored advice (Palmer & McCole, 1999). However, it should be noted that consumers tend to rely on the Internet or on a travel agent based on product characteristics. As far this aspect is concerned, previous studies have shown that consumers generally purchase low involvement products and services online (such as short trips), whereas they rely on the traditional intermediaries for the purchase of more complex or high involvement products, such as honeymoons or intercontinental travel (e.g., Prayag & Del Chiappa, 2014).

With the advent and development of user generated content, the relevance of the disintermediation phenomenon has been increasing. User Generated Content exerts a great influence in generating the idea of travelling, during the actual planning process and after tourists' decisions have been taken (e.g. Gretzel & Yoo, 2008; Hennig-Thurau, 2005; Law, Buhalis, & Cobanoglu, 2014). For instance, 77% of TripAdvisor users have been reported to check online reviews before choosing their accommodations (TripAdvisor, 2014). Gretzel and Yoo (2008) found that 33% of frequent travellers reported switching to a different hotel on the basis of online comments (Gretzel & Yoo, 2008). Del Chiappa (2011) showed that Italians change their hotel accommodation

- Sometimes (64.8%),
- Almost always (12%), or
- Always (0.5%) after having read reviews and comments posted online.

Recent studies have shown that tourists often prefer to use peer-to-peer travel applications (e.g., TripAdvisor, Booking.com, etc.), instead of traditional travel agencies to seek information and make hotel reservations (Sun, Nang Fong, Law, & Luk, 2015). Recent studies have provided further insights into the scientific debate on disintermediation for hotel booking, also considering the relative power of user generated content in influencing tourists' choices with respect to information delivered by travel agencies. Specifically, studies carried out in the context of Italy (Del Chiappa, 2013), France (Prayag & Del Chiappa, 2014), and Spain (Del Chiappa, Lorenzo-Romero, & Gallarza, 2014) have shown that consumers change the accommodation suggested by a travel agency based upon User Generated Content quite often.

In such a scenario, Gen Y, the cohort of people born between 1980 and 1999 (Lee, 2013),

- Is more heavily involved in travel planning (Xiang, Magnini, & Fesenmaier, 2015),
- Is technologically savvy, well engaged in online travel planning and booking (Bolton et al., 2013; Nusair, Bilgihan, & Okumus, 2013; Xiang et al., 2015), and
- Spends a considerable amount of time using online social networks for messaging, keeping in touch with friends and sharing travel experiences (Pempek, Yermolayeva, & Calvert, 2009).

This contributes to the explanation of the reason why Gen Y members have been considered expert social media users (Lester, Forman, & Lloyd, 2006), especially when online travel searching and purchasing behaviours are observed (Beldona, Morrison, & O'Leary, 2005). Jones and Fox (2009) showed that Gen Y members are more likely to read other people's blogs and to write their own than members of other generations. Richards (2007) pointed out increased usage of the Internet to book travel-related products among the Gen Y cohort; such evidence could be rooted in Gen Y's search for customised travel products and services that satisfy their needs, interests and tastes in a flexible and effective way (e.g., Sweeney, 2006). Several studies seemed to suggest a low dependence of Gen Y travellers on traditional travel agencies; for example, in the study carried out by Frost and Shanka (1999) with a sample of 1083 Asian Australian students, just 0.9% of the respondents reported being influenced in their destination choice by a traditional travel agent. However, other studies have shown that Gen Y travellers mostly tend to use the Internet when planning their trips and use travel agents when booking vacations (Beldona, Nusair, & Demicco, 2009). Similarly, Pizam et al. (2004) studied a sample of 1429 students at 11 universities located in 11 different countries and showed that 46.5% of the respondents booked and purchased their trips using a travel agent, especially respondents who scored low on risk-taking and sensation-seeking.

Gen Y is expected to reach its spending power peak in the future, especially for travel-related purchases (Lee, 2013). Therefore, Gen Y's perceptions, attitudes and online behaviours should constantly be monitored (e.g., Barton, Fromm, & Egan, 2012).

Several studies have shown that the influence of User Generated Content on tourism choices differs based on the socio-demographic characteristics of the travellers (e.g. Del Chiappa, 2013; Gretzel, Yoo, & Purifoy, 2007; Yoo & Gretzel, 2011). More specifically, significant differences have been reported for

- Gender (Del Chiappa, 2013; Gretzel et al., 2007; Yoo & Gretzel, 2011),
- Education (Yoo & Gretzel, 2011) and
- Income (Yoo & Gretzel, 2011).

Interestingly, with respect to age, young travellers (less than 35 years old) are influenced by User Generated Content as a travel information-seeking source (Gretzel et al., 2007; Ip, Lee, & Law, 2012). Broadly speaking, current studies on the influence of socio-demographics on online behaviours show findings that are sometimes contradictory. Gender generally has no influence on online behaviours (Ip et al., 2012; Kim & Kim, 2004), except in the study by Del Chiappa (2013). The education level influences online behaviours in some studies (Ip et al., 2012; Morrison, Jing, O'Leary, & Cai, 2001; Weber & Roehl, 1999), but not in others (Kim & Kim, 2004). Likewise, income level has no influence on online behaviour in some studies (Kim & Kim, 2004), but does have an influence in others (Del Chiappa, 2013; Ip et al., 2012). Further and deeper research is needed to investigate whether such differences exist within a seemingly heterogeneous population of Gen Y. Therefore, this study has been carried out to investigate Italian Gen Y travellers' views for and against disintermediation, and to analyse the extent to which their final choices are influenced by User Generated Content, rather than by information provided by high street travel agencies. Further, it aims at showing whether significant differences exist based on socio-demographic characteristics.

STUDY METHODS

For the purposes of this study, a questionnaire which includes four sections was developed based on a literature review (Del Chiappa, 2013; Prayag & Del Chiappa, 2014). In the first section, the respondents were first asked whether they belong to the 18-35 age bracket. Only people giving a positive answer to this question were allowed to complete the questionnaire. Hence, in the same section, Gen Y's usage patterns for both online hotel reservations and traditional travel agencies was investigated through questions such as "How do you usually book your accommodation?", "How often do you use the Internet for hotel reservations?", and "How often do you use traditional travel agencies to make hotel reservations?". In the second section, respondents were asked to express their level of agreement with a list of 13 statements used to investigate their views concerning the use of online booking and the use of traditional travel agencies when making hotel reservations; respondents gave their answers using a 5-point Likert scale (1= I completely disagree, 5 = I completely agree). The third section included several questions related to the use of the Internet for different types of holidays at different geographical scales (national, regional – Europe, and international) and the length of the trip (short vs. medium/long). Finally, socio-demographics such as age, gender, education level, and average monthly income were also measured.

The questionnaire was originally designed in English and then translated into Italian. The method of back-translation was used to ascertain the content of the translated version and was pre-tested on 20 Italians from the target population. An online survey using a snowball sampling technique was considered to be the best choice to address the difficulty of reaching participants in remote areas through face-to-face structured interviews, and with the financial constraints we faced in managing the research project (Wrenn, Stevens, & Loudon, 2007). The initial subjects were generated from the 2,500 contacts of an Italian tourism association based in the centre of Italy. These 2500 individuals, residing in all regions of Italy, received an e-mail inviting them to fill in the online questionnaire by clicking on a link provided in the e-mail. At the same time, they were invited to forward the survey to at least five of their friends, relatives and contacts (18 years of age or over). After a three-week survey period (during November 2013), a total of 1287 valid questionnaires were obtained; after having eliminated all of the surveys that had an unreasonable number of missing data, a total of 1131 completed questionnaires were obtained and used for the purposes of the study.

FINDINGS AND DISCUSSION

Table 2 summarises the demographic profile of the sample. More than half of the respondents reported being female (59.12%) and belonging to the 26-35 age bracket (61.17%). A significant percentage of the interviewees had a university degree (47.65%) or a high school diploma (39.95%). Further, 32.93% of the respondents reported having an average household monthly income below 2.000 € per month. The majority of the sample had previously made hotel reservations online (84.32%). Overall, 86.91% of the respondents reported having booked online one or two times per year.

Table 3 highlights that the respondents generally expressed very positive feelings toward the role of the Internet and technology in supporting tourists' choices.

Specifically, the respondents strongly agreed that the Information and Communication Technologies are useful to save money (M=3.95, SD=1.114) and time (M=3.65, SD=1.223) in making hotel room reservations, compared to using street travel agencies. Further, the respondents thought that the Internet and Information and Communication Technologies allowed them to book whenever they want – 24 hours a day (M=3.93, SD=1.114). Italian consumers from Gen Y reported searching for information on the Internet and checking User Generated Content (M=3.9, SD=1.028) when making hotel reservations, with User Generated Content being considered relatively trustworthy (M=3.55, SD=0.935). Moreover, the respondents agreed with the idea that websites are a more flexible channel to search for information and make reservations (M=3.87, SD=1.001). On the contrary, Italian travellers from Gen Y disagreed with the idea that consumers ultimately have to bear the cost of commissions to travel agents for their hotel rooms (M=2.64, SD=0979), and at the same time, they did not have a positive view about the convenience of seeking advice from travel agents and prefer the Internet (M=2.13, SD=1.059). Further, the respondents also thought that traditional travel agencies are not very effective in understanding the

Table 2. Socio-demographic characteristics of the sample

Variables	%
Gender	
Male	40.88
Female	59.12
Age	
≤ 25	32.63
26-35	67.37
Education	
Below secondary/high school	2.13
Secondary/high school	39.95
University degree	47.65
Post graduate degree (Master)/PhD	7.62
Other	2.65
Monthly Average Household Income €	
less than 1,000	7.01
1,001 -2,000	25.92
2,001 – 3,000	17.8
3.001 – 4.000	8.21
4.001 – 5.000	3.97
5.001 – 6.000	2.77
6.001 – 7.000	0.83
7.000 or above	8.39
I prefer not to answer	25.1
Have You Ever Used the Internet to Make Reservations for Hotel Rooms when You Travel?	
Yes	84.32
No	15.68
How Many Times a Year Do You Use the Internet to Make Hotel Reservations?	
1 – 2	73.90
3 – 4	18.05
> than 4	8.05
Have You Ever Changed the Hotel Accommodation Suggested by a Travel Agency Based on User Generated Content?	
Yes	30.19
No	69.81

needs and desires of their customers (M= 3.01, SD= 0.95) and disagreed with the fact that travel agencies can offer a human touch and interface with the hotel industry (M=2.73, SD=1.18). Despite this, Gen Y travellers showed a slightly positive attitude toward the idea that travel agencies are professional counsellors for hotel rooms (M=3.16, SD=0.993) and are able to personalise products and services (M=3.18, SD=0.96).

Table 3.Gen Y travellers' views for and against the hotel disintermediation

Items	M.	SD
The Internet allows people to save money when making hotel reservations	3.95	1.025
The Internet allows people to use their time in a very productive way as they can search for information and make reservations whenever they want (24 hours x 7 days)	3.93	1.114
When choosing hotels, I search for information through the Internet, and I check User Generated Content	3.9	1.028
Web sites (e.g., electronic intermediaries such as booking.com) are more flexible and can offer many more choices than travel agencies for hotel rooms	3.87	1.001
The Internet allows people to save a lot of time in making hotel room reservations compared to the use of traditional travel agencies	3.65	1.223
I trust the tourism information available online through User Generated Content	3.55	0.935
Travel agencies are able to personalise the products/services they provide to their customers	3.18	0.96
Travel agencies are professional counsellors for hotel rooms and offer valuable services and advice	3.16	0.993
The Internet provides tourist information in such a way that it is easy to choose hotels and spend free time online	3.11	1.195
Travel agencies are able to understand the needs and desires of their customers and to recommend the most suitable hotel accommodations	3.01	0.95
Travel agencies offer a human touch and interface with the hotel industry	2.73	1.185
Consumers ultimately have to bear the cost of commissions to travel agencies for their hotel rooms	2.64	0.979
It is more convenient to seek advice from travel agencies on hotel reservations than to use online technology only	2.13	1.059

To better examine the underlying dimensions of the disintermediation perceptions among Gen Y Italian travellers, a principal component factor analysis (PCA) with Varimax rotation was undertaken. The 13 factor items yielded three factors explaining 49.49% of the total variance (Table 4). The KMO-index (=0.831) and the Bartlett's test of sphericity (chi-square = 4721.056, p-value <0.0001) confirmed that the factor analysis was appropriate to explain the data (Hair, Black, Babin, & Anderson, 2014). The reliability of the three factors was calculated with Cronbach's alpha, and the results showed values higher than 0.7 for all factors, as recommended by Nunnally (1978). The three factors were labelled as follows:

- "F1 - Benefits of traditional travel agencies",
- "F2 - Benefits of online reservation", and
- "F3 - Online trust and search behaviour".

More specifically, F1 explained 27.76% of the total variance and was strongly related to items in favour of traditional travel agencies. Factor 2 explained 14.5% of the total variance and included items that were in favour of using the Internet for hotel room reservations. Finally, F3 explained 7.21% of the total variance depicting trust in online information and User Generated Content about hotels and the search behaviour of consumers. With the aim of verifying whether socio-demographics exert a significant influence on Gen Y perceptions, a series of statistical tests (T-test and Anova test) were performed. The findings revealed that age had a significant influence on F3 ($t = - 3.34$; $p < 0.05$), whereby the older Gen Y (26–25 years old) showed (M=3.72) marginally higher agreement levels, compared to younger Gen Y (18–25 years old) (M=3.52) on trust in online information. The level of education was reported

Table 4. Results of factor analysis

Item	Factor 1 Benefits of Traditional Travel Agencies	Factor 2 Benefits of Online Reservation	Factor 3 Online Trust and Search Behaviour
Travel agencies offer a human touch and interface with the hotel industry	0.58		
Travel agencies are professional counsellors for hotel rooms and offer valuable services and advice	0.737		
Consumers ultimately have to bear the cost of commissions to travel agencies for their hotel rooms	0.659		
It is more convenient to seek advice from travel agencies on hotel reservations than to use online technology only.	0.487		
Travel agencies are able to personalise the products/services they provide to their customers	0.782		
Travel agencies are able to understand the needs and desires of their customers and recommend the most suitable hotel accommodations	0.798		
Web sites (e.g., electronic intermediaries such as booking.com) are more flexible and can offer many more choices than travel agencies for hotel rooms		0.47	
The Internet allows people to save a lot of time in making hotel room reservations compared to the use of traditional travel agencies		0.809	
The Internet allows people to use their time in a very productive way as they can search for information and make reservations whenever they want (24 hours x 7 days)		0.811	
The Internet allows people to save money when making hotel reservations		0.613	
The Internet provides tourist information in such a way that it is easy to choose hotels and spend free time online		0.54	
When choosing hotels, I search for information through the Internet, and I check User Generated Content			0.854
I trust the tourism information available online through User Generated Content			0.817
Eigenvalues	4.43	2.32	1.15
% of variance	27.76	14.5	7.21
% cumulate	27.76	42.27	49.49
Alfa	0.795	0.754	0.775

to have a significant influence on F1 ($F=5.915$; $p< 0.05$) and on F3 ($F= 6.645$; $p< 0.05$). Contrariwise, income did not significantly influence the identified factors.

According to Law (2009) and Del Chiappa (2013), online users can be grouped into

- Occasional buyers (those who have used the Internet to make bookings one or two times),
- Moderate buyers (three to four times) and
- Frequent buyers (more than four times).

Hence, of the total sample, 73.90% were occasional buyers, 18.05% were moderate buyers and 8.05% were frequent buyers (Table 2). Interestingly, 30.19% of the respondents changed the hotel accommodation recommended by a traditional travel agency after reading reviews and comments posted online (Table 2).

Cross tabulations (Table 5) were used to evaluate whether offline buyers and online buyers were more or less likely to change the accommodation suggested by a travel agency based on UCG. Chi-square analyses indicated that online buyers were more likely to change the accommodation based on User Generated Content more often than offline buyers (χ^2=7.257, p<0,01).

Cross tabulations (Table 6) were used to evaluate whether occasional, moderate and frequent buyers were more or less likely to change the accommodation suggested by a travel agency based on UCG.

The findings indicated that significant differences did exist. In particular, frequent buyers changed their minds more often than moderate or occasional buyers ($\chi^2 = 15.569$, $p < 0.01$), thus disconfirming prior studies carried out in the French context (Prayag & Del Chiappa, 2014) that did not find any significant differences among these different groups; such evidence shows that differences based on cultural background still exist and should be further investigated, over time, via cross-cultural comparison. In line with prior studies, these findings could be explained by arguing that the extent to which travelers value User Generated Content, and are likely to be influenced by them, increases with the frequency of their traveling (Gretzel et al., 2007; Del Chiappa et al., 2016).

Previous studies have shown that the propensity to purchase travel products online is influenced by travel purpose and the duration of travel (short-haul vs. long-haul) (e.g. Del Chiappa, 2013; Law et al., 2004). Accordingly, findings of our study shows that respondents rely on the Internet for booking accommodations for domestic (95.6%) and European (95.3%) travel, rather than for intercontinental ones (50.4%) (table 7).

Table 5. The influence of user generated content on hotel choices: Comparing offline and online buyers

Have You Ever Changed the hotel accommodation Suggested by a Travel Agency Based on User Generated Content ?	Have You Ever Used the Internet to Make Hotel Reservations?					
	Yes		No		Tot.	χ^2
	N	%	N	%		
Yes	301	22.60	34	9.80	335	7.257*
No	650	77.40	127	90.20	777	
Tot.	951		161		1112	

*Significant at 0,1 level – missing data = 19.

Table 6. The influence of user generated content on tourist choices: Comparing occasional and moderate buyers

Have You Ever Changed the Accommodation Suggested by a Travel Agency Based on User Generated Content?	Occasional Buyers		Moderate Buyers		Frequent Buyers		Total		χ^2
	N	%	N	%	N	%	N	%	
Yes	221	28.0	70	36.2	40	46.5	331	30.9	15.569*
No	569	72.0	123	63.8	46	53.5	738	69.1	
Tot	790		193		86		1069		

*Significant at 01 level – missing data = 62.

Table 7. Types of travel and online booking of hotels

	Domestic	European	Intercontinental	Short-Term Journeys (4 Days or Less)	Medium-Long Term Journeys (5 Days or More)
Leisure/holidays	95.6%	95.3%	50.4%	95.6%	78.1%

Moreover, Gen Y travellers were found to book online more often for short-term journeys (95.6%) than for medium-long term journeys (78.1%). This confirms prior studies highlighting that short-haul travellers perceive traditional travel agents as biased operators (because they are commission oriented) and less flexible or able to offer a good variety of choices (Law et al., 2004). Meanwhile, the results seem to confirm, even in the context of Gen Y, that the greater the distance travelled and the longer the duration of the stay, the greater is the number of travellers using travel agencies (e.g., Del Chiappa, 2013; Prayag & Del Chiappa, 2014).

CONCLUSION

Young tourists' perceptions of online and offline hotel disintermediation are a prominent issue in the tourism literature (e.g., Barton et al., 2012). The findings of this paper provide further insights into the scientific debate on disintermediation and the perceived trustworthiness of User Generated Content for travellers from Gen Y. Given that the majority of studies on this topic are from the USA (McCubbrey, 1999) and Hong Kong (Law, 2009; Law & Lau, 2004), to our best knowledge, the only paper devoted to this topic in the European context is the one offered by Prayag and Del Chiappa (2014) in the French context.

Overall, the results reveal that the benefits offered by the Internet are rated more positively than the benefits offered by travel agencies, thus confirming prior research that shows the influence exerted by the Internet over holiday planning (e.g., Gretzel & Yoo, 2008; Lester et al., 2006; Nusair et al., 2013). Further, the study found three underlying dimensions of perceptions of disintermediation, namely:

- "Benefits of Traditional Travel Agency",
- "Benefits of Online Reservation",
- "Online Trust & Search Behaviour".

Further, our results highlight that age and education exert a significant influence on Gen Y perceptions, whereas income has no significant influence on the identified factors. This confirms Prayag and Del Chiappa's (2014) study on Gen Y in France, which highlighted the significant influence of age. At the same time, this study disconfirms Prayag and Del Chiappa (2014), which showed that education is not significantly influenced by Gen Y perceptions.

Online buyers were reported to change the accommodation suggested by a travel agent based on User Generated Content more often than offline buyers. Further, our findings showed that frequent buyers change their minds more often than moderate and occasional buyers (Table 6). This disconfirms prior studies carried out in the French context (Prayag & Del Chiappa, 2014), where such differences were not found when considering occasional, moderate and frequent buyers; further, it highlights that consumers from Gen Y who are more familiar with using the Internet to make hotel reservations tend to express a higher trustworthiness toward User Generated Content when compared with their counterparts, thus disconfirming prior studies (e.g., Yoo & Gretzel, 2010) in which the perceived trustworthiness of User Generated Content has been found to be higher for those who have less familiarity with the Internet. Further, the respondents reported preferring to use the Internet to make hotel reservations for short-term and short-haul trips, thus confirming prior studies (e.g., Prayag & Del Chiappa, 2014).

The study provides significant managerial implications. Based on the findings, it is clear that the benefits of the Internet for hotel booking outweigh the benefits offered by travel agencies. On the one hand, this suggests that hotel marketers could find it profitable to redistribute the hotel room quota, placing more hotel rooms on the Internet to increase their sales and their profits. Further, our findings indicate that User Generated Content has some influence on the hotel booking decision. In fact, 30.9% of online buyers in this study have changed the accommodation suggested by their travel agent based on User Generated Content. Hence, hospitality managers need to monitor their online reputations and engage, as much as possible, with their past and potential customers. Assuming that most of the booking services traditionally provided by travel agents can now be carried out over the Internet, our results suggested that travel agents should create and maintain a presence in the electronic marketplace, behave more as professional counsellors (Anckar, 2003), improve their customer service and ability to carve out Gen Y, thus developing marketing strategies that allow them to bind Gen Y in long-term relationships (Nusair, Parsa & Cobanoglu, 2011). Further, it should not be ignored that according to our findings Italian travelers belonging to Gen Y often change accommodation suggested by a travel agent based on User Generated Content. This suggests that travel agencies should identify the best offer for their customer complementing the information they can find in the catalogs offered by tour operators with the comments and review that travelers upload in peer-to-peer travel applications; by doing this, travel agents could show customers that the offer is the most suitable for their needs based on both offline (business-to-consumer) and online content (peer-to-peer). Broadly speaking, our results also support online marketers in their attempt to enhance their website design and promotional activities. For example, hotel marketers should design their websites to give Gen Y a sense of belonging to the hotel so that they can develop an emotional attachment toward the website and the hotel (Nusair et al., 2013; Nusair et al., 2011). Further, hoteliers could design their websites in a way that allows them to personalise their services for their market target and to offer their customers the possibility to be entertained, thus rendering the booking experience of Gen Y more enjoyable.

LIMITATIONS AND FUTURE RESEARCH

Although this study helps to fill a gap in the existing knowledge in the literature and proposes some implications for practitioners, several limitations still remain. First, the study used a convenience sample obtained on a web-based survey with a snowball sampling approach, and thus, the findings cannot be generalised. Second, the study did not investigate searching and booking behaviour pertaining to sub-groups of Italian consumers from Gen Y, and it did not explicitly consider the moderating effect that perceived risk and security exert on online tourist behaviour. Future research could be devoted to analysing these aspects. Third, the study is highly site-specific (i.e., a single country was investigated); hence, future research could replicate the study internationally, allowing for cross-cultural comparisons, to verify whether online buyers in different countries show different views in favour of and against hotel disintermediation. Future research is needed to understand whether the trustworthiness of UGC for online buyers is directly or indirectly related to their own experience using the Internet. According to some pioneering research (Grønflaten, 2011), future research around the topic of disintermediation in the context of Gen Y could be analyzed both as a choice between two information sources (travel agents vs. service providers), and as a choice between two information channels (face-to-face vs. the Internet), thus allowing us to consider more explicitly and in detail the relatively recent, and even growing, phenomenon of disintermediation in the online domain (hotel websites versus online travel agencies; e.g., Liu & Zhang, 2014). Finally, it could be interesting to investigate Gen Y online and offline behaviour by formulating a framework and/or setting up new hypotheses.

ACKNOWLEDGMENT

The authors would like to warmly thank the anonymous reviewers for their valuable help and suggestions in enhancing the quality of the manuscript.

REFERENCES

Anckar, B. (2003). Consumer intentions in terms of electronic travel distribution: Implications for future market structures. *E-Service Journal*, *2*(2), 68–86. doi:10.2979/ESJ.2003.2.2.68

Barton, C., Fromm, J., & Egan, C. (2012). *The millennial consumer, debunking stereotypes*. Retrieved January 12, 2016, from https://www.bcgperspectives.com/content/articles/consumer_insight_marketing_millennial_consumer/

Beldona, S., Morrison, A. M., & OLeary, J. (2005). Online shopping and pleasure travel products: A correspondence analysis. *Tourism Management*, *26*(4), 561–570. doi:10.1016/j.tourman.2004.03.008

Beldona, S., Nusair, K., & Demicco, F. (2009). Online travel purchase behavior of generational cohorts: A longitudinal study. *Journal of Hospitality Marketing & Management*, *18*(4), 406–420. doi:10.1080/19368620902799627

Benckendorff, P., Moscardo, G., & Pendergast, D. (Eds.). (2010). *Tourism and Generation Y*. Oxford, UK: CABI.

Berne, C., Garcia-Gonzalez, M., & Mugica, J. (2012). How ICT shifts the power balance of tourism distribution channels. *Tourism Management*, *33*(1), 205–214. doi:10.1016/j.tourman.2011.02.004

Bolton, R. N., Parasuraman, A., Hoefnagels, A., Migchels, N., Kabadayi, S., Gruber, T., & Solnet, D. et al. (2013). Understanding Generation Y and their use of social media: A review and research agenda. *Journal of Service Management*, *24*(3), 245–267. doi:10.1108/09564231311326987

Buhalis, D. (1998). Strategic use of information technologies in the tourism industry. *Tourism Management*, *19*(5), 409–421. doi:10.1016/S0261-5177(98)00038-7

Buhalis, D., & Law, R. (2008). Progress in information technology and tourism management: 20 years on and 10 years after the Internet – The state of eTourism research. *Tourism Management, 29*(4), 609–623. doi:10.1016/j.tourman.2008.01.005

Chen, S. L. (2001). Effects of value, affect, security, and web content on informational and transactional usage of the Internet. *Asia Pacific Journal of Tourism Research, 6*(81), 63–72. doi:10.1080/10941660108722089

Communify. (2015). *User-generated content and online reviews: Let your users do your marketing*. Retrieved February 12, 2016, from http://resources.communify.com/user-generated-content-and-online-reviews/

d'Italia, B. (2013). *Turismo internazionale dell'Italia* [Outbound tourism flows in Italy]. Retrieved September 29, 2013, from https://www.bancaditalia.it/statistiche/rapp_estero/turismo-int

Del Chiappa, G. (2011). Trustworthiness of Travel 2.0 applications and their influence on tourists' behaviour: An empirical investigation in Italy. In R. Law, M. Fuchis, & F. Ricci (Eds.), *Information and Communication Technologies in Tourism 2011 Proceedings of the International Conference in Innsbruck, Austria, 2011* (pp. 331–342). Vienna, Austria: Springer-Verlag.

Del Chiappa, G. (2013). Internet versus travel agencies: The perception of different groups of Italian online buyers. *Journal of Vacation Marketing, 19*(1), 1–12.

Del Chiappa, G., Lorenzo-Romero, C., & Alarcón-del-Amo, M. D. C. (2016). Internet and User-Generated Content Versus High Street Travel Agencies: A Latent Gold Segmentation in the Context of Italy. *Journal of Hospitality Marketing & Management, 25*(2), 197–217. doi:10.1080/19368623.2014.1001933

Del Chiappa, G., Lorenzo-Romero, C., & Gallarza, M. (2014). Attitude toward disintermediation in hotel reservations: Spanish travellers profile. *European Journal of Tourism Research, 9*, 129–143.

Djamasbi, S., Siegel, M., & Tullis, T. (2010). Generation Y, web design, and eye tracking. *International Journal of Human-Computer Studies, 68*(5), 307–323. doi:10.1016/j.ijhcs.2009.12.006

Frost, F. A., & Shanka, T. (1999). Asian Australian student travel preferences: An empirical study. *Asia Pacific Journal of Tourism Research, 4*(2), 19–26. doi:10.1080/10941669908722040

Gretzel, U., Fesenmaier, D., & O'Leary, J. (2006). The transformation of consumer behaviour. In D. Buhalis & C. Costa (Eds.), *Tourism business frontiers - Consumers, products and industry*. Oxford, UK: Elsevier. doi:10.1016/B978-0-7506-6377-9.50009-2

Gretzel, U., & Yoo, K. H. (2008). Use and impact of online travel reviews. In *Proceedings of the International Conference on Information and Communication Technologies in Tourism* (pp. 35-46). Innsbruck, Austria: Springer. doi:10.1007/978-3-211-77280-5_4

Gretzel, U., Yoo, K. H., & Purifoy, M. (2007). *Online travel review study: The role and impact of online travel reviews*. College Station, TX: Laboratory for Intelligent Systems in Tourism.

Grønflaten, Ø. (2011). Predicting travelers choice of information sources and information channels. *Journal of Travel Research, 48*(2), 230–244. doi:10.1177/0047287509332333

Guo, X., Ling, L., Dong, Y., & Liang, L. (2013). Optimal pricing strategy based on market segmentation for service products using online reservation systems: An application to hotel rooms. *International Journal of Hospitality Management*, *35*, 274–281. doi:10.1016/j.ijhm.2013.07.001

Hair, J. F., Black, W. C., Babin, B. J., & Anderson, R. E. (2014). Multivariate data analysis (7th ed.). Edinburgh Gate: Pearson Education Limited.

Hennig-Thurau, T. (2005). Word of mouse: Why consumers listen to each other on the internet. *Yearbook of Marketing and Consumer Research*, *3*, 37–58.

Hritz, N., Sidman, C. L., & DAbundo, M. (2014). Segmenting the college educated generation Y health and wellness traveler. *Journal of Travel & Tourism Marketing*, *31*(1), 132–145. doi:10.1080/10548408.2014.861727

Huang, L. (2011). Social media as a new play in a marketing channel strategy: Evidence from Taiwan travel agencies blogs. *Asia Pacific Journal of Tourism Research*, *17*(6), 615–634. doi:10.1080/10941665.2011.635664

International, I. P. K. (2013). *ITB world travel trends report*. Berlin, Germany: Messe Berlin GmbH.

Internet World Stats. (2016). *Internet usage statistics*. Retrieved January 30, 2016, from http://www.internetworldstats.com/stats.htm

Ip, C., Lee, H. A., & Law, R. (2012). Profiling the users of travel websites for planning and online experience sharing. *Journal of Hospitality & Tourism Research (Washington, D.C.)*, *36*(3), 418–426. doi:10.1177/1096348010388663

ITB. (2016). *World travel trends report 2015/2016*. Retrieved February 5, 2016, from http://www.itbberlin.de/media/itbk/itbk_dl_all/itbk_dl_all_itbkongress/itbk_dl_all_itbkongress_itbkongress365/itbk_dl_all_itbkongress_itbkongress365_itblibrary/itbk_dl_all_itbkongress_itbkongress365_itblibrary_studien/ITB_World_Travel_Trends_Report_2015_2016.pdf

Jin, T., Lin, V. S., & Hung, K. (2014). Chinas Generation Ys Expectation on Outbound group Package Tour. *Asia Pacific Journal of Tourism Research*, *19*(6), 617–644. doi:10.1080/10941665.2013.806939

Jones, A., & Fox, S. (2009). *Generations online 2009*. Retrieved September 25, 2014, from http://www.pewinternet.org/PPF/r/275/report_display.asp

Kim, W. G., & Kim, D. J. (2004). Factors affecting online hotel reservation intention between online and non-online customers. *Hospital Management*, *23*(4), 381–395. doi:10.1016/j.ijhm.2004.02.001

Law, R. (2009). Disintermediation of hotel reservations: The perception of different groups of online buyers in Hong Kong. *International Journal of Contemporary Hospitality Management*, *21*(6), 766–772. doi:10.1108/09596110910976007

Law, R., Buhalis, D., & Cobanoglu, C. (2014). Progress on information and communication technologies in hospitality and tourism. *International Journal of Contemporary Hospitality Management*, *26*(5), 727–750. doi:10.1108/IJCHM-08-2013-0367

Law, R., & Lau, W. (2004). A study of the perceptions of Hong Kong hotel managers on the potential disintermediation of travel agencies. *Journal of Travel & Tourism Marketing*, *17*(2-3), 117–131. doi:10.1300/J073v17n02_10

Law, R., Leung, K., & Wong, J. (2004). The impact of the internet on travel agencies. *International Journal of Contemporary Hospitality Management*, *16*(2), 100–107. doi:10.1108/09596110410519982

Lee, T. Y. (2013). *Top 10 trends of the next generation of travel: The millennials*. Retrieved January 10, 2016, from www.hvs.com

Lester, D. H., Forman, A. M., & Loyd, D. (2006). Internet shopping and buying behavior of college students. *Services Marketing Quarterly*, *27*(2), 123–138. doi:10.1300/J396v27n02_08

Liu, J. N. K., & Zhang, E. Y. (2014). An investigation of factors affecting customer selection of online hotel booking channels. *International Journal of Hospitality Management*, *39*, 71–83. doi:10.1016/j.ijhm.2014.01.011

Manjit, S. Y., & Pavlou, P. A. (2014). Marketing in computer-mediated environments: Research synthesis and new directions. *Journal of Marketing*, *78*(1), 20–40. doi:10.1509/jm.12.0020

McCubbrey, D. J. (1999). Disintermediation and reintermediation in the US air travel distribution study: A delphi study. *Communications of the Association for Information Systems*, *1*, 2–39.

Money, R., & Crotts, J. (2003). The effect of uncertainty avoidance on information search, planning and purchases of international vacations. *Tourism Management*, *24*(1), 191–202. doi:10.1016/S0261-5177(02)00057-2

Morrison, A., Jing, A., OLeary, J., & Cai, L. (2001). Predicting usage of the internet for travel bookings: An exploratory study. *Information Technology and Tourism*, *4*(1), 15–30. doi:10.3727/109830501108750868

Nunnally, J. C. (1978). *Psychometric theory* (2nd ed.). New York, NY: McGraw-Hill.

Nusair, K. K., Bilgihan, A., & Okumus, F. (2013). The role of online social network travel websites in creating social interaction for Generation Y travelers. *International Journal of Tourism Research*, *15*(5), 458–472. doi:10.1002/jtr.1889

Nusair, K. K., Parsa, H. G., & Cobanoglu, C. (2011). Building a model of commitment for Generation Y: An empirical study on e-travel retailers. *Tourism Management*, *32*(4), 833–843. doi:10.1016/j.tourman.2010.07.008

Palmer, A., & McCole, P. (1999). The virtual re-intermediation of travel services: A conceptual framework and empirical investigation. *Journal of Vacation Marketing*, *6*(1), 33–47. doi:10.1177/135676679900600105

Pempek, T. A., Yermolayeva, Y. A., & Calvert, S. L. (2009). College students social networking experiences on Facebook. *Journal of Applied Developmental Psychology*, *30*(3), 227–238. doi:10.1016/j.appdev.2008.12.010

Pizam, A., Jeong, G. H., Reichel, A., Van Boemmel, H., Lusson, J. M., Steynberg, L., & Montmany, N. et al. (2004). The relationship between risk-taking, sensation-seeking, and the tourist behavior of young adults: A cross-cultural study. *Journal of Travel Research*, *42*(3), 151–260. doi:10.1177/0047287503258837

Prayag, G., & Del Chiappa, G. (2014). Hotel disintermediation in France: Perceptions of students from Generation Y. *Anatolia – An International Journal of Tourism and Hospitality Research, 25*, 417–430.

Richards, G. (2007). *New horizons II: The young independent travellers 2008*. Amsterdam: WYSE Travel Confederation.

Richards, G., & Wilson, J. (2003). *New Horizons in Independent Youth and Student Travel, a Report for the International Student Travel Confederation (ISTC) and the Association of Tourism and Leisure Education (ATLAS)*. International Student Travel Confederation - ISTC, Amsterdam. Retrieved September 20, 2014, from http://www.atlas-euro.org/pages/pdf/FINAL_Full_Report.pdf

Sun, S., Nang Fong, L. H., Law, R., & Luk, C. (2015). An investigation of Gen-Ys online hotel information search: The case of Hong Kong. *Asia Pacific Journal of Tourism Research, 21*(4), 443–456. doi:10.1080/10941665.2015.1062405

Sweeney, R. (2006). *Millennial behaviors and demographics*. Retrieved October 29, 2015, from http://library1.njit.edu/staff-folders/sweeney/

TripAdvisor. (2014). *24 insights to shape your TripAdvisor strategy*. Retrieved February 15, 2016, from https://www.tripadvisor.com/TripAdvisorInsights/n2120/24-insights-shape-your-tripadvisor-strategy

Wan, C.-S. (2002). The web sites of international tourist hotels and tour wholesalers in Taiwan. *Tourism Management, 23*(2), 155–160. doi:10.1016/S0261-5177(01)00048-6

We Are Social. (2016). *Digital in 2016, We Are Social's compendium of global digital social and mobile data trends and statistics*. Retrieved January 30, 2016, from http://www.slideshare.net/wearesocialsg/digital-in-2016?from_action=save

Weber, K., & Roehl, W. S. (1999). Profiling people searching for and purchasing travel products on the world wide web. *Journal of Travel Research, 37*(3), 291–298. doi:10.1177/004728759903700311

Wrenn, B., Stevens, R. E., & Loudon, D. L. (2007). *Marketing research. Text and cases*. New York, NY: Haworth Press.

Xiang, Z., Magnini, V. P., & Fesenmaier, D. R. (2015). Information technology and consumer behavior in travel and tourism: Insights from travel planning using the Internet. *Journal of Retailing and Consumer Services, 22*, 244–249. doi:10.1016/j.jretconser.2014.08.005

Yoo, K. H., & Gretzel, U. (2010). Antecedents and impacts of trust in travel-related consumer-generated media. *Information Technology & Tourism, 12*(2), 139–152. doi:10.3727/109830510X12887971002701

Yoo, K. H., & Gretzel, U. (2011). Influence of personality on travel-related consumer-generated media creation. *Computers in Human Behavior, 27*(2), 609–621. doi:10.1016/j.chb.2010.05.002

Zehre, A., & Moschl, P. (2008). New distribution channels and business strategies for location-based travel agencies. In P. O'Connor, W. Höpken, & U. Gretzel (Eds.), *Proceedings of the International Conference on Information and Communication Technologies in Tourism* (pp. 359-382). Innsbruck, Austria: Springer.

ADDITIONAL READING

Abou-Shouk, M., Megicks, P., & Lim, W. M. (2013). Perceived Benefits and E-Commerce Adoption by SME Travel Agents in Developing Countries Evidence from Egypt. *Journal of Hospitality & Tourism Research (Washington, D.C.)*, *37*(4), 490–515. doi:10.1177/1096348012442544

Amaro, S., Duarte, P., & Henriques, C. (2016). Travelers use of social media: A clustering approach. *Annals of Tourism Research*, *59*, 1–15. doi:10.1016/j.annals.2016.03.007

Baka, V. (2016). The becoming of user-generated reviews: Looking at the past to understand the future of managing reputation in the travel sector. *Tourism Management*, *53*, 148–162. doi:10.1016/j.tourman.2015.09.004

Chen, S. L. (2001). Effects of value, affect, security, and web content on informational and transactions usage of the internet. *Asia Pacific Journal of Tourism Research*, *6*(1), 63–72. doi:10.1080/10941660108722089

Cheung, C. M., Lee, M. K., & Rabjohn, N. (2008). The impact of electronic word-of-mouth: The adoption of online opinions in online customer communities. *Internet Research*, *18*(3), 229–247. doi:10.1108/10662240810883290

Del Chiappa, G., Lorenzo-Romero, C., & Alarcón-del-Amo, M. D. C. (2015). The Influence of User-Generated Content on Tourists' Choices. *Market/Tržište*, *27*(2), 221-236.

Del Chiappa, G., Velčovská, S., & Atzeni, M. (2016). Hotel Disintermediation and User Generated Content in the Czech Republich: Perceptions of Students from Generation Y. In M. Kozack & N. Kozack (Eds.), *Tourist Behavior. International Perspectives*. Newcastle, UK: Cambridge Scholars Publishing.

Frost, F. A., & Shanka, T. (1999). Asian Australian student travel preferences: An empirical study. *Asia Pacific Journal of Tourism Research*, *4*(2), 19–26. doi:10.1080/10941669908722040

Heung, V. C., & Leong, J. S. (2006). Travel demand and behavior of university students in Hong Kong. *Asia Pacific Journal of Tourism Research*, *11*(1), 81–96. doi:10.1080/10941660500500766

Labrecque, L., & Mathwisk, C. (2013). Consumer power: Evolution in the digital age. *Journal of Interactive Marketing*, *27*(4), 257–269. doi:10.1016/j.intmar.2013.09.002

Lee, C. F., & King, B. (2016). International Students in Asia: Travel Behaviors and Destination Perceptions. *Asia Pacific Journal of Tourism Research*, *21*(4), 457–476. doi:10.1080/10941665.2015.1062786

Lu, Q. S., Yang, Y., & Yuksel, U. (2015). The impact of a new online channel: An empirical study. *Annals of Tourism Research*, *54*, 136–155. doi:10.1016/j.annals.2015.06.006

Peterson, R. A., & Merino, M. C. (2003). Consumer information search behavior and the Internet. *Psychology and Marketing*, *20*(2), 99–121. doi:10.1002/mar.10062

Seekings, J. (1998). The Youth Travel Market. *Travel and Tourism Analyst*, *5*, 37–55.

Sheahan, P. (2009). *Gen Y: Thriving and Surviving with Gen Y at Work*. New York: Hardie Grant Books.

Stangl, B., Inversini, A., & Schegg, R. (2016). Hotels dependency on online intermediaries and their chosen distribution channel portfolios: Three country insights. *International Journal of Hospitality Management*, *52*, 87–96. doi:10.1016/j.ijhm.2015.09.015

Sun, S., Fong, L. H. N., Law, R., & Luk, C. (2016). An Investigation of Gen-Ys Online Hotel Information Search: The Case of Hong Kong. *Asia Pacific Journal of Tourism Research*, *21*(4), 443–456. doi: 10.1080/10941665.2015.1062405

Tsai, H. T., Huang, L., & Lin, C. G. (2005). Emerging e-commerce development model for Taiwanese travel agencies. *Tourism Management*, *26*(5), 787–796. doi:10.1016/j.tourman.2004.04.009

Westerman, D., Spence, P. R., & Van Der Heide, B. (2014). Social media as information source: Recency of updates and credibility of information. *Journal of Computer-Mediated Communication*, *19*(2), 171–183. doi:10.1111/jcc4.12041

World Tourism Organization. (2008), Youth Travel Matter. Understanding the Global Phenomenon of Youth travel. Retrieved March 12, 2013, from http://www.siimt.com/work/sites/siimt/resources/LocalContent/1172/6/YouthTravel_protec.pdf (accessed 20 January 2013).

Yung, C. Y. (1997). Internet marketing: The perception of travel agencies in Taipei, Taiwan. *Asia Pacific Journal of Tourism Research*, *2*(1), 67–74. doi:10.1080/10941669808721986

KEY TERMS AND DEFINITIONS

ANOVA: A statistical test used to compare the means of more than two samples.

Disintermediation: The disintermediation hypothesis describes the idea that the role of tourism intermediaries can be eliminated.

Electronic Word of Mouth (E-WOM): All of the informal communications directed at consumers through Internet-based technology with regard to the consumption experience of products and services and their providers.

Factor Analysis: A statistical method used to describe the variability among observed variables in terms of a potentially lower number of unobserved variables called factors.

Gen Y: The cohort of people born between 1980 and 1999 (the so-called Millennials).

Tripographic Characteristics: Travel-related behavioural characteristics, such as travel party, the duration of the stay, means of transport and types of accommodation.

User Generated Content (UGC): All types of comments, reviews, photos and videos that consumers can upload on peer-to-peer applications as regards to products, services and their providers.

Chapter 19
Prey Predator Algorithm for Travelling Salesman Problem:
Application on the Ethiopian Tourism Sites

Surafel Luleseged Tilahun
University of KwaZulu-Natal, South Africa

Natnael Nigussie Goshu
Addis Ababa Science and Technology University, Ethiopia

Jean Medard T. Ngnotchouye
University of KwaZulu-Natal, South Africa

ABSTRACT

Visiting most, if not all, tourist destination of a country while visiting a country is an ideal plan of a tourist. In most cases if the tour is not carefully planned, it will be costly and time taking to travel between tourist destinations of a country. If we consider Ethiopia, a country which has been named as best tourism destination for 2015 by the European Council on Tourism and Trade (ECTT); there are many tourist destinations all over the country. The problem of determining the optimum route to visit all the tourist sites with minimum traveling time can be formulated as a travel salesman problem. In this study 17 of the famous tourist destination of Ethiopia will be selected and a travel salesman model will be formulated. Due to the NP hardness of the travel salesman problem, metaheuristic based algorithms are found to be more effective. Hence, a recently introduced swarm based metaheuristic algorithm, called prey predator algorithm will be used to deal with the formulated problem. The simulation result suggests that the best route to visit the selected destination is Addis Ababa → Sof Omar Caves → Bale Mountain National Park → Abijiata-Shalla Lakes → Netchisar National Park → Mago National Park → Omo National Park → Gambella National Park → Bahir Dar → Lalibela → Gonder → Semien Mountain National Park → Axum → Ertale → Yangudi Rassa National Park → Awash National Park → Harar → Addis Ababa. It should be noted in some of the sites there is no direct route and hence a route through other cities is used and hence it should be recomputed in the future when a direct route between these tourist destinations is constructed.

DOI: 10.4018/978-1-5225-1054-3.ch019

INTRODUCTION

Tourism can be recognized as long as people have traveled. Food, water, safety or acquisition of resources (trade) was the early travel motivations. But the idea of travel for pleasure or exploration soon emerged. Travel has always been depending on technology, to provide the means or mode of travel. The earliest travelers walked or rode domesticated animals. The invention of the wheel and the sail provided new modes of transportation. Each improvement in technology increased individuals' opportunities to travel. As roads were improved and governments stabilized, interest in travel increased for education, sightseeing, and religious purposes. One of the earliest travel guides was written by Pausanias, a Greek, which was a 10 volume guide to Greece, for Roman tourists in 170 A.D (Satta, 2005). However, there is no consensus concerning the definition of tourism it varies source by source, person by person.

The first definition of tourism was made by Guyer Feuler in 1905 (Bogahavatta, 2013). He defined Tourism as:

A phenomenon unique to modern time which is dependent on the people's increasing need for a change and relaxing, the wish of recognizing the beauties of nature and art and the belief that nature gives happiness to human beings and which helps nations and communities' approaching to each other thanks to the developments in commerce and industry and the communication and transportation tools' becoming excellent.

In 1976, the Tourism Society of England defined Tourism as "Tourism is the temporary, short-term movement of people to destination outside the places where they normally live and work and their activities during the stay at each destination. It includes movements for all purposes." (Bogahavatta, 2013). Mathieson and Wall (1982) created a good working definition of tourism as "the temporary movement of people to destinations outside their normal places of work and residence, the activities undertaken during their stay in those destinations, and the facilities created to cater to their needs." According to Macintosh and Goeldner (1986) tourism is "the sum of the phenomena and relationships arising from the interaction of tourists, business suppliers, host governments and host communities in the process of attracting and hosting these tourists and other visitors." But when it comes to explain it with the basic terms, we can sum it up as follows; Tourism is a collection of activities, services and industries which deliver a travel experience comprising

- Transportation,
- Accommodation,
- Eating and drinking establishments,
- Retail shops,
- Entertainment businesses and other hospitality services provided for individuals or groups traveling away from home not more than one consecutive year.

According to The World Tourism Organization (WTO) 2014 report, tourism is currently the world largest industry with annual revenues of over $1.5 trillion dollars (World Tourism Organization (UNWTO), 2016).

Tourism consists of certain components (dimensions), four of which may be considered as basic. These four basic component of tourism are;

1. **Attractions:** Usually focus on natural resources, culture, ethnicity or entertainment.
2. **Facilities:** When tourists arrive at attractions place they require facilities to provide services like lodging, food & beverage, support services, and communication infrastructure.
3. **Hospitality:** The community's attitude which permeates every tourism location that makes the tourist feels welcome and safe. It is the result of the interaction between the tourist and the local population.
4. Transportation.

Transportation is the critical component of tourism. The basic questions a traveler looks at are the time needed to travel to the destination and its corresponding traveling cost. Tourism developments are dependent on the ease of access and types of transportation available. Besides to this, the number of tourists has increased from time to time to visit different destination as the means of transportation improves with technology. However, most of the time tourist destination are not visited by minimizing either cost or time rather through suggestions without considering the minimum distance or cost due to lack of information about the destination and appropriate solution method to have minimum total distance or minimum total cost. For a tourist visiting a new country, visiting all major tourist destinations with minimum cost and minimum time is an ideal objective to achieve. Furthermore, this problem can be formulated as a travel salesman problem. Hence in this chapter, taking the tourism sites of Ethiopia as a cases study we will formulate the tourism visiting problem as a travel salesman problem and prey predator algorithm will be used to solve the problem.

Ethiopia is one of the oldest countries with intact cultural heritage as it never was colonized. Furthermore, it has been a center for trade for horn of Africa in old days and also an icon place for religion. That is in early ages Christianity has entered and the old tradition of that time Christianity remains and preserved from western culture. It has also mentioned many times in the bible. It has a big role in Islam as well, as the early Muslims fled to Ethiopia, the then called Abyssinia, to be saved. There are also Ethiopian Jews.

Prey predator is a recently introduced metaheuristic algorithm inspired by the interaction between a predator and its prey (Tilahun, 2013; Tilahun & Ong, 2015). The formulated problem will be solved using this algorithm. However, since the original algorithm is designed for continuous problem a proper modification to be used for travel salesman problem will be given and used to solve the problem. But before that brief explanation of Ethiopian and its tourism sites will be explained. After all of this, simulation result and conclusion will be presented.

TOURISM IN ETHIOPIA

Ethiopia has been voted as the world's best tourism destination for 2015 by the European Council on Tourism and Trade (ECTT), a non-profit group of European Union tourism trade organizations (ECTT, 2015). Government of Ethiopia have also plane to make this country one of the 5 destination in Africa. Of course, due to globalization and rapidly improving infrastructure, the tourism industry is showing a tremendous progress. Besides to this, the country makes commitment and willingness to develop tourism through a number of initiatives because tourism can tremendously contribute to the economy and improve gross domestic product (GDP).

Ethiopia is one of non-colonized countries in Africa along with Liberia and as large as France and Spain combined, above 1.1 million square kilometers, and has one of the richest histories on the African continent and located at 3-15 degree North, and 33-48 degree East. It has many unique identities like its own alphabet, calendar and time counting system. It is the source of many revers including Nile which is a backbone for Egypt's development. It is a big country with more than 100 million people speaking more than 80 languages with their own unique culture. It has a number of endemic animals and plants which attracted tourists all over the world. Nine entries are already registered under world heritage by UNESCO with many to be reviewed:

- 20 national parks,
- 4 wild life sanctuaries,
- 18 controlled hunting areas, and
- 2 pilot projects community conservation areas.

There are 861 species of Birds of which 17 are endemic to Ethiopia. There are 284 species of mammal of which 31 are endemic and 7 are large mammals. (FDREMCT, 2015)

Now a day's Ethiopia is drawn the attention of many tourists. Even though there are many destinations interesting for tourists, the major tourist destinations based on different websites and FDREMCT (2015) in the country are listed as follow:

1. **Addis Ababa (AA):** Addis Ababa is the capital city of the country, the country's commercial and cultural hub, and the sitting place for more than 115 embassies, Africa Union and many international offices. Its National Museum exhibits prehistoric fossils, including replicas of the famous early hominid known as Lucy, along with Ethiopian art and traditional crafts. See Figure 1.

Figure 1. Addis Ababa
Simien Thrills Tour and Travel (2016).

Figure 2. Lalibela
Chiu (2010).

Figure 3. Axum
Across Abyssinia Tour PLC (2014).

2. **Lalibela (La):** There are 11 incredible rock-hewn monolithic churches which were built by King Lalibela way back in the late 12th or early 13th century. The structures are noted for their artistic splendor and each of the buildings has distinctive architectural designs. The churches are actually dug into the ground, and St George's cross has become its most famous of all. See Figure 2.
3. **Axum (AX):** Axum happens to be one of the oldest cities in Africa and home of the ancient Aksumite Empire trading power that ruled the region from about 400 BC into the 10th century. Axum also is the supposed location of the Ark of the Covenant and Rock churches of Tigray. See Figure 3.
4. **Bahir Dar Lake Tana and Blue Nile Falls (BD):** Another place not to be missed place in Ethiopia is the Blue Nile, the longest river in Africa. It is locally known as Tis Isat, is 400m wide and 45m deep. When the water falls into a gorge, it creates the most spectacular view. In addition to this, more than 37 islands in the Lake Tana have historical monasteries. See Figure 4.
5. **Fassil Ghebbi, Gonder (Go):** Gondar happens to be the real life Camelot of Ethiopia. An entire castle complex that includes an entire complex of castles. There is also a small monastery in a nearby city called Gorgora which is very impressive. See Figure 5.

Figure 4. Bahir Dar
Gishen Tours (2015).

6. **Harar (Ha):** It is an ancient city which dates back to about 1520 and with the oldest Islamic city in the Continent. The city is famed for its ancient buildings and impressive city walls. Harar has approximately 99 mosques and it oozes old school islamic culture. A real highlight in Harar is the 'wild' hyenas though. At dusk you can see the local men attracting wild hyenas in the city. For generations these hyenas have scavenged within the city walls at night, through this they have become 'friendly' with a group of men who feed them and allow you to feed them. See Figure 6.
7. **Sof Omar Caves (SO):** According to tradition Sof Omar was the name of a Muslim holy man who lived in the area and Ayiew is the name of his daughter. Maco and Holuca are local names for 'name' and 'cave', respectively. Long a religious center, it is sacred both to Islam and the local Oromo traditional religion. The caves are known for their many pillars, particularly in the 'Chamber of Columns'. See Figure 7.
8. **Ert Ale or Danakil Depression (Er):** The Danakil depression shares the Eritrean border on the east of the Tigrian highlands in the North East of Ethiopia. It has been officially listed as one of the hottest and driest area on earth having an average temperature of 35+ degrees Celsius. It is an impressive area of geological fascination, a place dotted with active volcanoes, hot springs, Lava Lake, salted basins and above all lunar landscapes. It is a vast unpopulated region that lies below the sea level, it's the lowest point in all of Africa and the second lowest in the world (second to the Dead Sea in Jordan/Israel). Erta Ale is the most active volcano in Ethiopia. It is 613 metres (2,011 ft) high, with one or sometimes two active lava lakes at the summit which occasionally overflow on the south side of the volcano. It is notable for holding the longest-existing lava lake. Volcanoes with lava lakes are very rare: there are only six in the world. See Figure 8.
9. **Abijatta-Shalla Lakes (AS):** It covers 887 square kilometers including the Rift Valley lakes of Abijatta and Shalla. The two lakes are separated by three kilometers of hilly land. The altitude of the park ranges from 1540 to 2075 meters, the highest peak being Mount Fike, which is situated

Figure 5. Fasil Ghebbi
UNESCO (2006).

Figure 6. Harar
Brand Ethiopia (2016).

between the two lakes. Besides the two lakes, the primary attraction of this national park are a number of hot springs on the northeast corner of Lake Abijatta, and large numbers of flamingoes on the lake. See Figure 9.

10. **Awash National Park (AW):** Covers at least 756 square kilometers of acacia woodland and grassland was established in 1966. The Addis Ababa - Dire Dawa highway passes through this park, separating the Illala Saha Plains to the south from the Kudu Valley to the north. In the south of the park the Awash River gorge has amazing waterfalls. In the upper Kudu Valley at Filwoha are hot springs amid groves of palm trees. See Figure 10.
11. **Bale Mountain National Park (BM):** Is one of the highest incidences of animal endemicity of any terrestrial habitat in the world. The park encompasses an area of approximately 2,150 square kilometers, and is divided into five distinct and unique habitats:
    a. The Northern Grasslands (Gaysay Valley),
    b. Northern Woodlands (Park Headquarters),

Figure 7. Sof Omar
Sunny 13 Tour (2016).

Figure 8. Ert Ale
Safa (2015).

c. Afro-alpine Meadows (Sanetti Pleateau),
d. Erica Moorlands, and the
e. Harenna Forest.

The park is known for being home to the largest populations of both the endemic and endangered Ethiopian wolf (Canis simensis) and mountain nyala (Tragelaphus buxtoni), as well as the endemic Bale monkey (Chlorocebus djamdjamensis) and giant mole rat (Tachyoryctes macrocephalus). See Figure 11.

12. **Gambella National Park (GN):** Was established as a protected area in 1973 to conserve a diverse assemblage of wildlife and unique habitats. Although not technically in the Rift Valley, Gambella National Park lies along another of the country's important rivers: the Baro. Near the town of Gambella, Gambella National Park have large conservation area contains many species not found elsewhere in the country, such as the Nile lechwe and the white-eared kob. Roan antelope, topi, elephant, buffalo, giraffe, and the unusual whale-headed stork are also to be found here. See Figure 12.

Figure 9. Abijatta-Shalla
RIA Ethiopian Sports (2016).

Figure 10. Awash Park
Johnny Abyssinia (2016).

Figure 11. Bale Mountain
Aves (2016).

13. **Mago National park (MN):** Covers 2162 square kilometers. This park are divided by the Mago River, a tributary of the Omo, into two parts. To the west is the Tama Wildlife Reserve, with the Tama river defining the boundary between the two. To the south is the Murle Controlled Hunting Area, distinguished by Lake Dipa which stretches along the left side of the lower Omo. The major environments in and around the Park are the
    a. Rivers and riverine forest,
    b. The wetlands along the lower Mago and around Lake Dipa,
    c. The various grasslands on the more level areas, and
    d. Scrub on the sides of the hills. See Figure 13.
14. **Netchisar National Park (NN):** It contains 514 square kilometers of territory include the "Bridge of God" (an isthmus between Lakes Abaya and Chamo), and the Nechisar (English: white grass) plains east of the lakes. Park elevations range between 1108 and 1650 meters above sea level. Nechisar National Park was established in 1974. See Figure 14.
15. **Lower Valley of Omo and Omo National Park (ON):** Lower Omo valley especially tribal tour much like the Maasai in kenya this is one of the only places in the world where you can still find indigenous people that haven't been influenced by the outside and developed world. The south is also an excellent location for African Safaris. It is also contains a large national park with an area of 4,068 square kilometres. It is a vast expanse of true wilderness, adjacent to the Omo River, which flows southwards into Lake Turkana and is one of the richest and least-visited wildlife sanctuaries in eastern Africa.
    a. Eland,
    b. Oryx,
    c. Burchell's zebra,
    d. Lelwel hartebeest,
    e. Buffalo,
    f. Giraffe,
    g. Elephant,
    h. Waterbuck,
    i. Kudu,
    j. Lion,
    k. Leopard and
    l. Cheetah roam within the park's boundaries. See Figure 15.
16. **Semien Mountain National Park (SM):** A UNESCO World Heritages site, the Simien Mountains is often called God's playground because of their unique formations. There are treks available into the national park where you can find a unique combination of animals and incredible landscapes. It is home to a number of endangered species, including the Ethiopian wolf and the walia ibex, a wild goat found nowhere else in the world. The gelada baboon and the caracal, a cat, also occur within the Simien Mountains. More than 50 species of birds inhabit the park, including the impressive bearded vulture, or lammergeier, with its 10-foot (3m) wingspan. See Figure 16.
17. **Yangudi Rassa National Park (YR):** Covers 4730 square kilometers of territory include Mount Yangudi near the southern border and the surrounding Rassa Plains, with altitudes from 400 to 1459 meters above sea level. Sandy semi-desert and wooded grassland cover the majority of the park's area. This national park was proposed in 1977 in specific to protect the African wild ass. Large animals native to the park include Beisa oryx, Soemmering's gazelle, gerenuk and Grevy's zebra. Bird species of interest include Phoenicopterus minor, Petronia brachydactyla and Ardeotis arabs. See Figure 17.

Figure 12. Gambella Park
Tropic Air Kenya (2013).

Figure 13. Mago Park
Tana Travel and Tour (2016).

Figure 14. Nechisar Park
Wikipidia (2016).

Figure 15. Omo Park
Tana Travel and Tour (2016).

Figure 16. Semien Mountain
Schenck (2016).

Figure 17. Yangudi Rassa Park
Ethiopian Embassy, Turkey (2016).

TRAVEL SALESMAN PROBLEM

The objective of Traveling Salesman Problem (TSP) is to find a tour of a given number of n of cities, visiting each city exactly once and return back to the starting or initial city where the length of the tour is minimized. Currently the only known method guaranteed to the optimally solve the traveling salesman problem is by enumerating each possible $\frac{(n-1)!}{2}$ numbers of tour and search for the tour with smallest distance (Kumbharana & Pandey, 2013). However, this method of computing is impossible if the number of city n is large then the number of tours increases factorial times. Therefore, even though TSP is easy to understand, it is very difficult to solve. Researchers have proved that traveling salesman problem is Non-deterministic polynomial time complete problem (NPC) (Garey & David, 1979; Papadimitriou, 1977; Harp, 1972). A lot of researches have been done to investigate a method to solve TSP. One of them is Brute-force method, which is based on selecting the minimum distance out of all computed tours (Cook, 2010; Maredia, 2010). Using Brute-force method to solve TSP with a large number of cities can be frustrating and even take years or centuries to solve. In addition, it is inefficient as n gets large. Nearest Neighborhood approach is one of approximation method which is based on selecting the consequence nearest city (Cook, 2010; Maredia, 2010). However, even though it works well and gives a good approximation solution, it does not always give the best solution as brute-force method. Greedy approach is also another type of approximation approach which is based on choosing the n minimum distance and Hamiltonian cycle in order to find an optimal solution (Calgor, 2010; Maredia, 2010). If the tour doesn't have Hamiltonian cycle, then there is no optimal solution. This is one of the limitations of Greedy approach. Branch and bound method of solving based on dividing a problem into a number of sub problems (Calgor, 2010; Maredia, 2010). It is a system for solving a sequence of sub-problems each of which may have multiple possible solutions and where the solution chosen for one sub-problems. The algorithm depends on the efficient estimation of the lower and upper bounds of a region or branch of the search space and approaches enumeration as the size (n-dimensional volume) of the region tends zero (Land, 1960).

However, all the above methods are difficult to apply when the numbers of cities increase, because it becomes impossible to find the minimum distance of every tour in polynomial time. Therefore, a lot of studies are conducted to resolve this problem and most of them are concentrated using metaheuristic algorithm. Metaheuristic algorithm are becoming powerful in solving modern global optimization problems (Baeck, 1997; Bonabeau, 1999; Deb, 1995; Goldberg, 1989), especially for NP-hard optimization such as the traveling sales man problem. For example, Particle Swarm Optimization (Poli, 2007), Artificial Bee Colony Optimization (Baeck, 1997) and Anti Colony Optimization (Bonabeau, 1999) are some of swarm based Algorithms. Those algorithms are features of an approximation algorithm, which in less time will end up giving a good and acceptable result. In this paper, we will focus on prey predator algorithm (PPA) to solve TSP. Now let's formulate the problem of visiting all 17 well known tourist destination in Ethiopia including the capital city Addis Ababa by starting the journey from Addis Ababa as a starting point and back to the capital, as a travel salesman problem.

The objective will be minimizing travel distance. The decision variable x will be a vector of n dimension where n - 2 is the number of tourist destinations outside the capital. Hence, a given vector will be $x = [1, x_2, x_3, \ldots, x_{n-1}, n]$, where each entry, x_i, is an integer. 1 is assigned for Addis Ababa as a starting point and 18 is assigned again for Addis Ababa as a final destination. The distance between the major

Table 1. The distance between the sites in kilometers collected from the Ministry of Culture and Tourism of Ethiopia with additional computation using google map

	AA	La	AX	BD	Go	Ha	SO	Er	AS	AW	BM	GN	MN	NN	ON	SM	YR
AA	0	701	1024	563	738	525	576	983	200	215	382	714	800	510	590	890	520
La	701	0	390	312	357	940	1234	534	630	916	1083	1147	1500	1211	1292	494	430
AX	1024	390	0	540	350	1549	1600	500	1224	1240	1406	1330	1824	1534	1614	270	570
BD	563	312	540	0	176	1088	1139	846	763	778	945	790	1258	1074	1153	323	950
Go	738	357	350	176	0	1263	1314	891	938	853	1120	937	1538	1248	1328	147	717
Ha	525	940	1549	1088	1263	0	532	1300	1183	345	1365	1239	1160	870	1115	1415	608
SO	576	1234	1600	1139	1314	532	0	1559	384	598	200	1124	480	660	564	1466	861
Er	983	534	500	846	891	1300	1559	0	1183	1175	1365	1636	1783	1493	1573	770	427
AS	200	630	1224	763	938	1183	384	1183	0	274	184	550	410	246	441	1090	447
AW	215	916	1240	778	953	345	598	1175	274	0	398	929	685	570	712	1105	263
BM	382	1083	1406	945	1120	1365	200	1365	184	398	0	944	505	321	590	1272	661
GN	714	1147	1330	790	937	1239	1124	1636	550	929	944	0	664	555	560	1113	1234
MN	800	1500	1824	1258	1538	1160	480	1783	410	685	505	664	0	183	84	1690	1030
NN	510	1211	1534	1074	1248	870	660	1493	246	570	321	555	183	0	241	1400	880
ON	590	1292	1614	1153	1328	1115	564	1573	441	712	590	560	84	241	0	1480	960
SM	890	494	270	323	147	1415	1466	770	1090	1105	1272	1113	1690	1400	1480	0	650
YR	520	430	570	950	717	608	861	427	447	263	661	1234	1030	880	960	650	0

tourist sites listed in the previous section is presented in Table 1. In some of the tours, a direct route may not exit but the distance is computed through other cities with minimum distance to pass through.

Hence, the travel salesman problem formulation of the problem can be given as in Equation (1).

$$\begin{aligned} \min f(x) &= \sum_{i=1}^{18} d(x_i, x_{i+1}) \\ s.t.\ x &\in \aleph^{18} \\ x_1 &= 1 \\ x_{18} &= 18 \\ x_i &\neq x_j, \forall i \neq j \end{aligned} \qquad (1)$$

where $d(x_i, x_j)$ is the distance from destination x_i to destination x_j. For example if $x = [1, 4, 2, 6, 8, 3, 16, 11, 5, 7, 9, 12, 10, 13, 15, 14, 17, 1]$, then $d(x_1, x_2)$ is the distance between x_1 and x_2, that is between destination 1 (which is AA) and 4 (which is BD) and that is 563. In a similar way $d(x_2, x_3) = 312$. Hence $f(x) = 11864$.

PREY PREDATOR ALGORITHM

The Standard Prey Predator Algorithm

Prey predator algorithm is swarm based metaheuristic algorithm which is inspired by the interaction between a predator and its prey (Tilahun, 2013; Tilahun &Ong, 2015). The algorithm has been tested in different applications and compared with other algorithm providing a promising results (Tilahun & Ong, 2013; Hamadneh et al., 2013; Tilahun & Melesse, 2015; Tilahun, 2012; Bahmani-Firouzi et. al., 2015; Dai et. al., 2015Tilahun & Ngnotchouye, 2016a, Tilahun & Ngnotchouye, 2016b, Tilahun & Ngnotchouye, 2016c, Tilahun et. al., 2016). The algorithm works by categorizing the initial feasible solutions as predator and prey and updates the solutions mimicking the predator prey scenario. A solution is said to be predator if its performance is the least among current solution set, a solution is said to be best prey if its performance is better than all the current solution in the given iteration and the rest of the solutions are called ordinary prey. The predator works as an agent for exploration where it explores the solution space with bigger step length and also forces the ordinary prey to explore the solution space by scaring them. The best prey, in the other hand, works as an agent for exploitation. It is considered as a prey which found a hiding place and no more affected by the predator. Hence, it totally does an exploitation of its neighborhood with smaller step length and also attracts other ordinary prey to explore its neighborhood. The movement of the ordinary prey depends on the probability of follow-up, an algorithm parameter. If a random number is less than or equal to the probability of follow-up then the prey will follow other prey with better performance compared to itself and if the random number is greater than the probability of follow-up it will run randomly away from the predator. Prey predator algorithm is summarized in Table 2.

Table 2. Prey predator algorithm

1. Parameter setup
2. Generate random N solutions
3. Sort them based on their performance in the objective function, $f(x)$, as $x_1, x_2, \ldots, x_N$ from worst to best
4. Update the predator, $x_p = x_1$, using $x_p := x_p + l_{max} rand() + l_{min} rand(x_2 - x_1)$ where *rand*() and *rand* are random vector and random number
5. For i=1:N-2
6. If rand > Pf
7. U=rand()
8. If $\|x_p-(x_i+U)\| > \|x_p-(x_i-U)\|$
9. $x_i := x_i + l_{max} rand(U)$
10. else
11. $x_i := x_i + l_{max} rand(-U)$
12. end
13. else
14. $x_i := x_i + l_{max} \dfrac{\sum_{j=i+1}^{N}\left(x_j - x_i\right)}{\left\|\sum_{j=i+1}^{N}\left(x_j - x_i\right)\right\|} + l_{min} rand()$
15. end
16. end
17. randomly generate m unit directions, say $u_1, u_2, \ldots, u_m$ and set u_{m+1} a zero vector
18. chose the direction u which will improve x_N and move x_N in that direction using $x_N := x_N + l_{min} rand(u)$
19. if termination criterion is met stop else go back to step 3.

The prey predator algorithm is introduced for continuous problems. Its updating formulas use Euclidian distance to measure the distance between two solutions and it is not suitable for discrete problems. Hence, it needs to be modified to suit the problems as will be discussed in the next section.

Prey Predator Algorithm for Travel Salesman Problem

The decision variable for the travel salesman problem takes only integer and non-repeating values, as given in Equation (1). As in the standard firefly algorithm a random combination of N routs will be generated (N initial feasible solutions will be generated). For instance in MATLAB *1+randperm(16)* will be generated 16 entries from 2 to 17 by assigning this values from $x(2)$ to $x(17)$ with $x(1) = x(18)$, an initial solution can be generated.

After generating N initial solution, their performance will be measured according to the objective function $f(x)$. The best performing solution is called to be the best prey, the worst performing solution is called the predator and the rest ordinary prey. In the standard prey predator algorithm two step lengths, exploring and exploiting step length are used, however in this case it will be replaced by probability of change, P_c and p_c.

The predator, x_p, runs randomly. That is done by changing the entry of its components based on probability of change, P_c. That is demonstrated by Table 3:

In a similar manner, the updating of the best prey will be done with smaller probability of change, p_c.

The updating mechanism of ordinary prey, x_i, is done based on an algorithm parameter, called probability of follow-up, p_f. If a random number is less than or equal to the probability of follow-up, then it will follow better performing solutions and if it is greater than the probability of follow-up then the solution will randomly run away from the predator. A sample MATLAB code is given as follows (See Table 4), (suppose that the solutions are sorted in such a way that for a solution x_i all solution with higher index performs better and x_n is x_b):

The algorithm runs until a termination criterion is achieved. That can be maximum number of iteration, a certain level of tolerance or if no improvement is recorded in consecutive number of iterations. The flowchart of the algorithm is given in Figure 18.

SIMULATION

The formulated travel salesman problem is solved using the proposed modified prey predator approach.

Table 3. Changing of predator components on basis of probability of change

```
for i=1:16
if rand <= Pc
change xp(i) randomly
repair and make the solution feasible
end
end
```

Table 4. Sample MATLAB code

```
if rand<=p_f
for j=i+1:N
for k=2:16
if rand <= p_c
xi(k)=xi(k)+round(rand*(xj(k)-xi(k)))
repair xi for feasibility
end
end
end
else
for k=2:16
if rand <= Pc
xi(k)=xi(k)+round(rand*(xi(k)-xp(k)))
                              repair xi for feasibility
end
end
end
```

Figure 18. Prey predator algorithm for the travel salesman problem

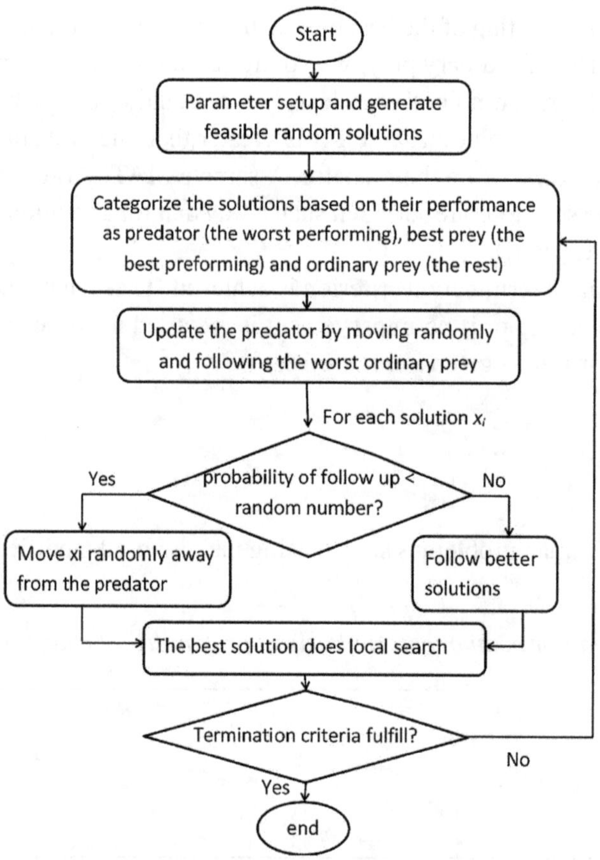

Simulation Setup

The simulation is done using MATLAB 7.10.0 (R2010a) on Intel(R) Core (TM) i3 - 3110M CPU 2.4 GHz; RAM 6GB and 64 bit operating system machine. The algorithm parameter P_c, p_c and p_f are set as 0.7, 0.4 and 0.6 respectively. The initial number of solutions, N, set to be 50. The termination criterion is maximum number of iterations and is set to be 500.

Simulation Results and Discussion

The simulation based on the setup given in the previous section is done 50 times. The best solution is found to be at x = [1 7 11 9 14 13 15 12 4 2 5 16 3 8 17 10 6 1], with $f(x)$ = 5969. In addition to the best solution, the mean and standard deviation is found to be 6587 and 272.2195, respectively. From the simulation, the best route found is:

Addis Ababa → Sof Omar Caves → Bale Mountain National Park → Abijiata-Shalla Lakes → Netchisar National Park → Mago National Park → Omo National Park → Gambella National Park → Bahir Dar → Lalibela → Gonder → Semien Mountain National Park → Axum → Ert Ale → Yangudi Rassa National Park → Awash National Park → Harar → Addis Ababa.

Figure 19 presented the functional value as a function of the number of iteration in one of the trials. The best rout on the map of the country is presented in Figure 20.

It should be noted that some places don't have a direct route, and hence even though they are close as can be seen on the map, the traveling distance can be longer. Hence the solution presented here is under

Figure 19. Simulation result of one of the trials

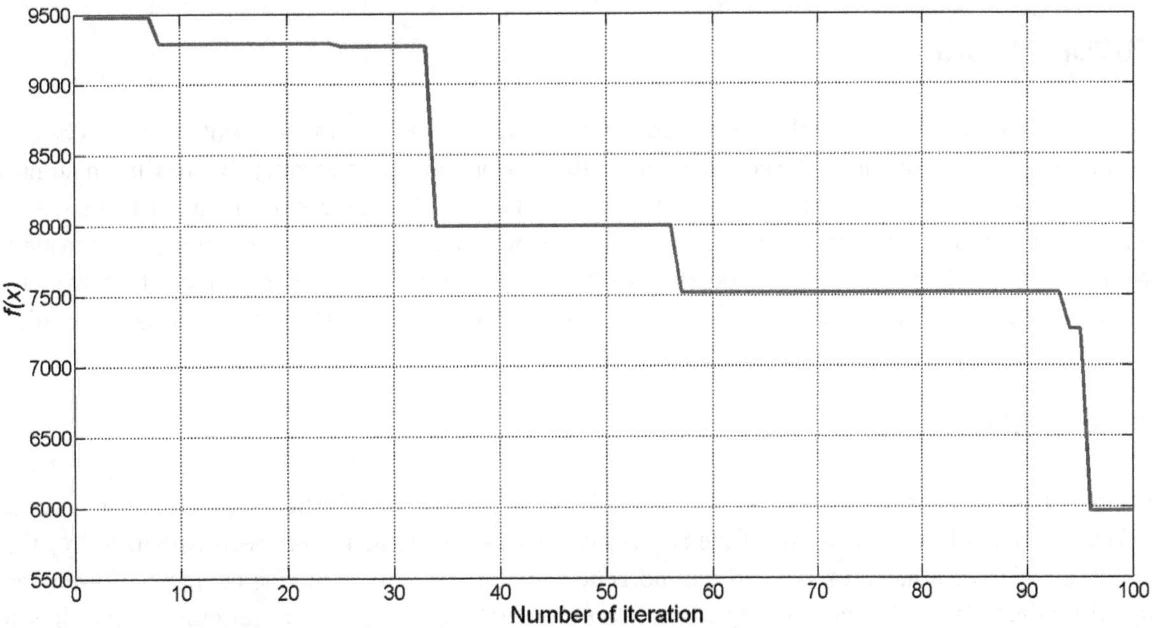

Figure 20. The best route found from the simulation

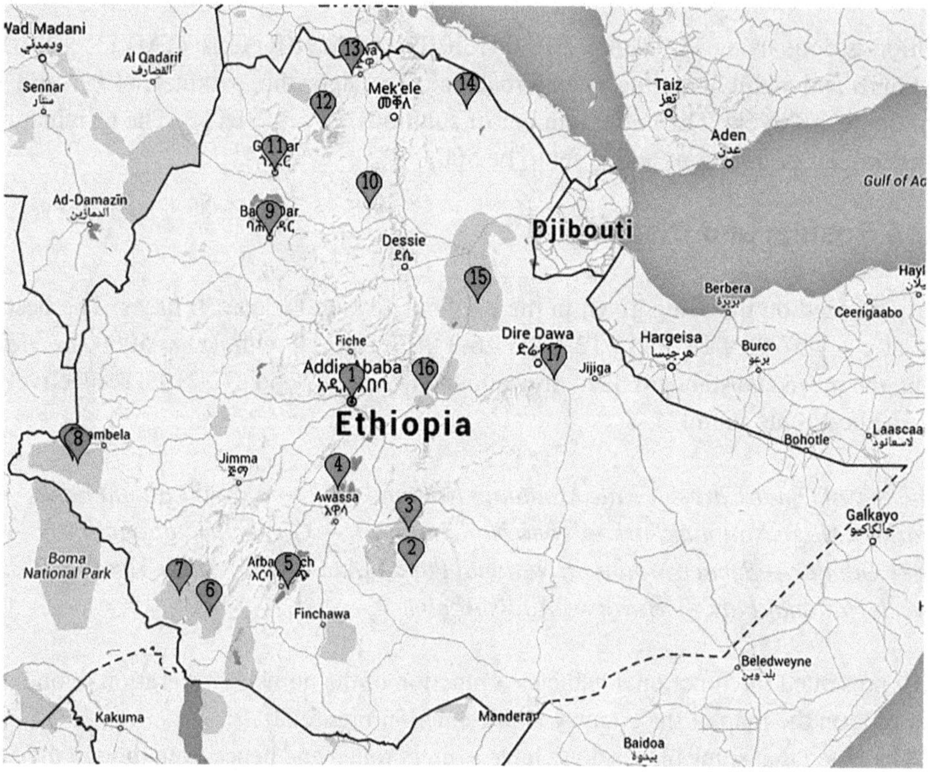

current situation. However, it should be noted that it may not remain to be the best solution as there are different ongoing road constructions in the country.

FUTURE WORKS

It should also need to be noted that for a tourist with limited time to stay in the country may not be able to explore all the destinations, hence one possible future work will be to identify the maximum number of tourist sites to be visited with a time constraint. In addition to the time constraint, a tourist may setup different criteria and hence incorporating the preference of the tourist may need further exploration along with a given time limitation. Another possible work is on the service provider parts. It can be considered as supply chain management and a study on making the stay of the tourist more favorable can be studied.

CONCLUSION

Ethiopia has been one of the oldest countries in history. It was a center for trade and exchange between Africa, Asia and Europe. It is one of the two countries in Africa which never been colonized by other countries. Hence, it has preserved all the indigenous culture as well as unique character from mixed up with others. It is religious country where more than 95% of its people are religious. According to central statistical agency of the country, based on its 2007 census, 62.8 are Christians with 43.5% being the Orthodox Christians and 33.9% of the population are Muslims. It has been mentioned a number of times in the Holy Bible and also have a very unique link with Islam. Other than that, it has a breath taking natural resources from the hottest place in the world to very top mountains suitable for hiking. A number of endemic animals in different parts of the country. The number of tourists visiting the country has increase from year to year. In this paper the major tourist destinations of the country is discussed and a travel salesman model of visiting all these sites with minimum traveling distance is presented. Prey predator algorithm is used to deal with the problem and the best route, starting from the capital and back to the capital is computed as:

Addis Ababa → Sof Omar Caves → Bale Mountain National Park → Abijiata-Shalla Lakes → Netchisar National Park → Mago National Park → Omo National Park → Gambella National Park → Bahir Dar → Lalibela → Gonder → Semien Mountain National Park → Axum → Ertale → Yangudi Rassa National Park → Awash National Park → Harar → Addis Ababa.

Even though in this study the tourist destinations are used from Ethiopia, it should be noted that it can been applied and used for any countries tourist sites.

REFERENCES

Across Abyssinia Tour, P. L. C. (2014). Retrieved from http://www.adventureabyssinia.com/wp-content/uploads/2014/08/axum1.jpg

Aves, E. (2016). *Exploring Bale Mountains National Park, Ethiopia, Rough Guides*. Accessed on Feb. 2016, from http://www.roughguides.com/wp-content/uploads/2014/09/95647017_14-660x431.jpg

Baeck, T., Fogel, D., & Michalewicz, Z. (1997). *Handbook of Evolutionary Computation*. Taylor & Francis. doi:10.1887/0750308958

Bogahavatta, C. (2013). *Introduction to tourism and cultural resources*. Retrieved Feb. 02, 2016 from http://si.archaeology.lk/materials/Introduction-to-Tourism-and-Cultural-Resources_2013_04_01.pdf

Bonabeau, E., Dorigo, M., & Theraulaz, G. (1999). *Swarm Intelligence: From Natural to Artificial Systems*. Oxford University Press.

Brand Ethiopia. (2016). Accessed on Feb. 2016, from http://www.brandethiopian.com/Content/uploads/tour-package/21/The_Gate_of__Harrar_jpg.jpg

Calgor, H. (n.d.). *The bruth Force Algorithm*. Retrieved January24, 2010, from http://www.ctl.ua.edu/math103/hamilton/hamilton.html

Chiu, V. K. H. (2010). Accessed from UNESCO website: http://whc.unesco.org/en/list/18/gallery/

Cook, W. (n.d.). *History of the Traveling Salesman Problem*. Retrieved January 22, 2010, from http://www.tsp.gatech.edu/index.html

Deb, K. (1995). *Optimization for Engineering Design*. New Delhi: Prentice-Hall.

Ethiopian Embassy in Ankara. (2016). *Natural Attractions*. Accessed Feb. 2016, from http://www.ethiopianembassyankara.org/wp-content/uploads/2015/10/yangud-national-park.gif

European Union Tourism Trade Organizations (ECTT). (2015). *European Union tourism trade organizations (ECTT) Best Tourist Destination Announcement*. Retrieved on Feb. 10, 2016 from http://ectt.webs.com/besttouristdestination.htm

FDREMCT. (2015). *Ethiopia, F. D. R. E. Ministry of Culture and Tourism*. Retrieved on Feb. 20, 2016 from http://www.moct.gov.et/index.php/en/about-ethopia

Garey, M., & David, J. (1979). *Computers and Intractability: A guide to the Theory of NP-Completeness*. Bell Laboratories Murray Hill.

Goldberg, D. (1989). *Genetic Algorithms in Search, Optimization and Machine Learning*. Reading, MA: Addison Wesley.

Gishen Tours. (2015). Accessed on Feb. 2016, from http://www.gishentours.com/wp-content/uploads/2015/12/BlueNileFalls1.jpg

Hamadneh, N., Tilahun, S. L., Sathasivam, S., & Ong, H. C. (2013). Prey-Predator Algrorithm as a New Optimization Technique Using in Radia Basis Function Neural Networks. *Research Journal of Applied Sciences*, *8*(7), 383–387.

Harp, R. (1972). *Reducibility among Combinatorial Problems*. University of California at Berkeley.

Johnny Abyssinia Tour and Travel. (2016). Accessed on Feb. 2016, http://www.johnnyabyssinia.com/images/imm38000_1%20%2813%29%20wm.jpg

Kumbharana, S., & Pandey, G. (2013). Solving Travelling Salesman Problem using Firefly Algorithm. *International Journal for Research in Science and Advanced Technologies, 2*(2), 53-57.

Land, A., & Doig, A. (1960). An outomotive method of solving discrete programming problems. *Econometrics*, *28*(3), 409–520.

Maredia, A. (2010). *History, Analysis, and Implementation of Traveling Salesman Problem (TSP) and Related Problems* (Unpublished doctoral dissertation). University of Houston-Downtown.

Mathieson, A., & Wall, G. (1982). *Tourism: economic, physical and social impacts*. Harlow, UK: Longman.

McIntosh, R. W., & Goeldner, C. R. (1986). *Tourism: principles, practices, philosophies*. New York: Wiley.

Papadimitriou, C. H. (1977). The Euclidean travelling salesman problem is NP-complete. *Theoretical Computer Science*, *4*(3), 237–244. doi:10.1016/0304-3975(77)90012-3

Poli, R., Kennedy, J., & Blackwell, T. (2007). *Particle swarm optimization: An overview*. Springer Science and Business Media.

RIA Ethiopian Sports. (2016). Accessed on Feb. 2016, http://www.ethiotrail.com/wp-content/uploads/2014/02/quees_02.jpg

Safa, H. (2015). *Dallol, Ethiopia, The Hottest Green Crater on Earth*. Accessed Feb. 2016, http://inspirationseek.com/wp-content/uploads/2015/01/Dallol-Green-Crater-in-Ethiopia.jpg

Satta, A. (2015). *Towards sustainable tourism in the Mediterranean: Situation and challenges at a regional level.* Retrieved January 28, 2016, from http://www.medpan.org/documents/ 10180/0/presentation_tourism_25nov15_satta/0c6a1ee4-075b-4eea-9e75-052db62e9208

Schenck, C. (2016). *Simien National Park.* UNESCO. Accessed Feb. 2106, http://whc.unesco.org/?cid=31&l=en&id_site=9&gallery=1&&maxrows=46

Simien Thrills Tour and Travel. (2016). Retrieved from http://simienthrills.com/621-2/

Sunny 13 Tour and Travel. (2016). Accessed on Feb. 2016, http://sunny13.com/images/Attractions/Natural/Sof-Omara.jpg

Tana Travel and Tour. (2016a). Accessed Feb. 2016, http://www.tanatravelandtour.com/image/OmoNationalPark.jpg

Tana Travel and Tour. (2016b). Accessed Feb. 2016, http://www.tanatravelandtour.com/image/ MAGO-NATIONALPARK.jpg

Tilahun, S. L. (2012). Fuzzy graph representation of bus timetabling problem and its solution method using Prey-Predator algorithm. *International Workshop on Optimal Network Topologies (IWONT) 2012.* Institut Teknologi Bandung.

Tilahun, S. L. (2013). *Prey predator algorithm: A new metaheuristic optimization approach* (PhD thesis). School of Mathematical Sciences, Universiti Sains Malaysia, Malaysia.

Tilahun, S. L., & Ong, H. C. (2013). Comparison between Genetic Algorithm and Prey-Predator Algorithm. *Malaysian Journal of Fundamental and Applied Sciences*, *9*(4), 167–170.

Bahmani-Firouzi, B., Sharifinia, S., Azizipanah-Abarghooee, R., & Niknam, T. (2015). Scenario-Based Optimal Bidding Strategies of GENCOs in the Incomplete Information Electricity Market Using a New Improved Prey Predator Optimization Algorithm. *IEEE Systems Journal, 9*(4), 1485 - 1495.

Dai, W., Liu, Q., & Chai, T. (2015). Particle size estimate of grinding processes using random vector functional link networks with improved robustness. *Neurocomputing*, *169*, 361–372. doi:10.1016/j.neucom.2014.08.098

Tilahun, S. L., & Ong, H. C. (2015). Prey predator algorithm: A new metaheuristic optimization algorithm. *International Journal of Information Technology & Decision Making*, *14*(6), 1331–1352. doi:10.1142/S021962201450031X

Tilahun, S. L., & Melese, S. (2015). *Prey predator with adaptive step length.* The 8th International Congress of Industrial and Applied Mathematics, Beijing, China.

Tilahun, S. L., & Ngnotchouye, J. M. T. (2016a). Prey predator algorithm with adaptive step length. *International Journal of Bio-inspired Computation*, *8*(4), 195–204.

Tilahun, S. L., & Ngnotchouye, J. M. T. (2016b). *Two stage travel salesman model of world tourism.* The 14th Pacific Rim International Conference on Artificial Intelligence (PRICAI 2016), Phuket, Thailand.

Tilahun, S. L., & Ngnotchouye, J. M. T. (2016c). *Examination timetabling using prey predator algorithm.* The 14th Pacific Rim International Conference on Artificial Intelligence (PRICAI 2016), Phuket, Thailand.

Tilahun, S. L., Ong, H. C., & Ngnotchouye, J. M. T. (2016). Extended Prey Predator Algorithm with a Group Hunting Scenario. *Advances in Operations Research, Volume 2016*, Article ID 7325263, 14 pages

Tropic Air Kenya. (2013). *Ethiopia's richest wildlife area – Gambela.* Accessed on Feb. 2016, http://www.tropicairkenya.com/wp-content/uploads/Gambela-giraffe.jpg

UNESCO. (2006). Accessed on Feb. 2016, http://whc.unesco.org/ en/list/19/gallery/

Wikipedia. (2016). Accessed Feb. 2016, https://upload.wikimedia.org/wikipedia/commons/ thumb/0/00/LakeChamoCrocodile.jpg/1024px-LakeChamoCrocodile.jpg

World Tourism Organization (UNWTO). (2015). *World Tourism Organization (UNWTO) Annual Report 2014.* Retrieved Feb. 26, 2016 from http://dtxtq4w60xqpw.cloudfront.net/sites/all/files/pdf/unwto_annual _report_2014.pdf

KEY TERMS AND DEFINITION

Algorithm: (Pronounced AL-go-rith-um) Is a procedure or formula for solving a problem. The word derives from the name of the mathematician, Mohammed ibn-Musa al-Khwarizmi, who was part of the royal court in Baghdad and who lived from about 780 to 850.

Ark of the Covenant: Also known as the Ark of the Testimony, was a wooden chest clad with gold containing the two stone tablets of the Ten Commandments as well as, according to various texts within the Hebrew Bible, Aaron's rod and a pot of manna.

Lava Lakes: Large volumes of molten lava, usually basaltic, contained in a volcanic vent, crater, or broad depression.

Lucy: The common name of AL 288-1, several hundred pieces of bone fossils representing 40 percent of the skeleton of a female of the hominin species *Australopithecus afarensis*. In Ethiopia, the assembly is also known as Dinkinesh, which means "you are marvelous" in the Amharic language.

Metaheuristic: A higher-level procedure or heuristic designed to find, generate, or select a heuristic (partial search algorithm) that may provide a sufficiently good solution to an optimization problem, especially with incomplete or imperfect information or limited computation capacity.

NP-Hardness: (Non-deterministic polynomial-time hard), in computational complexity theory, is a class of problems that are, informally, "at least as hard as the hardest problems in NP".

Rift Valley: A linear-shaped lowland between several highlands or mountain ranges created by the action of a geologic rift or fault.

UNESCO World Heritage Site: A place (such as a forest, mountain, lake, island, desert, monument, building, complex, or city) that is listed by the United Nations Educational, Scientific and Cultural Organization (UNESCO) as of special cultural or physical significance.

Chapter 20
Modeling Tourists' Opinions Using RIDIT Analysis

Subhajit Bhattacharya
Xavier Institute of Social Service, India

Rohit Vishal Kumar
International Management Institute, Bhubaneswar, India

ABSTRACT

In this chapter we have attempted to use "Relative to an Identified Distribution" (RIDIT) algorithms based modelling for analysing real-time empirical data relating to tourists' attitude and preference for a better understanding of the tourists' motivation and behaviour. RIDIT approach for evaluating the factors that influence tourist behaviour is not a very common approach in tourism sector. This chapter on modelling tourists' opinions and perceptions with RIDIT analysis would try to guide the empirical research in the domains of hospitality, tourism and travel research and analytics process in generating Optimized research outcomes.

INTRODUCTION

Travel and tourism is considered as one of the largest and fastest growing industries (Ninemeier & Perdue, 2008; Cooper & Hall, 2008) and it is also strongly influencing the world GDP, employment, exports and taxes (Kay, 2003; Koc, 2004; Ninemeier & Perdue, 2008). World Travel & Tourism industry has contributed US$7. 6 trillion (9.8% of global GDP) and created 277 million jobs (1 in 11 jobs) directly or indirectly in 2014 (UNWTO, 2015). According to the latest UNWTO World Tourism Barometer, it has been expressed that international tourists (overnight visitors) reached 1,138 million in 2014 with the growth rate of 4.7%, which is 51 million more than in 2013. Tourism is not only a social phenomenon it is also a big business (Cohen, 1979; Krippendorf 1986). Hence, growing popularity of global tourism demands proper destination marketing (Echtner & Ritchie, 1993; Murphy et al., 2007). To achieve this, tourism marketers nowadays need to apply modern branding techniques and effective positioning strategies (Hosany et al., 2006; Skinner, 2008; Hankinson, 2009; Hanna & Rowley, 2011) for motivating the prospective tourists and that it would also be helpful to cope up with current unstable and changing

DOI: 10.4018/978-1-5225-1054-3.ch020

competitive tourism marketing environment (Ekinci & Hosany, 2006; Pike et al., 2004). Pike (2009) established that 70% of worldwide travellers visit only 10 countries and the remaining countries are struggling to attract the outstanding 30% of the total international arrivals (Morgan et al., 2003). This is not just the problem of destination marketing but also the problem of understanding the tourists' insights and preferences. Not only in the product market but also in the tourism sector tastes and preferences vary. Tourists are not completely the similar in nature; they have different set of needs, motivations and preferences relating to their ideal vacation. Tourists are heterogeneous in nature (Dolnicar, 2008), so it is very essential to project the different tourists' destination brands with proper tourism marketing initiatives to attract the different prospective tourist segments (Echtner & Ritchie, 1993; Bhat & Reddy, 1998; Hankinson, 2005; Mowle & Merrilees, 2005; Wood, 2007; Campo-Martinez et al., 2010). In this context the judgement of tourists' preferences and attitudes (George, 2003; March & Woodside, 2005) are necessary to achieve the equilibrium in tourism marketing. In better understanding the travel behaviour, it is important to know how the key attributes and factors relating to destination brand influence tourist choices (March & Woodside, 2005; Holloway, 2004). Various researchers have argued that tourist motivations analysis helps in better understanding of travel behaviour (Huang & Xiao, 2000; Lee, Lee & Wicks, 2004; Law, Cheung & Lo, 2004; Correia et al., 2006; Saayman, et al., 2009). Hence travel motivation and behaviour are considered as the important fields in tourism marketing research literature (Huang & Xiao, 2000; Lam & Hsu, 2006). The prediction of travel motivation and behaviour play a significant role in tourism marketing domain, in order to generate more demand and also this to help in tourists marketing decisions (Holloway, 2004; March & Woodside, 2005; Decrop, 2006; Smallman et al., 2010). "In order to understand tourists' behaviour, researchers have considered the primary data relating to tourists' assertiveness, attitudes, interest perceptions, behavioural changes, intentions, and knowledge. These types of tourists' data and information are frequently measured and gathered by using Likert scale (as cited by Wu, 2007 in Likert, 1932; Fink, 1995; Fink et al., 1998; Peterson, 2000; Alexandrov, 2010; Boone et al., 2012)." Instead of commonly used statistical approaches for Likert scale data examining, we have used RIDIT algorithms based modelling in analysing the tourists' opinions and preferences.

The chapter is divided into the following sections:

- Section-II deals with brief review of literature.
- Section-III deals with the Existing technique of Likert scale data analysis.
- Section-IV deals with the use of RIDIT analysis in optimizing empirical research outcome. Section-V is concerned with RIDIT Algorithm.
- Section-VI deals with data analysis with RIDIT model.
- In Section-VII results and discussion are being discussed.
- Section-VIII deals issues with conclusion and recommendations.

It is expected that this chapter would motivate the budding researchers in applying the RIDIT algorithms based modelling for getting deeper understanding of the minds of the prospective tourists.

BACKGROUND

One of the key requirements in tourism research is to identify components which are having high individual contributions to tourists' perceptions and destination selection behaviour. Here the concept of ranking of the multivariate components comes under consideration. Few of the resent tourism researchers have examined the multi-attribute preferences in selection of a tourists' destination in doing so some researchers have used weighted average scores (Oh et al., 2009; Maciejewski et al., 2014), few researchers have studied tourist motivation and preferences by using factor analysis and t-test (Tsaur et al., 1996; Das et al., 2007; Salman et al., 2012; Shuzhen et al., 2011; Tomiš et al., 2014). Several researchers have used some highly complex techniques like 4-level AHP model and TOPSIS in measuring and ranking the tourism attributes or components (Chen, 2006; Hsu et al., 2009; Zhang et al., 2011; Lai et al., 2013; Wang et al.,2013). Bat all of the aforesaid studies were limited to a maximum sample size of 500 along with the specific assumptions about the sample distributions. In the domain of tourism research sometime the research may conduct with huge data gathered form different source and that is not also may not follow any specific distribution, in that situation determining the perfect tool of data analysis is very difficult. To counter with this problem RIDIT analysis may be seen as one of the possible tools which may rank-wise arrange the tourists' preference related components in an easier way without the prior assumption about the sampling process.

In 1958, Irwin D. J. Bross had propounded the use of RIDIT (Relative to an Identified Distribution) analysis for dealing with ordinal variables. The measurement system relies heavily on the skills of the researcher involved in the measurement of the variable. According to Bross, these variables may not be effectively analysed by Chi Square and t-test as they might not be proper due to assumption of a parametric distribution used by the Chi Square and t-test (Bross, 1958). RIDIT analyses the probability that a given observation is better than a visible observation from a reference distribution. "The outcomes from the RIDIT analysis can be used to arrange Likert scale items either in an ascending or in a descending order based on importance. This can also examine the relations among scale items in terms of degrees of significance or agreement (Wu, 2007, p. 676)."

RIDIT analysis has been comparatively less used in literature because of the complexity of interpreting the results. There seems to be two distinctive trends in the development of RIDIT literature. First we have reviewed some papers which have tried to improve the RIDIT analysis and remove its inherent shortcomings; while on the other there are papers which directly use the technique in analysis. In 1974, Jairus D. Flora Jr (1974), working at University of Michigan, published a technical work circulated to the Motor Vehicle Manufacturer Organization of USA in which she pointed out that the problem with RIDIT analysis. She showed that the choice of reference population can lead to rejection or selection of null hypothesis that the two populations (reference and comparison) are equivalent or not. This is because the choice of reference population impacts the calculation of standard deviation used in the denominator for testing of hypothesis. Similarly other researchers had compared RIDIT to quartile ranks (Kantor, Winkelstein, & Ibrahim, 1968; Selvin, 1977). WJ Conover (1973) noticed similarities between RIDIT analysis and the Wilcoxon Rank Sum Test. Some researchers compared Raw, RIDIT, Normal Median and Normal Mean scoring methods and found that RIDIT scoring methods perform better than the Normal Mean and Raw scoring system. He also showed that under assumption of a uniform distribution, RIDIT scoring became the optimal scoring system (Brockett, 1981). The work on RIDIT analysis continues with other researcher contributing to the development of RIDIT in recent times (Oyeka, 1992; Uwawunkonye & Anaene, 2013)

The initial use of RIDIT analysis was in the field of epidemiology where it received a lot of attention (Dhillon, Goldstein, & Schiffman, 1978; Forthofer & Koch, 1973; Williams & Grizzle, 1972). Subsequently there was a lull in the usage of the technique as other techniques were developed. But use of RIDIT has been seen sporadically in some places. It was used to study atmospheric visibility in Britain (Craig & Faulkenberry, 1979; Doyle & Dorling, 2002). Lieberthal (2008) used a modified form of RIDIT analysis to Hospital Quality. The modification was that in the first step the RIDIT scores were calculated and in the second step the principal component analysis approach was used on RIDIT scores to extract factors for analysis. This approach – termed as PRIDIT approach – is gaining currency amongst researchers. In a study, the RIDIT analysis along with Grey Relational Analysis was used to evaluate the factors which influenced the purchase of NANO cars in the Indian markets (Bikash, Pravat, & Sreekumar, 2010). RIDIT approach has also been used to prioritize agribusiness issues in the Indian scenario (Panda & Sreekumar, 2012). In another work, Mahapatra and Sreekumar (2012) have used RIDIT analysis to identify the importance of factors which influence online shopping behaviour. Mohanty and Gahan (2012) have used RIDIT analysis to understand the Buyer Supplier Relationship in the manufacturing sector in India. Similarly RIDIT analysis has been used to understand the importance of attributes for apparel buying amongst teenaged buyers in Lucknow region of India (Kaushal, 2013). There are less evidences of RIDIT analysis in the domain of tourism research in understanding the mind of the prospective tourists. Most of the statistical ranking or ordering techniques is based either on simple mean ranks which can be affected by presence of skewed data. In case of factor analysis the objective it to group the number of scales into a smaller (artificially created) subsets – which, more often than not, are hypothetical in nature. The reduced scales so created are difficult to implement in the real world and the artificiality makes it difficult to work with. More so, both factor analysis and cluster analysis are highly sensitive to variable selection and different choices of variables may produce different results. To counter the problem of ranking of items in ascending or descending order RIDIT analysis can be used.

STEPS OF RIDIT ANALYSIS

The use of RIDIT calculation and analysis have been recommended for the processing of ordinal data or the data which are in ordered scale, but are not on an interval scale. The procedure for the RIDIT Analysis as follows; from a reference population with the same categories one determines a "RIDIT" or score for each category. This score for each category is the percentile rank of an item in the reference population and is equal to the number of items in all lower categories plus one-half the number of items in the subject category, divided by the population size. Once the RIDIT's for each category have been determined, they are taken as values of a dependent variable for the comparison groups and the normal distribution is applied (Flora Jr, 1974). In simpler terms, suppose that there are m items and n ordered categories listed from the most favoured to the least favoured in the scale, then using the steps as outlined by Chien-Ho Wu (2007) RIDIT analysis goes as follows (See Figure 1):

1. Calculate RIDITs for the accumulated reference dataset.
    a. Select a population set to serve as a reference sample data set. In case the population cannot be easily recognised then the total response set for the survey might be considered as the reference data set.
    b. Calculate frequency f_j for each group of responses, where j = 1,…, n.
    c. Calculate the mid-point collected frequency F_j for each group-wise of responses.

Figure 1. Block diagram showing RIDIT analysis process

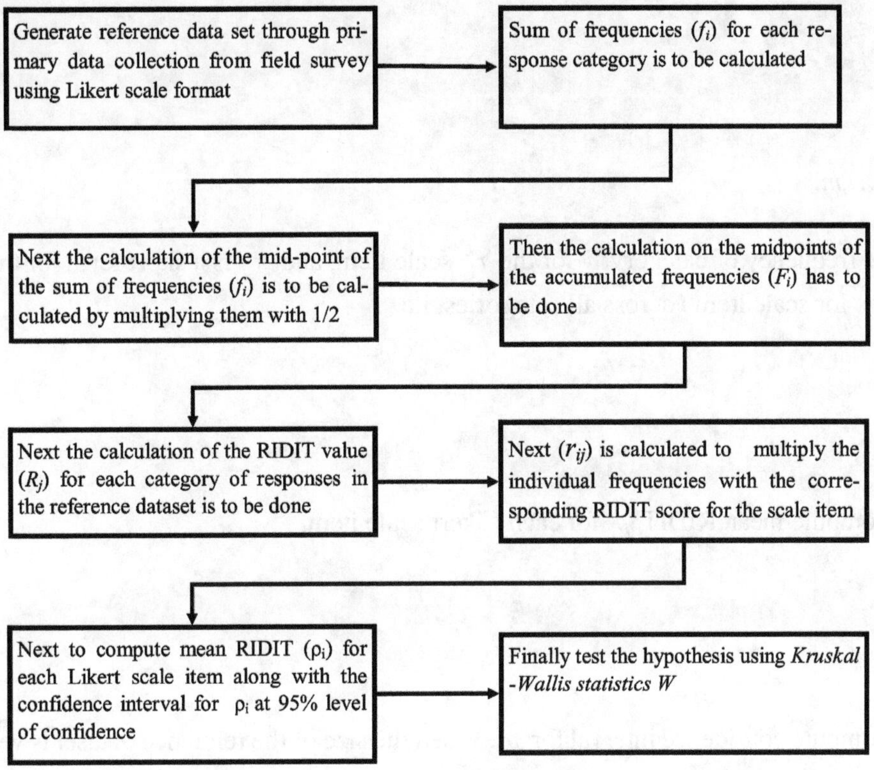

$$F_1 = \frac{1}{2} f_1$$

$$F_j = \frac{1}{2} f_j + \sum_{K=1}^{J-1} f_k \text{, where j=2 …n}$$

d. Calculate RIDIT value R_j for each group-wise of responses in the reference dataset.

$$R_j = \frac{F_j}{N},$$

where $j = 1\dots n$.

Here 'N' is the total count of sample survey responses in ordinal scale. By description, the expected value of 'R' for the existing reference dataset is always 0.5 (Agresti, 1984; Bross, 1958; Wu, 2007).

2. Compute RIDITs and mean RIDITs for comparison datasets. Note that a comparison dataset is comprised of the frequencies of responses for each category of a Likert scale item. Since there are *m* Likert scale items in this illustration, there will be *m* comparison datasets.

a. Compute RIDIT value r_{ij} for each category of scale items.

$$\pi_i = R_j \times \frac{\pi_{ij}}{\pi_i}$$

where $j = 1,\ldots,m$.

π_{ij} is the frequency of category j for the ith scale item, and π_{ij} is a short form for the summation of frequencies for scale item i across all categories, i.e.,

$$\pi_i = \sum_{k=1}^{n} \pi_{ik}$$

b. Compute mean RIDIT ρ_i for each Likert scale item.

$$\rho_i = \sum_{k=1}^{n} r_{ik}$$

c. Compute confidence interval for ρ_i. When the size of the reference dataset is very large relative to that of any comparison dataset, the 95% confidence interval of any ρ_i is:

$$\rho \pm {}^{1}\!\!/\!{}_{\sqrt{3}}\,\pi_i$$

d. Test the following hypothesis using *Kruskal-Wallis statistics W*.

$$\begin{cases} H_0 : \forall_i, \rho_i = 0.5 \\ H_1 : \exists_i, \rho_i \neq 0.5 \end{cases}$$

$$W = 12\sum_{i=1}^{m} \pi_i(\rho_i - 0.5)^2$$

W follows a χ^2 distribution with (m – 1) degree of freedom. If H_0 cannot be accepted, examine the relationships among confidence intervals of ρ.The general rules for interpreting the values of ρ are shown below.

A scale item with its ρ_i value statistically deviate from 0.5 implies a significant difference in the response patterns between the reference dataset and the comparison dataset for the particular scale item. If the confidence interval of a ρ_i contains 0.5, then it is accepted that the ρ_i value is not significantly deviated from 0.5.

A low value of ρ_i is preferred over a high value of ρ_i because a low value of ρ_i indicates a low probability of being in a negative propensity.

The response patterns of scale items with overlapped confidence intervals of ρ are considered, among the respondents, to be statistically indifferent from each other.

Above block diagram showing that how the execute RIDIT analysis stepwise basis for placing and arranging the scale items in the order of priority. Above diagram can help the researchers in understanding the RIDIT algorithm in a simple way. Firstly there is need of reference dataset (in Likert scale format) based on primary survey. Then the frequencies (f_i) for each response category is to be added. Next mid-point of the sum of frequencies (f_i) is to be calculated, followed by the midpoints of the accumulated frequencies (F_j). After that the calculation of the RIDIT value for each category of responses existing in the reference dataset is to be done. Then r_{ij} is calculated with the objective of multiplying the separate frequencies with the corresponding R_j score for the scale item. Next to ρ_i for each Likert scale item is to be calculated at 95% confidence. At end of the process *Kruskal-Wallis statistics W* test has to perform for the testing of hypothesis. The next part of the book chapter has critically demonstrated how to use Microsoft Excel functions to perform RIDIT analysis.

Simple Process of Using RIDIT with Microsoft Excel Functions

Firstly there is a need of a specific target population to generate reference data set through primary data collection in Likert scale format. Then the count and tabulate the frequency of responses in the dataset has to be meticulously written that can be simply arrange by considering the total response gathered from the field survey, this data set is called the reference dataset as shown in Table 1in this current chapter. RIDIT analysis requires a simple Microsoft Excel based programming or use of $SUMIF()$ function in Microsoft Excel. It can be visible that sum of frequencies (f_i) for each response category is done. Next the calculation of the mid-point of the sum of frequencies (f_i) is done by multiplying them with ½. Then the calculation on the midpoints of the accumulated frequencies (F_j) has to be done. For F_1 is calculated as $0.5 * f_1$. F_2 is calculated as $0.5 * f_2$ plus the sum of frequencies of the previous class i.e. f_1. Similarly F_3 is calculated as $0.5 * f_3$ plus the sum of frequencies of the previous classes i.e. f_1 and f_2 and so on. Next there is a need of compute the RIDIT value R_j for each category of responses in the reference dataset. This is simply done by dividing the midpoint of F_j by N – the total number of responses in the reference dataset. In this example N is equal to 2240. After this part of the algorithm we have to compute the r_{ij}. In calculation of r_{ij}, it is needed to multiply the individual frequencies with the corresponding RIDIT score for the scale item. For example to get *rij* for "opinion or view 1 response or reply category 1" we need to multiply the frequency (response category) with the computed RIDIT score of 0.19 and divide the resultant by the row total (2240). The results of the computations are shown in Table 2. The mean RIDIT score (denoted by ρ_i) is the sum of *rij* for across the response categories for a scale item. Once ρ_i are calculated, then it is need to calculate the 95% upper and lower bound for by using *step- ii-c* which has been shown in the above algorithm section. Next test the null hypothesis has to be done, where all the ρ_i are equal to 0.5 versus the hypothesis that at least one of the ρ_i is different. This is done by calculating the Kruskal-Wallis Statistics (W) as shown in *step- ii-d*. Kruskal-

Wallis statistics W follows the χ^2 distribution with (m – 1) degrees of freedom. In this case the value of χ^2 for (26-1) degrees of freedom is 37.65248 and calculated Kruskal-Wallis Statistics (W) is 4912.857 which is considerably higher. Therefore, we can reject the null hypothesis.

Data Analysis with RIDIT Model

Here we have taken 26 (twenty six) components or items which may be considered by the tourists in selection of a holiday destination such components are as follows:

- *Experimentation,*
- *Destination Promotion,*
- *Cultural elements,*
- *Information sharing with Family,*
- *Uniqueness,*
- *Local artefact,*
- *Liveliness, Pleasure,*
- *Weather,*
- *Cost,*
- *Food & Beverage,*
- *Value addition,*
- *Peace & relaxation,*
- *Hospitality,*
- *Logistic Support,*
- *Safety and Hassles free Services,*
- *Exploring New Things,*
- *Nightlife,*
- *Travel Insurance Service,*
- *Self-concept,*
- *Opinion sharing at Social Networking platform,*
- *Online Reviews,*
- *Natural beauty,*
- *Across Group Experience sharing,*
- *Fun,* and
- *Adventure.*

We have use a real time data base relating to tourists preference, the empirical dataset of 2240 responders has been taken into consideration. For execution for RIDIT algorithm no such high-end statistical software is required. Here we have used Microsoft office excel 2007 to test the RIDIT model. The RIDITs for the reference dataset is shown in Table 1, based on the survey conducted relating to destination brand choices behaviour pertaining to Indian tourism sector. For the need of normality requirement we can execute the skewness and kurtosis analysis which will help us in justifying that our research sample is representing the population characteristics or not. Hear I this chapter we have used Microsoft office excel 2007 to perform skewness and kurtosis tests to judge the normality of our real-time dataset.

Results of skewness and kurtosis tests have depicted that their values are very near to zero. The skewness index ranges from -0.183 to 0.028 and kurtosis index ranges from - 0.623 to -0.375, which is signifying that the hypothesis of normality seems to be valid (Groeneveld et al., 1984; Horswell et al., 1993; Arnold et al., 1995; Joanes et al., 1998; Kline, 2010). After getting the empirical data we have prepared a Microsoft office excel dataset representing frequency pertaining to the different items along with their preferences. This dataset is being considered as the reference dataset for the RIDIT calculation. Here f_i is the sum of each response category in Table 1on basis of f_i the all F_j values have been calculated where the mid-point analysis has been instead simple mean consideration.

It is always true that the tourist are different in nature but their preference and liking types may be similar or diverse. From the Table 1it can be visible that the opinions of tourist respondents in the current book chapter about scale items are significantly dissimilar from each other. RIDITs reference dataset (in Table 1) has confirmed that the component (in Likert scale) relating to tourists' destination marketing are meaningfully carrying the different likings and preferences. Above tables shows that for the same item the similarity in preference it is very less that has also be visible in cross items too. Computed mid-point accumulated frequency F_j for each category of responses follows:

$$F_1 = \frac{1}{2} f_1$$

$$F_j = \frac{1}{2} f_j + \sum_{K=1}^{J-1} f_k ,$$

where j=2 …n

Which actually indicates that similarity and dissimilarity each category of responses (see Table 1). Computed RIDIT value R_j for each category of responses in the reference dataset.

$$R_j = \frac{F_j}{N} ,$$

where j =1… n

N is the total number of responses from the Likert scale survey of interest. By definition, the expected value of R for the reference dataset is always 0.5 (Agresti, 1984; Bross, 1958; Wu, 2007). From the RIDITs reference dataset in Table 1we can see that the tourist are having heterogeneous preferences in each category of responses.

The Kruskal-Wallis W is calculated as follows:

$$W = 12\sum_{i=1}^{m} \pi_i (\rho_i - 0.5)^2$$

Kruksal Wallis – W 4912.857

Chi-Square (26-1 = 25 df) 37.65248

Table 1. RIDITs calculation from the tourists' reference dataset

Item	V.I.(5)	I (4)	N (3)	U(2)	V.U.(1)	π_i
Experimentation	750	910	250	195	135	2240
Destination Promotion	830	815	295	190	110	2240
Cultural element	1020	755	175	160	130	2240
Information sharing with Family, Friends, Relatives	960	795	190	170	125	2240
Uniqueness	1110	730	95	170	135	2240
Local artefact	705	1045	155	215	120	2240
Liveliness	765	915	245	190	125	2240
Pleasure	825	915	210	185	105	2240
Weather	950	755	175	215	145	2240
Cost	965	725	205	190	155	2240
Food & Beverages	930	760	260	165	125	2240
Value addition	645	915	330	185	165	2240
Peace and relaxation	670	980	250	205	135	2240
Hospitality	735	850	330	215	110	2240
Logistic Support	750	880	290	205	115	2240
Safety and Hassles free Services	735	705	340	290	170	2240
Exploring New Things	700	855	345	185	155	2240
Nightlife	695	815	325	260	145	2240
Travel Insurance Service	620	945	345	205	125	2240
Self-concept	850	830	245	205	110	2240
Opinion sharing at Social Networking platform	770	710	355	250	155	2240
Online Reviews	990	765	190	160	135	2240
Natural beauty	1015	845	120	160	100	2240
Across Group Experience sharing	1075	765	125	165	110	2240
Fun	1005	755	160	180	140	2240
Adventure	605	970	285	200	180	2240
f_i	21675	21709	6293	5117	3461	58240
$1/2 * f_i$	10837.5	10854.5	3146.5	2558.5	1730.5	
F_j	10837.5	32529.5	46530.5	52235.5	56524.5	
R_j	0.19	0.56	0.8	0.9	0.97	

Notes: 1 – very unimportant; 2 – unimportant; 3 – Neutral; 4 – important; 5 – very important.

From the Table 2 it can be observed that the tourists are having higher probability to less emphasize and overlooked the factors ranked 26th, 25th and 24th in selection of the holiday destination the reason being that the upper bound value of these components more than 0.50 (the threshold limit), this indicates that the factors ranked 26th, 25th and 24th are having higher probability to be ignored by the prospective tourist. The components that are within the threshold limits will be considered as the potential item by the potential tourist. Table 2 has depicted that scale component with its ρ_i value statistically diverge from 0.5 infers a significant inequality in the opinions of the responders and the comparative data set for the specific scale component. A relatively less value of ρ_i is favoured over a greater value of ρ_i because a less value of ρ_i specifies a high probability of being considered as a very important item in the preference set relating consumer preference. The opinion patterns of scale components with coincide confidence intervals of ρ indicates the indifferent property statistically. Above analysis in Table 2 has critically demonstrated that higher the *calculated W* is indicating higher degree of probability to be considered as the important components which influence the tourists' behaviour. The closeness of the *calculated W* of the individual factors shows the probability led familiarity of the items. Since the overall Kruskal-Wallis W is significantly large, so it can be said that the views and attitudes about the scale items among the respondents are statistically different.

Table 3 is properly showing the tourism marketing mix components in an order of priority that indicates the tourists' likings and preferences, upon which the tourists select any tourism destination brand. In one hand components with higher order of priorities are having higher impacts in tourism destination brand selection behaviour of the tourists. On the other hand components with lower order of priorities are having less impacts on tourists' destination brand selection behaviour.

RESULTS AND DISCUSSION

From the above analysis it can be observed that a RIDITs ranks of the separate tourism marketing mix items in terms of the probability of being considered as important or most significant item according to the preferences of the potential tourist. The different tourism marketing mix components have got there unique rank which has been shown in Table 3. The RIDITs ranks is arranged in following sequence:

Uniqueness > Across Group Experience sharing> Natural beauty> Cultural element > Fun>Online Reviews >Information sharing with Family, Friends, Relatives >Cost > Weather >Food & Beverages >Pleasure >Self-concept >Destination Promotion >Liveliness >Local artefact >Logistic Support >Hospitality >Opinion sharing at Social Networking platform >Peace and relaxation >Exploring New Things >Nightlife >Safety and Hassles free Services >Value addition >Travel Insurance Service >Adventure >Experimentation.

The RIDIT analysis have helped the research in understanding the tourists' preferences and behaviour with the RIDIT algorithm based management model. The RIDIT model has specified the precise location of the individual tourism marketing mix components within the entire series of the items which are having influences on tourists' behaviour. This analysis has demonstrated that most important factors influencing tourist holiday destination selection behaviour includes:

Table 2. Calculation of RIDITs comparison on the tourists' reference dataset

Item	5	4	3	2	1	ρ_i	Rank	LB	UB	W-Calculated
Experimentation	0.06	0.23	0.09	0.08	0.06	0.51	26	0.50	0.53	0.50
Destination Promotion	0.00	0.20	0.11	0.08	0.05	0.43	13	0.42	0.44	10.30
Cultural element	0.00	0.19	0.06	0.06	0.06	0.37	4	0.36	0.38	37.24
Information sharing with Family, Friends, Relatives	0.00	0.20	0.07	0.07	0.05	0.39	7	0.38	0.40	27.98
Uniqueness	0.00	0.18	0.03	0.07	0.06	0.34	1	0.33	0.35	55.59
Local artefact	0.00	0.26	0.06	0.09	0.05	0.45	15	0.44	0.47	4.75
Liveliness	0.00	0.23	0.09	0.08	0.05	0.45	14	0.43	0.46	6.59
Pleasure	0.00	0.23	0.07	0.07	0.05	0.42	11	0.41	0.43	13.41
Weather	0.00	0.19	0.06	0.09	0.06	0.40	9	0.39	0.41	22.58
Cost	0.00	0.18	0.07	0.08	0.07	0.40	8	0.38	0.41	23.70
Food & Beverages	0.00	0.19	0.09	0.07	0.05	0.40	10	0.39	0.41	21.31
Value addition	0.00	0.23	0.12	0.07	0.07	0.49	23	0.48	0.50	0.16
Peace and relaxation	0.00	0.24	0.09	0.08	0.06	0.47	19	0.46	0.49	1.50
Hospitality	0.00	0.21	0.12	0.09	0.05	0.46	17	0.45	0.48	3.00
Logistic Support	0.00	0.22	0.10	0.08	0.05	0.45	16	0.44	0.47	4.58
Safety and Hassles free Services	0.00	0.18	0.12	0.12	0.07	0.49	22	0.47	0.50	0.39
Exploring New Things	0.00	0.21	0.12	0.07	0.07	0.48	20	0.47	0.49	1.14
Nightlife	0.00	0.20	0.12	0.10	0.06	0.49	21	0.47	0.50	0.43
Travel Insurance Service	0.00	0.24	0.12	0.08	0.05	0.49	24	0.48	0.51	0.06
Self-concept	0.00	0.21	0.09	0.08	0.05	0.42	12	0.41	0.44	12.91
Opinion sharing at Social Networking platform	0.00	0.18	0.13	0.10	0.07	0.47	18	0.46	0.48	1.89
Online Reviews	0.00	0.19	0.07	0.06	0.06	0.38	6	0.37	0.39	31.68
Natural beauty	0.00	0.21	0.04	0.06	0.04	0.36	3	0.35	0.37	43.35
Across Group Experience sharing	0.00	0.19	0.04	0.07	0.05	0.35	2	0.34	0.36	51.03
Fun	0.00	0.19	0.06	0.07	0.06	0.38	5	0.37	0.39	33.31
Adventure	0.00	0.24	0.10	0.08	0.08	0.50	25	0.49	0.51	0.01

Notes: LB: lower bound of the 95% confidence interval of mean RIDIT ρ_i ; UB: upper bound of the 95% confidence interval of mean RIDIT ρ_i,

Table 3. Rank wise significance of tourism marketing mix components in the order of priority

Items	Rank
Uniqueness	1
Across Group Experience sharing	2
Natural beauty	3
Cultural element	4
Fun	5
Online Reviews	6
Information sharing with Family, Friends, Relatives	7
Cost	8
Weather	9
Food & Beverages	10
Pleasure	11
Self-concept	12
Destination Promotion	13
Liveliness	14
Local artefact	15
Logistic Support	16
Hospitality	17
Opinion sharing at Social Networking platform	18
Peace and relaxation	19
Exploring New Things	20
Nightlife	21
Safety and Hassles free Services	22
Value addition	23
Travel Insurance Service	24
Adventure	25
Experimentation	26

Uniqueness; Across Group Experience sharing; Natural beauty; Cultural element; Fun; Online Reviews; Information sharing with Family, Friends, Relatives; Cost; Weather; Food & Beverages; Pleasure; Self-concept; Destination Promotion; Liveliness; Local artefact; Logistic Support; Hospitality; Opinion sharing at Social Networking platform; Peace and relaxation; Exploring New Things, while Nightlife; Safety and Hassles free Services; Value addition; Travel Insurance Service; Adventure; Experimentation

These have been figured out not to be significantly important components to influence tourists behaviour in connection with holiday destination selection behaviour.

FUTURE RESEARCH DIRECTIONS

RIDIT is very useful data analytical technique which can provide sufficient help the budding researchers to process a large number of tourists' data in a simplified way. Further the scholars in the domain of tourism research can think for the greater use of the RIDIT analysis in determining the tourism marketing mix strategies. Researchers can also add value to the existing RIDIT analysis technique for enhanced outcomes. The budding researchers can use this data analysis technique for processing the large size tourists' data in optimising marketing strategic planning in the domain of hospitality, tourism and travel industry. We are also suggesting that the future studies can also think for the fusion technique of tourism large size tourists' data analysis based on RIDIT algorithm for higher degree of optimisation. This analytical technique could also be used in service quality gap analysis or in service quality benchmarking in hospitality, tourism and travel industry. We are hopeful the management and tourism student can also use RIDIT analysis technique in their master thesis to improve the tourists' primary data analysis with high degree of robustness and accuracy.

CONCLUSION

After going through the above discussion, it is clear that RIDIT provides researchers a simplified means of ranking the items measured on Likert scale items. This provides a major advantage in calculating the importance of items on the Likert scale as the researchers need not be bothered in assuming a distribution (mostly normal) and trying to test whether the data points conform to the distribution assumption. This analysis requires a reference data series which needs to be generated cautiously based on the type of items being provided in the Likert scale. RIDIT analysis can process a large number of samples with commonly available software, Microsoft office excel 2007. RIDIT analysis should be used when the data series are known to be ordinal in nature. The RIDIT value represents the probabilities of the scale items being in better off or worse off conditions than the reference data series. Calculation is much more complex and there are issues in correctly interpreting the RIDIT values (because of the overlap of the confidence intervals). It requires a representative reference dataset to calculate reliable values and is highly sensitive to the choice of the reference data set. In case of non-availability of the reference dataset – the entire response set can be used as the reference dataset. An advantage of RIDIT analysis is that it allows testing of hypothesis using the Kruskall-Wallis test statistics. Budding researchers can use this analytical tool whenever there is need of find the rank-wise positions of the preferences or likings of the tourist, it will also be used in judging the motivational level of the tourist. RIDIT analysis can also be used in analysing the tourist behaviour relating to selection of the tourists' destination options. Another advantage of RIDIT analysis is that this analytical tool can be applied any type of sample size either small or large, this can be considered as a data examining tool that can easily determine the large variances in the response data set or the more homogeneity in the response data set.

In this chapter, we found that the rankings derived from the RIDIT model based analysis are very much consistent with each other. It is hoped that this chapter has been able to throw some light on the application of the RIDIT algorithms based modelling for getting a deeper understanding of the minds of the prospective tourists.

LIMITATION

The factors that have been taken into consideration in finding the tourists' preferences are not exhaustive. The RIDIT analysis is shown in this book chapter has been restricted within the use of Microsoft Excel functions, no such other software has been used. The current research oriented book chapter is restricted to the primary tourists'survey data collected in Indian. The outputs of RIDIT analysis may vary with the different sample sizes. The different country related effects are not discussed in this research based book chapter relating to holistic optimisation techniques in hospitality, tourism and travel research.

MOTIVATION

Inspiration to write this chapter was to academically contribute in the domain of tourism research. We were trying to introduce the 'Relative to an Identified Distribution' (RIDIT) analysis in tourism research to help the budding researcher with an alternative analytical methodology.

ACKNOWLEDGMENT

We thank all the tourists who have given their valuable responses that have helped us in preparing a sizable tourists' preference related database.

REFERENCES

Alexandrov, A. (2010). Characteristics of Single-Item Measures in Likert Scale Format. *Electronic Journal of Business Research Methods*, *8*(1), 1–12.

Bhat, S., & Reddy, S. K. (1998). Symbolic and Functional Positioning of Brands. *Journal of Consumer Marketing*, *15*(1), 32–43. doi:10.1108/07363769810202664

Bikash, R. D., & Pravat, S. K., & Sreekumar. (2010). Factors Influencing Purchase of 'NANO' the Innovative Car from India - An Empirical Study. *Asian Journal of Business Management*, *2*(3), 48–56.

Boone, H. N. J., & Boone, D. a. (2012). Analyzing Likert Data. *Journal of Extension*, *50*(2), 30.

Brockett, P. L. (1981). A note on Numerical Assignment of Scores to Ranked Categorical Data. *The Journal of Mathematical Sociology*, *8*(1), 91–101. doi:10.1080/0022250X.1981.9989917

Bross, I. D. J. (1958). How to use Ridit Analysis. *Biometrics*, *14*(1), 18–38. doi:10.2307/2527727

Campo-Martinez, S., Garau-Vadell, J. B., & Martinez-Ruiz, M. P. (2010). Factors influencing repeat visits to a destination: The influence of group composition. *Tourism Management*, *31*(6), 862–870. doi:10.1016/j.tourman.2009.08.013

Chen, C.-F. (2006). Applying the analytical hierarchy process (AHP) approach to convention site selection. *Journal of Travel Research, 45*(2), 167–174. doi:10.1177/0047287506291593

Cohen, E. (1979). A phenomenology of tourist experiences. Sociology. *The Journal of the British Sociological Association, 13*(2), 179–201.

Conover, W. (1973). Rank Tests for One Sample, Two Samples and K Samples without the Assumptions of a Continuous Distribution Function. *Annals of Statistics, I*(6), 1105–1125. doi:10.1214/aos/1176342560

Cooper, C., & Hall, M. (2008). *Contemporary Tourism: An International Approach.* London: Butterworth Heinemann.

Correia, A., Oom Do Valle, P., & Moço, C. (2006). Why People Travel to Exotic Places, *International Journal of Culture. Tourism and Hospitality, 1*(1), 45–61.

Craig, C. D., & Faulkenberry, G. D. (1979). The application of ridit analysis to detect trends in visibility. *Atmospheric Environment, 13*(12), 1617–1622. doi:10.1016/0004-6981(79)90319-6

Das, D., Sharma, S. K., Mohapatra, P. K. J., & Sarkar, A. (2007). Factors influencing the attractiveness of a tourist destination: A case study. *Journal of Service Research, 7*, 103–134.

Decrop, A. (2006). *Vacation Decision-Making.* Wallingford, UK: CABI-Publishing. doi:10.1079/9781845930400.0000

Dhillon, W. R., Goldstein, M., & Schiffman, L. (1978). On the appropriateness of Linear Discriminant and Multinomial Classification Analyisis in Marketing Research. *JMR, Journal of Marketing Research, 15*(1), 103–112. doi:10.2307/3150405

Dolnicar, S., Crouch, G. I., Devinney, T., Huybers, T., Louviere, J. J., & Opperwal, H. (2008). Tourism and discretionary income allocation. Heterogeneity among households. *Tourism Management, 29*(1), 44–52. doi:10.1016/j.tourman.2007.04.008

Doyle, M., & Dorling, S. (2002). Visibility Trends in UK 19501007. *Atmospheric Environment, 36*(19), 3161–3172. doi:10.1016/S1352-2310(02)00248-0

Echtner, C. M., & Ritchie, J. R. B. (1993). The Measurement of Destination Image: An Empirical Assessment. *Journal of Travel Research, 31*(4), 3–13. doi:10.1177/004728759303100402

Ekinci, Y., & Hosany, S. (2006). Destination personality: An application of brand personality to tourism destinations. *Journal of Travel Research, 45*(2), 127–139. doi:10.1177/0047287506291603

Fink, A. (1995). *How to Design Surveys.* London: SAGE Publications.

Fink, A., & Kosecoff, J. (1998). *How to Conduct Survey-a Step by Step Guide.* London: SAGE Publications.

Flora, J. D. Jr. (1974). *A note on Ridit Analysis.* Ann Arbor, MI: University of Michigan.

Forthofer, R. N., & Koch, G. (1973). An analysis for compounded functions of categorical data. *Biometrics, 29*(1), 143–157. doi:10.2307/2529683 PMID:4691050

George, R. (2003). Tourists perceptions of safety and security while visiting Cape Town. *Tourism Management, 24*(3), 575–585. doi:10.1016/S0261-5177(03)00003-7

Goldstein, G., & Hersen, M. (1984). *Handbook of Psychological Assessment*. Pergamon Press.

Hankinson, G. (2005). Destination brand image: A business tourism perspective. *Journal of Services Marketing, 19*(1), 24–33. doi:10.1108/08876040510579361

Hankinson, G. (2009). Managing destination brands: Establishing a theoretical foundation. *Journal of Marketing Management, 25*(1-2), 97–115. doi:10.1362/026725709X410052

Hanna, S., & Rowley, J. (2011). Towards a strategic place brand-management model. *Journal of Marketing Management, 27*(5-6), 458–476. doi:10.1080/02672571003683797

Holloway, J. C. (2004). *Marketing for Tourism*. New York: Prentice Hall.

Hosany, S., Ekinci, Y., & Uysal, M. (2006). Destination image and destination personality: An application of branding theories to tourism places. *Journal of Business Research, 59*(5), 638–642. doi:10.1016/j.jbusres.2006.01.001

Hsu, T. K., Tsai, Y. F., & Wu, H. H. (2009). The preference analysis for tourist choice of destination: A case study of Taiwan. *Tourism Management, 30*(2), 288–297. doi:10.1016/j.tourman.2008.07.011

Huang, A., & Xiao, H. (2000). Leisure-Based Tourist Behavior: A Case Study of Changchun. *International Journal of Contemporary Hospitality Management, 12*(3), 210–214. doi:10.1108/09596110010320850

Kantor, S., Winkelstein, W., & Ibrahim, M. A. (1968). A note on Interpretetion of the ridit as quantile ranks. *American Journal of Epidemiology, 87*, 609–615. PMID:5654949

Kaushal, S. K. (2013). The Importance of Apparel Product Attributes for Teenaged Buyers. *NMIMS Management Review, 23*, 45–64.

Kay, H. K. (2003). *Selling Tourism*. New York: Delmar Learning.

Koc, E. (2004). The role of family members in the family holiday purchase decision-making process. *International Journal of Hospitality & Tourism Administration, 5*(2), 85–101. doi:10.1300/J149v05n02_05

Krippendorf, J. (1986). The new tourist: Turning point for leisure and travel. *Tourism Management, 7*(2), 131–135. doi:10.1016/0261-5177(86)90025-7

Lai, W. H., & Vinh, N. Q. (2013). An application of ahp approach to investigate tourism promotional effectiveness. *Tourism and Hospitality Management, 19*(1), 1–22.

Lam, T., & Hsu, C. H. C. (2006). Predicting behavioral intention of choosing a travel destination. *Tourism Management, 27*(4), 589–599. doi:10.1016/j.tourman.2005.02.003

Law, R., Cheung, C., & Lo, A. (2004). The Relevance of Profiling, Travel activities for Improving Destination Marketing Strategies. *International Journal of Contemporary Hospitality Management, 16*(6), 355–362. doi:10.1108/09596110410550798

Lee, C. K., Lee, Y. K., & Wicks, B. E. (2004). Segmentation of Festival Motivation by Nationality and Satisfaction. *Tourism Management*, *25*(1), 61–70. doi:10.1016/S0261-5177(03)00060-8

Lieberthal, R. D. (2008). Hospital Quality: A PRIDIT Approach. *Health Services Research*, *43*(3), 18. doi:10.1111/j.1475-6773.2007.00821.x PMID:18454777

Likert, R. (1932). A Technique for the Measurement of Attitudes. *Archives de Psychologie*, *140*, 1–55.

Litwin, M. (1995). *How to Measure Survey Reliability and Validity*. London: SAGE Publication. doi:10.4135/9781483348957

Maciejewski, K., & Kerley, G. I. H. (2014). Understanding tourists' preference for mammal species in private protected areas: Is there a case for extralimital species for ecotourism? *PLoS ONE*, *9*(2).

Mahapatra, S., & Sreekumar, N. A. (2012). A RIDIT apprach to evaluate factors influencing online shopping behaviour. *International Journal of Internet Marketing and Advertising*, *7*(1), 83–98. doi:10.1504/IJIMA.2012.044936

March, R., & Woodside, A. (2005). Testing theory of planned versus realized behaviour. *Annals of Tourism Research*, *32*(4), 905–924. doi:10.1016/j.annals.2004.07.012

Mohanty, M. K., & Gahan, P. (2012). Buyer Supplier Relationship in Manufacturing Industry - Findings from Indian Manufacturing Sector. *Business Intelligence Journal*, *5*(1), 319–333.

Morgan, N., Pritchard, A., & Pride, R. (2003). Destination branding and the role of the stakeholders: The case of New Zealand. *Journal of Vacation Marketing*, *9*(3), 285–299. doi:10.1177/135676670300900307

Mowle, J., & Merrilees, B. (2005). A functional and symbolic perspective to branding Australian SME wineries. *Journal of Product and Brand Management*, *14*(4), 220–227. doi:10.1108/10610420510609221

Murphy, L., Moscardo, G., & Benckendorff, P. (2007). Using brand personality to differentiate regional tourism destinations. *Journal of Travel Research*, *46*(1), 5–14. doi:10.1177/0047287507302371

Ninemeier, J. D., & Perdue, J. (2008). *Discovering Hospitality and Tourism. In The World's Greatest Industry* (2nd ed.). Pearson Prentice Hall.

Oyeka, I. C. A. (1992). A Modified Ridit Analysis. *Journal of Nigerian Statistical Association*, *8*(1), 45–69.

Panda, R. K., & Sreekumar. (2012). Prioritizing Issues in Agribusiness Using Ridit Analysis: An Indian Perspective. *World Journal of Agricultural Sciences*, *8*(1), 80–88.

Peterson, R. (2000). *Constructing Effective Questionnaires*. London: SAGE Publications. doi:10.4135/9781483349022

Pike, S. (2009). Destination brand positions of a competitive set of near-home destinations. *Tourism Management*, *30*(6), 857–866. doi:10.1016/j.tourman.2008.12.007

Pike, S., & Ryan, C. (2004). Destination Positioning Analysis through a Comparison of Cognitive, Affective, and Conative Perceptions. *Journal of Travel Research*, *42*(May), 333–342. doi:10.1177/0047287504263029

Saayman, M., Slabbert, E., & Van Der Merwe, P. (2009). Travel Motivation: A Tale of Two Marine Destinations in South Africa. *South African Journal for Research in Sport. Physical Education and Recreation, 31*(1), 81–94.

Selvin, S. (1977). A Further Note on the Interpretetion of ridit analysis. *American Journal of Epidemiology, 105*, 16–20. PMID:831461

Shuzhen, P., Leiting, W., Feifei, Q., & Wei, Z. (2011). Factor Analysis of College students' Travel Motivations in the Tourist City: A case Study for College Students in Taiwan City. In M. Zhou (Ed.), *Advances in Education and Management: International Symposium* (Vol. 211). Springer Science & Business Media.

Skinner, H. (2008). The emergence and development of place marketings confused identity. *Journal of Marketing Management, 24*(9-10), 915–928. doi:10.1362/026725708X381966

Smallman, C., & Moore, K. (2010). Process studies of tourists' decision-making. *Annals of Tourism Research, 37*(2), 397–422. doi:10.1016/j.annals.2009.10.014

Tsaur, S.-H., & Tzeng, G.-H. (1996). Multiattribute Decision Making Analysis for Customer Preference of Tourist Hotels. *Journal of Travel & Tourism Marketing, 4*(4), 55–69. doi:10.1300/J073v04n04_04

UNWTO. (2015). *UNWTO Tourism Highlights 2015 Edition*. Author.

Uwawunkonye, E. G., & Anaene, O. I. C. (2013). A comparative study between ridit and modified ridit analysis. *American Journal of Theoretical and Applied Statistics, 2*(6), 248–254. doi:10.11648/j.ajtas.20130206.23

Wang, Y., Guo, L., & Huang, L. (2013). A sports tourism resources comprehensive evaluation model based on AHP-FUZZY. *Journal of Theoretical and Applied Information Technology, 48*(3), 1473–1479.

Williams, P. D., & Grizzle, J. E. (1972). Analysis of Contingency Table having Ordered Response Categories. *American Statistical Association, 67*(337), 55–63. doi:10.1080/01621459.1972.10481205

Wood, L. (2007). Functional and symbolic attributes of product selection. *British Food Journal, 109*(2), 108–118. doi:10.1108/00070700710725482

Wu, C. H. (2007). On the application of grey relational analysis and RIDIT analysis to Likert scale surveys.*International Mathematical Forum, 2*(14), 675-687. doi:10.12988/imf.2007.07059

Zhang, H., Gu, C., Gu, L., & Zhang, Y. (2011). The evaluation of tourism destination competitiveness by TOPSIS & information entropy - A case in the Yangtze River Delta of China. *Tourism Management, 32*(2), 443–451. doi:10.1016/j.tourman.2010.02.007

ADDITIONAL READING

Dolnicar, S., & Ring, A. (2014). Tourism marketing research: Past, present and future. *Annals of Tourism Research, 47*, 31–47. doi:10.1016/j.annals.2014.03.008

Fuchs, M., Hopken, W., & Lexhagen, M. (2014). Big data analytics for knowledge generation in tourism destinations - A case from Sweden. *Journal of Destination Marketing and Management*, *3*(4), 198–209. doi:10.1016/j.jdmm.2014.08.002

Kotler, P., Bowen, J. T., & Makens, J. C. (2010). *Marketing for Hospitality and Tourism* (5th ed.). London: Pearson.

Lew, A., & McKercher, B. (2006). Modeling tourist movements: A local destination analysis. *Annals of Tourism Research*, *33*(2), 403–423. doi:10.1016/j.annals.2005.12.002

McCabe, S. (2005). Who is a tourist? A critical review. *Tourist Studies*, *5*(1), 85–106. doi:10.1177/1468797605062716

Mckercher, B., & Lau, G. (2008). Movement Patterns of Tourists within a Destination. *Tourism Geographies*, *10*(3), 355–374. doi:10.1080/14616680802236352

Middleton, V. T. C., & Clarke, J. (2001). *Marketing in Travel and Tourism* (3rd ed.). Oxford: Butterworth Heinemann.

Morrison, A. (2002). *Hospitality and Travel Marketing* (3rd ed.). Albany: Delmar, Thomson Learning.

Tsiotsou, R., & Ratten, V. (2010). Future research directions in tourism marketing. *Marketing Intelligence & Planning*, *28*(4), 533–544. doi:10.1108/02634501011053702

Vuuren, C. V., & Slabbert, E. (2011). Travel Motivation and Behaviour of Tourists to a South African Resort.*International Conference on Tourism and Management Studies* - Algarve2011, I, 295-304.

Williams, A. (2006). Tourism and hospitality marketing: Fantasy, feeling and fun. *International Journal of Contemporary Hospitality Management*, *18*(6), 482–495. doi:10.1108/09596110610681520

KEY TERMS AND DEFINITIONS

Kruskal-Wallis Statistics W: Kruskal-Wallis statistics W is a hypothesis testing tool which follows the χ^2 distribution with $(m - 1)$ degrees of freedom. Here 'm' is the total number of variables.

Likert Scale: A Likert scale is a psychometric scale commonly involved in research and it is the most widely used approach to scaling responses in survey research, such that the term is often used interchangeably with rating scale, or more accurately the Likert-type scale, even though the two are not synonymous.

RIDIT Analysis: RIDIT (Relative to an Identified Distribution) is a very efficient technique that can be used to examine the Likert scale data. The outcomes from the RIDIT analysis can be used to arrange Likert scale items either in an ascending or in a descending order based on importance.

Tourism: The World Tourism Organization defines tourism as "Tourism is a social, cultural and economic phenomenon which entails the movement of people to countries or places outside their usual environment for personal or business/professional purposes. These people are called visitors (which may be either tourists or excursionists; residents or non-residents) and tourism has to do with their activities, some of which imply tourism expenditure".

Tourists: The World Tourism Organization defines Tourists as people who "travel to and stay in places outside their usual environment for more than twenty-four (24) hours and not more than one consecutive year for leisure, business and other purposes not related to the exercise of an activity remunerated from within the place visited".

Travel Behaviour: This refers to the way in which tourists behave according to their attitudes before, during and after travelling.

Travel Motivation: Travel motivation is the inner state of a person, or certain needs and wants of the tourists that can be considered as one of the most important psychological influences of tourist behaviour.

Compilation of References

Ab Hamid, N. R., Ahmad, F., Tahir Shas, S. N., & Arshad, N. H. (2010). Evaluating the adoption level of electronic customer relationship management by telecommunication companies. *WSEAS Transactions on Systems*, *9*(2), 115–124.

Abbatista, F., Licchelli, M. D. O., Lops, P., Semerao, G., & Zambetta, F. (1999). Improving the usability of an e-commerce web site through personalization. In *Proceedings of theWorkshop on Recommendation and Personalization in Electronic Commerce, 2nd International Conference on Adaptative Hypermedia and Adaptative Web Based Systems*. University of Bari.

Abbott, D. (2014). *Applied predictive analytics: Principles and techniques for the professional data analyst*. Indianapolis, IN: Wiley.

Abdel-Azim, T. S. (2010). The Relationship between the perception of Risk and the Decision Making process of Travel of French Tourists. The case of Egypt. *Tourismos: An International Multidisplinary Journal of Tourism, 2*, 29-47.

Abeles, E. G., White, M. D., & Hahn, K. (1999). A user-based design process for web sites. *OCLC Systems & Services*, *15*(1), 35–44. doi:10.1108/10650759910257850

Abou-Shouk, M., Megicks, P., & Lim, W. M. (2013). Perceived benefits and e- commerce adoption by SME travel agents in developing countries: Evidence from Egypt. *Journal of Hospitality & Tourism Research (Washington, D.C.)*, *37*(4), 490–515. doi:10.1177/1096348012442544

Accinelli, E., Brida, J. G., & Carrera, E. (2008). A good policy of sustainable tourism. *Revista de Administración, Finanzas y Economía, 2*(2), 150–161.

Across Abyssinia Tour, P. L. C. (2014). Retrieved from http://www.adventureabyssinia.com/wp-content/uploads/2014/08/axum1.jpg

Adolphs, C., & Winkelmann, A. (2010). Personalization research in e-commerce a state of the art review (2000-2008). *Journal of Electronic Commerce Research*, *11*(4), 326–341.

Agresti, A. (1990). *Categorical data analysis*. New York, NY: John Wiley & Sons.

Ahmed, E. (2012). Network analysis. In L. Dwyer, A. Gill, & N. Seetaram (Eds.), *Handbook of research methods in tourism: Quantitative and qualitative approaches* (pp. 472–494). Cheltenham, UK: Edward Elgar. doi:10.4337/9781781001295.00033

Akehurst, G. (2009). User-generated content: The use of blogs for tourism organizations and tourism consumers. *Service Business*, *3*(1), 51–61. doi:10.1007/s11628-008-0054-2

Akgũl, Y., & Vatansever, K. (2016). Web accessibility evaluation of government websites for people with disabilities in Turkey. *Journal of Advance Management*, *4*(3), 201–210.

Alba, E. (2005). *Parallel Metaheuristics: A New Class of Algorithms*. Wiley-Interscience. doi:10.1002/0471739383

Alba, E., Luque, G., & Nesmachnow, S. (2013). Parallel metaheuristics: Recent advances and new trends. *International Transactions in Operational Research*, *20*(1), 1–48. doi:10.1111/j.1475-3995.2012.00862.x

Al-Debei, M., Ashouri, M., & Akroush, M. (2015). Consumer attitudes towards online shopping: The effects of trust, perceived benefits, and perceived web quality. *Internet Research*, *25*(5), 707–733. doi:10.1108/IntR-05-2014-0146

Aledo, A., García, H., & Ortíz, G. (2010). Análisis de mapas causales de impactos del turismo residencial. *Empiria: Revista De Metodología. Ciencia y Sociedad*, (20), 2061–2086.

Alexandrov, A. (2010). Characteristics of Single-Item Measures in Likert Scale Format. *Electronic Journal of Business Research Methods*, *8*(1), 1–12.

Algieri, B., & Kanellopoulou, S. (2009). Determinants of demand for exports of tourism: An unobserved component model. *Tourism and Hospitality Research*, *9*(1), 9–19. doi:10.1057/thr.2008.39

Alshawi, S., Themistocleous, M., & Almadani, R. (2004). Integrating diverse ERP systems: A case study. *The Journal of Enterprise Information Management*, *17*(6), 454–462. doi:10.1108/17410390410566742

Alvarez, L., Ruiz, F., & Solana, P. (2008, September). *Yield management under the real options scheme for optimal decision making in hotels*. Paper presented at the II International Conference on Industrial Engineering and Industrial Management, Burgos, Spain.

Amadeus. (2011). *The always-connected traveller: How mobile will transform the future of air travel.* Retrieved March 15, 2015 from http://www.amadeus.com/airlineit/the-always-connected-traveller/docs/amadeus-the-always-connected-traveller-2011-en.pdf

Amdahl, G. M. (1967). Validity of the single processor approach to achieving large scale computing capabilities. *Proceedings of the April 18-20, 1967, spring joint computer conference.* ACM. doi:10.1145/1465482.1465560

Amorin, E., Gandara, J., & Korsanje, M. E. (2012). Seguridad percibida en la ciudad de Curitiba. Un sondeo exploratorio en residentes locales y turistas. *Revista de análisis turístico*, *14*, 1-9.

Amram, M., & Kulatilaka, N. (1999). *Real options: managing strategic investment in an uncertain world.* Oxford Press.

Amran, M., & Kulatilaka, N. (1999). *Real options—Managing strategic investment in an uncertain world.* Boston, MA: Harvard Business School Press.

Anckar, B. (2003). Consumer intentions in terms of electronic travel distribution: Implications for future market structures. *e-Service Journal*, *2*(2), 68–86. doi:10.2979/ESJ.2003.2.2.68

Andergassen, R., Candela, G., & Figini, P. (2013). An economic model for tourism destinations: Product sophistication and price coordination. *Tourism Management*, *37*, 86–98. doi:10.1016/j.tourman.2012.10.013

Andersen, L. (2002). How options analysis can enhance managerial performance. *European Management Journal*, *20*(5), 505–511. doi:10.1016/S0263-2373(02)00093-2

Anderson, C. K. (2009). The billboard effect: Online travel agent impact on non-OTA reservation volume. CHR Report. *The Center for Hospitality Research at Cornell University, 9*(16).

Anderson, C. K. (2011). Search, OTAs, and online booking: An expanded analysis of the billboard effect. CHR Report. *The Center for Hospitality Research at Cornell University, 11*(8).

Anderson, C. K., Davison, M., & Rasmussen, H. (2004). Revenue management: A real options approach. *Naval Research Logistics*, *51*(5), 686–703. doi:10.1002/nav.20026

Anderson, D. R., Sweeney, D. J., & Williams, T. A. (2001). *Quantitative methods for business.* Cincinnati, OH: South Western College Publishing.

Anderson, W., Juaneda, C., & Sastre, F. (2009). Influences of pro-all-inclusive Travel Decisions. *Tourism Review, 64*(2), 4–18. doi:10.1108/16605370910963482

Andrés, M. E., Gómez, M. A., & Mondéjar, J. A. (2014). A model to evaluate the effects of price fairness perception in online hotel booking. *Electronic Commerce Research, 14*(2), 171–187. doi:10.1007/s10660-014-9137-4

Andrew, P., & Schmidgall, R. (1993). *Financial Management for the Hospitality Industry.* Washington, DC: AH&MA.

Angelou, G. N., & Economides, A. A. (2009). A multi-criteria game theory and real-options model for irreversible ICT investment decisions. *Telecommunications Policy, 33*(10-11), 686–705. doi:10.1016/j.telpol.2009.07.005

Araujo, L., & Merelo, J. (2011). Diversity through multiculturality: Assessing migrant choice policies in an island model. Evolutionary Computation. *IEEE Transactions on, 15*(4), 456–469. doi:10.1109/TEVC.2010.2064322

Arndt, O., & Sternberg, R. (2000). Do manufacturing firms profit from intraregional innovation linkages? An empirical based answer. *European Planning Studies, 8*(4), 465–485. doi:10.1080/713666423

Arsal, I., Woosnam, K. M., Baldwin, E. D., & Backman, S. J. (2010). Residents as travel destination information providers: An online community perspective. *Journal of Travel Research, 49*(4), 400–413. doi:10.1177/0047287509346856

Asero, V., D'Agata, R., & Tomaselli, V. (2013). Tourism mobility among destinations: An application of network analysis. In A. M. Oliveri & S. De Cantis (Eds.), *Analysing local tourism. A statistical perspective* (pp. 309–324). Maidenhead, UK: McGraw-Hill Education.

Asero, V., Gozzo, S., & Tomaselli, V. (2016). Building tourism networks through tourist mobility. *Journal of Travel Research*, 55(6), 751–763. doi:10.1177/0047287515569777

Ashill, N. J., Rod, M., & Carruthers, J. (2008). The effect of management commitment to service quality on frontline employees' job attitudes, turnover intentions and service recovery performance in a new public management context. *Journal of Strategic Marketing, 16*(5), 437–462. doi:10.1080/09652540802480944

Au, N., Law, R., & Buhalis, D. (2010). The impact of Culture on eComplaints: Evidence from Chinese Consumers in Hospitality Organizations. In U. Gretzel R. Law, & M Fuchs (Eds.), Information and Communication Technologies in Tourism 2010 (pp. 285-296). Vienna, Austria: Springer.

Aves, E. (2016). *Exploring Bale Mountains National Park, Ethiopia, Rough Guides.* Accessed on Feb. 2016, from http://www.roughguides.com/wp-content/uploads/2014/09/95647017_14-660x431.jpg

Bachelier, L. (1900). *The'orie de la Speculation.* Paris: Gauthier-Villars.

Back, T., Fogel, B., & Michalewicz, Z. (1997). *Handbook of Evolutionary Computation.* London: Institute of Physics. doi:10.1887/0750308958

Badea, M. (2014). Customer orientation in the local media. Case study: The audiovisual in Dambovita county. *Procedia: Social and Behavioral Sciences, 149*, 76–80. doi:10.1016/j.sbspro.2014.08.333

Baggio, R. (2008). Symptoms of complexity in a tourism system. *Tourism Analysis, 13*(1), 1–20. doi:10.3727/108354208784548797

Baggio, R. (2014). Complex tourism systems: A visibility graph approach. *Kybernetes, 43*(3/4), 445–461. doi:10.1108/K-12-2013-0266

Baggio, R., & Cooper, C. (2010). Knowledge transfer in a tourism destination: The effects of a network structure.The-*Service Industries Journal*, *30*(10), 1–15. doi:10.1080/02642060903580649

Baggio, R., & Del Chiappa, G. (2014). Real and virtual relationships in tourism digital ecosystems. *Information Technology & Tourism*, *14*(1), 3–19. doi:10.1007/s40558-013-0001-5

Baggio, R., & Sainaghi, R. (2011). Complex and chaotic tourism systems: Towards a quantitative approach. *International Journal of Contemporary Hospitality Management*, *23*(6), 840–861. doi:10.1108/09596111111153501

Baggio, R., Scott, N., & Cooper, C. (2010). Network science - A review focused on tourism. *Annals of Tourism Research*, *37*(3), 802–827. doi:10.1016/j.annals.2010.02.008

Baggio, R., Scott, N., & Cooper, C. (2011). Design of tourism governance networks. In E. Laws, H. Richins, J. Agrusa, & N. Scott (Eds.), *Tourism destination governance: Practice, theory and issues* (pp. 150–171). Wallingford, UK: CAB International. doi:10.1079/9781845937942.0159

Bahmani-Firouzi, B., Sharifinia, S., Azizipanah-Abarghooee, R., & Niknam, T. (2015). Scenario-Based Optimal Bidding Strategies of GENCOs in the Incomplete Information Electricity Market Using a New Improved Prey Predator Optimization Algorithm. *IEEE Systems Journal, 9*(4), 1485 - 1495.

Baker, M., & Wurgler, J. (2000). The equity share in new issues and aggregate stock returns. *The Journal of Finance*, *55*(5), 2219–2257. doi:10.1111/0022-1082.00285

Baker, M., & Wurgler, J. (2002). Market timing and capital structure. *The Journal of Finance*, *57*(1), 1–32. doi:10.1111/1540-6261.00414

Baldursson, F. M. (1998). Irreversible Investment under Uncertainty in Oligopoly. *Journal of Economic Dynamics & Control*, *22*(4), 627–644. doi:10.1016/S0165-1889(97)00070-5

Banati, H., Bedi, P., & Grover, P. S. (2006). Evaluating web usability from the users perspective. *Journal of Computer Science*, *2*(4), 314–317. doi:10.3844/jcssp.2006.314.317

Banker, R. D., Field, J. M., Schroeder, R. G., & Sintia, K. K. (1996). Impact of work teams on manufacturing performance: A longitudinal field study. *Academy of Management Journal*, *39*(4), 867–890. doi:10.2307/256715

Barnes, S. J., & Vidgen, R. (2003). Measuring website quality improvements: A case study of the Forum on Strategic Management Knowledge Exchange. *Industrial Management & Data Systems*, *103*(5), 297–309. doi:10.1108/02635570310477352

Barr, A. (1995). Drink: An informal social history. London, UK: Bantham.

Barro, R. J. (1976). The Loan Market, Collateral, and Rates of Interest. *Journal of Money, Credit and Banking*, *8*(4), 439–456. doi:10.2307/1991690

Barth, J. E. (2011). A model for wine list and wine inventory yield management. *International Journal of Hospitality Management*, *30*(3), 701–707. doi:10.1016/j.ijhm.2010.12.004

Barton, C., Fromm, J., & Egan, C. (2012). *The millennial consumer, debunking stereotypes*. Retrieved January 12, 2016, from https://www.bcgperspectives.com/content/articles/consumer_insight_marketing_millennial_consumer/

Batabyal, A. A., & Beladi, H. (2011). An alternate approach to modeling the slack season provision of guided tours to tourists. *Tourism Management*, *32*(5), 1047–1049. doi:10.1016/j.tourman.2010.09.001

Baudrillard, J. (2006). Virtuality and Events: The hell of power. *Baudrillard Studies*, *3*(2), 1–15.

Bauman, Z. (2008). *Liquid fear*. Buenos Aires: Paidos.

Bauman, Z. (2009). Freudian Civilization revisited – or whatever happened to the reality principie? *Anthopological Psychology*, *21*, 1–9.

Bauman, Z. (2013a). *Consuming life*. New York, NY: John Wiley & Sons.

Bayless, M., & Chaplinsky, S. (1996). Is there a Window of Opportunity for Seasoned Equity Issuance? *The Journal of Finance*, *51*(1), 253–278. doi:10.1111/j.1540-6261.1996.tb05209.x

Beaumont, N., & Dredge, D. (2010). Local tourism governance: A comparison of three network approaches. *Journal of Sustainable Tourism*, *18*(1), 7–28. doi:10.1080/09669580903215139

Beck, U. (2006). *Risk society, towards a new modernity*. Buenos Aires: Paidos.

Beldona, S., Cai, L., & Pearson, T. (2001). Impact of the Internet on the hospitality value chain – towards an information sharing network. Advances in Hospitality and Tourism Research.*Proceedings of the Sixth Annual Conference and Graduate Students Research in Hospitality and Tourism*, 6, 25-29.

Beldona, S., Morrison, A. M., & OLeary, J. (2005). Online shopping motivations and pleasure travel products: A correspondence analysis. *Tourism Management*, *26*(4), 561–570. doi:10.1016/j.tourman.2004.03.008

Beldona, S., Nusair, K., & Demicco, F. (2009). Online travel purchase behavior of generational cohorts: A longitudinal study. *Journal of Hospitality Marketing & Management*, *18*(4), 406–420. doi:10.1080/19368620902799627

Beldona, S., Racherla, P., & Mundhra, G. D. (2011). To buy or not to buy: Indian consumers choices of online versus offline channels for air travel purchase. *Journal of Hospitality Marketing and Management*, *20*(8), 831–854. doi:10.1080/19368623.2011.577693

Belediye Belgeli Tesislerde Konaklama İstatistikleri. (2013). Retrieved December 15, 2015, from http://www.tuik.gov.tr/PreTablo.do?alt_id=1072

Bell, E. (2009). *Glass Half Empty? On the State of Restaurant Wine Sales in the Recession*. Retrieved from http://www.starchefs.com/features/editors_dish/wine-survey/html/index.shtml

Benckendorff, P., Moscardo, G., & Pendergast, D. (Eds.). (2010). *Tourism and Generation Y*. Oxford, UK: CABI.

Bendoly, E. (2013). Real-time feedback and booking behavior in the hospitality industry: Moderating the balance between imperfect judgment and imperfect prescription. *Journal of Operations Management*, *31*(1), 62–71. doi:10.1016/j.jom.2012.06.003

Benjamin, D. K. (1978). The use of collateral to enforce debt contracts. *Economic Inquiry*, *16*(3), 333–359. doi:10.1111/j.1465-7295.1978.tb00507.x

Bennet, M. M., & Buhalis, D. (2003). *Tourism and travel distribution, the travel agency perspective. Insights, D25-30*. English Tourism Council.

Bennett, M. M., & Lai, C. (2005). The impact of the internet on travel agencies in Taiwan. *Tourism and Hospitality Research*, *6*(1), 8–23. doi:10.1057/palgrave.thr.6040041

Berenguer, G., Gil, I., & Ruiz, M. E. (2009). Do upscale restaurant owners use wine lists as a differentiation strategy? *International Journal of Hospitality Management*, *28*(1), 86–95. doi:10.1016/j.ijhm.2008.04.003

Beritelli, P. (2011). Do actors really agree on strategic issues? Applying consensus analysis of stakeholder perceptions in tourist destination communities. *Tourism Analysis*, *16*(4), 219–241. doi:10.3727/108354211X13110944386924

Beritelli, P., Bieger, T., & Laesser, C. (2007). The impact of the Internet on information source portfolios: Insight from a mature market. *Journal of Travel & Tourism Marketing*, *22*(1), 63–80. doi:10.1300/J073v22n01_05

Beritelli, P., Bieger, T., & Laesser, C. (2014). The new frontiers of destination management: Applying variable geometry as a function-based approach. *Journal of Travel Research*, *53*(4), 403–417. doi:10.1177/0047287513506298

Berman, A. R. (2005). "Once a Mortgage, Always a Mortgage" - The Use (and Misuse of) Mezzanine Loans and Preferred Equity Investments. *Stanford Journal of Law. Business & Finance*, *11*, 76.

Berman, B., & Evans, J. (1995). *Principles of marketing*. Englewood Cliffs, NJ: Prentice Hall.

Berne, C., Garcia-Gonzalez, M., & Mugica, J. (2012). How ICT shifts the power balance of tourism distribution channels. *Tourism Management*, *33*(1), 205–214. doi:10.1016/j.tourman.2011.02.004

Berry, L. L., Carbone, L. P., & Haeckel, S. H. (2002). Managing the total customer experience. *MIT Sloan Management Review*, *43*(3), 85–89.

Berry, L. L., Shankar, V., Parish, J. T., Cadwallader, S., & Dotzel, T. (2006). Creating new markets through service innovation. *MIT Sloan Management Review*, *47*(2), 56–63.

Bertoluci, J. (2007). Search Engine Shoot-out. *PC World*, *25*(6), 86–96.

Bettencourt, L. A., & Brown, S. W. (1997). Contact employees: Relationships among workplace fairness, job satisfaction and prosocial service behaviors. *Journal of Retailing*, *73*(1), 39–61. doi:10.1016/S0022-4359(97)90014-2

Bevan, A. A., & Danbolt, J. (2004). Testing for Inconsistencies in the Estimation of UK Capital Structure Determinants. *Applied Financial Economics*, *14*(1), 55–66. doi:10.1080/0960310042000164220

Bhat, S. S., & Milne, S. (2008). Network effects on cooperation in destination website development. *Tourism Management*, *29*(6), 1131–1140. doi:10.1016/j.tourman.2008.02.010

Bhat, S., & Reddy, S. K. (1998). Symbolic and Functional Positioning of Brands. *Journal of Consumer Marketing*, *15*(1), 32–43. doi:10.1108/07363769810202664

Bianchi, R. (2007). Tourism and The Globalization of Fear: Analyzing the politics of risk and (in)security in global travel. *Tourism and Hospitality Research*, *7*(1), 64–74. doi:10.1057/palgrave.thr.6050028

Bianchi, R., & Stephenson, M. (2014). *Tourism and citizenship: Rights, freedoms and responsibilities in the global order* (Vol. 11). Abingdon: Routledge.

Bianchi, R., & Stephenson, M. (2014). *Tourism and Citizenship: rights, freedoms and responsibilities in the global Order*. Abingdon: Routledge.

Bickerdyke, I. (1996). Australia: The evolving structure and strategies of business networks. In OECD (Ed.), *Networks of enterprises and local development: Competing and co-operating in local productive systems* (pp. 203–216). Paris: OECD.

Bieger, T. (2004). SMEs and cooperation. In P. Keller & T. Bieger (Eds.), *The future of small and medium sized enterprises in tourism* (pp. 141–150). St. Gallen, Switzerland: AIEST.

Bieger, T., Beritelli, P., & Laesser, C. (2009). Size matters! Increasing DMO effectiveness and extending tourism destination boundaries. *Tourism Review*, *57*(3), 309–327.

Bikash, R. D., & Pravat, S. K., & Sreekumar. (2010). Factors Influencing Purchase of 'NANO' the Innovative Car from India - An Empirical Study. *Asian Journal of Business Management*, *2*(3), 48–56.

Bitner, M. J., Booms, B. H., & Mohr, L. A. (1994). Critical service encounters: The employee's viewpoint. *Journal of Marketing*, *58*(4), 95–106. doi:10.2307/1251919

Bitran, G. R., & Mondschein, S. V. (1995). An application of yield management to the hotel industry considering multiple day stays. *Operations Research*, *43*(3), 427–443. doi:10.1287/opre.43.3.427

Black, F., & Scholes, M. S. (1973). The pricing of options and corporate liabilities. *Journal of Political Economy*, *81*(3), 637–659. doi:10.1086/260062

Blake, A., & Sinclair, M. T. (2003). Tourism crisis management: US response to September 11. *Annals of Tourism Research*, *30*(4), 813–832. doi:10.1016/S0160-7383(03)00056-2

Blocker, C. P., Flint, D. J., Myers, M. B., & Slater, S. F. (2011). Proactive customer orientation and its role for creating customer value in global markets. *Journal of the Academy of Marketing Science*, *39*(2), 216–233. doi:10.1007/s11747-010-0202-9

Block, S. (2011). Does the Weighted Average Cost of Capital Describe the Real-World Approach to the Discount Rate? *The Engineering Economist*, *56*(2), 170–180. doi:10.1080/0013791X.2011.573618

Bogahavatta, C. (2013). *Introduction to tourism and cultural resources*. Retrieved Feb. 02, 2016 from http://si.archaeology.lk/materials/Introduction-to-Tourism-and-Cultural-Resources_2013_04_01.pdf

Bolino, M. C., & Turnley, W. H. (2009). Relative deprivation among employees in lower-quality leader-member exchange relationships. *The Leadership Quarterly*, *20*(3), 276–286. doi:10.1016/j.leaqua.2009.03.001

Bolton, R. N., Parasuraman, A., Hoefnagels, A., Migchels, N., Kabadayi, S., Gruber, T., & Solnet, D. et al. (2013). Understanding Generation Y and their use of social media: A review and research agenda. *Journal of Service Management*, *24*(3), 245–267. doi:10.1108/09564231311326987

Bonabeau, E., Dorigo, M., & Theraulaz, G. (1999). *Swarm Intelligence: From Natural to Artificial Systems*. Oxford University Press.

Boone, H. N. J., & Boone, D. a. (2012). Analyzing Likert Data. *Journal of Extension*, *50*(2), 30.

Borgatti, S. P., & Halgin, D. S. (2011). On network theory. *Organization Science*, *22*(5), 1168–1181. doi:10.1287/orsc.1100.0641

Brakus, J. J., Schmitt, B. H., & Zarantonello, L. (2009). Brand experience: What is it? How is it measured? Does it affect loyalty? *Journal of Marketing*, *73*(3), 52–68. doi:10.1509/jmkg.73.3.52

Brand Ethiopia. (2016). Accessed on Feb. 2016, from http://www.brandethiopian.com/Content/uploads/tour-package/21/The_Gate_of__Harrar_jpg.jpg

Breiger, R. L., Boorman, S. A., & Arabie, P. (1975). An algorithm for clustering relational data with applications to social network analysis and comparison with multidimensional scaling. *Journal of Mathematical Psychology*, *12*(3), 328–383. doi:10.1016/0022-2496(75)90028-0

Briedenhann, J., & Wickens, E. (2004). Tourism routes as a tool for the economic development of rural areas - vibrant hope or impossible dream? *Tourism Management*, *25*(1), 71–79. doi:10.1016/S0261-5177(03)00063-3

Bring, S., & Page, L. (1998). The anatomy of a large-scale hypertextual web search engine. *Computer Networks*, *30*, 107–117.

Brockett, P. L. (1981). A note on Numerical Assignment of Scores to Ranked Categorical Data. *The Journal of Mathematical Sociology*, *8*(1), 91–101. doi:10.1080/0022250X.1981.9989917

Brohman, M. K., Piccoli, G., Martin, P., Zulkernine, F., Parasuraman, A., & Watson, R. T. (2009). A design theory approach to building strategic network-based customer service systems. *Decision Sciences*, *40*(3), 403–430. doi:10.1111/j.1540-5915.2009.00242.x

Brooks, R. (2009). Researching democracy and terrorism: How political access affects militant activity. *Security Studies*, *18*(4), 756–788. doi:10.1080/09636410903369027

Bross, I. D. J. (1958). How to use Ridit Analysis. *Biometrics*, *14*(1), 18–38. doi:10.2307/2527727

Brotherton, B. (2004). Critical success factors in the UK budget hotel operations. *International Journal of Operations & Production Management*, *24*(9/10), 944–969. doi:10.1108/01443570410552135

Brown, G., Boya, U. O., Humphreys, N., & Widing, R. E. (1993). Attributes and behaviors of salespeople preferred by buyers: High socializing vs. low socializing industrial buyers. *Journal of Personal Selling & Sales Management*, *13*(1), 25–33.

Bruhn, M., Schnebelen, S., & Schäfer, D. (2014). Antecedents and consequences of the quality of e-customer-to-customer interactions in B2B brand communities. *Industrial Marketing Management*, *43*(1), 164–176. doi:10.1016/j.indmarman.2013.08.008

Brunori, G., & Rossi, A. (2000). Collective action: Some insights from wine routes in Tuscany. *Sociologia Ruralis*, *40*(4), 409–423. doi:10.1111/1467-9523.00157

Bruwer, J. (2003). South African wine routes: Some perspectives on the wine tourism industrys structural dimensions and wine tourism product. *Tourism Management*, *24*(4), 423–435. doi:10.1016/S0261-5177(02)00105-X

Bucklin, L. P., & Sengupta, S. (1989). Organizing successful co-marketing alliances. *Journal of Marketing*, *57*(2), 32–46. doi:10.2307/1252025

Buhalis, D. (1998). Strategic use of information technologies in the tourism industry. *Tourism Management*, *19*(5), 409–421. doi:10.1016/S0261-5177(98)00038-7

Buhalis, D. (2000). Marketing the competitive destination of the future. *Tourism Management*, *21*(1), 97–116. doi:10.1016/S0261-5177(99)00095-3

Buhalis, D., & Darcy, S. (2010). *Accessible Tourism. Concept and Issues*. Bristol: Channel View Publications.

Buhalis, D., Darcy, S., & Ambrose, I. (2012). *Best Practice in Accessible Tourism: Inclusion, Disability, Ageing*. Bristol: Channel View Publications.

Buhalis, D., & Law, R. (2008). Progress in information technology and tourism management: 20 years on and 10 years after the Internet – The state of eTourism research. *Tourism Management*, *29*(4), 609–623. doi:10.1016/j.tourman.2008.01.005

Burns, P., & Novelli, M. (Eds.). (2008). *Tourism and mobilities: Local-global connections*. Wallingford, UK: CAB International. doi:10.1079/9781845934040.0000

Burt, R. S. (1976). Positions in networks. *Social Forces*, *55*(1), 93–122. doi:10.1093/sf/55.1.93

Cachon, G. P. (2001). Stock wars: Inventory competition in a two-echelon supply chain with multiple retailers. *Operations Research*, *49*(5), 658–674. doi:10.1287/opre.49.5.658.10611

Calero, C., Ruiz, J., & Piattini, M. (2005). Classifying web metrics using the web quality model. *Online Information Review*, *29*(3), 227–248. doi:10.1108/14684520510607560

Calgor, H. (n.d.). *The bruth Force Algorithm.* Retrieved January24, 2010, from http://www.ctl.ua.edu/math103/hamilton/hamilton.html

Calle, A. M., & Tamayo, V. M. (2009). Decisiones de inversión a través de opciones reales. *Estudios Gerenciales, 25*(111), 107–126. doi:10.1016/S0123-5923(09)70073-7

Campo-Martinez, S., Garau-Vadell, J. B., & Martinez-Ruiz, M. P. (2010). Factors influencing repeat visits to a destination: The influence of group composition. *Tourism Management, 31*(6), 862–870. doi:10.1016/j.tourman.2009.08.013

Candela, G., & Cellini, R. (2006). Investment in tourism market: A dynamic model of differentiated oligopoly. *Environmental and Resource Economics, 35*(1), 41–58. doi:10.1007/s10640-006-9005-3

Cao, X., & Mokhtarian, P. L. (2005). *The intended and actual adoption of online purchasing a brief review of recent literature.* Working paper Institute of Transportation Studies (ITS). 2007. Department of Civil and Environmental Engineering, University of California. Retrieved from http://www.tc.umn.edu/~cao/UCD-ITS-RR-05-07.pdf

Capello, R. (2009). Space, growth and development. In R. Capello & P. Nijkamp (Eds.), *Handbook of regional growth and development theories* (pp. 33–52). Cheltenham, UK: Edward Elgar. doi:10.4337/9781848445987.00008

Card, J., Chen, C., & Cole, S. (2003). Online travel product shopping: Differences between shoppers and non shoppers. *Journal of Travel Research, 42*(2), 133–139. doi:10.1177/0047287503257490

Carlsen, J. (1999). A systems approach to island tourism destination management. *Systems Research and Behavioral Science, 16*(4), 321–327. doi:10.1002/(SICI)1099-1743(199907/08)16:4<321::AID-SRES255>3.0.CO;2-5

Carlshamre, P. (2001). *A usability perspective on requirements engineering – From methodology to product development.* Linköping: Linköping University Electronic Press.

Carlson, J., & OCass, A. (2011). Developing a framework for understanding e-service quality, its antecedents, consequences, and mediators. *Managing Service Quality: An International Journal, 21*(3), 264–286. doi:10.1108/09604521111127965

Casanovas, I. (2010). Exploring the current theoretical background about adoption until institutionalization of online education in universities: Needs for further research. *Electronic Journal of E-learning, 8*(2), 73–84.

Cavalletti, A. (2010). *Mitología de la seguridad: la ciudad biopolítica.* Madrid: Adriana Hidalgo.

Cerny, V. (1985). Thermodynamical approach to the traveling salesman problem: An efficient simulation algorithm. *Journal of Optimization Theory and Applications, 45*(1), 41–51. doi:10.1007/BF00940812

Chakraborty, G., Lala, V., & Warren, D. (2003). What do customers consider important in B2B websites. *Journal of Advertising Research, 43*(1), 50–61. doi:10.2501/JAR-43-1-50-61

Chang, S., Gong, Y., & Shum, C. (2011). Promoting innovation in hospitality companies through human resource management practices. *International Journal of Hospitality Management, 30*(4), 812–818. doi:10.1016/j.ijhm.2011.01.001

Chapman, P., Clinton, J., Kerber, R., Khabaza, T., Reinartz, T., Shearer, C., & Wirth, R. (2000). *CRISP-DM 1.0: Step-by-step data mining guide.* Retrieved from https://the-modeling-agency.com/crisp-dm.pdf

Charters, S. (2006). *Wine and Society: The social and cultural context of a drink.* Burlington, MA: Elsevier Butterworth-Heinemann.

Chathoth, P., Altinay, L., Harrington, R. J., Okumus, F., & Chan, E. S. (2013). Co-production versus co-creation: A process based continuum in the hotel service context. *International Journal of Hospitality Management, 32*, 11–20. doi:10.1016/j.ijhm.2012.03.009

Chen, A. H., Peng, N., & Hackley, C. (2008). Evaluating service marketing in airline industry and its influence on student passengers purchasing behaviour using Taipei–London route as an example. *Journal of Travel & Tourism Marketing*, *25*(2), 149–160. doi:10.1080/10548400802402503

Chen, C. F., & Myagmarsuren, O. (2013). Exploring the moderating effects of value offerings between market orientation and performance in tourism industry. *International Journal of Tourism Research*, *15*(6), 595–610. doi:10.1002/jtr.1900

Chen, C. M., & Lin, Y. C. (2013). The influence of uncertain demand on hotel capacity. *International Journal of Hospitality Management*, *34*, 462–465. doi:10.1016/j.ijhm.2012.11.008

Chen, C.-F. (2006). Applying the analytical hierarchy process (AHP) approach to convention site selection. *Journal of Travel Research*, *45*(2), 167–174. doi:10.1177/0047287506291593

Cheng, J., Grossman, M. & McKercher, T. (2014). *Professional CUDA C Programming*. Wrox.

Chen, J. J., & Dimou, I. (2005). Expansion strategy of international hotel firms. *Journal of Business Research*, *58*(12), 1730–1740. doi:10.1016/j.jbusres.2004.11.001

Chen, L., & Holsapple, C. (2012). E-business adoption research: Analysis and structure. *Proceedings of AMCIS*, *2012*(August), 9–12.

Chen, L., & Holsapple, C. (2013). E-Business adoption research: State of the art. *Journal of Electronic Commerce Research*, *14*(3), 261–286.

Chen, S. C., & Lin, C. P. (2015). The impact of customer experience and perceived value on sustainable social relationship in blogs: An empirical study. *Technological Forecasting and Social Change*, *96*, 40–50. doi:10.1016/j.techfore.2014.11.011

Chen, S. L. (2001). Effects of value, affect, security, and web content on informa- tional and transactional usage of the Internet. *Asia Pacific Journal of Tourism Research*, *6*(81), 63–72. doi:10.1080/10941660108722089

Cheyne, J., Downes, M., & Legg, S. (2006). Travel agent vs. Internet: What influences travel consumer choices? *Journal of Vacation Marketing*, *12*(1), 41–57. doi:10.1177/1356766706059307

Chiang, W.-C., Chen, J. C. H., & Xu, X. (2007). An overview of research on revenue management: Current issues and future research. *International Journal of Revenue Management*, *1*(1), 97–128. doi:10.1504/IJRM.2007.011196

Chitura, T., Mupemhi, S., Dube, T., & Bolongkikit, J. (2008). Barriers to electronic commerce adoption in small and medium enterprise: A critical literature review. *Journal of Internet Banking and Commerce*, *13*(2), 1–13.

Chiu, V. K. H. (2010). Accessed from UNESCO website: http://whc.unesco.org/en/list/18/gallery/

Choe, H., Masulis, R. W., & Nanda, V. (1993). Common stock offerings across the business cycle: Theory and evidence. *Journal of Empirical Finance*, *1*(1), 3–31. doi:10.1016/0927-5398(93)90003-A

Choi, S. C. (1991). Price competition in a channel structure with a common retailer. *Marketing Science*, *10*(4), 271–296. doi:10.1287/mksc.10.4.271

Choudhury, M. M., & Harrigan, P. (2014). RM to social CRM: The integration of new technologies into customer relationship management. *Journal of Strategic Marketing*, *22*(2), 149–176. doi:10.1080/0965254X.2013.876069

Chow, C. W. C., Lai, J. Y. M., & Loi, R. (2015). Motivation of travel agents' customer service behavior and organizational citizenship behavior: The role of leader-member exchange and internal marketing orientation. *Tourism Management*, *48*, 362–369. doi:10.1016/j.tourman.2014.12.008

Christodoulides, G., de Chernatony, L., Furrer, O., Shiu, E., & Abimbola, T. (2006). Conceptualising and measuring the equity of online brands. *Journal of Marketing Management*, *22*(7/8), 799–825. doi:10.1362/026725706778612149

Chu, P. (1997). *A Genetic Algorithm Approach for Combinatorial Optimization Problems* (Ph.D. Thesis). The Management School Imperial College of Science, Technology and Medicine, London, UK

Chung, J. Y., & Buhalis, D. (2008). Web 2.0: A study of online travel community. In P. O'Connor, W. Höpken and U. Gretzel (Eds.), Information and Communication Technologies in Tourism 2008 (pp. 70-81). Vienna, Austria: Springer.

Chu, R. (2001). What online Hong Kong travelers look for on airline/travel website? *International Journal of Hospitality Management*, *20*(1), 95–100. doi:10.1016/S0278-4319(00)00046-3

Clarke, J. (2005). Effective marketing for rural tourism. In D. Hall, I. Kirkpatrick, & M. Mitchell (Eds.), *Rural tourism and sustainable business* (pp. 87–102). Clevedon, UK: Channel View.

Claveria, O., Monte, E., & Torra, S. (2015). A new forecasting approach for the hospitality industry. *International Journal of Contemporary Hospitality Management*, *27*(7), 1520–1538. doi:10.1108/IJCHM-06-2014-0286

Clemes, M. D., Gan, C., & Ren, M. (2011). Synthesizing the effects of service quality, value, and customer satisfaction on behavioral intentions in the motel industry: An empirical analysis. *Journal of Hospitality & Tourism Research (Washington, D.C.)*, *35*(4), 530–568. doi:10.1177/1096348010382239

Cobanoglu, C., Bill, W., & Patrick, J. M. (2001). A comparison of mail, fax and web-based survey methods. *International Journal of Market Research*, *43*(4), 441–452.

Cohen, E. (1979). A phenomenology of tourist experiences. Sociology. *The Journal of the British Sociological Association*, *13*(2), 179–201.

Cohen, E. H. (2014). Tourism and Terror: a Case Study: Israel 1948-2012. *International Journal of Religious Tourism and Pilgrimage*, *2*(1), 13–26.

Cohen, E., & Neal, M. (2010). Coinciding crises and tourism in contemporary Thailand. *Current Issues in Tourism*, *13*(5), 455–475. doi:10.1080/13683500.2010.491898

Communify. (2015). *User-generated content and online reviews: Let your users do your marketing*. Retrieved February 12, 2016, from http://resources.communify.com/user-generated-content-and-online-reviews/

Connoly, D., Olsen, M., & Moore, R. (1998). The Internet as a Distribution Channel. *The Cornell Hotel and Restaurant Administration Quarterly*, *98*(4), 42–54. doi:10.1177/001088049803900408

Conover, W. (1973). Rank Tests for One Sample, Two Samples and K Samples without the Assumptions of a Continuous Distribution Function. *Annals of Statistics*, *1*(6), 1105–1125. doi:10.1214/aos/1176342560

Contractor, J. F., & Kundu, K. S. (1998). Modal choice in the world of alliances: Analysing organizational forms in the international hotel sector. *Journal of International Business Studies*, *29*(2), 325–358. doi:10.1057/palgrave.jibs.8490039

Cook, W. (n.d.). *History of the Traveling Salesman Problem*. Retrieved January 22, 2010, from http://www.tsp.gatech.edu/index.html

Cook, S. (2012). *CUDA Programming: A Developer's Guide to Parallel Computing with GPUs (Applications of Gpu Computing)*. Morgan Kaufmann.

Cooper, C., & Hall, M. (2008). *Contemporary Tourism: An International Approach*. London: Butterworth Heinemann.

Cooper, C., & Hall, M. (2013). *Contemporary tourism: An international approach* (2nd ed.). Oxford, UK: Goodfellow Publishers.

Correia, A., Oom Do Valle, P., & Moço, C. (2006). Why People Travel to Exotic Places, *International Journal of Culture. Tourism and Hospitality*, *1*(1), 45–61.

Cox, J. C., Ross, S., & Rubinstein, M. (1979, September). Option pricing: A simplified approach. *Journal of Financial Economics*, *7*(3), 229–263. doi:10.1016/0304-405X(79)90015-1

Craig, C. D., & Faulkenberry, G. D. (1979). The application of ridit analysis to detect trends in visibility. *Atmospheric Environment*, *13*(12), 1617–1622. doi:10.1016/0004-6981(79)90319-6

Cremer, H., Marchand, M., & Thisse, J. (1989). The Public Firm as an Instrument for Regulating an Oligopolistic Market. *Oxford Economic Papers*, *41*(2), 283–301.

Crnojevac, I. H., Gugié, J., & Karlovčan, S. (2010). eTourism: A comparison of online and offline bookings and the importance of hotel attributes. *Journal of Information and Organizational Sciences*, *34*(1), 41–54.

Cross, R. L., Borgatti, S. P., & Parker, A. (2002). Making invisible work visible: Using social network analysis to support human networks. *California Management Review*, *44*(2), 25–46. doi:10.2307/41166121

Crouch, G. I., & Ritchie, J. B. (1999). Tourism, competitiveness, and societal prosperity. *Journal of Business Research*, *44*(3), 137–152. doi:10.1016/S0148-2963(97)00196-3

Cunill, O. M., Forteza, C. M., & Miralles, M. R. (2008). Valuing growth strategy management by hotel chains based on the real options approach. *Tourism Economics*, *14*(3), 511–526. doi:10.5367/000000008785633640

Czakon, W., Mucha-Kus, K., & Rogalski, M. (2014). Coopetition Research Landscape – A Systematic Literature Review 1997-2010. *Journal of Economics & Management*, *17*, 122–150.

Czernek, K. (2013). Determinants of cooperation in a tourist region. *Annals of Tourism Research*, *40*(1), 83–104. doi:10.1016/j.annals.2012.09.003

d'Italia, B. (2013). *Turismo internazionale dell'Italia* [Outbound tourism flows in Italy]. Retrieved September 29, 2013, from https://www.bancaditalia.it/statistiche/rapp_estero/turismo-int

Dai, W., Liu, Q., & Chai, T. (2015). Particle size estimate of grinding processes using random vector functional link networks with improved robustness. *Neurocomputing*, *169*, 361–372. doi:10.1016/j.neucom.2014.08.098

Dalbor, M. C., & Upneja, A. (2004). The Investment Opportunity Set and the Long-Term Debt Decision of U.S. Lodging Firms. *Journal of Hospitality & Tourism Research (Washington, D.C.)*, *28*(3), 346–355. doi:10.1177/1096348004265024

Dallen, J. T., & Stephen, W. B. (2003). *Heritage Tourism*. Harlow, UK: Prentice Hall.

Dammert, L., Malone, M. F. T., & Dammert, L. (2002). Inseguridad y temor en la Argentina: El impacto de la confianza en la policía y la corrupción sobre la percepción ciudadana del crimen. *Desarrollo Economico*, *42*(166), 285–301. doi:10.2307/3455944

Darcy, S. (2010). Inherent complexity: Disability, accessible tourism and accommodation information preferences. *Tourism Management Journal, Elsevier.*, *31*(6), 816–826. doi:10.1016/j.tourman.2009.08.010

Darcy, S., & Buhalis, D. (2011). *Accessible tourism: concepts and issues. Aspects of Tourism.* Bristol: Channel View Publications Ltd.

Darcy, S., & Dickson, T. (2009). A Whole – of – Life Approach to Tourism: The Case of Accessible Tourism Experiences. *Journal of Hospitality and Tourism Management*, *16*(1), 32–44. doi:10.1375/jhtm.16.1.32

Das, D., Sharma, S. K., Mohapatra, P. K. J., & Sarkar, A. (2007). Factors influencing the attractiveness of a tourist destination: A case study. *Journal of Service Research*, *7*, 103–134.

Davendra, D. D., & Zelinka, I. (2016). *Self-Organizing Migrating Algorithm - Methodology and Implementation*. Springer. doi:10.1007/978-3-319-28161-2

Davendra, D., Zelinka, I., Bialic-Davendra, M., Senkerik, R., & Jasek, R. (2013). Discrete self-organizing migrating algorithm for flow-shop scheduling with no-wait makespan. *Mathematical and Computer Modelling*, *57*(1-2), 100–110. doi:10.1016/j.mcm.2011.05.029

Davenport, T. (2000). *Mission Critical: Realizing the Promise ok Enterprise Systems*. Boston, MA: Harvard Business School Press.

Davidavičienė, V., & Tolvaišas, J. (2011). Measuring quality of e-commerce web sites: Case of Lithuania. *Economics and Management*, *16*, 723–729.

Davis, F. D., Bagozzi, R. P., & Warshaw, P. R. (1989). User acceptance of computer technology: A comparison of two theoretical models. *Management Science*, *35*(8), 982–1003. doi:10.1287/mnsc.35.8.982

Davis, L. (1996). *Handbook of Genetic Algorithms*. Berlin: Van Nostrand Reinhold.

Day, G. S. (1994). The capabilities of market-driven organizations. *Journal of Marketing*, *58*(4), 37–52. doi:10.2307/1251915

De Jong, A., Wetzels, M., & de Ruyter, K. (2008). Linking employee perceptions of collective efficacy in self-managing service teams with customer-perceived service quality: A psychometric assessment. *International Journal of Service Industry Management*, *19*(3), 353–378. doi:10.1108/09564230810875011

De Luca, L. M., Verona, G., & Vicari, S. (2010). Market orientation and R&D effectiveness in high-technology firms: An empirical investigation in the biotechnology industry. *Journal of Product Innovation Management*, *27*(3), 299–320. doi:10.1111/j.1540-5885.2010.00718.x

De Pablo, R. (2004). *Las nuevas tecnologías aplicadas al turismo*. Madrid: Editorial Universitaria Ramón Areces.

De Villes, R. F. (1991). *Scale development: theory and applications. Applied Social Research Series 26*. Newbury Park: Sage.

Deb, K. (1995). *Optimization for Engineering Design*. New Delhi: Prentice-Hall.

Deb, K. (2001). *Multi-objective optimization using evolutionary algorithms*. John Wiley & Sons.

Decrop, A. (2006). *Vacation Decision-Making*. Wallingford, UK: CABI-Publishing. doi:10.1079/9781845930400.0000

Deighton, J. (1996). The future of interactivity marketing. *Harvard Business Review*, *74*(6), 151–160.

Del Chiappa, G. (2011). Trustworthiness of Travel 2.0 applications and their influence on tourists' behaviour: An empirical investigation in Italy. In R. Law, M. Fuchis, & F. Ricci (Eds.), *Information and Communication Technologies in Tourism 2011 Proceedings of the International Conference in Innsbruck, Austria, 2011* (pp. 331–342). Vienna, Austria: Springer-Verlag.

Del Chiappa, G. (2011b). *Trustworthiness of Travel 2.0 applications and their influence on tourist behaviour*. Retrieved August 27, 2015, from http://tourismintelligence.ca/2011/06/17/trustworthiness-of-travel-2-0-applications-and-their-influence-on-tourist-behaviour/

Del Chiappa, G. (2011a). Trustworthiness of Travel 2.0 applications and their influence on tourist behaviour: An empirical investigation in Italy. In *Proceedings of TTRA Europe and AFM Conference, Creativity and Innovation in Tourism.* doi:10.1007/978-3-7091-0503-0_27

Del Chiappa, G. (2012). User-generated content and its influence on tourists' choices: How do tourists make hotel reservations online? *Turistica: Italian Journal of Tourism, 1*(1), 39–50.

Del Chiappa, G. (2013). Internet versus travel agencies: The perception of different groups of Italian online buyers. *Journal of Vacation Marketing, 19*(1), 1–12.

Del Chiappa, G., Alarcón-del-Amo, M., & Lorenzo-Romero, C. (2016). Internet and user-generated content versus high-street travel agencies: A latent gold segmentation in the context of Italy. *Journal of Hospitality Marketing and Managament, 25*(2), 197–217. doi:10.1080/19368623.2014.1001933

Del Chiappa, G., Lorenzo-Romero, C., & Alarcón-del-Amo, M. D. C. (2015). Profiling tourists based on their perceptions of the trustworthiness of different types of peer-to-peer applications. *Current Issues in Tourism*, 1–18. doi:10.1080/13683500.2015.1077208

Del Chiappa, G., Lorenzo-Romero, C., & Gallarza, M. (2014). Attitude toward disintermediation in hotel reservations: Spanish travellers profile. *European Journal of Tourism Research, 9*(1), 129–143.

Dhar, V. (2013). Data science and prediction. *Communications of the ACM, 56*(12), 64–73. doi:10.1145/2500499

Dhillon, W. R., Goldstein, M., & Schiffman, L. (1978). On the appropriateness of Linear Discriminant and Multinomial Classification Analyisis in Marketing Research. *JMR, Journal of Marketing Research, 15*(1), 103–112. doi:10.2307/3150405

Dix, A., Finlay, J. E., Abowd, G. D., & Beale, R. (2003). *Human-computer interaction* (3rd ed.). New York: Prentice Hall.

Dixit, A. K. (1989). Entry and Exit Decisions under Uncertainty. *Journal of Political Economy, 97*(3), 620–638. doi:10.1086/261619

Dixit, A. K. (1994). *Investment under Uncertainty*. Princeton, NJ: Princeton University Press.

Djamasbi, S., Siegel, M., & Tullis, T. (2010). Generation Y, web design, and eye tracking. *International Journal of Human-Computer Studies, 68*(5), 307–323. doi:10.1016/j.ijhcs.2009.12.006

Doi, H. (2003). *Discounted Cash Flow: Business Analysis and Valuation. Toyo Keizai Inc.* (in Japanese)

Dollinger, M. J. (1990). The evolution of collective strategies in fragmented industries. *Academy of Management Review, 15*(2), 266–285.

Dolnicar, S. (2005a). Fear Segment in tourism. In *Proceedings of the 14 International Research Conference of the Council for Australian University and Hospitality Education.* CAUTHE.

Dolnicar, S. (2005b). Understanding barriers to leisure travel, tourists fears as marketing basis. *Journal of Vacation Marketing, 11*(3), 197–208. doi:10.1177/1356766705055706

Dolnicar, S., Crouch, G. I., Devinney, T., Huybers, T., Louviere, J. J., & Opperwal, H. (2008). Tourism and discretionary income allocation. Heterogeneity among households. *Tourism Management, 29*(1), 44–52. doi:10.1016/j.tourman.2007.04.008

Domínguez, P., Burguette, E., & Bernard, A. (2003). Efectos del 11 de Septiembre en la hotelería Mexicana: reflexión sobre la mono-dependencia turística. *Estudios y Perspectivas en Turismo, 12*(3-4), 335-348.

Dorigo, M., & Suttzle, T. (2004). *Ant Colony Optimization.* MIT Press.

Dostál, P. (2014). The use of soft computing in management. In P. Vasant (Ed.), *Handbook of research on novel soft computing intelligent algorithms: Theory and practical applications* (pp. 294–326). Hershey, PA: IGI Global. doi:10.4018/978-1-4666-4450-2.ch010

Doyle, M., & Dorling, S. (2002). Visibility Trends in UK 19501007. *Atmospheric Environment*, *36*(19), 3161–3172. doi:10.1016/S1352-2310(02)00248-0

Dredge, D., & Moore, S. (1992). A methodology for the integration of tourism in town planning. *Journal of Tourism Studies*, *3*(3-1), 8-27.

Dredge, D. (1999). Destination place planning and design. *Annals of Tourism Research*, *26*(4), 772–791. doi:10.1016/S0160-7383(99)00007-9

Dredge, D. (2006). Policy networks and the local organisation of tourism. *Tourism Management*, *27*(2), 269–280. doi:10.1016/j.tourman.2004.10.003

Dredge, D., & Jamal, T. (2013). Mobilities on the Gold Coast, Australia: Implications for destination governance and sustainable tourism. *Journal of Sustainable Tourism*, *21*(4), 557–579. doi:10.1080/09669582.2013.776064

Driver, J. (1999). Developments in airline marketing practice. *Journal of Marketing Practice: Applied Marketing Science*, *5*(5), 134–150. doi:10.1108/EUM0000000004571

Drost, A. (1996). Developing sustainable tourism for world heritage sites. *Annals of Tourism Research*, *23*(2), 479–484. doi:10.1016/0160-7383(96)83345-7

e-Business W@tch. (2006). *ICT and e-Business in the Tourism Industry: ICT adoption and e-business activity in 2006*. The European Commission, Enterprise & Industry Directorate General.

Echtner, C. M., & Ritchie, J. R. B. (1993). The Measurement of Destination Image: An Empirical Assessment. *Journal of Travel Research*, *31*(4), 3–13. doi:10.1177/004728759303100402

Economics, T. (2013). The impact of online content on European Tourism. Tourism Economics and Oxford Economics Company.

Econsultancy. (2011). *Three quarters of travel research takes place online: Survey*. Retrieved December 30, 2015, from http://econsultancy.com/uk/blog/8275-three-quarters-of-travel-research-takes-place-online-survey

Edwards, B., Goodwin, M., Pemberton, S., & Woods, M. (2000). *Partnership working in rural regeneration: Governance and empowerment?* Bristol, UK: The Policy Press and the Joseph Rowntree Foundation.

Eighmey, J. (1997). Profiling user responses to commercial websites. *Journal of Advertising Research*, *37*(3), 59–66.

Ekinci, Y., & Hosany, S. (2006). Destination personality: An application of brand personality to tourism destinations. *Journal of Travel Research*, *45*(2), 127–139. doi:10.1177/0047287506291603

Elgonemy, A. R. (2002). Debt-financing Alternatives: Refinancing and Restructuring in the Lodging Industry. *The Cornell Hotel and Restaurant Administration Quarterly*, *43*, 7–21.

eMarket. (2014, July 23). Worldwide Ecommerce Sales to Increase Nearly 20% in 2014. *eMarket*.

Emarketer. (2007). *Niche sites invigorate online travel*. Retrieved April 11, 2015, from http://www.eMarketer.com

Ernst, H., Hoyer, W. D., Krafft, M., & Krieger, K. (2011). Customer relationship management and company performance: The mediating role of new product performance. *Journal of the Academy of Marketing Science*, *39*(2), 290–306. doi:10.1007/s11747-010-0194-5

Ertek, G., & Griffin, P. M. (2002). Supplier- and buyer-driven channels in a two-stage supply chain. *IIE Transactions*, *34*(8), 691–700. doi:10.1080/07408170208928905

Ertuna, C., & Ertuna, Z. I. (2009). The Sensitivity of German and British tourists to new shocks. *Tourism Review.*, *64*(3), 19–27. doi:10.1108/16605370910988809

Ethiopian Embassy in Ankara. (2016). *Natural Attractions*. Accessed Feb. 2016, from http://www.ethiopianembassyankara.org/wp-content/uploads/2015/10/yangud-national-park.gif

Eubank, W. L., & Weinberg, L. (1994). Does democracy encourage terrorism? *Terrorism and Political Violence*, *6*(4), 417–435. doi:10.1080/09546559408427271

European Commission. Enterprise and Industry. (2015). *Accessible tourism.* Retrieved December 15, 2015 from: http://www.europa.eu

European Commission. Regional Policy. (2011). *Cities of tomorrow. Challenges, visions, ways forward.* Retrieved February 29, 2016 from: http://www.europa.eu

European Network for Accessible Tourism (ENAT). (n.d.). Retrieved November 10, 2015 from: http:// www.accessibletourism.org

European Union Tourism Trade Organizations (ECTT). (2015). *European Union tourism trade organizations (ECTT) Best Tourist Destination Announcement*. Retrieved on Feb. 10, 2016 from http://ectt.webs.com/besttouristdestination.htm

Fama, E. F., & French, K. R. (2004). The Capital Asset Pricing Model: Theory and Evidence. *The Journal of Economic Perspectives*, *18*(3), 25–40. doi:10.1257/0895330042162430

Faulkner, B. (2001). Towards a framework for tourism disaster management. *Tourism Management*, *22*(2), 135–147. doi:10.1016/S0261-5177(00)00048-0

Faust, K., & Wasserman, S. (1992). Blockmodels: Interpretation and evaluation. *Social Networks*, *14*(1-2), 5–61. doi:10.1016/0378-8733(92)90013-W

Fawcett, P., Mcleish, R. E., & Ogden, I. D. (1992). *Logistics management*. London, UK: Pitman Publishing.

Fazlollahi, B. (2002). *Strategies for e-commerce success*. IRM Press. doi:10.4018/978-1-93177-708-7

FDREMCT. (2015). *Ethiopia, F. D. R. E. Ministry of Culture and Tourism*. Retrieved on Feb. 20, 2016 from http://www.moct.gov.et/index.php/en/about-ethopia

Fennell, D. A. (2014). Foreword. In C. Weeden & K. Boluk (Eds.), *Managing Ethical Consumption in Tourism*. New York, NY: Routledge, xv-xvii. doi:10.1016/B978-0-12-411577-4.09985-8

Fernández-Delgado, M., Cernadas, E., Barro, S., & Amorim, D. (2014). Do we need hundreds of classifiers to solve real world classification problems? *Journal of Machine Learning Research*, *15*(1), 3133–3181.

Fesenmaier, D. R., Xiang, Z., Pan, B., & Law, R. (2011). A framework of search engine use for travel planning. *Journal of Travel Research*, *50*(6), 587–601. doi:10.1177/0047287510385466

Fine, C. H. (1999). *Clockspeed: Winning industry control in the age of temporary advantage*. New York, NY: Basic Books.

Fink, A. (1995). *How to Design Surveys*. London: SAGE Publications.

Fink, A., & Kosecoff, J. (1998). *How to Conduct Survey-a Step by Step Guide*. London: SAGE Publications.

Finn, M., Elliott-White, M., & Walton, M. (2000). *Tourism and Leisure Research Methods. Data Collection, Analysis and Interpretation.* Harlow, UK: Pearson – Longman.

Fisher, I. (1907). *The Rate of Interest.* New York: The Macmillan Company.

Flaherty, T. B., Dahlstrom, R., & Skinner, S. J. (1999). Organizational values and role stress as determinants of customer-oriented selling performance. *Journal of Personal Selling & Sales Management, 19*(2), 1–18.

Flavigny, C., & Pascal, M. (1995). Tourism among the disabled: Dissatisfied clients. Espaces (Paris), (134), 36-42.

Floh, A., & Madlberger, M. (2013). The role of atmospheric cues in online impulse-buying behavior. *Electronic Commerce Research and Applications, 12*(6), 425–439. doi:10.1016/j.elerap.2013.06.001

Flora, J. D. Jr. (1974). *A note on Ridit Analysis.* Ann Arbor, MI: University of Michigan.

Floyd, M. F., Gibson, H., Pennington-Gray, L., & Thapa, B. (2004). The effect of risk perceptions on intentions to travel in the aftermath of September 11, 2001. *Journal of Travel & Tourism Marketing, 15*(2-3), 2–3, 19–38. doi:10.1300/J073v15n02_02

Forbes. (2015, January 5). Competition is shaking up the online travel market. *Forbes.*

Forthofer, R. N., & Koch, G. (1973). An analysis for compounded functions of categorical data. *Biometrics, 29*(1), 143–157. doi:10.2307/2529683 PMID:4691050

Foucault, M. (2001). *Defender la Sociedad.* Buenos Aires: Fondo de Cultura Económica.

Foucault, M. (2006). *Seguridad, Territorio, Población.* Buenos Aires: Fondo de Cultura Económica.

Framke, W. (2002). The destination as a concept: A discussion of the business-related perspective versus the socio-cultural approach in tourism theory. *Scandinavian Journal of Hospitality and Tourism, 2*(2), 92–108. doi:10.1080/15022250216287

Franklin, A. (2003). *Tourism: An Introduction.* London: SAGE Publications Ltd. doi:10.4135/9781446220108.n2

Freisleben, B., & Gleichmann, G. (1993). Controlling airline seat allocations with neural networks.*Proceeding of the Twenty-Sixth Hawaii International Conference on System Sciences.* doi:10.1109/HICSS.1993.284243

Froehle, C. M. (2006). Service personnel, technology, and their interaction in influencing customer satisfaction. *Decision Sciences, 37*(1), 5–38. doi:10.1111/j.1540-5414.2006.00108.x

Frost, F. A., & Shanka, T. (1999). Asian Australian student travel preferences: An empirical study. *Asia Pacific Journal of Tourism Research, 4*(2), 19–26. doi:10.1080/10941669908722040

Frow, P., & Payne, A. (2007). Towards the "perfect" customer experience. *Journal of Brand Management, 15*(2), 89–101. doi:10.1057/palgrave.bm.2550120

Fuchs, G. & Reichel, A. (2010). Health Tourists Visiting a Highly Volatile Destination. *Anatolia: An International Journal of Tourism and Hospitality Research, 21*(2), 205-226.

Fuchs, G. & Reichel, A. (2010). Health Tourists Visiting a Highly Volatile Destination. *Anatolia: An International Journal of Tourism and Hospitality Research,21*, 205-226.

Fuchs, G., & Pizam, A. (2011). *The Importance of Safety and Security for Tourism Destinations. Tourism Destination Marketing and Management: Collaborative Strategies, 300.* CABI.

Fuchs, G., & Reichel, A. (2004). Cultural Differences in tourist destination risk perception: An exploratory study. *Tourism, 52*(1), 21–37.

Fuchs, G., & Reichel, A. (2004). Cultural Differences in tourist destination risk perception: An exploratory study. *Tourism, 52*, 21–37.

Fudenberg, D., & Tirole, J. (1985, July). Pre-emption and Rent Equalisation in the Adoption of New Technology. *The Review of Economic Studies, 52*(3), 383–401. doi:10.2307/2297660

Fuller, G. (2010). *A World Without Islam.* New York, NY: Back Bay Books.

Gale, T. (2008). The end of tourism or endings of tourism? In P. Burns & M. Novelli (Eds.), *Local-Global Connections.* Wallingford: CABI. doi:10.1079/9781845934040.0001

Gallego, G., Kou, S. G., & Phillips, R. (2008). Revenue management of callable products. *Management Science, 54*(3), 550–564. doi:10.1287/mnsc.1070.0782

Garcia de Soto Camacho, E. (2015). Mode of entry, real options and strategic flexibility in internationalization processes of hotel chains. Doctoral Dissertation Summary. *European Journal of Tourism Research, 10*, 131-135.

García de Soto-Camacho, E., & Vargas-Sánchez, A. (2015). Choice of entry mode, strategic flexibility and performance of international strategy in hotel chains: An approach based on real options. *European Journal of Tourism Research, 9*, 92–114.

Garcia, D., & Tugores, M. (2006). Optimal choice of quality in hotel services. *Annals of Tourism Research, 33*(2), 456–469. doi:10.1016/j.annals.2005.12.010

Garey, M., & David, J. (1979). *Computers and Intractability: A guide to the Theory of NP-Completeness.* Bell Laboratories Murray Hill.

Garrow, L., & Ferguson, M. (2008). Revenue management and the analytics explosion: Perspectives from industry experts. *Journal of Revenue and Pricing Management, 7*(2), 219–229. doi:10.1057/rpm.2008.3

Gatignon, H., & Xuereb, J. M. (1997). Strategic orientation of the firm and new product performance. *JMR, Journal of Marketing Research, 34*(1), 77–90. doi:10.2307/3152066

Gentile, C., Spiller, N., & Noci, G. (2007). How to sustain the customer experience: An overview of experience components that co-create value with the customer. *European Management Journal, 25*(5), 395–410. doi:10.1016/j.emj.2007.08.005

George, B, Inbakaran, R. & Poyyamoli, G. (2010). To Travel or Not to travel: towards understanding the theory of nativistic motivation. *Tourism: An International Interdisciplinary Journal, 58*(4), 395-407.

George, B. P., Inbakaran, R., & Poyyamoli, G. (2010). To Travel or Not to travel: towards understanding the theory of nativistic motivation. *Turizam: znanstveno-stručni časopis, 58*, 395-407.

George, R. (2003). Tourists perceptions of safety and security while visiting Cape Town. *Tourism Management, 24*(3), 575–585. doi:10.1016/S0261-5177(03)00003-7

Gertner, R., & Rosenfield, A. (1999). *How Real Options Lead to Better Decisions.* Retrieved from http://www.ftmastering.com/risk/sr3.html

Getz, D. (1987). Capacity to absorb tourism: Concepts and implications for strategic planning. *Annals of Tourism Research, 10*(2), 239–261. doi:10.1016/0160-7383(83)90028-2

Ghose, A. K., & Billiau, G. (2011). The optimizing WEB: A green ICT research perspective. In B. Unhelkar (Ed.), *Handbook of research on green ICT: Technology, business and social perspectives* (pp. 184–196). Hershey, PA: IGI Global. doi:10.4018/978-1-61692-834-6.ch012

Gibson, L., Lynch, P. A., & Morrison, A. (2005). The local destination tourism network: Development issues. *Tourism and Hospitality Planning & Development*, *2*(2), 87–99. doi:10.1080/14790530500171708

Giddens, A. (1991). *Modernity and Self-Identity: Self and Society in the Late Modern Age*. Stanford University Press.

Giddens, A. (1999). *The Consequences of Modernity*. Madrid: Alianza Editorial.

Giddens, A. (2000). *Run Away World: how globalization is reshaping our lives*. Madrid: Taurus.

Gilmore, J. H., & Pine, B. J. (2002). Differentiating hospitality operations via experiences: Why selling services is not enough. *The Cornell Hotel and Restaurant Administration Quarterly*, *43*(3), 87–96. doi:10.1016/S0010-8804(02)80022-2

Gishen Tours. (2015). Accessed on Feb. 2016, from http://www.gishentours.com/wp-content/uploads/2015/12/BlueNileFalls1.jpg

Glover, F., & Laguna, M. (1997). Tabu Search. Springer.

Go, F., & Pine, R. (1995). *Globalization strategy in the hotel industry*. New York, NY: Routledge.

Goh, C., Ong, Y., & Tan, K. (2009). *Multi-Objective Memetic Algorithms*. Springer-Verlag. doi:10.1007/978-3-540-88051-6

Goldberg, D. (1989). *Genetic Algorithms in Search, Optimization and Machine Learning*. Reading, MA: Addison Wesley.

Goldberg, D. (1989). *Genetic Algorithms in Search, Optimization, and Machine Learning*. Addison-Wesley Publishing Company Inc.

Goldhar, J. D., & Lei, D. (1991). The shape of twenty-first century global manufacturing. *The Journal of Business Strategy*, *12*(2), 37–41. doi:10.1108/eb039401 PMID:10111375

Goldstein, G., & Hersen, M. (1984). *Handbook of Psychological Assessment*. Pergamon Press.

Gopalani, A., & Shick, K. (2011). The service-enabled customer experience: A jump-start to competitive advantage. *The Journal of Business Strategy*, *32*(3), 4–12. doi:10.1108/02756661111121947

Gorin, T., Brunger, W. G., & White, M. M. (2006). No-show forecasting: A blended cost-based, PNR-adjusted approach. *Journal of Revenue and Pricing Management*, *5*(3), 188–206. doi:10.1057/palgrave.rpm.5160039

Govindarajan, V., Kopalle, P. K., & Danneels, E. (2011). The effects of mainstream and emerging customer orientations on radical and disruptive innovations. *Journal of Product Innovation Management*, *28*(s1), 121–132. doi:10.1111/j.1540-5885.2011.00865.x

Granados, N. F., Kauffman, R. J., & King, B. (2008). The emerging role of vertical search engines in travel distribution: a newly-vulnerable electronic markets perspective.*Proceedings of the 41st Hawaii International Conference on System Sciences*. doi:10.1109/HICSS.2008.435

Greene, W. H. (2012). *Econometric analysis* (7th ed.). Boston, MA: Pearson Education.

Green, Y. N., & Weaver, P. A. (2008). Approaches, Techniques, and Information Technology Systems in the Restaurants and Foodservice Industry: A Qualitative Study in Sales Forecasting. *International Journal of Hospitality & Tourism Administration*, *9*(2), 164–191. doi:10.1080/15256480801909071

Greer, C. R., & Lei, D. (2012). Collaborative innovation with customers: A review of the literature and suggestions for future research. *International Journal of Management Reviews*, *14*(1), 63–84. doi:10.1111/j.1468-2370.2011.00310.x

Grenadier, S. R. (1996). The Strategic Exercise of Options: Development Cascades and Overbuilding in Real Estate Markets. *The Journal of Finance*, *51*(5), 1653–1679. doi:10.1111/j.1540-6261.1996.tb05221.x

Gretzel, U., Fesenmaier, D., & O'Leary, J. (2006). The transformation of consumer behaviour. In D. Buhalis & C. Costa (Eds.), *Tourism business frontiers - Consumers, products and industry*. Oxford, UK: Elsevier. doi:10.1016/B978-0-7506-6377-9.50009-2

Gretzel, U., & Yoo, K. H. (2008). Use and impact of online travel reviews. In P. O'Connor, W. Höpken, & U. Gretzel (Eds.), *Information and communication technologies in tourism 2008 proceedings of the international conference in Innsbruck, Austria, 2008* (pp. 35–46). Vienna: Springer-Verlag. doi:10.1007/978-3-211-77280-5_4

Gretzel, U., Yoo, K. H., & Purifoy, M. (2007). *Online travel review study: The role and impact of online travel reviews*. College Station, TX: Laboratory for Intelligent Systems in Tourism.

Grissemann, U., Plank, A., & Brunner-Sperdin, A. (2013). Enhancing business performance of hotels: The role of innovation and customer orientation. *International Journal of Hospitality Management*, *33*, 347–356. doi:10.1016/j.ijhm.2012.10.005

Gronfeldt, S., & Strother, J. (2006). *Service leadership: The quest for competitive advantage*. London, UK: Sage Publications.

Grønflaten, Ø. (2011). Predicting travelers choice of information sources and information channels. *Journal of Travel Research*, *48*(2), 230–244. doi:10.1177/0047287509332333

Grunewald, L. (2012). La Seguridad en el marco de la competitividad de los destinos turísticos. *Revista de Ciencias Sociales*, *21*, 141–158.

Guenzi, P., Pardo, C., & Georges, L. (2007). Relational selling strategy and key account managers' relational behaviors: An exploratory study. *Industrial Marketing Management*, *36*(1), 121–133. doi:10.1016/j.indmarman.2005.03.014

Gulati, R. (1998). Alliances and networks. *Strategic Management Journal*, *19*(4), 293–317. doi:10.1002/(SICI)1097-0266(199804)19:4<293::AID-SMJ982>3.0.CO;2-M

Guo, X., Ling, L., Dong, Y., & Liang, L. (2013). Cooperation contract in tourism supply chains: The optimal pricing strategy of hotels for cooperative third party strategic websites. *Annals of Tourism Research*, *41*, 20–41. doi:10.1016/j.annals.2012.11.009

Guo, X., Ling, L., Dong, Y., & Liang, L. (2013). Optimal pricing strategy based on market segmentation for service products using online reservation systems: An application to hotel rooms. *International Journal of Hospitality Management*, *35*, 274–281. doi:10.1016/j.ijhm.2013.07.001

Guo, X., Zheng, X., Ling, L., & Yang, C. (2014). Online coopetition between hotels and online travel agencies: From the perspective of cash back after stay. *Tourism Management Perspectives*, *12*, 104–112. doi:10.1016/j.tmp.2014.09.005

Guttentag, D. A. (2010). Virtual reality: Applications and implications for tourism. *Tourism Management*, *31*(5), 637–651. doi:10.1016/j.tourman.2009.07.003

Guyon, I., & Elisseeff, A. (2003). An introduction to variable and feature selection. *Journal of Machine Learning Research*, *3*, 1157–1182.

Gu, Z. (2003). Analysis of Las Vegas Strip casino hotel capacity: An inventory model for optimization. *Tourism Management*, *24*(3), 309–314. doi:10.1016/S0261-5177(02)00073-0

Hair, J. F., Black, W. C., Babin, B. J., & Anderson, R. E. (2014). Multivariate data analysis (7th ed.). Edinburgh Gate: Pearson Education Limited.

Håkansson, H., & Snehota, I. (Eds.). (1995). *Developing relationships in business networks*. London: Routledge.

Haley, G., & Goldberg, S. (1995). Net Present Value Techniques and Their Effects on New Product Research. *Industrial Marketing Management*, *24*(3), 177–190. doi:10.1016/0019-8501(94)00044-W

Hall, C. M. (2008). *Tourism planning: Policies, processes and relationship*. Harlow, UK: Pearson-Prentice Hall.

Hall, C. M. (2008). *Tourism planning: Policies, processes and relationships* (2nd ed.). Harlow, UK: Pearson Prentice Hall.

Hall, C. M., Sharples, L., Cambourne, B., & Macionis, N. (Eds.). (2000). *Wine tourism around the world: Development, management and markets*. Oxford, UK: Butterworth-Heinemann.

Hall, M. (2002). Travel Safety, terrorism and the media the significance circle of the issue attention cycle. *Current Issues in Tourism*, *5*(5), 458–466. doi:10.1080/13683500208667935

Hall, M. (2010). Crisis Events in Tourism: Subjects of crisis in Tourism. *Current Issues in Tourism*, *13*(5), 401–417. doi:10.1080/13683500.2010.491900

Hall, M. C. (2000). *Tourism Planning*. Harlow, UK: Prentice Hall.

Hamadneh, N., Tilahun, S. L., Sathasivam, S., & Ong, H. C. (2013). Prey-Predator Algrorithm as a New Optimization Technique Using in Radia Basis Function Neural Networks. *Research Journal of Applied Sciences*, *8*(7), 383–387.

Han, J. K., Namwoon, K., & Srivastava, R. K. (1998). Market orientation and organizational performance: Is innovation a missing link? *Journal of Marketing*, *62*(4), 30–45. doi:10.2307/1252285

Hankinson, G. (2005). Destination brand image: A business tourism perspective. *Journal of Services Marketing*, *19*(1), 24–33. doi:10.1108/08876040510579361

Hankinson, G. (2009). Managing destination brands: Establishing a theoretical foundation. *Journal of Marketing Management*, *25*(1-2), 97–115. doi:10.1362/026725709X410052

Hanna, S., & Rowley, J. (2011). Towards a strategic place brand-management model. *Journal of Marketing Management*, *27*(5-6), 458–476. doi:10.1080/02672571003683797

Hanson, W. (2000). *Principles of internet marketing*. Cincinnati, OH: South-Western Collage Publishing.

Harper, D. (2008). *Valuation of Hotels for Investors*. Estates Gazette.

Harp, R. (1972). *Reducibility among Combinatorial Problems*. University of California at Berkeley.

Harrington, R. J. (2007). *Food and Wine Pairing: A Sensory Experience*. Wiley and Sons, Publishers.

Harris, F.W. (1913). How many parts to make at once. *Factory: The Magazine of Management, 10*(2), 135-136, 152.

Harris, E. G., Mowen, J. C., & Brown, T. J. (2005). Re-examining salesperson goal orientations: Personality influencers, customer orientation, and work satisfaction. *Journal of the Academy of Marketing Science*, *33*(1), 19–35. doi:10.1177/0092070304267927

Harrison, D. M., Panasian, C. A., & Seiler, M. J. (2011). Further Evidence on the Capital Structure of REITs. *Real Estate Economics*, *39*(1), 133–166. doi:10.1111/j.1540-6229.2010.00289.x

Harrison, D., & Hitchcock, M. (2005). *The politics of world heritage. Negotiating tourism and conservation*. Paris: UNESCO World Heritage Centre.

Hartman, K. B., Meyer, T., & Scribner, L. L. (2009). Retail and service encounters: The inter-cultural tourist experience. *Journal of Hospitality Marketing & Management*, *18*(2/3), 197–215. doi:10.1080/19368620802591934

Hart, W., Krasnogor, N., & Smith, J. (2005). *Recent Advances in Memetic Algorithms*. SpringerVerlag. doi:10.1007/3-540-32363-5

Hayes, D. K., & Miller, A. A. (2011). *Revenue management for the hospitality industry*. Hoboken, NJ: Wiley.

Hayes, R., & Abernathy, W. (1980). Managing our way to economic decline. *Harvard Business Review*, *58*(4), 66–77.

Heart, T., Pliskin, N., Schechtman, E., & Reichel, A. (2001). Information Technology in the hospitality Industry: The Israeli Scene and Beyond. *Information Technology & Tourism*, *4*(1), 41–64. doi:10.3727/109830501108750886

Heath, D. B. (2000). *Drinking Occasions: Comparative Perspectives on Alcohol and Culture*. Ann Arbor, MI: Taylor and Francis.

Hemmington, N. (2007). From service to experience: Understanding and defining the hospitality business. *Service Industries Journal*, *27*(6), 747–755. doi:10.1080/02642060701453221

Hendershott, P. H., & Ward, C. W. R. (2000). Incorporating option-like features in the valuation of shopping centers. *Real Estate Finance*, (Winter), 31–36.

Henderson, J. C. (2003). Managing tourism and Islam in peninsular Malaysia. *Tourism Management*, *24*(4), 447–456. doi:10.1016/S0261-5177(02)00106-1

Henderson, J. C. (2008). Managing Crises: UK Civil Aviation, BAA airpots and the August 2006 terrorist Threat. *Tourism and Hospitality Research*, *8*(2), 125–136. doi:10.1057/thr.2008.12

Hennig-Thurau, T. (2005). Word of mouse: Why consumers listen to each other on the internet. *Yearbook of Marketing and Consumer Research*, *3*, 37–58.

Henry, S. L. (n.d.). *Introduction to web accessibility*. World Wide Web Consortium. Retrieved from www.w3.org/WAI/intro/accessibility.php

Herath, H. S. B., & Barth, J. R. (2001). *A Nested Option Model for Multi-stage Capital Investment Decisions*. Eighth Asia Pacific Finance Association (APFA) Annual Conference 2001, Bangkok, Thailand.

Herr, P. M., Kardes, F. R., & Kim, J. (1991). Effects of word-of-mouth and product-attribute information on persuasion: An accessibility-diagnosticity perspective. *The Journal of Consumer Research*, *17*(4), 454–462. doi:10.1086/208570

Hirschleifer, J. (1958, August). On the theory of optimal investment decision. *Journal of Political Economy*.

Ho, C. I., Lin, M. H., & Chen, H. M. (2012). Web users behavioural patterns of tourism information search: From online to offline. *Tourism Management*, *33*(6), 1468–1482. doi:10.1016/j.tourman.2012.01.016

Hodder, J., & Riggs, H. (1985). Pitfalls in evaluating risky projects. *Harvard Business Review*, *63*(1), 128–135.

Hofstede, G. (1980). *Culture's Consequences*. Beverly Hills, CA: Sage Publications.

Holcomb, J., & Pizam, A. (2006). *Do incidents of theft at tourist destinations have a negative effect on tourists' decisions to travel to affected destinations? In Tourism, Security & Safety: From Theory to Practice* (pp. 105–124). Oxford, UK: Butterworth-Heinemann.

Holland, A. S., Ott, S. H., & Riddough, T. J. (2000). The role of uncertainty in investment: An examination of competing investment models using commercial real estate data. *Real Estate Economics*, *28*(1), 33–64. doi:10.1111/1540-6229.00792

Holland, J. (1975). *Adaptation in natural and artificial systems*. Ann Arbor, MI: Univ. of Michigan Press.

Holloway, J. C. (2004). *Marketing for Tourism*. New York: Prentice Hall.

Homburg, C., Müller, M., & Klarmann, M. (2011). When should the customer really be king? On the optimum level of salesperson customer orientation in sales encounters. *Journal of Marketing*, *75*(2), 55–74. doi:10.1509/jmkg.75.2.55

Hong, T., Ma, T., & Huan, T. C. T. (2015). Network behavior as driving forces for tourism flows. *Journal of Business Research*, *68*(1), 146–156. doi:10.1016/j.jbusres.2014.04.006

Hosany, S., Ekinci, Y., & Uysal, M. (2006). Destination image and destination personality: An application of branding theories to tourism places. *Journal of Business Research*, *59*(5), 638–642. doi:10.1016/j.jbusres.2006.01.001

Hosmer, D. W., & Lemeshow, S. (2000). *Applied logistic regression.* New York, NY: John Wiley & Sons. doi:10.1002/0471722146

Hritz, N., Sidman, C. L., & DAbundo, M. (2014). Segmenting the college educated generation Y health and wellness traveler. *Journal of Travel & Tourism Marketing*, *31*(1), 132–145. doi:10.1080/10548408.2014.861727

Hsu, H. Y., & Tsou, H. T. (2011). Understanding customer experiences in online blog environments. *International Journal of Information Management*, *31*(6), 510–523. doi:10.1016/j.ijinfomgt.2011.05.003

Hsu, T. K., Tsai, Y. F., & Wu, H. H. (2009). The preference analysis for tourist choice of destination: A case study of Taiwan. *Tourism Management*, *30*(2), 288–297. doi:10.1016/j.tourman.2008.07.011

Huang, A., & Xiao, H. (2000). Leisure-Based Tourist Behavior: A Case Study of Changchun. *International Journal of Contemporary Hospitality Management*, *12*(3), 210–214. doi:10.1108/09596110010320850

Huang, C. W., Ho, F. N., & Chiu, Y. H. (2014). Measurement of tourist hotels productive efficiency, occupancy, and catering service effectiveness using a modified two-stage DEA model in Taiwan. *Omega*, *48*, 49–59. doi:10.1016/j.omega.2014.02.005

Huang, G. Q., Chen, W., Song, H., & Zhang, X. (2010). Game-theoretic study of the dynamics of tourism supply chains for package holidays under quantity competition. *Tourism Economics*, *16*(1), 197–216. doi:10.5367/000000010790872150

Huang, H.-C., Chang, A. Y., & Ho, C.-C. (2013). Using artificial neural networks to establish a customer-cancellation prediction model. *Przeglad Elektrotechniczny*, *89*(1b), 178–180.

Huang, L. (2011). Social media as a new play in a marketing channel strategy: Evidence from Taiwan travel agencies blogs. *Asia Pacific Journal of Tourism Research*, *17*(6), 615–634. doi:10.1080/10941665.2011.635664

Hua, W., Chan, A., & Mao, Z. (2009). Critical success factors and customer expectation in budget hotel segment: A case study of China. *Journal of Quality Assurance in Hospitality & Tourism*, *10*(1), 59–74. doi:10.1080/15280080802713702

Hueglin, C., & Vannotti, F. (2001). Data mining techniques to improve forecast accuracy in airline business.*Proceedings of the seventh ACM SIGKDD international conference on knowledge discovery and data mining*, (pp. 438–442). doi:10.1145/502512.502578

Huisman, K. J. M., & Kort, P. M. (1999). *Effects of Strategic Interactions on the Option Value of Wating*. Discussion Paper 1999-92.

Huisman, K. J. M. (2001). *Technology and Investment: A Game Theoretic Real Options Approach*. Kluwer Academic Publishers. doi:10.1007/978-1-4757-3423-2

Huntington, S. P. (1997). *The clash of civilizations and the remaking of world order*. New Dehli, Penguin Books India.

Hu, T., Marchiori, E., Kalbaska, N., & Cantoni, L. (2014). Online representation of Switzerland as a tourism destination: An exploratory research on a Chinese microblogging plataform. *Studies in Communication Sciences*, *14*(2), 136–143. doi:10.1016/j.scoms.2014.12.001

Hwang, Y.-H., & Fesenmaier, D. R. (2003). Multidestination pleasure travel patterns: Empirical evidence from the American Travel Survey. *Journal of Travel Research*, *42*(2), 166–171. doi:10.1177/0047287503253936

Hwang, Y.-H., Gretzel, U., & Fesenmaier, D. R. (2006). Multicity trip patterns. Tourist to the United States. *Annals of Tourism Research*, *33*(4), 1057–1078. doi:10.1016/j.annals.2006.04.004

Ilfinedo, P., & Nahar, N. (2006). Prioritization of *Enterprise Resource Planning* (ERP) Systems Success Measures: Viewpoints of Two Organizationaal Stakeholder Groups.*Proceedings of the 2006 ACM symposium on Applied computing*, (pp. 1554-1560). doi:10.1145/1141277.1141636

Iliescu, D. C., Garrow, L. A., & Parker, R. A. (2008). A hazard model of US airline passengers refund and exchange behavior. *Transportation Research Part B: Methodological*, *42*(3), 229–242. doi:10.1016/j.trb.2007.10.005

Im, S., & Workman, J. P. Jr. (2004). Market orientation, creativity, and new product performance in high-technology firms. *Journal of Marketing*, *68*(2), 114–132. doi:10.1509/jmkg.68.2.114.27788

Inkpen, A. C., & Tsang, E. W. K. (2005). Social capital, networks, and knowledge transfer. *Academy of Management Review*, *30*(1), 146–165. doi:10.5465/AMR.2005.15281445

Inskeep, E. (1991). *Tourism planning: an integrated and sustainable development approach*. New York: Van Nostrand Reinhold.

Instituto Nacional de Estadística. (2012). *Survey on equipment and use of information and communication technologies (ICT) in households year 2012*. Retrieved April 12, 2012, from http://www.ine.es/en/prensa/np738_en.pdf

International Civil Aviation Organization. (2010). *Guidelines on Passenger Name Record (PNR) data*. Retrieved from https://www.iata.org/iata/passenger-data-toolkit/assets/doc_library/04-pnr/New%20Doc%209944%201st%20Edition%20PNR.pdf

International, I. P. K. (2011). *Global travel trends 2011/2012*. Retrieved September 20, 2013, from http://www.ipkinternational.com/uploads/media/IPK_PR_ITB_2012.pdf

International, I. P. K. (2013). *ITB world travel trends report*. Berlin, Germany: Messe Berlin GmbH.

Internet Word Stats. (2011). *Top 20 countries with the highest number of internet users*. Retrieved August 31, 2015, from http://www.internetworldstats.com/top20.htm

Internet Word Stats. (2013). *Top 20 countries with the highest number of internet users*. Retrieved December 31, 2014, from http://www.internetworldstats.com/top20.htm

Internet World Stats. (2016). *Internet usage statistics*. Retrieved January 30, 2016, from http://www.internetworldstats.com/stats.htm

Inversini, A., & Masiero, L. (2014). Selling rooms online: The use of social media and online travel agencies. *International Journal of Contemporary Hospitality Management*, *26*(2), 272–292. doi:10.1108/IJCHM-03-2013-0140

Ip, C., Lee, H. A., & Law, R. (2012). Profiling the users of travel wesbites for planning and online experience sharing. *Journal of Hospitality & Tourism Research (Washington, D.C.)*, *36*(3), 418–426. doi:10.1177/1096348010388663

Ipsos Media CT & Google Travel Study. (2014). *The 2013 traveler's road to decision: affluent insights*. Thinkinsights Google. Retrieved from https://think.storage.googleapis.com/docs/travelers-road-to-decision-affluent-insights_research-studies.pdf

Ireland, R. D., Hitt, M. A., & Vaidyanath, D. (2002). Alliance management as a source of competitive advantage. *Journal of Management*, *28*(3), 413–446. doi:10.1177/014920630202800308

ITB. (2016). *World travel trends report 2015/2016.* Retrieved February 5, 2016, from http://www.itbberlin.de/media/itbk/itbk_dl_all/itbk_dl_all_itbkongress/itbk_dl_all_itbkongress_itbkongress365/itbk_dl_all_itbkongress_itbkongress365_itblibrary/itbk_dl_all_itbkongress_itbkongress365_itblibrary_studien/ITB_World_Travel_Trends_Report_2015_2016.pdf

Ivanov, S. (2014). *Hotel revenue management: From theory to practice.* Varna, Bulgary: Zangador.

Ivanov, S., & Zhechev, V. (2012). Hotel revenue management—A critical literature review. *Turizam: Znanstveno-Strucnicasopis, 60*(2), 175–197.

Jagels, M. G., & Coltman, M. M. (2007). *Hospitality management accounting* (9th ed.). Hoboken, NJ: John Wiley & Sons.

Jamal, T. B., & Getz, D. (1995). Collaboration theory and community tourism planning. *Annals of Tourism Research, 22*(1), 186–204. doi:10.1016/0160-7383(94)00067-3

Jamal, T., & Jamrozy, U. (2006). Collaborative networks and partnerships for integrated destination management. In D. Buhalis & C. Costa (Eds.), *Tourism management dynamics: Trends, management and tools* (pp. 164–172). Oxford, UK: Elsevier Butterworth-Heinemann. doi:10.1016/B978-0-7506-6378-6.50028-7

James, W. N. (2006). *Game Theory Decisions, Interaction and Evolution.* Springer-Verlag.

Jamshidi, B. (2008). *Web usability in B2B websites: Users' perspective* (M. S. Thesis). Lulea University of Technology, Sweden.

Janda, S., & Ybarra, A. (2005). Do product and consumer characteristics affect the relationship between online experience and customer satisfaction? *Journal of Internet Commerce, 4*(4), 133–151. doi:10.1300/J179v04n04_09

Jang, S., Tang, C. H., & Chen, M. H. (2008). Financing behaviors of hotel companies. *International Journal of Hospitality Management, 27*, 478–487. doi:10.1016/j.ijhm.2007.08.010

Jayachandran, S., Sharma, S., Kaufman, P., & Raman, P. (2005). The role of relational information processes and technology use in customer relationship management. *Journal of Marketing, 69*(4), 177–192. doi:10.1509/jmkg.2005.69.4.177

Jeng, J., & Fesenmaier, D. R. (1998). Destination compatibility in multidestination pleasure travel. *Tourism Analysis, 3*(3), 77–87.

Jenkins, J., Dredge, D., & Taplin, J. (2011). Destination planning and policy: Process and practice. In Y. Wang & A. Pizam (Eds.), *Destination marketing and management: Theories and applications* (pp. 21–38). Wallingford, UK: CAB International. doi:10.1079/9781845937621.0021

Jensen, B. M. C. (1986). Agency Costs of Free Cash Flow, Corporate Finance, and Takeovers. *The American Economic Review, 76*, 323–329.

Jin, T., Lin, V. S., & Hung, K. (2014). Chinas Generation Ys Expectation on Outbound group Package Tour. *Asia Pacific Journal of Tourism Research, 19*(6), 617–644. doi:10.1080/10941665.2013.806939

Johnny Abyssinia Tour and Travel. (2016). Accessed on Feb. 2016, http://www.johnnyabyssinia.com/images/imm38000_1%20%2813%29%20wm.jpg

Johnson, P. (1993). The notion of capacity in tourism: A review of the issues. In C. P. Cooper & A. Lockwood (Eds.), *Proceedings in Tourism, Recreation, and Hospitality Management* (Vol. 5, pp. 297–308). New York, NY: John Wiley & Sons.

Johnston, R., & Kong, X. (2011). The customer experience: A road-map for improvement. *Managing Service Quality, 21*(1), 5–24. doi:10.1108/09604521111100225

Jones, A., & Fox, S. (2009). *Generations online 2009*. Retrieved September 25, 2014, from http://www.pewinternet.org/PPF/r/275/report_display.asp

Joshi, S. (2014). Customer experience management: An exploratory study on the parameters affecting customer experience for cellular mobile services of a telecom company. *Procedia: Social and Behavioral Sciences*, *133*, 392–399. doi:10.1016/j.sbspro.2014.04.206

Jun, S. H., Vogt, C. A., & MacKay, K. J. (2010). Online information search strategies: A focus on flights and accommodations. *Journal of Travel & Tourism Marketing*, *27*(8), 579–595. doi:10.1080/10548408.2010.507151

Kalakota, R., & Robinson, M. (1999). *E-business: roadmap for success*. Addison Wesley Publishing Company.

Kale, J. R., & Shahrur, H. (2007). Corporate capital structure and the characteristics of suppliers and customers. *Journal of Financial Economics*, *83*(2), 321–365. doi:10.1016/j.jfineco.2005.12.007

Kamis, A., & Frank, J. (2014). Advice shearing between paired users in online travel planning. *Journal of Theoretical and Applied Electronic Commerce Research*, *9*(1), 1–15. doi:10.4067/S0718-18762014000100002

Kantor, S., Winkelstein, W., & Ibrahim, M. A. (1968). A note on Interpretetion of the ridit as quantile ranks. *American Journal of Epidemiology*, *87*, 609–615. PMID:5654949

Kao, Y. F., Huang, L. S., & Yang, M. H. (2007). Effects of experiential elements on experiential satisfaction and loyalty intentions: A case study of the super basketball league in Taiwan. *International Journal of Revenue Management*, *1*(1), 79–96. doi:10.1504/IJRM.2007.011195

Kasavana, M. L., & Smith, D. I. (1982). *Menu Engineering: A Practical Guide to Menu Analysis*. Okemos, MI: Hospitality Publications.

Kasemsap, K. (2013). Innovative human resource practice: A synthesized framework and causal model of leader-member exchange, organizational justice, job satisfaction, and organizational citizenship behavior. *International Journal of e-Education, e-Business, e-Management Learning*, *3*(1), 13–17. doi:10.7763/IJEEEE.2013.V3.185

Kasemsap, K. (2014). Developing a unified framework and a causal model of transformational leadership, empowerment, innovation support, and organizational innovation. In N. Erbe (Ed.), *Approaches to managing organizational diversity and innovation* (pp. 280–303). Hershey, PA: IGI Global. doi:10.4018/978-1-4666-6006-9.ch014

Kasemsap, K. (2015a). The role of customer relationship management in the global business environments. In T. Tsiakis (Ed.), *Trends and innovations in marketing information systems* (pp. 130–156). Hershey, PA: IGI Global. doi:10.4018/978-1-4666-8459-1.ch007

Kasemsap, K. (2015b). The role of marketing strategies in the tourism industry. In N. Ray (Ed.), *Emerging innovative marketing strategies in the tourism industry* (pp. 174–194). Hershey, PA: IGI Global. doi:10.4018/978-1-4666-8699-1.ch010

Kasemsap, K. (2015c). The roles of cross-cultural perspectives in global marketing. In J. Alcántara-Pilar, S. del Barrio-García, E. Crespo-Almendros, & L. Porcu (Eds.), *Analyzing the cultural diversity of consumers in the global marketplace* (pp. 37–59). Hershey, PA: IGI Global. doi:10.4018/978-1-4666-8262-7.ch003

Kasemsap, K. (2016a). Promoting service quality and customer satisfaction in global business. In U. Panwar, R. Kumar, & N. Ray (Eds.), *Handbook of research on promotional strategies and consumer influence in the service sector* (pp. 247–276). Hershey, PA: IGI Global. doi:10.4018/978-1-5225-0143-5.ch015

Kasemsap, K. (2016b). Promoting leadership development and talent management in modern organizations. In U. Aung & P. Ordoñez de Pablos (Eds.), *Managerial strategies and practice in the Asian business sector* (pp. 238–266). Hershey, PA: IGI Global. doi:10.4018/978-1-4666-9758-4.ch013

Kasemsap, K. (2016c). The roles of information technology and knowledge management in global tourism. In A. Nedelea, M. Korstanje, & B. George (Eds.), *Strategic tools and methods for promoting hospitality and tourism services* (pp. 109–138). Hershey, PA: IGI Global. doi:10.4018/978-1-4666-9761-4.ch006

Kasemsap, K. (2016d). Encouraging supply chain networks and customer loyalty in global supply chain. In N. Kamath & S. Saurav (Eds.), *Handbook of research on strategic supply chain management in the retail industry* (pp. 87–112). Hershey, PA: IGI Global. doi:10.4018/978-1-4666-9894-9.ch006

Kasemsap, K. (2016e). Implementing electronic commerce in global marketing. In I. Lee (Ed.), *Encyclopedia of e-commerce development, implementation, and management* (pp. 591–602). Hershey, PA: IGI Global. doi:10.4018/978-1-4666-9787-4.ch043

Kasemsap, K. (2016f). Utilizing social media in modern business. In I. Lee (Ed.), *Encyclopedia of e-commerce development, implementation, and management* (pp. 2171–2182). Hershey, PA: IGI Global. doi:10.4018/978-1-4666-9787-4.ch156

Kasemsap, K. (2016g). The roles of social media marketing and brand management in global marketing. In W. Ozuem & G. Bowen (Eds.), *Competitive social media marketing strategies* (pp. 173–200). Hershey, PA: IGI Global. doi:10.4018/978-1-4666-9776-8.ch009

Kaushal, S. K. (2013). The Importance of Apparel Product Attributes for Teenaged Buyers. *NMIMS Management Review*, *23*, 45–64.

Kay, H. K. (2003). *Selling Tourism*. New York: Delmar Learning.

Kennedy, K. N., Goolsby, J. R., & Arnould, E. J. (2003). Implementing a customer orientation: Extension of theory and application. *Journal of Marketing*, *67*(4), 67–81. doi:10.1509/jmkg.67.4.67.18682

Khan, I., Garg, R. J., & Rahman, Z. (2015). Customer service experience in hotel operations: An empirical analysis. *Procedia: Social and Behavioral Sciences*, *189*, 266–274. doi:10.1016/j.sbspro.2015.03.222

Kijima, M., & Shibata, T. (2002). *Real Options in a Duopoly Market with General Volatility Structure*. Discussion Paper, Kyoto University

Kim, D., & Perdue, R. R. (2013). The effects of cognitive, affective, and sensory attributes on hotel choice. *International Journal of Hospitality Management*, *35*(1), 246–257. doi:10.1016/j.ijhm.2013.05.012

Kimes, S. (1999). Implementing Restaurant Revenue Management: A Five-step Approach. *The Cornell Hotel and Restaurant Administration Quarterly*, *40*(3), 16–21. doi:10.1177/001088049904000315

Kimes, S. E. (2010). The future of hotel revenue management. *Cornell Hospitality Report*, *10*(14), 6–15.

Kimes, S. E., & Wirtz, J. (2003). Has revenue management become acceptable? Findings from an international study on the perceived fairness of rate fences. *Journal of Service Research*, *6*(2), 125–135. doi:10.1177/1094670503257038

Kim, H. R. (2005). Developing an index of online customer satisfaction. *Journal of Financial Services Marketing*, *10*(1), 49–64. doi:10.1057/palgrave.fsm.4770173

Kim, S., & Fesenmaier, D. R. (1990). Evaluating spatial structure effects in recreational travel. *Leisure Sciences*, *12*(4), 367–381. doi:10.1080/01490409009513115

Kim, W. G., & Kim, D. J. (2004). Factors affecting online hotel reservation intention between online and non-online customers. *Hospital Management*, *23*(4), 381–395. doi:10.1016/j.ijhm.2004.02.001

Kindratenko, V. (Ed.). (2014). *Numerical Computations with GPUs*. Springer. doi:10.1007/978-3-319-06548-9

Kirk, D. B., Mei, W., & Hwu, W. (2012). Programming Massively Parallel Processors: A Hands-on Approach (2nd ed.). Morgan Kaufmann.

Kirk, D., & Pine, R. (1998). Research in Hospitality and Technology. *Hospital Management, 17*(2), 203–217. doi:10.1016/S0278-4319(98)00016-4

Kirkpatrick, S., Gelatt, C. Jr, & Vecchi, M. (1983). Optimization by Simulated Annealing. *Science, 220*(4598), 671–680. doi:10.1126/science.220.4598.671 PMID:17813860

Kittner, C. C. B. (2007). The role of safe havens in Islamist terrorism. *Terrorism and Political Violence, 19*(3), 307–329. doi:10.1080/09546550701246791

Klaus, P., & Maklan, S. (2013). Towards a better measure of customer experience. *International Journal of Market Research, 55*(2), 227–246. doi:10.2501/IJMR-2013-021

Klepers, A., & Rozite, M. (2010). Administrative border as a challenge in spatial aspects of tourism competitiveness. In A. Clarke (Ed.), *Constructing central Europe: Tourism competitiveness* (pp. 23–33). Veszprém, HU: University Press, University of Pannonia.

Knutson, B. J., Beck, J. A., Kim, S., & Cha, J. (2009). Identifying the dimensions of the guests' hotel experience. *Cornell Hospitality Quarterly, 50*(1), 44–55. doi:10.1177/1938965508326305

Kobayashi, K., Matsushima, K., & Hishida, K. (2008). Economic benefit evaluation of reservation system. *JapanSociety of Civil Engineering, Transaction D, 64*(2), 299-318. (in Japanese)

Koc, E. (2004). The role of family members in the family holiday purchase decision-making process. *International Journal of Hospitality & Tourism Administration, 5*(2), 85–101. doi:10.1300/J149v05n02_05

Koh, K. Y., & Hatten, T. S. (2002). The tourism entrepreneur. The overlooked player in tourism development studies. *International Journal of Hospitality & Tourism Administration, 3*(1), 21–48. doi:10.1300/J149v03n01_02

Kong, J. J., & Kwok, Y. K. (2007). Real options in strategic investment games between two asymmetric firms. *European Journal of Operational Research, 181*(2), 967–985. doi:10.1016/j.ejor.2006.07.006

Korajczyk, R. A., Lucas, D. J., & McDonald, R. L. (1992). Equity issues with time-varying asymmetric information. *Journal of Financial and Quantitative Analysis, 27*(3), 397–417. doi:10.2307/2331327

Korstanje, M & Tarlow, P. (2013) Disasters, tourism and mobility: the case of Japan's earthquake. *Pasos: revista de turismo y patrimonio cultural,11*, 17-32

Korstanje, M. E. (2009). Miedos democráticos: nuevas expresiones del temor en Latinoamérica. *Eikasia: revista de filosofía, 28*, 109-123.

Korstanje, M. E. (2011b). Why risk why now? Conceptual problems around the risk perception in tourism industry. *RBTur, 5*(1), 4-22.

Korstanje, M. E. (2013c). Estar protegido: el rol de las compañías de asistencia al viajero en el sistema turístico. *El Periplo Sustentable: revista de turismo, desarrollo y competitividad, 25*, 43-76.

Korstanje, M. E., & George, B. P. (2015). Understanding the Sociological Roots of'No-Shows' in Technology Mediated Transactions in Tourism. *E-Review of Tourism Research*, 11.

Korstanje, M. (2009). Re-visiting risk perception theory in the context of travel. *E-Review of Tourism Research, 7*(4), 68–81.

Korstanje, M. E. (2011). Influence of history in the encounter of guests and hosts. *Anatolia*, *22*(2), 282–285. doi:10.1080/13032917.2011.597991

Korstanje, M. E. (2011). La fobología,¿ ciencia o forma de entretenimiento? *Nómadas. Revista Crítica de Ciencias Sociales y Jurídicas*, *31*, 175–189.

Korstanje, M. E. (2011a). The fear of traveling: A new perspective for tourism and hospitality. *Anatolia*, *22*(2), 222–233. doi:10.1080/13032917.2011.597935

Korstanje, M. E. (2012). ¿Se puede ponderar la seguridad turística? Un ensayo conceptual. *Turismo e Sociedade*, *5*(2), 368–390. doi:10.5380/tes.v5i2.26332

Korstanje, M. E. (2013a). Empire and Democracy. A critical reading of Michael Ignatieff. *Nómadas. Revista Crítica de Ciencias Sociales y Jurídicas*, *38*, 69–78.

Korstanje, M. E. (2013b). Preemption and Terrorism. When the Future Governs. *Cultura (Asociación de Docentes de la Universidad de San Martín de Porres)*, *10*, 167–184.

Korstanje, M. E. (2014). Towards an Index of Fear: The Role of Capital in Risk' s Construction. *International Journal of Cyber Warfare & Terrorism*, *4*(1), 19–26. doi:10.4018/ijcwt.2014010103

Korstanje, M. E. (2015). *A Difficult World, examining the roots of Capitalism*. New York, NY: Nova Science Publishers.

Korstanje, M. E., & Olsen, D. H. (2011). The discourse of risk in horror movies post 9/11: Hospitality and hostility in perspective. *International Journal of Tourism Anthropology*, *1*(3), 304–317. doi:10.1504/IJTA.2011.043712

Korstanje, M. E., & Skoll, G. (2013). An Essay on the Social Costs and Benefits of Technology Evolution. *Human & Social Studies. Research and Practice*, *2*(2), 13–39.

Korstanje, M. E., & Tarlow, P. (2012). Being lost: Tourism, risk and vulnerability in the post-9/11entertainment industry. *Journal of Tourism and Cultural Change*, *10*(1), 22–33. doi:10.1080/14766825.2011.639455

Koza, J. R., Andre, D., Bennett, F. H., & Keane, M. A. (1999). *Genetic Programming III: Darwinian Invention and Problem Solving* (1st ed.). San Francisco, CA: Morgan Kaufmann Publishers Inc.

Kozak, M., Crotts, J., & Law, R. (2007). The Impact of the pereption of risk on international Travellers. *International Journal of Tourism Research*, *9*(4), 233–242. doi:10.1002/jtr.607

Kracht, J., & Wang, Y. (2010). Examining the tourism distribution channel: Evolution and transformation. *International Journal of Contemporary Hospitality Management*, *22*(5), 736–757. doi:10.1108/09596111011053837

Krippendorf, J. (1986). The new tourist: Turning point for leisure and travel. *Tourism Management*, *7*(2), 131–135. doi:10.1016/0261-5177(86)90025-7

Ku, E. C. S., Wu, W., & Lin, A. (2011). Strategic alignment leverage between hotels and companies: The buyer-supplier relationship perspective. *International Journal of Hospitality Management*, *30*(3), 735–745. doi:10.1016/j.ijhm.2010.10.006

Kulatilaka, N., & Perotti, E. C. (1998, August). Strategic Growth Options. *Management Science*, *44*(8), 1021–1031. doi:10.1287/mnsc.44.8.1021

Kumar, V., & Raheja, G. (2012). Business To Business (B2b) and Business to Consumer (B2c) Management. *International Journal of Computer & Technology*, *3*(3), 447–451.

Kumbharana, S., & Pandey, G. (2013). Solving Travelling Salesman Problem using Firefly Algorithm. *International Journal for Research in Science and Advanced Technologies*, *2*(2), 53-57.

Kushida, J. I., Hara, A., Takahama, T., & Kido, A. (2013). Island-based differential evolution with varying subpopulation size. In *Computational Intelligence & Applications (IWCIA), 2013 IEEE Sixth International Workshop on*, (pp. 119–124). IEEE doi:10.1109/IWCIA.2013.6624798

Kwortnik, R. J. Jr, & Ross, W. T. Jr. (2007). The role of positive emotions in experiential decisions. *International Journal of Research in Marketing, 24*(4), 324–335. doi:10.1016/j.ijresmar.2007.09.002

Lafferty, G., & van Fossen, A. (2001). Integrating the tourism industry: Problems and strategies. *Tourism Management, 22*(1), 11–19. doi:10.1016/S0261-5177(00)00021-2

Lages, C. R., & Piercy, N. F. (2012). Key drivers of frontline employee generation of ideas for customer service improvement. *Journal of Service Research, 15*(2), 215–230. doi:10.1177/1094670511436005

Lai, R. N., Ng, L. H. M., & Zhou, Y. (2001, January). *Analysis of Hotel Overbuilding Based on Real Options Exercise Strategies*. Paper presented at the sixth annual meeting of the Asian Real Estate Society, Tokyo, Japan.

Lai, W. H., & Vinh, N. Q. (2013). An application of ahp approach to investigate tourism promotional effectiveness. *Tourism and Hospitality Management, 19*(1), 1–22.

Lambert, D., & Stocs, J. (1993). *Strategic logistics management*. New York, NY: McGraw–Hill.

Laming, C., & Mason, K. (2014). Customer experience: An analysis of the concept and its performance in airline brands. *Research in Transportation Business & Management, 10*, 15–25. doi:10.1016/j.rtbm.2014.05.004

Lämmel, R. (2008). Googles MapReduce programming model—Revisited. *Science of Computer Programming, 70*(1), 1–30. doi:10.1016/j.scico.2007.07.001

Lam, T., & Hsu, C. H. C. (2006). Predicting behavioral intention of choosing a travel destination. *Tourism Management, 27*(4), 589–599. doi:10.1016/j.tourman.2005.02.003

Land, A., & Doig, A. (1960). An outomotive method of solving discrete programming problems. *Econometrics, 28*(3), 409–520.

Larsen, S. (2007). Aspects of a Psychology of the Tourist Experience. *Scandinavian Journal of Hospitality and Tourism, 7*(1), 7-18.

Larsen, S., Brun, W., & Øgaard, T. (2009). What Tourists worry about: Construction of a scale measuring tourist worries. *Tourism Management, 30*(2), 260–265. doi:10.1016/j.tourman.2008.06.004

Lash, S. M., Urry, S. L. J., & Urry, J. (1993). *Economies of signs and space* (Vol. 26). London: Sage.

Lau, A., & Lau, H.-S. (2002). Effects of a Demand-Curve"s Shape on The Optimal Solutions of a MultiEchelon Inventory/Pricing-model. *European Journal of Operational Research, 147*(3), 530–548. doi:10.1016/S0377-2217(02)00291-6

Law, R. (2009). Disintermediation of hotel reservations: The perception of different groups of online buyers in Hong Kong. *International Journal of Contemporary Hospitality Management, 21*(6), 766–772. doi:10.1108/09596110910976007

Law, R., Buhalis, D., & Cobanoglu, C. (2014). Progress on information and communication technologies in hospitality and tourism. *International Journal of Contemporary Hospitality Management, 26*(5), 727–750. doi:10.1108/IJCHM-08-2013-0367

Law, R., Cheung, C., & Lo, A. (2004). The Relevance of Profiling, Travel activities for Improving Destination Marketing Strategies. *International Journal of Contemporary Hospitality Management, 16*(6), 355–362. doi:10.1108/09596110410550798

Law, R., & Lau, W. (2004). A study of the perceptions of Hong Kong hotel managers on the potential disintermediation of travel agencies. *Journal of Travel & Tourism Marketing*, *17*(2-3), 117–131. doi:10.1300/J073v17n02_10

Law, R., Leung, K., & Wong, J. (2004). The impact of the internet on travel agencies. *International Journal of Contemporary Hospitality Management*, *16*(2), 100–107. doi:10.1108/09596110410519982

Lawrence, R. D. (2003). A machine-learning approach to optimal bid pricing. In H. K. Bhargava & N. Ye (Eds.), *Computational modeling and problem solving in the networked world: Interfaces in computer science and operations research* (pp. 97–118). New York, NY: Springer. doi:10.1007/978-1-4615-1043-7_5

Lawson-Body, A., & Limayem, M. (2004). The impact of customer relationship management on customer loyalty: The moderating role of web site characterisitcs. *Journal of Computer-Mediated Communication*, *9*(4), 153–174.

Leahy, J. V. (1993). Investment in Competitive Equilibrium: The Optimality of Myopic Behavior. *The Quarterly Journal of Economics*, *108*(4), 1105–1133. doi:10.2307/2118461

Lee, J., Siau, K. e Hong, S. (2003). Enterprise Integration with ERP and EAI. *Communications of the ACM*, *468*(2), 54-60.

Lee, T. Y. (2013). *Top 10 trends of the next generation of travel: The millennials*. Retrieved January 10, 2016, from www.hvs.com

Lee, C. K., Lee, Y. K., & Wicks, B. E. (2004). Segmentation of Festival Motivation by Nationality and Satisfaction. *Tourism Management*, *25*(1), 61–70. doi:10.1016/S0261-5177(03)00060-8

Lee, E., & Staelin, R. (1997). Vertical Strategic Interaction : Implications for Channel Pricing. *Marketing Science*, *16*(3), 185–207. doi:10.1287/mksc.16.3.185

Lee, S. M., Hwang, T., & Kim, J. (2007). An analysis of diversity in electronic commerce research. *International Journal of Electronic Commerce*, *12*(1), 31–67. doi:10.2753/JEC1086-4415120102

Lemke, C., Riedel, S., & Gabrys, B. (2009). Dynamic combination of forecasts generated by diversification procedures applied to forecasting of airline cancellations. In *IEEE Symposium on Computational Intelligence for Financial Engineering* (pp. 85–91). New York, NY: Institute of Electrical and Electronics Engineers. doi:10.1109/CIFER.2009.4937507

Lemke, F., Clark, M., & Wilson, H. (2011). Customer experience quality: An exploration in business and consumer contexts using repertory grid technique. *Journal of the Academy of Marketing Science*, *39*(6), 846–869. doi:10.1007/s11747-010-0219-0

Lepp, A., & Gibson, H. (2008). Sensation Seeking and Tourism: Tourist role, perception of risk and Destination Choice. *Tourism Management*, *29*(4), 740–750. doi:10.1016/j.tourman.2007.08.002

Lester, D. H., Forman, A. M., & Loyd, D. (2006). Internet shopping and buying behavior of college students. *Services Marketing Quarterly*, *27*(2), 123–138. doi:10.1300/J396v27n02_08

Leuschner, R., Charvet, F., & Rogers, D. S. (2013). A meta-analysis of logistics customer service. *Journal of Supply Chain Management*, *49*(1), 47–63. doi:10.1111/jscm.12000

Lewis, R. C., & Chambers, R. E. (2000). *Marketing leadership in hospitality, foundations and practices*. New York: Wiley.

Liaw, Y. J., Chi, N. W., & Chuang, A. (2010). Examining the mechanisms linking transformational leadership, employee customer orientation, and service performance: The mediating roles of perceived supervisor and coworker support. *Journal of Business and Psychology*, *25*(3), 477–492. doi:10.1007/s10869-009-9145-x

Li, B. (2014). A Cruise Line Dynamic Overbooking Model with Multiple Cabin Types from the View of Real Options. *Cornell Hospitality Quarterly*, *55*(2), 197–209. doi:10.1177/1938965513507126

Li, C., & Bernoff, J. (2008). *Groundswell: Winning in a world transformed by social technologies*. New York: Harvard Business Press.

Lieberthal, R. D. (2008). Hospital Quality: A PRIDIT Approach. *Health Services Research*, *43*(3), 18. doi:10.1111/j.1475-6773.2007.00821.x PMID:18454777

Likert, R. (1932). A Technique for the Measurement of Attitudes. *Archives de Psychologie*, *140*, 1–55.

Lin, C. F. (2014). Exploring the hotel service personnels' cognitive implications toward service attributes and ethics. *Human Factors and Ergonomics in Manufacturing & Service Industries*, *24*(1), 14–28. doi:10.1002/hfm.20349

Lings, I. N., & Greenley, G. E. (2009). The impact of internal and external market orientations on firm performance. *Journal of Strategic Marketing*, *17*(1), 41–53. doi:10.1080/09652540802619251

Linneman, P. (2011). Real Estate Finance and Investments - Risk and Opportunities. Linneman Associates, Ed.

Lin, T. C., & Huang, S. L. (2014). Understanding the determinants of consumers switching intentions in a standards war. *International Journal of Electronic Commerce*, *19*(1), 163–189. doi:10.2753/JEC1086-4415190105

Liou, J. J. H., Hsu, C., Yeh, W., & Lin, R. (2011). Using a modified grey relation method for improving airline service quality. *Tourism Management*, *32*(6), 1381–1388. doi:10.1016/j.tourman.2011.01.013

Li, Q. (2005). Does democracy promote or reduce transnational terrorist incidents? *The Journal of Conflict Resolution*, *49*(2), 278–297. doi:10.1177/0022002704272830

Litvin, S. W., Goldsmith, R. E., & Pan, B. (2008). Electronic word-of-mouth in hospitality and tourism management. *Tourism Management*, *29*(3), 458–468. doi:10.1016/j.tourman.2007.05.011

Litwin, M. (1995). *How to Measure Survey Reliability and Validity*. London: SAGE Publication. doi:10.4135/9781483348957

Liu, F., Zhang, J., Zhang, J., Chen, D., Liu, Z., & Lu, S. (2012). Roles and functions of tourism destinations in tourism region of South Anhui: A tourist flow network perspective. *Chinese Geographical Science*, *22*(6), 755–764. doi:10.1007/s11769-012-0557-6

Liu, J. N. K., & Zhang, E. Y. (2014). An investigation of factors affecting customer selection of online hotel booking channels. *International Journal of Hospitality Management*, *39*, 71–83. doi:10.1016/j.ijhm.2014.01.011

Liu, P. H. (2004). Hotel demand/cancellation analysis and estimation of unconstrained demand using statistical methods. In I. Yeoman & U. McMahon-Beattie (Eds.), *Revenue management and pricing: Case studies and applications* (pp. 91–101). London, England: Thomas Learning.

Loiacono, E. T., Watson, R. T., & Goodhue, D. L. (2007). WebQual: An instrument for consumer evaluation of web sites. *International Journal of Electronic Commerce*, *11*(3), 51–87. doi:10.2753/JEC1086-4415110302

Lorrain, F., & White, H. C. (1971). Structural equivalence of individuals in social networks. *The Journal of Mathematical Sociology*, *1*(1), 49–80. doi:10.1080/0022250X.1971.9989788

Lovelock, B., & Lovelock, K. M. (2013). *The Ethics of Tourism. Critical and Applied Perspectives*. New York: Routledge.

Lu, W., & Stepchenkova, S. (2014). User-generated content as a research mode in tourism and hospitality applications: Topics, methods, and software. *Journal of Hospitality Marketing & Management*. doi: .10.1080/19368623.2014.907758

Lu, A. C. C., & Chen, B. T. (2014). Information search behavior of independent Guests' Complaints Online. *Journal of Hospitality Marketing & Management*, *19*(7), 797–818.

Lue, C. C., Crompton, J. L., & Stewart, W. P. (1996). Evidence of cumulative attraction in multidestination recreational trip decision. *Journal of Travel Research*, *35*(1), 41–49. doi:10.1177/004728759603500107

Luhmann, N. (2006). *Sociología del Riesgo [Sociology of Risk]*. Universidad Iberoamericana.

Luo, M. M., Chen, J. S., Ching, R. K., & Liu, C. C. (2011). An examination of the effects of virtual experiential marketing on online customer intentions and loyalty. *Service Industries Journal*, *31*(13), 2163–2191. doi:10.1080/02642069.2010.503885

Lu, P. (2012). *Optimizing hotel pricing: A new approach to hotel reservations. Cornell Hospitality Reports, 12(10).*

MacGillavry, K., & Wilson, A. (2014). Delivering loyalty via customer experience management at DHL freight. *Global Business and Organizational Excellence*, *33*(6), 6–20. doi:10.1002/joe.21569

Maciejewski, K., & Kerley, G. I. H. (2014). Understanding tourists' preference for mammal species in private protected areas: Is there a case for extralimital species for ecotourism? *PLoS ONE*, *9*(2).

MacKie-Mason, J. (1990). Do taxes affect corporate financing decisions? *The Journal of Finance*, *45*(5), 1471–1493. doi:10.1111/j.1540-6261.1990.tb03724.x

Magun, S. (1996). *The development of strategic alliances in Canadian industries: A micro-analysis*. Working paper No. 13. Industry Canada.

Mahapatra, S., & Sreekumar, N. A. (2012). A RIDIT apprach to evaluate factors influencing online shopping behaviour. *International Journal of Internet Marketing and Advertising*, *7*(1), 83–98. doi:10.1504/IJIMA.2012.044936

Manjit, S. Y., & Pavlou, P. A. (2014). Marketing in computer-mediated environments: Research synthesis and new directions. *Journal of Marketing*, *78*(1), 20–40. doi:10.1509/jm.12.0020

Mansfeld, Y., & Pizam, A. (Eds.). (2006). *Tourism, security and safety: from theory to practice*. Oxford, UK: Elsevier.

March, R., & Woodside, A. (2005). Testing theory of planned versus realized behaviour. *Annals of Tourism Research*, *32*(4), 905–924. doi:10.1016/j.annals.2004.07.012

Maredia, A. (2010). *History, Analysis, and Implementation of Traveling Salesman Problem (TSP) and Related Problems* (Unpublished doctoral dissertation). University of Houston-Downtown.

Markus, L., Petrie, D., & Axline, S. (2000). Bucking the Trends: What the Future May Hold for ERP Packages. *Information Systems Frontiers*, *2*(2), 181–193. doi:10.1023/A:1026548107263

Marsden, P. V. (1989). Methods for the characterization of role structures in network analysis. In L. C. Freeman, D. R. White, & A. K. Romney (Eds.), *Research methods in social network analysis* (pp. 489–530). Fairfax, VA: George Mason University Press.

Martínez, J., Majó, J., & Casadesús, M. (2006). *El uso de las Tecnologias de la Información en el Sector Hotelero*. Academic Press.

Martin, J., Mortimer, G., & Andrews, L. (2015). Re-examining online customer experience to include purchase frequency and perceived risk. *Journal of Retailing and Consumer Services*, *25*, 81–95. doi:10.1016/j.jretconser.2015.03.008

Marzano, G., & Scott, N. (2009). Power in destination branding. *Annals of Tourism Research*, *36*(2), 247–267. doi:10.1016/j.annals.2009.01.004

Massey, D. (1994). *Space, place and gender*. Cambridge, UK: Polity Press.

Mathieson, A., & Wall, G. (1982). *Tourism: economic, physical and social impacts*. Harlow, UK: Longman.

Mattson, T. G., Sanders, B. A., & Massingill, B. L. (2004). *Patterns for Parallel Programming*. Addison-Wesley Professional.

Mauri, A. G., & Minazzi, R. (2013). Web review influence on expectations and purchasing intentions of hotel potential customers. *International Journal of Hospitality Management*, *34*(1), 99–107. doi:10.1016/j.ijhm.2013.02.012

Maurya, U. K., Mishra, P., Anand, S., & Kumar, N. (2015). Corporate identity, customer orientation and performance of SMEs: Exploring the linkages. *IIMB Management Review*, *27*(3), 159–174. doi:10.1016/j.iimb.2015.05.001

McCarthy, L., Stock, D., & Verma, R. (2010). How travelers use online and social media channels to make hotel-choice decisions. *Cornell Hospitality Report*, *10*(18), 4–18.

Mccartney, G. (2008). Does one culture all think the same? An investigation of destination image perceptions from several origins. *Tourism Review*, *63*(4), 13-26.

McCauley, C., & Moskalenko, S. (2008). Mechanisms of political radicalization: Pathways toward terrorism. *Terrorism and Political Violence*, *20*(3), 415–433. doi:10.1080/09546550802073367

McCool, M., Reinders, J., & Robison, A. (2012). *Structured Parallel Programming: Patterns for Efficient Computation*. Morgan Kaufmann.

McCool, M., Robison, A. D., & Reinders, J. (2012). *Structured Parallel Programming*. Morgan Kaufmann.

McCubbrey, D. J. (1999). Disintermediation and reintermediation in the US air travel distribution study: A delphi study. *Communications of the Association for Information Systems*, *1*, 2–39.

McDonald, R., & Siegel, D. (1986). The Value of Waiting to Invest. *The Quarterly Journal of Economics*, *101*(4), 707–728. doi:10.2307/1884175

McIntosh, R. W., & Goeldner, C. R. (1986). *Tourism: principles, practices, philosophies*. New York: Wiley.

McIntosh, R. W., Goeldner, C. R., & Ritchie, J. R. B. (1995). *Tourism, principles, practices, philosophies* (2nd ed.). New York, NY: John Wiley & Sons.

Mead, G. H. (2009). Mind, self, and society: From the standpoint of a social behaviourist (Vol. 1). Chicago: University of Chicago Press.

Mehrotra, R., & Ruttley, J. (2006). *Revenue management* (2nd ed.). Washington, DC: American Hotel & Lodging Association.

Mei, H., & Zhan, Z. (2013). An analysis of customer room choice model and revenue management practices in the hotel industry. *International Journal of Hospitality Management*, *33*, 178–183. doi:10.1016/j.ijhm.2012.08.001

Melovic, B., Mitrovic, S., Djokaj, A., & Vatin, N. (2015). Logistics in the function of customer service: Relevance for the engineering management. *Procedia Engineering*, *117*, 802–807. doi:10.1016/j.proeng.2015.08.255

Merinero-Rodríguez, R., & Pulido-Fernández, J. I. (2016). Analysing relationships in tourism: A review. *Tourism Management*, *54*, 122–135. doi:10.1016/j.tourman.2015.10.010

Messinger, P. R., & Narasimhan, C. (1995). Has power shifted in the grocery channel? *Marketing Science*, *14*(2), 189–223. doi:10.1287/mksc.14.2.189

Meyer, C., & Schwager, A. (2007). Understanding customer experience. *Harvard Business Review*, *85*(2), 117–126. PMID:17345685

Meyer-Cech, K. (2005). Regional cooperation in rural theme trails. In D. Hall, I. Kirkpatrick, & M. Mitchell (Eds.), *Rural tourism and sustainable business* (pp. 137–148). Clevedon, UK: Channel View.

Michalewicz, Z. (1996). *Genetic Algorithms + Data Structures = Evolution Programs*. Berlin: Springer-Verlag. doi:10.1007/978-3-662-03315-9

Michalewicz, Z., & Fogel, D. B. (2000). *How to Solve It: Modern Heuristics*. Berlin: Springer-Verlag. doi:10.1007/978-3-662-04131-4

Mich, L., Franch, M., & Cilione, G. (2003). The 2QCV3Q quality model for the analysis of web site requirements. *Journal of Web Engineering*, *2*(1), 105–127.

Michopoulou, E. E., & Moisa, D. (2016). The Role of Culture on Online Search Behaviour: A Comparative Study Between British and Chinese Travellers. In A. Inversini & R. Schegg (Eds.), Information and Communication Technologies in Tourism 2016 (pp. 765-777). Vienna: Springer International Publishing.

Microsoft. (n.d.a). *Load the Solver Add-in*. Retrieved July 8, 2016, from https://support.office.com/en-us/article/Load-the-Solver-Add-in-612926fc-d53b-46b4-872c-e24772f078ca

Microsoft. (n.d.b). *Solver Uses Generalized Reduced Gradient Algorithm*. Retrieved May 5, 2016, from https://support.microsoft.com/en-us/kb/82890

Miguel, M., Fernández, E., Olmeda, I., & Segui, M. (2000). Aplicación de las Nuevas Tecnologías de la Informatión al Sector Turístico. *Estudios y Perspectivas en Turísmo*, *9*, 5–23.

Milano, R., Baggio, R., & Piattelli, R. (2011, January). The effects of online social media on tourism websites. In R. Law, M. Fuchs, & F. Ricci (Eds.), *Information and Communication Technologies in Tourism 2011* (pp. 343–353). Vienna, Austria: Springer. doi:10.1007/978-3-7091-0503-0_38

Ministry of Infrastructure. (2002). *Ministry of Infrastructure Recommendations on Building Design and Practices*. Warsaw, Poland: Ministry of Infrastructure.

Mitsche, N. (2005). Understanding the information search process within a tourism domain specific search engine. In *Information and communication technologies in tourism* (pp. 183–193). Springer. doi:10.1007/3-211-27283-6_17

Modigliani, F., & Miller, M. (1958). The cost of capital, corporation finance, and the theory of investment. *The American Economic Review*, *48*(6), 261–297.

Modigliani, F., & Miller, M. H. (1963). Corporate income taxes and the cost of capital — A correction. *The American Economic Review*, *53*(3), 433–443.

Mohamad, R., & Ismail, N. A. (2009). Electronic commerce adoption in SME: The trend of prior studies. *Journal of Internet Banking and Commerce*, *14*(2), 1–16.

Mohanty, M. K., & Gahan, P. (2012). Buyer Supplier Relationship in Manufacturing Industry - Findings from Indian Manufacturing Sector. *Business Intelligence Journal*, *5*(1), 319–333.

Mohanty, P. (2008). *Hotel industry and Tourism in India*. APH Publishing.

Molina, A., Martín, V. J., Santos, J., & Aranda, E. (2009). Consumer service and loyalty in Spanish grocery store retailing: An empirical study. *International Journal of Consumer Studies*, *33*(4), 477–485. doi:10.1111/j.1470-6431.2009.00780.x

Molz, J. G. (2012). *Travel connections: Tourism, technology, and togetherness in a mobile world*. New York: Routledge.

Money, R., & Crotts, J. (2003). The effect of uncertainty avoidance on information search, planning and purchases of international vacations. *Tourism Management*, *24*(1), 191–202. doi:10.1016/S0261-5177(02)00057-2

Morales, D. R., & Wang, J. (2010). Forecasting cancellation rates for services booking revenue management using data mining. *European Journal of Operational Research, 202*(2), 554–562. doi:10.1016/j.ejor.2009.06.006

Morgan, N., Pritchard, A., & Pride, R. (2003). Destination branding and the role of the stakeholders: The case of New Zealand. *Journal of Vacation Marketing, 9*(3), 285–299. doi:10.1177/135676670300900307

Morgeson, F. P., DeRue, D. S., & Karam, E. P. (2010). Leadership in teams: A functional approach to understanding leadership structures and processes. *Journal of Management, 36*(1), 5–39. doi:10.1177/0149206309347376

Morrison, A., Jing, A., OLeary, J., & Cai, L. (2001). Predicting usage of the internet for travel bookings: An exploratory study. *Information Technology and Tourism, 4*(1), 15–30. doi:10.3727/109830501108750868

Mowle, J., & Merrilees, B. (2005). A functional and symbolic perspective to branding Australian SME wineries. *Journal of Product and Brand Management, 14*(4), 220–227. doi:10.1108/10610420510609221

Muhtaseb, R., Lakiotaki, K., & Matsatsinis, N. (2012). Applying a multicriteria satisfaction analysis approach based on user preferences to rank usability attributes in e-tourism websites. *Journal of Theoretical and Applied Electronic Commerce Research, 7*(3), 28–48. doi:10.4067/S0718-18762012000300004

Murphy, L., Moscardo, G., & Benckendorff, P. (2007). Using brand personality to differentiate regional tourism destinations. *Journal of Travel Research, 46*(1), 5–14. doi:10.1177/0047287507302371

Murrey, M., & Sproats, J. (1990). The disabled traveller: Tourism and disability in Australia. *Journal of Tourism Studies, 19*(1), 9–14.

Murto, P. (2004). Exit in Duopoly under Uncertainty. *The Rand Journal of Economics, 35*(1), 111–127. doi:10.2307/1593732

Murto, P., Nasakkala, E., & Keppo, J. (2004). Timing of investments in oligopoly under uncertainty: A framework for numberical analysis. *European Journal of Operational Research, 157*(2), 486–500. doi:10.1016/S0377-2217(03)00234-0

Muscatello, J., & Chen, I. (2008). Enterprise Resource Planning (ERP) Implementations: Theory and Practice. *International Journal of Enterprise Information Systems, 4*(1), 63–83. doi:10.4018/jeis.2008010105

Mutingi, M., & Mbohwa, C. (2014). Fuzzy system dynamics: An application to supply chain management. In P. Vasant (Ed.), *Handbook of research on novel soft computing intelligent algorithms: Theory and practical applications* (pp. 234–257). Hershey, PA: IGI Global. doi:10.4018/978-1-4666-4450-2.ch008

Myers, S. C. (1977, November). Determinants of corporate borrowing. *Journal of Financial Economics, 5*(2), 147–175. doi:10.1016/0304-405X(77)90015-0

Myers, S. C. (1984). Finance Theory and Financial Strategy. *Interfaces, 14*(1), 126–137. doi:10.1287/inte.14.1.126

Myers, S. C. (2001). Capital Structure. *The Journal of Economic Perspectives, 15*(2), 81–102. doi:10.1257/jep.15.2.81

Myers, S. C., & Majluf, N. (1984). Corporate Financing and Investment Decisions when Firms have Information that Investors do not Have. *Journal of Financial Economics, 13*(2), 187–221. doi:10.1016/0304-405X(84)90023-0

Nadiri, H., Hussain, K., Ekiz, E. H., & Erdoğan, Ş. (2008). An investigation on the factors influencing passengers' loyalty in the North Cyprus national airline. *Total Quality Management Journal, 20*(3), 265–280.

Narver, J. C., & Slater, S. F. (1990). The effect of a market orientation on business profitability. *Journal of Marketing, 54*(4), 20–35. doi:10.2307/1251757

Nash, J. F. Jr. (1950, January). Equilibrium points in n-person games. *Procs.of the National Academy of Sciences, 36*(1), 48–49. doi:10.1073/pnas.36.1.48 PMID:16588946

Nasution, R. A., Sembada, A. Y., Miliani, L., Resti, N. D., & Prawono, D. A. (2014). The customer experience framework as baseline for strategy and implementation in services marketing. *Procedia: Social and Behavioral Sciences*, *148*, 254–261. doi:10.1016/j.sbspro.2014.07.041

Neuling, R., Riedel, S., & Kalka, K.-U. (2004). New approaches to origin and destination and no-show forecasting: Excavating the passenger name records treasure. *Journal of Revenue and Pricing Management*, *3*(1), 62–72. doi:10.1057/palgrave.rpm.5170094

Newman, P., & Thornley, A. (2011). *Planning World Cities. Globalisation and Urban Politics.* New York: Palgrave - Macmillan.

Ngai, E. W. T., & Wat, F. K. T. (2002). A literature review and classification of electronic commerce research. *Information & Management*, *39*(5), 415–429. doi:10.1016/S0378-7206(01)00107-0

Nielsen, J. (1993). *Usability engineering*. San Francisco: Academic Press.

Niemann, M., Mochol, M., & Tolksdorf, R. (2008). Enhancing hotel research with semantic web technologies. *Journal of Theoretical and Applied Electronic Commerce Research*, *3*(2), 82–96.

Nilashi, M., bin Ibrahim, O., Ithnin, N., & Sarmin, N. H. (2015). A multi-criteria collaborative filtering recommender system for the tourism domain using Expectation Maximization (EM) and PCA-ANFIS. *Electronic Commerce Research and Applications*, *14*(6), 542–562. doi:10.1016/j.elerap.2015.08.004

Ninemeier, J. D., & Perdue, J. (2008). *Discovering Hospitality and Tourism. In The World's Greatest Industry* (2nd ed.). Pearson Prentice Hall.

Noone, B. M., & Lee, C. H. (2011). Hotel overbooking: The effect of overcompensation on customers reactions to denied service. *Journal of Hospitality & Tourism Research (Washington, D.C.)*, *35*(3), 334–357. doi:10.1177/1096348010382238

Noro, R., Dobashi, Y., Wang, Z. Y., Takahashi, T., Okamoto, T., Tanaka, T., & Watada, J. (2006, April). Real Options Approach to Life Cycle Management of Power Equipments. *Proceedings, CMD 2006, International Conference on Condition Monitoring and Diagnosis Changwon.*

Novelli, M., Schmitz, B., & Spencer, T. (2006). Networks, clusters and innovation in tourism: A UK experience. *Tourism Management*, *27*(6), 1141–1152. doi:10.1016/j.tourman.2005.11.011

Núñez, J. D. (2012). Centrales de reservas: Sentido y sensibilidad. *Tecnohotel*, *454*, 8–10.

Nunkesser, M., & Sawitzki, D. (2005). Blockmodels. In U. Brandes & T. Erlebach (Eds.), *Network analysis: Methodological foundation* (pp. 253–292). Berlin: Springer-Verlag. doi:10.1007/978-3-540-31955-9_10

Nunkoo, R., Gursoy, D., & Ramkissoon, H. (2013). Developments in hospitality marketing and management: Social network analysis and research themes. *Journal of Hospitality Marketing & Management*, *22*(3), 269–288. doi:10.1080/19368623.2013.753814

Nunnally, J. C. (1978). *Psychometric theory* (2nd ed.). New York, NY: McGraw-Hill.

Nuri, J., & Archer, S. (2001). Target adjustment model against pecking order model of capital structure. In *European Financial Management Association Annual Meeting*.

Nusair, K. K., Bilgihan, A., & Okumus, F. (2013). The role of online social network travel websites in creating social interaction for Generation Y travelers. *International Journal of Tourism Research*, *15*(5), 458–472. doi:10.1002/jtr.1889

Nusair, K. K., Parsa, H. G., & Cobanoglu, C. (2011). Building a model of commitment for Generation Y: An empirical study on e-travel retailers. *Tourism Management*, *32*(4), 833–843. doi:10.1016/j.tourman.2010.07.008

O'Connor, P. (2008). User-Generated Content and Travel: A Case Study on Tripadvisor.com. In P. O'Connor, W. Höpken & U. Gretzel (Eds.), Information and Communication Technologies in Tourism 2008 (pp. 47-58). Vienna, Austria: Springer.

O'Neil, C., & Schutt, R. (2014). *Doing data science: Straight talk from the frontline*. Sebastopol, CA: O'Reilly Media.

O'Neill, M., & Brabazon, A. (2006), Grammatical differential evolution. In *Proceedings of the 2006 International Conference on Artificial Intelligence*. Las Vegas, NV: CSREA Press.

O'Reilly, K., & Paper, D. (2012). CRM and retail service quality: Front-line employee perspectives. *International Journal of Retail & Distribution Management*, *40*(11), 865–881. doi:10.1108/09590551211267610

Observatorio Nacional de las Telecomunicaciones y de la Sociedad de la Información. (2011). *Estudio sobre Comercio Electrónico B2C 2011 Edición 2012*. Retrieved March 12, 2013, from http://www.ontsi.red.es/ontsi/es/estudios-informes/estudio-b2c-2011-edición-2012

Oflaç, B. S., Sullivan, U. Y., & Baltacioğlu, T. (2012). An attribution approach to consumer evaluations in logistics customer service failure situations. *Journal of Supply Chain Management*, *48*(4), 51–71. doi:10.1111/j.1745-493X.2012.03280.x

Ohmae, K. (1989). The global logic of strategic alliances. *Harvard Business Review*, *67*(2), 143–154.

Ojiako, U., & Maguire, S. (2009). Seeking the perfect customer experience: A case study of British Telecom. *Strategic Change*, *18*(5/6), 179–193. doi:10.1002/jsc.846

Olsen, N. V., & Sallis, J. (2006). Market scanning for new service development. *European Journal of Marketing*, *40*(5/6), 466–484. doi:10.1108/03090560610657796

Onwubolu, G., & Babu, B. (2004). *New Optimization Techniques in Engineering*. New York: Springer-Verlag. doi:10.1007/978-3-540-39930-8

Ooi, J. T. L. (2000). An empirical investigation on the incidence of secured debt. *Journal of Property Research*, *17*(3), 185–201. doi:10.1080/09599910050119976

Opler, T., Pinkowitz, L., Stulz, R., & Williamson, R. (1999). The determinants and implications of corporate cash holdings. *Journal of Financial Economics*, *52*(1), 3–46. doi:10.1016/S0304-405X(99)00003-3

Ordover, J. A., & Willig, R. D. (1981). An Economic Definition of Predation : Pricing and product innovation. *The Yale Law Journal*, *91*(1), 8–53. doi:10.2307/795848

OReilly, A. (1986). Tourism carrying capacity: Concepts and issues. *Tourism Management*, *7*(4), 254–258. doi:10.1016/0261-5177(86)90035-X

Orfila-Sintes, F., Crespí-Cladera, R., & Martínez-Ros, E. (2005). Innovation activity in the hotel industry: Evidence from Balearic Islands. *Tourism Management*, *26*(6), 851–865. doi:10.1016/j.tourman.2004.05.005

Organ, D. W. (1988). *Organizational citizenship behavior: The good soldier syndrome*. Lexington, MA: Lexington Books.

Ottenbacher, M., & Gnoth, J. (2005). How to develop successful hospitality innovation. *Cornell Hospitality Quarterly*, *46*(2), 205–222. doi:10.1177/0010880404271097

Oyeka, I. C. A. (1992). A Modified Ridit Analysis. *Journal of Nigerian Statistical Association*, *8*(1), 45–69.

Pacheco, P. (2011). *An Introduction to Parallel Programming*. Morgan Kaufmann.

Packer, T.L., Small, J., & Darcy, S. (2008). *Tourist Experiences of Individuals with Vision Impairment*. Gold Coast Australia, Sustainable Tourism CRC Tech Report.

Page, S. J., & Hall, M. C. (2003). *Managing Urban Tourism*. Harlow, UK: Prentice Hall.

Pallud, J., & Straub, D. W. (2014). Effective Website Design for Experience-Influenced Environments: The Case of High Culture Museums. *Information & Management*, *51*(3), 359–373. doi:10.1016/j.im.2014.02.010

Palmer, A., & McCole, P. (1999). The virtual re-intermediation of travel services: A conceptual framework and empirical investigation. *Journal of Vacation Marketing*, *6*(1), 33–47. doi:10.1177/135676679900600105

Panda, R. K., & Sreekumar. (2012). Prioritizing Issues in Agribusiness Using Ridit Analysis: An Indian Perspective. *World Journal of Agricultural Sciences*, *8*(1), 80–88.

Panorama Consulting Group. (2009). *2009 ERP Report: Hospitality and Entertainment*. Relatório.

Papadimitriou, C. H. (1977). The Euclidean travelling salesman problem is NP-complete. *Theoretical Computer Science*, *4*(3), 237–244. doi:10.1016/0304-3975(77)90012-3

Parasuraman, A., Zeithaml, V. A., & Malhotra, A. (2007). E-S-QUAL: A multiple-item scale for assessing electronic servisse quality. *Journal of Service Research*, *7*(3), 13–33.

Parasuraman, A., Zeithaml, V., & Malhotra, N. (2005). E-S-QUAL: A multiple-item scale for assessing electronic service quality. *Journal of Service Research*, *7*(3), 213–233. doi:10.1177/1094670504271156

Paredes, M. R., Barrutia, J. M., & Echebarria, C. (2014). Resources for value co-creation in e-commerce: A review. *Electronic Commerce Research*, *14*(2), 111–136. doi:10.1007/s10660-014-9135-6

Parker, C. M., & Castleman, T. A. (2006). A preliminary meta-analysis of SEM ebusiness journal publications: current trends and future research opportunities moving beyond adoption factor studies. In *Collecter 2006.Proceedings of the 2006 Collecter E-Commerce Conference*.

Park, J.-W., Robertson, R., & Wu, C.-L. (2006). Modelling the impact of airline service quality and marketing variables on passengers future behavioural intentions. *Transportation Planning and Technology*, *29*(5), 359–381. doi:10.1080/03081060600917686

Park, K., & Reisinger, Y. (2010). Differences in the perceived influence of natural disasters and travel risk on international travels. *Tourism Geographies*, *12*(1), 1–24. doi:10.1080/14616680903493621

Park, Y. A., & Gretzel, U. (2007). Success factors for destination marketing websites: A Qualitative Meta-Analysis. *Journal of Travel Research*, *46*(1), 26–35. doi:10.1177/0047287507302381

Parlar, M., & Weng, Z. K. (2006). Coordinating Pricing and Production Decisions in The Presence of Price Competition. *European Journal of Operational Research*, *170*(1), 211–227. doi:10.1016/j.ejor.2004.06.026

Parolo, G., Ferrarini, A., & Rossi, G. (2009). Optimization of tourism impacts within protected areas by means of genetic algorithms. *Ecological Modelling*, *220*(8), 1138–1147. doi:10.1016/j.ecolmodel.2009.01.012

Paswan, A., D'Souza, D., & Zolfagharian, M. A. (2009). Toward a contextually anchored service innovation typology. *Decision Sciences*, *40*(3), 513–540. doi:10.1111/j.1540-5915.2009.00239.x

Pavlovich, K. (2003). The evolution and transformation of a tourism destination network: The Waitomo Caves, New Zealand. *Tourism Management*, *24*(2), 203–216. doi:10.1016/S0261-5177(02)00056-0

Pawlikowska-Piechotka, A. (2013). *Planowanie przestrzeni turystycznej* [Planning Tourism]. Gdynia: Novae Res.

Pawlina, G., & Kort, P. M. (2006). Real options in an asymmetric duopoly: Who benefits from your competitive disadvantage? *Journal of Economics & Management Strategy*, *15*(1), 1–35. doi:10.1111/j.1530-9134.2006.00090.x

Pearce, D. G. (2014). Toward an integrative conceptual framework of destinations. *Journal of Travel Research, 53*(2), 141–153. doi:10.1177/0047287513491334

Pearce, D. G., & Schänzel, H. A. (2013). Destinations: Tourists' perspectives from New Zealand. *International Journal of Tourism Research.* doi:10.1002/jtr.1955

Pearce, D. G., Tan, R., & Schott, C. (2007). Distribution channels in international markets: A comparative analysis of the distribution of New Zealand tourism in Australia, Great Britain and the USA. *Current Issues in Tourism, 10*(1), 33–60. doi:10.2167/cit290.0

Pedemonte, M., Nesmachnow, S., & Cancela, H. (2011). A survey on parallel ant colony optimization. *Applied Soft Computing, 11*(8), 5181–5197. doi:10.1016/j.asoc.2011.05.042

Pempek, T. A., Yermolayeva, Y. A., & Calvert, S. L. (2009). College students social networking experiences on Facebook. *Journal of Applied Developmental Psychology, 30*(3), 227–238. doi:10.1016/j.appdev.2008.12.010

Pennington-Gray, L., London, B., Cahyanto, I., & Klages, W. (2011). Expanding the tourism crisis management planning framework to include social media: Lessons from the Deepwater Horizon Oil Spill 2010. *International Journal of Tourism Anthropology, 1*(3/4), 239–253. doi:10.1504/IJTA.2011.043708

Peterson, R. (2000). *Constructing Effective Questionnaires.* London: SAGE Publications. doi:10.4135/9781483349022

Pforr, C., & Hosie, P. (Eds.). (2009). *Crisis management in the tourism industry: beating the odds?* London: Ashgate Publishing, Ltd.

Phillimore, J., & Goodson, L. (2009). *Qualitative Research in Tourism.* London: Routledge.

Phillips, R. L. (2005). *Pricing and revenue optimization.* Stanford, CA: Stanford University Press.

PhoCusWright. (2009). *Consumer Travel Report.* Report. Author.

PhoCusWright. (2011). *Social Media in Travel 2011: Traffic, Activity & Sentiment.* Report. Author.

PhoCusWright. (2012a). European Consumer Travel Report (3rd ed.). Report. Author.

PhoCusWright. (2012b). *Asia Pacific Online Travel Overview.* Report. Author.

PhoCusWright's. (2012, September 22). The global travel e-commerce. *PhoCusWright's.*

Piazza, J. A. (2006). Rooted in Poverty? Terrorism, Poor Economic Development, and Social Cleavages. *Terrorism and Political Violence, 18*(1), 159–177. doi:10.1080/095465590944578

Piazza, J. A. (2007). Draining the swamp: Democracy promotion, state failure, and terrorism in 19 Middle Eastern countries. *Studies in Conflict and Terrorism, 30*(6), 521–539. doi:10.1080/10576100701329576

Pike, S. (2008). *Destination marketing. An integrated marketing communication approach.* Oxford, UK: Elsevier Butterworth-Heinemann.

Pike, S. (2009). Destination brand positions of a competitive set of near-home destinations. *Tourism Management, 30*(6), 857–866. doi:10.1016/j.tourman.2008.12.007

Pike, S., & Ryan, C. (2004). Destination Positioning Analysis through a Comparison of Cognitive, Affective, and Conative Perceptions. *Journal of Travel Research, 42*(May), 333–342. doi:10.1177/0047287504263029

Pine, B. J., & Gilmore, J. H. (2011). *The experience economy.* Boston, MA: Harvard Business Review Press.

Pizam, A. (1999). A comprehensive approach to classifying acts of crime and violence at tourism destinations. *Journal of Travel Research, 38*(1), 5–12. doi:10.1177/004728759903800103

Pizam, A. (2002). Tourism and terrorism. *International Journal of Hospitality Management, 21*(1), 1–3. doi:10.1016/S0278-4319(01)00037-8

Pizam, A., Jeong, G. H., Reichel, A., Van Boemmel, H., Lusson, J. M., Steynberg, L., & Montmany, N. (2004). The relationship between risk-taking, sensation-seeking, and the tourist behavior of young adults: A cross-cultural study. *Journal of Travel Research, 42*(3), 151–260. doi:10.1177/0047287503258837

Pizam, A., & Mansfeld, Y. (1996). *Tourism, crime and international security issues*. New York, NY: John Wiley & Sons.

Pizam, A., Tarlow, P. E., & Bloom, J. (1997). Making tourists feel safe: Whose responsibility is it? *Journal of Travel Research, 36*(1), 23–28. doi:10.1177/004728759703600104

Polatoglu, L. K. (1991). Optimal order quantity and pricing decisions in single-inventory inventory systems. *International Journal of Production Economics, 23*(1–3), 175–185. doi:10.1016/0925-5273(91)90060-7

Poli, R., Kennedy, J., & Blackwell, T. (2007). *Particle swarm optimization: An overview*. Springer Science and Business Media.

Porter, M. (1998), Clusters and the new economics of competition. *Harvard Business Review, 76*(6), 77-90.

Porter, M. (1998). Clusters and the new economics of competition. *Harvard Business Review, 76*(6), 77–90. PMID:10187248

Prayag, G., & Del Chiappa, G. (2014). Hotel disintermediation in France: Perceptions of students from Generation Y. *Anatolia – An International Journal of Tourism and Hospitality Research, 25*(3), 417-430.

Prayag, G., & Del Chiappa, G. (2014). Hotel disintermediation in France: Perceptions of students from Generation Y. *Anatolia – An International Journal of Tourism and Hospitality Research, 25*, 417–430.

Preiser, W. F. E., & Ostroff, E. (2001). *Universal Design Handbook*. New York: McGraw – Hill.

Prell, C. (2012). *Social Network Analysis: History, theory and methodology*. London: Sage.

Price, K. (1999). An introduction to differential evolution. In D. Corne, M. Dorigo, & F. Glover (Eds.), *New Ideas in Optimisation* (pp. 79–108). McGraw Hill, International.

Prideaux, B., Laws, E., & Faulkner, B. (2003). Events in Indonesia: Exploring the limits to formal tourism trends forecasting methods in complex crisis situations. *Tourism Management, 24*(4), 475–487. doi:10.1016/S0261-5177(02)00115-2

Pullman, M. E., & Gross, M. A. (2004). Ability of experience design elements to elicit emotions and loyalty behaviours. *Decision Sciences, 35*(3), 551–578. doi:10.1111/j.0011-7315.2004.02611.x

Qi, C. X., Gibson, H. J., & Zhang, J. J. (2009). Perceptions of risk and travel intentions: The case of China and the Beijing Olympic Games. *Journal of Sport & Tourism, 14*(1), 43–67. doi:10.1080/14775080902847439

Quan, D. C. (2002). The price of a reservation. *The Cornell Hotel and Restaurant Administration Quarterly, 43*(3), 77–86. doi:10.1016/S0010-8804(02)80021-0

Quarantelli, E. (1975). *Panic Behavior: Some empirical observations*. Disaster Research Center, Preliminary Paper 20.

Quarantelli, E. (1982). *People´s reactions to emergency warnings*. Disaster Research Center, Preliminary Paper 75.

Quarantelli, E. (1990). *The Mass media in disasters in the United States*. Disaster Research Center, Preliminary Paper.

Quarantelli, E. (2001). *The Sociology of Panic.* Disaster Research Center, Preliminary Paper 283.

Quigg, L. (1993). Empirical Testing of Real Option-Pricing Models. *The Journal of Finance*, *48*(2), 621–641. doi:10.1111/j.1540-6261.1993.tb04730.x

Quigg, L. (1995). Optimal land development. In L. Trigeorgis (Ed.), *Real Options in Capital Investment: Models, Strategies and Applications* (pp. 265–280). Westport, CT: Praeger.

Quin, Z., & Zhang, L. (2001). *Customizable e-commerce: introduction, application and proposition.* Xi'an Jiaotong University.

Rabianski, J. S. (2003). Primary and secondary data: Concepts, concerns, errors, and issues. *The Appraisal Journal*, *71*(1), 43–55.

Racela, O. C. (2014). Customer orientation, innovation competencies, and firm performance: A proposed conceptual model. *Procedia: Social and Behavioral Sciences*, *148*, 16–23. doi:10.1016/j.sbspro.2014.07.010

Rappaport, A. (1986). Creating Shareholder Value. The Free Press.

Rayport, J. F., Jaworski, B. J., & Kyung, E. J. (2005). Best face forward: Improving companies' service interfaces with customers. *Journal of Interactive Marketing*, *19*(4), 67–80. doi:10.1002/dir.20050

Rechenberg, I. (1971). *Evolutionsstrategie - Optimierung technischer Systeme nach Prinzipien der biologischen Evolution* (PhD thesis). Fromman-Holzboog.

Reeves, C. (1993). *Modern Heuristic Techniques for Combinatorial Problems.* Oxford, UK: Blackwell Scientific Publications.

Reichel, A., Fuchs, G., & Uriely, N. (2007). Perceived risk and the non-institutionalized tourist role: The case of Israeli student ex-backpackers. *Journal of Travel Research*, *46*(2), 217–226. doi:10.1177/0047287507299580

Reichhart, F. (2009). Means of access to tourism activities for disabled people. *Teoros Journal. Revue de Recherche en Tourisme*, (2), 73-81.

Reid, L. J., Smith, S. L. J., & McCloskey, R. (2008). The effectiveness of regional marketing alliances: A case study of the Atlantic Canada Tourism Partnership 2000–2006. *Tourism Management*, *29*(3), 581–593. doi:10.1016/j.tourman.2007.05.005

Reinganum, J. F. (1981). On the diffusion of new technology: A game theoretic approach. *The Review of Economic Studies*, *48*(3), 395–405. doi:10.2307/2297153

Ren, L., Qiu, H., Wang, P., & Lin, P. M. C. (2016). Exploring customer experience with budget hotels: Dimensionality and satisfaction. *International Journal of Hospitality Management*, *52*, 13–23. doi:10.1016/j.ijhm.2015.09.009

RIA Ethiopian Sports. (2016). Accessed on Feb. 2016, http://www.ethiotrail.com/wp-content/uploads/2014/02/quees_02.jpg

Ribalaygua, L. (2000). Estrategia Tecnológica en ele Sector Hotelero. *Estudios y Perspectivas en Turísmo*, *92*, 99–111.

Richards, G., & Wilson, J. (2003). *New Horizons in Independent Youth and Student Travel, a Report for the International Student Travel Confederation (ISTC) and the Association of Tourism and Leisure Education (ATLAS).* International Student Travel Confederation - ISTC, Amsterdam. Retrieved September 20, 2014, from http://www.atlas-euro.org/pages/pdf/FINAL_Full_Report.pdf

Richards, G. (2007). *New horizons II: The young independent travellers 2008.* Amsterdam: WYSE Travel Confederation.

Rifai, T. (2013, August). *Accessible Tourism for All.* UN Secretary General speech, UN General Assembly Resolution of August 13, Zimbabwe (A/RES/637(XX).

Ritchie, B. W. (2004). Chaos, crises and disasters: A strategic approach to crisis management in the tourism industry. *Tourism Management, 25*(6), 669–683. doi:10.1016/j.tourman.2003.09.004

Ritchie, B. W. (2009). *Crisis and disaster management for tourism.* Bristol: Channel View Publications.

Ritchie, J. R. B., & Crouch, G. I. (2003). *The competitive destination. A sustainable tourism perspective.* Wallingford, UK: CAB International. doi:10.1079/9780851996646.0000

Rodríguez, I. (2002). *Marketing.com y comercio electrónico en la sociedad de la información.* Madrid: Pirámide.

Roehl, W., & Fesenmaier, D. (1992). Risk Perceptions and Pleasure Travel: An exploratory analysis. *Journal of Travel Research, 30*(4), 17–29. doi:10.1177/004728759203000403

Ronald, W. M., & Edgar, A. N. (2008). *Introduction to Finance; Markets, Investments, and Financial Management.* Thirteenth.

Rondán, F. J., Arenas, J., & Ramierez, P. (2015). Travel buying behavior in Social Network site users: To buy online vs. offline. *Journal of Theoretical and Applied Electronic Commerce Research, 10*(1), 49–62. doi:10.4067/S0718-18762015000100005

Rose, S., Hair, N., & Clark, M. (2011). Online customer experience: A review of the business-to-consumer online purchase context. *International Journal of Management Reviews, 13*(1), 24–39. doi:10.1111/j.1468-2370.2010.00280.x

Ross, J., & Vitale, M. (2000). The ERP Revolution: Surviving vs Thiring. *Information Systems Frontiers, 2*(2), 233–241. doi:10.1023/A:1026500224101

Rubock, D. B. (2007). *US CMBS and CRE CDO : Moody' s Approach to Rating Commercial Real Estate Mezzanine Loans.* Academic Press.

Ruciński, M., Izzo, D., & Biscani, F. (2010). On the impact of the migration topology on the island model. *Parallel Computing, 36*(10), 555–571. doi:10.1016/j.parco.2010.04.002

Rus, R. (2009). The use of Information systems in Hotel Market of Cluj-Napoca. *Studia universitatis babes-bolyai, 3*, 51-59.

Russell, R., & Faulkner, B. (2004). Entrepreneurship, chaos and the tourism area lifecycle. *Annals of Tourism Research, 31*(3), 556–579. doi:10.1016/j.annals.2004.01.008

Russell, S. J., & Norvig, P. (2003). *Artificial Intelligence: A Modern Approach* (2nd ed.). Upper Saddle River, NJ: Prentice Hall.

Ryan, C. (1993). Crime, violence, terrorism and tourism: An accidental or intrinsic relationship? *Tourism Management, 14*(3), 173–183. doi:10.1016/0261-5177(93)90018-G

Saarinen, J. (2014). Transforming destinations: A discoursive approach to tourist destinations and development. In A. Viken & B. Granås (Eds.), *Tourism destination development: Turns and tactics* (pp. 47–62). Farnham, UK: Ashgate Publishing Limited.

Saayman, M., Slabbert, E., & Van Der Merwe, P. (2009). Travel Motivation: A Tale of Two Marine Destinations in South Africa. *South African Journal for Research in Sport. Physical Education and Recreation, 31*(1), 81–94.

Sackett, H. & Botterill, D. (2006). Perception of international travel risk: an exploratory study of the influence of proximity to terrorist attack. *E-Review of Tourism Research, 4*(2), 44-49.

Safa, H. (2015). *Dallol, Ethiopia, The Hottest Green Crater on Earth*. Accessed Feb. 2016, http://inspirationseek.com/wp-content/uploads/2015/01/Dallol-Green-Crater-in-Ethiopia.jpg

Saha, S., & Yap, G. (2014). The moderation effects of political instability and terrorism on tourism development: A cross-country panel analysis. *Journal of Travel Research, 53*(4), 509–521. doi:10.1177/0047287513496472

Sanders, J., & Kandrot, E. (2010). *CUDA by Example: An Introduction to General-Purpose GPU Programming, Portable Documents*. Addison-Wesley Professional.

Sanders, J., & Kandrot, E. (2010). *CUDA by Example: An Introduction to General-Purpose GPU Programming*. Addison-Wesley Professional.

Sandler, T. (1995). On the relationship between democracy and terrorism. *Terrorism and Political Violence*, *7*(4), 1–9. doi:10.1080/09546559508427315

Santiago, R. (2011). *Yield Management aplicado a la medición de la performance de la carta de vinos del restaurante Mercado Central* (Unpublished undergraduate thesis). Universidad de San Andrés, Buenos Aires, Argentina.

Santos, J. (2003). E-service quality: A model of virtual service quality dimensions. *Managing Service Quality: An International Journal*, *13*(3), 233–246. doi:10.1108/09604520310476490

Saraniemi, S., & Kylänen, M. (2011). Problematizing the concept of tourism destination: An analysis of different theoretical approaches. *Journal of Travel Research, 50*(2), 133–143. doi:10.1177/0047287510362775

Sarkar, S. (2000). On the investment–uncertainty relationship in a real options model. *Journal of Economic Dynamics & Control*, *24*(2), 219–225. doi:10.1016/S0165-1889(99)00005-6

Satta, A. (2015). *Towards sustainable tourism in the Mediterranean: Situation and challenges at a regional level.* Retrieved January 28, 2016, from http://www.medpan.org/documents/ 10180/0/presentation_tourism_25nov15_satta/0c6a1ee4-075b-4eea-9e75-052db62e9208

Saxe, R., & Weitz, B. A. (1982). The SOCO scale: A measure of the customer orientation of salespeople. *JMR, Journal of Marketing Research, 19*(3), 343–351. doi:10.2307/3151568

Schenck, C. (2016). *Simien National Park*. UNESCO. Accessed Feb. 2106, http://whc.unesco.org/?cid=31&l=en&id_site=9&gallery=1&&maxrows=46

Schibrowsky, J. A., Peltier, J. W., & Nill, A. (2007). The state of internet marketing research: A review of the literature and future research directions. *European Journal of Marketing*, *41*(7/8), 722–733. doi:10.1108/03090560710752366

Schmallegger, D., & Carson, D. (2008). Blogs in tourism: Changing approaches to information exchange. *Journal of Vacation Marketing, 14*(2), 99-110.

Schmidgall, R. (2011). *Hospitality industry managerial accounting* (7th ed.). Lansing, MI: The Educational Institute of the American Hotel & Lodging Association.

Schmitt, B. H. (1999). *Experiential marketing: How to get customers to sense, feel, think, act and relate to your company and brand.* New York, NY: The Free Press.

Schonberger, J. (2005). *Operational Freight Carrier Planning, Basic Concepts, Optimization Models and Advanced Memetic Algorithms*. Springer-Verlag.

Schwartz, J. E. (1977). An Examination of *CONCOR* and related methods for blocking sociometric data. In D. R. Heise (Ed.), *Sociological Methodology* (pp. 255–282). San Francisco, CA: Jossey-Bass. doi:10.2307/270759

Schwefel, H. (1974). *Numerische Optimierung von Computer-Modellen* (PhD thesis). Birkhuser.

Schwepker, C. H. Jr. (2003). Customer-oriented selling: A review, extension, and directions for future research. *Journal of Personal Selling & Sales Management*, *23*(2), 151–171.

Scott, J. (2000). *Social network analysis. A handbook* (2nd ed.). London: Sage Publications.

Scott, J. H. Jr. (1977). Bankruptcy, Secured Debt, and Optimal Capital Structure. *The Journal of Finance*, *32*(1), 1–19. doi:10.1111/j.1540-6261.1977.tb03237.x

Scott, N., Baggio, R., & Cooper, C. (2008a). *Network analysis and tourism: From theory to practice*. Clevedon, UK: Channel View.

Scott, N., Cooper, C., & Baggio, R. (2008b). Destination networks: Four Australian cases. *Annals of Tourism Research*, *35*(1), 169–188. doi:10.1016/j.annals.2007.07.004

Sellers-Rubio, R., & Más-Ruiz, F. J. (2009). Technical efficiency in the retail food industry: The influence of inventory investment, wage levels, and age of the firm. *European Journal of Marketing*, *43*(5/6), 652–669. doi:10.1108/03090560910946981

Selvin, S. (1977). A Further Note on the Interpretetion of ridit analysis. *American Journal of Epidemiology*, *105*, 16–20. PMID:831461

Serra, F. (2013). Estratégias de negócio: Os conceitos de yield e de revenue management e a sua aplicação às operações hoteleiras. In C. Henriques, I. Monteiro, F. Serra, J. Santos, & P. Águas (Eds.), *Inovação e Qualidade na Hotelaria* (pp. 161–172). Faro, Portugal: Universidade do Algarve, Escola Superior de Gestão, Hotelaria e Turismo.

Shah, S., & Thakor, A. V. (1987). Optimal capital structure and project financing. *Journal of Economic Theory*, *42*(2), 209–243. doi:10.1016/0022-0531(87)90086-X

Sharpley, R., & Telfer, D. (2015) Tourism & Development: concepts and Issues (2nd ed.). Bristol: Channel View Publications.

Shaw, C. (2005). *Revolutionize your customer experience*. New York, NY: Palgrave Macmillan. doi:10.1057/9780230513457

Shaw, G., Bailey, A., & Williams, A. (2011). Aspects of service-dominant logic and its implications for tourism management: Examples from the hotel industry. *Tourism Management*, *32*(2), 207–214. doi:10.1016/j.tourman.2010.05.020

Shaw, G., & Coles, T. (2004, June). Disability, holiday making and the tourism industry in the UK: A preliminary survey. *Tourism Management*, *25*(3), 397–403. doi:10.1016/S0261-5177(03)00139-0

Sheehan, L. R., & Ritchie, J. R. B. (2005). Destination stakeholders. Exploring identity and salience. *Annals of Tourism Research*, *32*(3), 711–734. doi:10.1016/j.annals.2004.10.013

Sheller, M. (2009). The new Caribbean complexity: Mobility systems, tourism and spatial rescaling. *Singapore Journal of Tropical Geography*, *30*(2), 189–203. doi:10.1111/j.1467-9493.2009.00365.x

Shelton, E., & Tucker, H. (2005). Tourism and disability: Issues beyond access. *Tourism Review International*, *8*(3), 211–219. doi:10.3727/154427205774791528

Sheng, L. (2011). Regional Competition and Sustainable Development: A Game Theory Model for Tourism Destinations. *European Planning Studies*, *19*(4), 669–681. doi:10.1080/09654313.2011.548470

Shigeo, M. (2001). *Introduction to Game Theory*. Nikkei Publishing Inc.

Shih, H. Y. (2006). Network characteristics of drive tourism destinations: An application of network analysis in tourism. *Tourism Management*, *27*(5), 1029–1039. doi:10.1016/j.tourman.2005.08.002

Shuzhen, P., Leiting, W., Feifei, Q., & Wei, Z. (2011). Factor Analysis of College students' Travel Motivations in the Tourist City: A case Study for College Students in Taiwan City. In M. Zhou (Ed.), *Advances in Education and Management: International Symposium* (Vol. 211). Springer Science & Business Media.

Sigala, M., & Leslie, D. (2006). *International Cultural Tourism. Management, Implication and Cases*. Oxford: Elsevier.

Simien Thrills Tour and Travel. (2016). Retrieved from http://simienthrills.com/621-2/

Sin, L. Y. M., Tse, A. C. B., Heung, V. C. S., & Yim, F. H. K. (2005). An analysis of the relationship between market orientation and business performance in the hotel industry. *International Journal of Hospitality Management*, *24*(6), 555–577. doi:10.1016/j.ijhm.2004.11.002

Sirmans, C. F. (1997). Research on discounted cash flow models. *Real Estate Finance*, *13*(4), 93–95.

Skift. (2014, June 17). The most popular online booking sites in travel, 2014 Edition. *Skift*.

Skinner, H. (2008). The emergence and development of place marketings confused identity. *Journal of Marketing Management*, *24*(9-10), 915–928. doi:10.1362/026725708X381966

Skoll, G. R., & Korstanje, M. E. (2014a). Points of Discussion around 9/11, terrorism and tourism revisited. *e-Review of Tourism Research, 11*(1-2), 1-17.

Skoll, G. R., & Korstanje, M. E. (2012). Risks, totems, and fetishes in Marx and Freud. *Sincronía*, *2*, 11–27.

Skoll, G. R., & Korstanje, M. E. (2014b). Terrorism, homeland safety and event management. *International Journal of Hospitality and Event Management*, *1*(1), 95–110. doi:10.1504/IJHEM.2014.062869

Skyscanner.com. (2011). *Facebook factor inspires*. Retrieved April 22, 2014, from http://www.skyscanner.net/news/facebook-factor-inspires-52-book-holiday

Slater, S. F., Reddy, V. K., & Zwirlein, T. J. (1998). Evaluating strategic investments: Complementing discounted cash flow analysis with options analysis. *Industrial Marketing Management*, *27*(5), 447–458. doi:10.1016/S0019-8501(97)00097-7

Small, J., Darcy, S., & Packer, T. (2012, August). The embodied tourist experiences of people with vision impairment: Management implications beyond the visual gaze. *Tourism Management*, *33*(4), 941–950. doi:10.1016/j.tourman.2011.09.015

Smallman, C., & Moore, K. (2010). Process studies of tourists' decision-making. *Annals of Tourism Research*, *37*(2), 397–422. doi:10.1016/j.annals.2009.10.014

Smit, H. T. J. (2003). Infrastructure investment as a real options game: The case of European airport expansion. *Financial Management*, *32*(4), 27–57. doi:10.2307/3666135

Smit, H. T. J., & Trigeorgis, L. (2006, January). Real options and games: Competition, alliances and other applications of valuation and strategy. *Review of Financial Economics*, *15*(2), 95–112. doi:10.1016/j.rfe.2005.12.001

Smith, K. M. (2003). *Issues in Cultural Tourism Studies*. London: Routledge. doi:10.4324/9780203402825

Smith, S. J., Parsa, H. G., Bujisic, M., & van der Rest, J.-P. (2015). Hotel cancelation policies, distributive and procedural fairness, and consumer patronage: A study of the lodging industry. *Journal of Travel & Tourism Marketing*, *32*(7), 886–906. doi:10.1080/10548408.2015.1063864

Smith, S., & Wheeler, J. (2002). *Managing the customer experience: Turning customers into advocates*. Upper Saddle River, NJ: Prentice Hall.

Song, H., Dwyer, L., Li, G., & Cao, Z. (2012). Tourism economics research: A review and assessment. *Annals of Tourism Research*, *39*(3), 1653–1682. doi:10.1016/j.annals.2012.05.023

Song, H., Yang, S., & Huang, G. Q. (2009). Price interactions between theme park and tour operator. *Tourism Economics, 15*(4), 813–824. doi:10.5367/000000009789955099

Sönmez, S. F. (1998). Tourism, terrorism, and political instability. *Annals of Tourism Research, 25*(2), 416–456. doi:10.1016/S0160-7383(97)00093-5

Sönmez, S. F., Apostolopoulos, Y., & Tarlow, P. (1999). Tourism in crisis: Managing the effects of terrorism. *Journal of Travel Research, 38*(1), 13–18. doi:10.1177/004728759903800104

Sorooshian, S., Salimi, M., Salehi, M., Nia, N. B., & Asfaranjan, Y. S. (2013). Customer experience about service quality in online environment: A case of Iran. *Procedia: Social and Behavioral Sciences, 93*, 1681–1695. doi:10.1016/j.sbspro.2013.10.101

Srinivasan, R. (2004). Role of trust in e-business success. *Information Management & Computer Security, 12*(1), 66–72. doi:10.1108/09685220410518838

Stackelberg, H. V. (1934). Marktform und Gleichgewicht. Springer Verlag.

Steiner, C., & Al-Hamarneh, A. (2005). Islamic tourism: Rethinking the strategies of tourism development in the Arab world after September 11, 2001. *Comparative Studies of South Asia, Africa and the Middle East, 24*(1), 173–182.

Stenbacka, R., & Tombak, M. M. (1994, September). Strategic Timing of Adoption of New Technologies under Uncertainty. *International Journal of Industrial Organization, 12*(3), 387–411. doi:10.1016/0167-7187(94)90040-X

Stienmetz, J. L., & Fesenmaier, D. R. (2013a). Assessing the economic value of tourist attraction using a network approach. In *Proceedings of the third international conference on the measurement and economic analysis of regional tourism.*

Stienmetz, J. L., & Fesenmaier, D. R. (2013b). Traveling the network: A proposal for destination performance metrics. *International Journal of Tourism Sciences, 13*(2), 57–75. doi:10.1080/15980634.2013.11434673

Stienmetz, J. L., Maxcy, J. G., & Fesenmaier, D. R. (2015). Evaluating destination advertising. *Journal of Travel Research, 54*(1), 22–35. doi:10.1177/0047287513514295

Stock, R. M., & Hoyer, W. D. (2005). An attitude-behavior model of salespeoples' customer orientation. *Journal of the Academy of Marketing Science, 33*(4), 536–552. doi:10.1177/0092070305276368

Strobl, A., & Peters, M. (2013). Entrepreneurial reputation in destination networks. *Annals of Tourism Research, 40*, 59–82. doi:10.1016/j.annals.2012.08.005

Strombeck, S. D., & Wakefield, K. L. (2008). Situational influences on service quality evaluations. *Journal of Services Marketing, 22*(5), 409–419. doi:10.1108/08876040810889175

Stulz, R., & Johnson, H. (1985). An analysis of secured debt. *Journal of Financial Economics, 14*(4), 501–521. doi:10.1016/0304-405X(85)90024-8

Subramanian, J., Stidham, S. Jr, & Lautenbacher, C. J. (1999). Airline yield management with overbooking, cancellations, and no-shows. *Transportation Science, 33*(2), 147–167. doi:10.1287/trsc.33.2.147

Suh, J. W., & Kim, Y. (2013). *Accelerating MATLAB with GPU Computing: A Primer with Examples.* Morgan Kaufmann.

Sunny 13 Tour and Travel. (2016). Accessed on Feb. 2016, http://sunny13.com/images/Attractions/Natural/Sof-Omara.jpg

Sun, S., Nang Fong, L. H., Law, R., & Luk, C. (2015). An investigation of Gen-Ys online hotel information search: The case of Hong Kong. *Asia Pacific Journal of Tourism Research, 21*(4), 443–456. doi:10.1080/10941665.2015.1062405

Sunstein, C. R. (2005). Laws of fear: Beyond the precautionary principle (Vol. 6). Cambridge, MA: Cambridge University Press.

Sunstein, C. (2006). *Risk and Reason. Safety, law and environment.* Buenos Aires: Katz.

Sunstein, C. R. (1990). *After the Rights Revolution. Reconceiving the regulatory state.* Cambridge, MA: Harvard University Press.

Suzuki, K. (Ed.). (1999). *MBA Game Theory.* Diamond Ltd. (in Japanese)

Svaton, M., Slaninova, & Bures. (2014). Improving strategy in robot soccer game by sequence extraction. In *18th International Conference in Knowledge Based and Intelligent Information and Engineering Systems, KES 2014.*

Svaton, V., Martinovic, J., Slaninova, K., & Snasel, V. (2014). Improving rule selection from robot soccer strategy with substrategies. In *Computer Information Systems and Industrial Management - 13th IFIP TC8 International Conference.* doi:10.1007/978-3-662-45237-0_9

Swarbrooke, J., & Horner, S. (2007). *Consumer behaviour in tourism.* Elsevier Ltd.

Swart, W. W., Var, T., & Gearing, C. E. (1978). Operations research applications to tourism. *Annals of Tourism Research, 5*(4), 414–428. doi:10.1016/0160-7383(78)90321-3

Sweeney, R. (2006). *Millennial behaviors and demographics.* Retrieved October 29, 2015, from http://library1.njit.edu/staff-folders/sweeney/

Taft, M. (2014, October 13). The definition of online booking tool in business travel. *ClickTravel.*

Tajeddini, K. (2010). Effect of customer orientation and entrepreneurial orientation on innovativeness: Evidence from the hotel industry in Switzerland. *Tourism Management, 31*(2), 221–231. doi:10.1016/j.tourman.2009.02.013

Tajeddini, K., & Trueman, M. (2012). Managing Swiss hospitality: How cultural antecedents of innovation and customer-oriented value systems can influence performance in the hotel industry. *International Journal of Hospitality Management, 31*(4), 1119–1129. doi:10.1016/j.ijhm.2012.01.009

Talluri, K. T., & Van Ryzin, G. J. (2004). *The theory and practice of revenue management.* New York, NY: Springer. doi:10.1007/b139000

Tana Travel and Tour. (2016a). Accessed Feb. 2016, http://www.tanatravelandtour.com/image/OmoNationalPark.jpg

Tana Travel and Tour. (2016b). Accessed Feb. 2016, http://www.tanatravelandtour.com/image/ MAGONATIONAL-PARK.jpg

Tang, C. H., & Jang, S. (2007). Revisit to the determinants of capital structure: A comparison between lodging firms and software firms. *International Journal of Hospitality Management, 26*(1), 175–187. doi:10.1016/j.ijhm.2005.08.002

Tang, L., & Jang, S. (2014). Information value and destination image: Investigating the moderating role of processing fluency. *Journal of Hospitality Marketing & Management, 23*(7), 790–814. doi:10.1080/19368623.2014.883585

Tang, T. W. (2014). Becoming an ambidextrous hotel: The role of customer orientation. *International Journal of Hospitality Management, 39,* 1–10. doi:10.1016/j.ijhm.2014.01.008

Tantawy, A., Kim, W. G., & Pyo, S. S. (2004). Evaluation of hotels to accommodate disabled visitors. *Journal of Quality Assurance in Hospitality & Tourism, 5*(1), 91–101. doi:10.1300/J162v05n01_07

Tarlow, P. & Korstanje M. (2013). Disasters, tourism, and mobility: The case of Japan Earthquake. *Pasos, revista de Turismo y Patrimonio Cultural, 11*(3), 17-32.

Tarlow, P. (2011b). Tourism Disaster Management in a age of Terrorism. Int. Journal of Tourism Anthropology, 1, 254-272. doi:10.1504/IJTA.2011.043709

Tarlow, P. (2012). Rosh Ha` Shanah Day. *International Journal of Safety and Security in Tourism and Hosp*, *3*, 1–19.

Tarlow, P. (2012). Rosh Ha` Shanah Day. *International Journal of Safety and Security in Tourism and Hosp.*, *3*, 1–19.

Tarlow, P. (2014). *Tourism Security*. Oxford, UK: Elsevier.

Tarlow, P. E. (2006). Terrorism and tourism. In J. Wilks, D. Pendergast, & P. Leggat (Eds.), *Tourism in turbulent times: Towards safe experiences for visitors* (pp. 79–92). New York, NY: Routledge. doi:10.1016/B978-0-08-044666-0.50014-2

Tarlow, P. E. (2011a). Tourism risk management in an age of terrorism. *Economía Autónoma*, *4*, 18–30.

Tayler, P., & McRae-Williams, P. (2005). *Internal versus external economies of scale*. Working Paper, Centre for Regional Innovation & Competitiveness.

Telfar, G. (1996). *Acceleration Techniques for Simulated Annealing* (MSc Thesis). Victoria University of Wellington, New Zealand.

Telfer, D. J. (2001). Strategic alliances along the Niagara Wine Route. *Tourism Management*, *22*(1), 21–30. doi:10.1016/S0261-5177(00)00033-9

Thakor, M. V., & Joshi, A. W. (2005). Motivating salesperson customer orientation: Insights from the job characteristics model. *Journal of Business Research*, *58*(5), 584–592. doi:10.1016/j.jbusres.2003.10.002

Thakran, K., & Verma, R. (2013). The emergence of hybrid online distribution channels in travel, tourism and hospitality. *Cornell Hospitality Quarterly*, *54*(3), 240–247. doi:10.1177/1938965513492107

Themistocleous, M., Irani, Z., O'Keefe, R., & Paul, R. (2001). ERP Problems and Application Integration Issues.*Proceedings of the 34th Hawaii International Conference on System Sciences: An Empirical Survey (HICSS-34).*

Thorelli, H. B. (1986). Networks: Between markets and hierarchies. *Strategic Management Journal*, *7*(1), 37–51. doi:10.1002/smj.4250070105

Tideswell, C., & Faulkner, W. (1999). Multidestination travel patterns of international visitors to Queensland. *Journal of Travel Research*, *37*(4), 364–374. doi:10.1177/004728759903700406

Tierney, K. (1994). *Sociology´s Unique Contributions to the Study of Risk*. Disaster Research Center, Preliminary Paper, 204.

Tilahun, S. L. (2012). Fuzzy graph representation of bus timetabling problem and its solution method using Prey-Predator algorithm. *International Workshop on Optimal Network Topologies (IWONT) 2012*. Institut Teknologi Bandung.

Tilahun, S. L. (2013). *Prey predator algorithm: A new metaheuristic optimization approach* (PhD thesis). School of Mathematical Sciences, Universiti Sains Malaysia, Malaysia.

Tilahun, S. L., & Melese, S. (2015). *Prey predator with adaptive step length*. The 8th International Congress of Industrial and Applied Mathematics, Beijing, China.

Tilahun, S. L., & Ngnotchouye, J. M. T. (2016b). *Two stage travel salesman model of world tourism*. The 14th Pacific Rim International Conference on Artificial Intelligence (PRICAI 2016), Phuket, Thailand.

Tilahun, S. L., & Ngnotchouye, J. M. T. (2016c). *Examination timetabling using prey predator algorithm*. The 14th Pacific Rim International Conference on Artificial Intelligence (PRICAI 2016), Phuket, Thailand.

Tilahun, S. L., Ong, H. C., & Ngnotchouye, J. M. T. (2016). Extended Prey Predator Algorithm with a Group Hunting Scenario. *Advances in Operations Research, Volume 2016*, Article ID 7325263, 14 pages

Tilahun, S. L., & Ngnotchouye, J. M. T. (2016a). Prey predator algorithm with adaptive step length. *International Journal of Bio-inspired Computation, 8*(4), 195–204.

Tilahun, S. L., & Ong, H. C. (2013). Comparison between Genetic Algorithm and Prey-Predator Algorithm. *Malaysian Journal of Fundamental and Applied Sciences, 9*(4), 167–170.

Tilahun, S. L., & Ong, H. C. (2015). Prey predator algorithm: A new metaheuristic optimization algorithm. *International Journal of Information Technology & Decision Making, 14*(6), 1331–1352. doi:10.1142/S021962201450031X

Timothy, D. J., & Nyaupane, G. P. (2009). *Cultural heritage and tourism: a regional perspective*. New York: Routledge.

Tinsley, R., & Lynch, P. (2001). Small tourism business networks and destination development. *International Journal of Hospitality Management, 20*(4), 367–378. doi:10.1016/S0278-4319(01)00024-X

Tisdell, C. A. (2013). *Handbook of Tourism Economics: Analysis, New Applications, and Case Studies*. Singapore: World Scientific Publishing Company. doi:10.1142/7956

Titman, S., & Wessels, R. (1988). The Determinants of Capital Structure Choice. *The Journal of Finance, 43*(1), 1–19. doi:10.1111/j.1540-6261.1988.tb02585.x

Torres, E. N., Fu, X., & Lehto, X. (2014). Examining key drivers of customer delight in a hotel experience: A cross-cultural perspective. *International Journal of Hospitality Management, 36*, 255–262. doi:10.1016/j.ijhm.2013.09.007

Tortosa, V., Moliner, M. A., & Sanchez, J. (2009). Internal market orientation and its influence on organisational performance. *European Journal of Marketing, 43*(11/12), 1435–1456. doi:10.1108/03090560910989975

Tourassi, G. D., Frederick, E. D., Markey, M. K., & Floyd, C. E. (2001). Application of the mutual information criterion for feature selection in computer-aided diagnosis. *Medical Physics, 28*(12), 2394–2402. doi:10.1118/1.1418724 PMID:11797941

Tourism Business Portal. (2014). EU Virtual tourism observatory. *European Tourism Forum*. Retrieved from https://ec.europa.eu/growth/tools-databases/vto/

Tourism Exchange. (2011). *Friends' holiday pictures on Facebook inspires*. Retrieved August 31, 2014, from http://www.tourismexchange.com/exchange/en/newsroom/home/getArticle.jsp?articleID=13738

Tremblay, P. (1998). The economic organization of tourism. *Annals of Tourism Research, 24*(4), 837–859. doi:10.1016/S0160-7383(98)00028-0

Trigeorgis, L. (1991). Anticipated Competitive Entry and Early Preemptive Investment in Deferrable Projects. *Journal of Economics and Business, 43*(2), 143–156. doi:10.1016/0148-6195(91)90014-N

Trigeorgis, L. (1996). *Real options: Managerial flexibility and strategy in resource allocation*. Cambridge, MA: MIT Press.

Trigeorgis, L. (1996). *Real Options: Managerial Flexibility and Strategy n Resource Allocation*. The MIT Press.

TripAdvisor. (2014). *24 insights to shape your TripAdvisor strategy*. Retrieved February 15, 2016, from https://www.tripadvisor.com/TripAdvisorInsights/n2120/24-insights-shape-your-tripadvisor-strategy

Tropic Air Kenya. (2013). *Ethiopia's richest wildlife area – Gambela*. Accessed on Feb. 2016, http://www.tropicairkenya.com/wp-content/uploads/Gambela-giraffe.jpg

Tsai, H., & Gu, Z. (2007). Optimizing room capacity and profitability for Hong Kong hotels. *Journal of Travel & Tourism Marketing, 29*(1), 57–68. doi:10.1080/10548408.2012.638563

Tsai, H., & Gu, Z. (2012). Institutional ownership and firm performance: Empirical evidence from US-based publicly traded restaurant firms. *Journal of Hospitality & Tourism Research (Washington, D.C.), 31*(1), 19–38. doi:10.1177/1096348006296056

Tsai, W. H., & Hung, S. J. (2009, April). Dynamic pricing and revenue management process in Internet retailing under uncertainty: An integrated real options approach. *Omega, 37*(2), 471–481. doi:10.1016/j.omega.2007.07.001

Tsaur, S.-H., & Tzeng, G.-H. (1996). Multiattribute Decision Making Analysis for Customer Preference of Tourist Hotels. *Journal of Travel & Tourism Marketing, 4*(4), 55–69. doi:10.1300/J073v04n04_04

Tseng, M. L., Chiu, A. S., & Vo, M. P. N. (2011). Evaluating the tourists demand to develop Vietnamese tourism performance. *Procedia: Social and Behavioral Sciences, 25*, 311–326. doi:10.1016/j.sbspro.2011.10.550

Tucker, A. W., & Luce, R. D. (Eds.). (1959). *Contributions to the theory of games* (Vol. 4). Princeton: Princeton University Press.

UN ENABLE. (2013). *Promoting Accessible Tourism for All. Development and human rights for all.* New York: UN ENABLE.

UN WHO. (2007). *UN WHO Global Age-friendly Cities: A Guide.* Retrieved February 29, 2016 from: http://www.who.int

UN WHO. (2011). *UN WHO World report on disability.* New York: Routledge.

UN WTO. (2013). *UN WTO Recommendations on Accessible Tourism for All.* New York: UN WTO.

UN WTO. (2014). *San Marino Declaration on Accessible Tourism.* San Marino: UN WTO.

UN WTO. (2015). *Manual on Accessible Tourism for All: Public – Private Partnership and Good Practice.* New York: UN WTO.

UN. (2006). *The Convention on the Rights of Persons with Disabilities (CRPD).* New York: UN.

UNESCO ICOMOS. (2011). *The Valetta Principles for the Safeguarding and Management of Historic Cities, Towns and Urban Areas.* Paris: UNESCO.

UNESCO. (2006). Accessed on Feb. 2016, http://whc.unesco.org/ en/list/19/gallery/

United Nations. (2015). *Information economy report 2015. Unlocking the potential of e-commerce for developing countries.* United Nations Publication, UNCTAD/IER/2015.

UNWTO. (2015). *UNWTO Tourism Highlights 2015 Edition.* Author.

Urbaczewski, A., Jessup, L. M., & Wheeler, B. (2002). Electronic commerce research: A taxonomy and synthesis. *Journal of Organizational Computing and Electronic Commerce, 12*(4), 263–305. doi:10.1207/s15327744joce1204_1

Urban, T. L. (2008). Coordinating pricing and inventory decisions under reference price effects. *International Journal of Manufacturing Technology and Management, 13*(1), 78–94. doi:10.1504/IJMTM.2008.015975

Uwawunkonye, E. G., & Anaene, O. I. C. (2013). A comparative study between ridit and modified ridit analysis. *American Journal of Theoretical and Applied Statistics, 2*(6), 248–254. doi:10.11648/j.ajtas.20130206.23

Vael, A. J. (2002). *Leisure and Tourism. Policy and Planning.* London: Routledge; doi:10.1079/9780851995465.0000

Valenzuela, L. M., Mulki, J. P., & Jaramillo, J. F. (2010). Impact of customer orientation, inducements and ethics on loyalty to the firm: Customers' perspective. *Journal of Business Ethics*, *93*(2), 277–291. doi:10.1007/s10551-009-0220-z

Vallespín, M., & Molinillo, S. (2014). El futuro de la intermediación en el sector turístico. *Revista de Análisis Turístico*, *17*, 13–25.

Vanharanta, H., Kantola, J., & Seikola, S. (2015). Customers' conscious experience in a coffee shop. *Procedia Manufacturing*, *3*, 618–625. doi:10.1016/j.promfg.2015.07.283

Van, T., Taillard, E., Melab, N., & Talbi, E.-G. (2012). Parallelization strategies for hybrid metaheuristics using a single GPU and multi-core resources. In *Parallel Problem Solving from Nature-PPSN XII* (pp. 368–377). Springer.

Vargas, A. (2014, April 5). Los metabuscadores aumentan su influencia en 2014. *Hosteltur*.

Varian, H. R. (1992). *Microeconomic Analysis* (3rd ed.). New York, NY: Norton & C.

Vellas, F., & Becherel, L. (1999). *The international marketing of travel and tourism: A strategic approach*. London: Macmillan. doi:10.1007/978-1-349-27486-4

Vera, J., & Trujillo, A. (2013). Service quality dimensions and superior customer perceived value in retail banks: An empirical study on Mexican consumers. *Journal of Retailing and Consumer Services*, *20*(6), 579–586. doi:10.1016/j.jretconser.2013.06.005

Verhoef, P. C., Lemon, K. N., Parasuraman, A., Roggeveen, A., Tsiros, M., & Schlesinger, L. A. (2009). Customer experience creation: Determinants, dynamics and management strategies. *Journal of Retailing*, *85*(1), 31–41. doi:10.1016/j.jretai.2008.11.001

Vernon, J., Essex, S., Pinder, D., & Curry, K. (2005). Collaborative policymaking: Local sustainable projects. *Annals of Tourism Research*, *32*(2), 325–345. doi:10.1016/j.annals.2004.06.005

Vijayasarathy, L. R. (2004). Predicting consumer intentions to use on-line shopping: The case for an augmented technology acceptance model. *Information & Management*, *41*(6), 747–762. doi:10.1016/j.im.2003.08.011

Visinescu, L. L., Sidorova, A., Jones, M. C., & Prybutok, V. R. (2015). The influence of website dimensionality on customer experiences, perceptions and behavioral intentions: An exploration of 2D vs. 3D web design. *Information & Management*, *52*(1), 1–17. doi:10.1016/j.im.2014.10.005

Volo, S. (2008). Bloggers reported tourist experiences: Their utility as a tourism data source and their effect on prospective tourists. *Journal of Vacation Marketing*, *16*(4), 297–311. doi:10.1177/1356766710380884

von Neumann, J. (1928, June). Sur la theorie des jeux. *Comptes Rendus delAcademie des Sciences, 186*(25), 1689-1691.

von Neumann, J. (1928, December). Zur Theorie der Gesellschaftsspiele. *Mathematische Annalen*, *100*(1), 295–320. doi:10.1007/BF01448847

von Neumann, J., & Morgenstern, O. (1947). *Theory of Games and Economic Behavior* (2nd ed.). Princeton, NJ: Princeton University Press.

von Wallpach, S., & Kreuzer, M. (2013). Multi-sensory sculpting (MSS): Eliciting embodied brand knowledge via multisensory metaphors. *Journal of Business Research*, *66*(9), 1325–1331. doi:10.1016/j.jbusres.2012.02.032

Vora, M., & Mirnalinee, T. T. (2015). From optimization to clustering: A swarm intelligence approach. In P. Vasant (Ed.), *Handbook of research on artificial intelligence techniques and algorithms* (pp. 594–619). Hershey, PA: IGI Global. doi:10.4018/978-1-4666-7258-1.ch019

Vukonic, B. (2002). Religion, tourism and economics: A convenient symbiosis. *Tourism Recreation Research*, *27*(2), 59–64. doi:10.1080/02508281.2002.11081221

Vukonic, B. (2010). Do we always understand each other? In *Tourism and Muslim World* (pp. 31–45). West Yorkshire: Emerald Group Publishing. doi:10.1108/S2042-1443(2010)0000002006

Waddock, S. A. (1989). Understanding social partnerships: An evolutionary model of partnership organizations. *Administration & Society*, *21*(1), 78–100. doi:10.1177/009539978902100105

Wald, J. K., & Long, M. S. (2007). The effect of state laws on capital structure. *Journal of Financial Economics*, *83*(2), 297–319. doi:10.1016/j.jfineco.2005.10.008

Walker, P. (2005, October). A Chronology of Game Theory. *History of Game Theory.*

Wall, G. (1997). Tourism attractions: Points, lines and areas. *Annals of Tourism Research*, *24*(1), 240–243. doi:10.1016/S0160-7383(96)00039-4

Walls, A. R. (2013). A cross-sectional examination of hotel consumer experience and relative effects on consumer values. *International Journal of Hospitality Management*, *32*, 179–192. doi:10.1016/j.ijhm.2012.04.009

Wan, C.-S. (2002). The web sites of international tourist hotels and tour wholesalers in Taiwan. *Tourism Management*, *23*(2), 155–160. doi:10.1016/S0261-5177(01)00048-6

Wang, B., Wang, S. M., & Watada, J. (2009, July). Improved real option analysis based on fuzzy random variables. *Machine Learning and Cybernetics, 2009 International Conference on*, (pp. 694- 699). doi:10.1109/ICMLC.2009.5212448

Wang, B., Wang, S. M., & Watada, J. (2010, June). Real Options Analysis Basedon Fuzzy Random Variables. *International Journal of Innovative Computing, Information, & Control*, *6*(4), 1689–1698.

Wang, C. C., & Chen, C. C. (2010). Electronic commerce research in latest decade: A literature review. *International Journal of Electronic Commerce Studies*, *1*(1), 1–14.

Wang, C. H., Chen, K. Y., & Chen, S. C. (2012). Total quality management, market orientation and hotel performance: The moderating effects of external environmental factors. *International Journal of Hospitality Management*, *31*(1), 119–129. doi:10.1016/j.ijhm.2011.03.013

Wang, Q., & Parlar, M. (1994). A three-person game theory model arising in stochastic inventory control theory. *European Journal of Operational Research*, *76*(1), 83–97. doi:10.1016/0377-2217(94)90008-6

Wang, Y. (2011). Destination marketing and management: Scope, definition and structures. In Y. Wang & A. Pizam (Eds.), *Destination marketing and management: Theories and applications* (pp. 1–20). Wallingford, UK: CAB International. doi:10.1079/9781845937621.0001

Wang, Y., & Fesenmaier, D. R. (2007). Collaborative destination marketing: A case study of Elkhart county, Indiana. *Tourism Management*, *28*(3), 863–875. doi:10.1016/j.tourman.2006.02.007

Wang, Y., Guo, L., & Huang, L. (2013). A sports tourism resources comprehensive evaluation model based on AHP-FUZZY. *Journal of Theoretical and Applied Information Technology*, *48*(3), 1473–1479.

Wang, Y., & Krakover, S. (2008). Destination marketing: Competition, cooperation or coopetition? *International Journal of Contemporary Hospitality Management*, *20*(2), 126–141. doi:10.1108/09596110810852122

Ward, J., Griffiths, P., & Whitmore, P. (1990). *Strategic planning for Information Systems.* John Wiley & Sons.

Ward, J., Peppard, J., & Daniel, E. (2008). Building Better Business Cases for IT Investments. *MIS Quarterly Executive*, *7*(1), 1–15.

Wareham, J., Zheng, J. G., & Straub, D. (2005). Critical themes in electronic commerce research: A meta-analysis. *Journal of Information Technology*, *20*(1), 1–19. doi:10.1057/palgrave.jit.2000034

Wasserman, S., & Faust, K. (1994). *Social network analysis: Methods and applications*. Cambridge, UK: Cambridge University Press. doi:10.1017/CBO9780511815478

Watada, J. (2008, May). *A Real Options Approach to Project Evaluation for MIS*. The 19th International Conference on Information Management.

Watada, J., Waripan, T., & Wu, B. (2014). Optimal decision methods in two-echelon logistic models. *Journal of Management History*, *52*(7), 1273 – 1287.

Watada, J., & Song, W. (2010, December). Strategy Building for Foreign Exchange Exposure. *ICIC Express Letters*, *1*(1), 1–17.

We Are Social. (2016). *Digital in 2016, We Are Social's compendium of global digital social and mobile data trends and statistics*. Retrieved January 30, 2016, from http://www.slideshare.net/wearesocialsg/digital-in-2016?from_action=save

Weber, K., & Roehl, W. S. (1999). Profiling people searching for and purchasing travel products on the world wide web. *Journal of Travel Research*, *37*(3), 291–298. doi:10.1177/004728759903700311

Weeden, C., & Boluk, K. (2014). *Managing Ethical Consumption in Tourism*. New York, NY: Routledge.

Weiss, A. M., Lurie, N. H., & MacInnis, D. J. (2008). Listening to strangers: Whose responses are valuable, how valuable are they, and why? *JMR, Journal of Marketing Research*, *45*(4), 425–436. doi:10.1509/jmkr.45.4.425

Weissenberg, A. F., Cohen, J., & Culliney, P. (2003). Mezzanine financing for hospitality property owners. *Journal of Retail & Leisure Property*, *3*(2), 115–121. doi:10.1057/palgrave.rlp.5090168

Wertener, H., & Klein, S. (1999). *Information technology and tourism: A challenging relationship*. Wien: Springer-Verlag. doi:10.1007/978-3-7091-6363-4

Werthner, H., & Ricci, F. (2004). E-commerce and tourism. *Communications of the ACM*, *47*(12), 101–105. doi:10.1145/1035134.1035141

White, T. (2012). Hadoop: The Definitive Guide (3rd ed.). Yahoo Press.

Wie, B. W. (2005). A dynamic game model of strategic capacity investment in the cruise line industry. *Tourism Management*, *26*(2), 203–217. doi:10.1016/j.tourman.2003.09.020

Wieseke, J., Ullrich, J., Christ, O., & van Dick, R. (2007). Organizational identification as a determinant of customer orientation in service organizations. *Marketing Letters*, *18*(4), 265–278. doi:10.1007/s11002-007-9021-2

Wikipedia. (2016). Accessed Feb. 2016, https://upload.wikimedia.org/wikipedia/commons/ thumb/0/00/LakeChamo-Crocodile.jpg/1024px-LakeChamoCrocodile.jpg

Wilkinson, P. (2011). *Terrorism versus democracy: The liberal state response*. London: Routledge.

Williams, J. T. (1993). Equilibrium and Options on Real Assets. *Review of Financial Studies*, *6*(4), 825–850. doi:10.1093/rfs/6.4.825

Williams, P. D., & Grizzle, J. E. (1972). Analysis of Contingency Table having Ordered Response Categories. *American Statistical Association*, *67*(337), 55–63. doi:10.1080/01621459.1972.10481205

Wilson, R. H. (1934). A Scientific Routine for Stock Control. *Harvard Business Review*, *13*, 116–128.

Wilt, N. (2013). *CUDA Handbook: A Comprehensive Guide to GPU Programming*. Addison-Wesley Professional.

Winston, W. L. (1991). *Operations Research Applications and Algorithms*. Duxbury Press.

Wöber, K. W. (2003). Domain-specific search engines. In Destination recommendation systems: behavioral foundations and applications. CABI.

Wong, A., Liu, Y., & Tjosvold, D. (2015). Service leadership for adaptive selling and effective customer service teams. *Industrial Marketing Management*, *46*, 122–131. doi:10.1016/j.indmarman.2015.01.012

Wood, D. J., & Gray, B. (1991). Toward a comprehensive theory of collaboration. *The Journal of Applied Behavioral Science*, *27*(2), 139–162. doi:10.1177/0021886391272001

Wood, L. (2007). Functional and symbolic attributes of product selection. *British Food Journal*, *109*(2), 108–118. doi:10.1108/00070700710725482

Woodside, A. G., & Dubelaar, C. (2002). A general theory of tourism consumption systems: A conceptual framework and an experimental exploration. *Journal of Travel Research*, *41*(2), 120–132. doi:10.1177/004728702237412

Workman, J. P. Jr, Homburg, C., & Jensen, O. (2003). Intraorganizational determinants of key account management effectiveness. *Journal of the Academy of Marketing Science*, *31*(1), 3–21. doi:10.1177/0092070302238599

World Tourism Organization (UNWTO). (2014). *Tourism highlights 2014 Edition*. UNWTO.

World Tourism Organization (UNWTO). (2015). *Tourism highlights 2015 Edition*. UNWTO.

World Tourism Organization (UNWTO). (2015). *World Tourism Organization (UNWTO) Annual Report 2014*. Retrieved Feb. 26, 2016 from http://dtxtq4w60xqpw.cloudfront.net/sites/all/files/pdf/unwto_annual _report_2014.pdf

Wrenn, B., Stevens, R. E., & Loudon, D. L. (2007). *Marketing research. Text and cases*. New York, NY: Haworth Press.

Wrenn, B., Stevens, R. E., & Loudon, D. L. (2007). *Marketing research: Text and cases*. Binghamton: Haworth Press.

Wu, C. C., & Lo, Y. H. (2012). Customer reactions to encountering consecutive service failures. *Journal of Consumer Behaviour*, *11*(3), 217–224. doi:10.1002/cb.1376

Wu, C. H. (2007). On the application of grey relational analysis and RIDIT analysis to Likert scale surveys.*International Mathematical Forum*, *2*(14), 675-687. doi:10.12988/imf.2007.07059

Wu, W. Y., Tsai, C. C., & Fu, C. S. (2013). The relationships among internal marketing, job satisfaction, relationship marketing, customer orientation, and organizational performance: An empirical study of TFT-LCD companies in Taiwan. *Human Factors and Ergonomics in Manufacturing & Service Industries*, *23*(5), 436–449. doi:10.1002/hfm.20329

Xiang, Z., Wang, D., & Fesenmaier, D. R. (2014). Adaptive strategies to technological change: understanding travellers using the internet for trip planning. In Information and communication technologies in toursm (pp. 411–424).

Xiang, Z. (2011). *Dynamic social media in online travel information search: A preliminary analysis* (pp. 343–353). Vienna: Springer.

Xiang, Z., & Gretzel, U. (2010). Role of social media in online travel information search. *Tourism Management*, *31*(2), 179–188. doi:10.1016/j.tourman.2009.02.016

Xiang, Z., Magnini, V. P., & Fesenmaier, D. R. (2015). Information technology and consumer behavior in travel and tourism: Insights from travel planning using the Internet. *Journal of Retailing and Consumer Services*, *22*, 244–249. doi:10.1016/j.jretconser.2014.08.005

Xiang, Z., Wöber, K., & Fesenmaier, D. R. (2008). Representation of the online tourism domaine in search engines. *Journal of Travel Research*, *47*(2), 137–150. doi:10.1177/0047287508321193

Xie, S., & Zhang, Z. (2012). Models and optimization techniques of machining parameters in turning operations. In J. Davim (Ed.), *Computational methods for optimizing manufacturing technology: Models and techniques* (pp. 162–192). Hershey, PA: IGI Global. doi:10.4018/978-1-4666-0128-4.ch007

Yaakob, S.B., & Watada, J. (2010, July). *Power System Equipments Investment Decision-Making under Uncertainty: A Real Options Approach.* KES-IDT2010, Baltimore, MD.

Yang, S., & Lynn, M. (2009). Wine List Characteristics Associated with Greater Wine Sales. *Cornell Hospitality Reports*, *9*(11). Retrieved from http://www.hotelschool.cornell.edu/research/chr/pubs/reports/abstract-15089.html

Yang. (2010). A New Metaheuristic Bat-Inspired Algorithm. In *Nature Inspired Cooperative Strategies for Optimization* (NISCO 2010). Springer.

Yang, C., & Chang, T. (2011). Binomial real option pricing for restaurant menu analysis. *Cornell Hospitality Quarterly*, *52*(3), 273–282. doi:10.1177/1938965511410054

Yang, S. L., & Zhou, Y.-W. (2006). Two-echelon supply chain models: Considering duopolistic retailers different competitive behaviors. *International Journal of Production Economics*, *103*(1), 104–116. doi:10.1016/j.ijpe.2005.06.001

Yang, X. S. (2009). Firefly algorithms for multimodal optimization. Stochastic Algorithms: Foundations and Applications, SAGA 2009. *Lecture Notes in Computer Science*, *5792*, 169178.

Yang, X.-S., & Deb, S. (2009). Cuckoo search via Lvy flights.*World Congress on Nature and Biologically Inspired Computing (NaBIC 2009)*. IEEE Publications.

Yangyong, Z., & Yun, X. (2011). *Dataology and data science: Up to now*. Retrieved from http://iwdds.fudan.edu.cn/second/abstract/Dataology%20and%20DataScienceUp%20to%20Now_ZYYFudan.pdf

Yan, T., Ulanova, L., Ouyang, Y., & Xu, F. (2014). Data mining in time series: Current study and future trend. *Journal of Computer Science*, *10*(12), 2358–2359. doi:10.3844/jcssp.2014.2358.2359

Ye, Q., Law, R., Gu, B., & Chen, W. (2011). The influence of user-generated content on traveler behavior: An empirical investigation on the effects of e-word-of-mouth to hotel online bookings. *Computers in Human Behavior*, *27*(2), 634–639. doi:10.1016/j.chb.2010.04.014

Yılı İşletme Belgeli Konaklama Tesislerine Giriş ve Geceleme Sayıları. (2013). Retrieved December 15, 2015, from http://www.tuik.gov.tr/PreTablo.do?alt_id=1072

Yoo, K. H., & Gretzel, U. (2010). Antecedents and impacts of trust in travel-related consumer-generated media. *Information Technology & Tourism*, *12*(2), 139–152. doi:10.3727/109830510X12887971002701

Yoo, K. H., & Gretzel, U. (2011). Influence of personality on travel-related consumer-generated media creation. *Computers in Human Behavior*, *27*(2), 609–621. doi:10.1016/j.chb.2010.05.002

Yoo, K. H., Lee, Y., Gretzel, U., & Fesenmaier, D. R. (2009). Trust in travel-related consumer-generated media. In W. Höpken, U. Gretzel, & R. Law (Eds.), *Information and Communication Technologies in Tourism 2009* (pp. 49–59). Vienna, Austria: Springer. doi:10.1007/978-3-211-93971-0_5

Yoon, M. G., Lee, H. Y., & Song, Y. S. (2012). Linear approximation approach for a stochastic seat allocation problem with cancellation & refund policy in airlines. *Journal of Air Transport Management*, *23*, 41–46. doi:10.1016/j.jairtraman.2012.01.013

Yoon, M. H., Seo, J. H., & Yoon, T. S. (2004). Effects of contact employee supports on critical employee responses and customer service evaluation. *Journal of Services Marketing*, *18*(5), 395–412. doi:10.1108/08876040410548311

Yuan, M. (2005). After September 11: determining its Impacts on Rural Canadians travel to U.S. *E-Review of Tourism Research*, *3*(5), 103-108.

Yuan, M. (2005). After September 11: determining its Impacts on Rural Canadians travel to U.S. *E-Review of Tourism Research*,*3*, 103-108.

Yuan, Y. H. E., & Wu, C. K. (2008). Relationships among experiential marketing, experiential value, and customer satisfaction. *Journal of Hospitality & Tourism Research (Washington, D.C.)*, *32*(3), 387–410. doi:10.1177/1096348008317392

Yuksel, A., & Yuksel, F. (2002). Measurement of tourist satisfaction with restaurant services: A segment-based approach. *Journal of Vacation Marketing.*, *9*(1), 52–68. doi:10.1177/135676670200900104

Zach, F., & Gretzel, U. (2012). Tourist-activated networks: Implications for dynamic bundling and en route recommendations. *Journal of Information Technology and Tourism*, *13*(3), 239–257.

Zaharias, P., & Poylymenakou, A. (2009). Developing a usability evaluation method for e-learning applications: Beyond functional usability. *International Journal of Human-Computer Interaction*, *25*(1), 75–98. doi:10.1080/10447310802546716

Zehre, A., & Moschl, P. (2008). New distribution channels and business strategies for location-based travel agencies. In P. O'Connor, W. Höpken, & U. Gretzel (Eds.), *Proceedings of the International Conference on Information and Communication Technologies in Tourism* (pp. 359-382). Innsbruck, Austria: Springer.

Zelinka, I. (2004). SOMA - Self Organizing Migrating Algorithm. In B. B. Onwubolu (Ed.), *New Optimization Techniques in Engineering* (pp. 167–218). New York: Springer-Verlag. doi:10.1007/978-3-540-39930-8_7

Zelinka, I., Celikovský, S., Richter, H., & Chen, G. (2010). *Evolutionary Algorithms and Chaotic Systems.* Springer. doi:10.1007/978-3-642-10707-8

Zelinka, I., Davendra, D., Senkerik, R., Jasek, R., & Oplatkova, Z. (2011). *Analytical Programming - a Novel Approach for Evolutionary Synthesis of Symbolic Structures.* InTech. doi:10.5772/16166

Zelinka, I., Oplatkova, Z., & Nolle, L. (2005). Analytic programming - symbolic regression by means of arbitrary evolutionary algorithms. International Journal of Simulation Systems. *Science and Technology*, *6*(9), 44–56.

Zhang, H., Gu, C., Gu, L., & Zhang, Y. (2011). The evaluation of tourism destination competitiveness by TOPSIS & information entropy - A case in the Yangtze River Delta of China. *Tourism Management*, *32*(2), 443–451. doi:10.1016/j.tourman.2010.02.007

Zhang, J. (2010). Employee orientation and performance: An exploration of the mediating role of customer orientation. *Journal of Business Ethics*, *91*(S1s1), 111–121. doi:10.1007/s10551-010-0570-6

Zhang, J., Cai, L. A., & Kavanaugh, R. R. (2008). Dimensions in building brand experience for economy hotels: A case of emerging market. *Journal of China Tourism Research*, *4*(1), 61–77. doi:10.1080/19388160802099782

Zhang, X., Song, H., & Huang, G. Q. (2009). Tourism supply chain management: A new research agenda. *Tourism Management*, *30*(3), 345–358. doi:10.1016/j.tourman.2008.12.010

Zheng, T., & Gu, Z. (2011). Overcapacity in Shanghais high-end hotel sector: Analysis based on an inventory model. *Journal of Convention & Event Tourism, 12*(4), 253–270. doi:10.1080/15470148.2011.621585

Zhou, K. Z., Brown, J. R., & Dev, C. S. (2009). Market orientation, competitive advantage, and performance: A demand-based perspective. *Journal of Business Research, 62*(11), 1063–1070. doi:10.1016/j.jbusres.2008.10.001

Zhou, L., Dai, L., & Zhang, D. (2007). Online shopping acceptance model-A critical survey of consumer factors in online shopping. *Journal of Electronic Commerce Research, 8*(1), 41–62.

Zhu, M. Y., & Watada, J. (2008). A Real Option Approach to Pharmaceutical R&D Project Valuation. *Proceedings of, Int. Symposium on Management Engineering 2008.*

Zhu, W., Zeng, N., & Wang, N. (2010). Sensitivity, specificity, accuracy, associated confidence interval and ROC analysis with practical SAS implementations. *NESUG Proceedings: Health Care and Life Sciences.*

Ziakas, V., & Costa, C. A. (2010). Explicating inter-organizational linkages of a host communitys events network. *International Journal of Event and Festival Management, 1*(2), 132–147. doi:10.1108/17852951011056919

Zingales, L., & Rajan, R. (1995). What Do We Know about Capital Structure? Some Evidence from International Data. *The Journal of Finance, 50*(5), 1421–1460. doi:10.1111/j.1540-6261.1995.tb05184.x

Zinn, J. O. (2010). Biography, Risk and Uncertainty – Is there common ground for biographical Research and Risk Research. *FQS, Forum, Qualitative Social Research, 11*(1), 1-19.

Zornoza, C. (1995). Dirección hotelera y tecnologías de la información: Un Análisis Empírico de la Empresa Hotelera Valenciana. *Estudios turisticos, 125*, 19-52.

About the Contributors

Pandian Vasant is a senior lecturer at Department of Fundamental and Applied Sciences, Faculty of Science and Information technology, Universiti Teknologi PETRONAS in Malaysia. He holds PhD in Computational Intelligence, MSc (Engineering Mathematics) and BSc (Hons) in Mathematics. His research interests include Soft Computing, Hybrid Optimization, Holistic Optimization, innovative computing and Applications. He has co-authored research papers and articles in national journals, international journals, conference proceedings, conference paper presentation, and special issues lead guest editor, lead guest editor for book chapters' project, conference abstracts, edited books, keynote lecture and book chapters (150 publications indexed in SCOPUS). In the year 2009, Dr. Pandian Vasant was awarded top reviewer for the journal Applied Soft Computing (Elsevier) and awarded outstanding reviewer in the year 2015 for ASOCO (Elsevier) journal. He has 25 years of working experience at the various universities from 1989-2016. Currently he is Editor-in-Chief of IJCO, IJSIEC, IEM, IJEOE and Editor of GJTO. H-Index SCOPUS Citations = 33, H-Index Google Scholar = 22.

Kalaivanthan M. or well known as Kay originally from Kedah, Malaysia. He is currently appointed as a Director of Food & Beverage at one of Shangri-La's Hotels & Resorts property. He was an officer with Royal Malaysian Navy prior to his career in the hospitality industry which started in the early 1998. His education background includes Executive Diploma Management in University Malaya and Degree in Business Management at Open University. With over 16 years' experience in the hospitality industry, Kay also has overseas exposure in Maldives for more than five years. He worked for a number of accoladed hotel chains such as Pan Pacific Kuala Lumpur, Hyatt Subang & Kuantan, Centara Grand and Jumeirah Dhevanafushi Maldives. It will come as no surprise that travelling is one of his favorite hobbies, through which he develops the new innovation and trend in Food & Beverage, deep understanding of culture, local tradition and the art of spices. Kay philosophy is "Vision without action is a daydream; action without vision is a nightmare. Patience, persistence and perspiration make an unbeatable combination for success".

* * *

Elisa Alén-González is a Lecturer of Marketing at Faculty of Turism and Business Administration (University of Vigo). PhD in Business Administration. Her main research interests focus on senior tourism, e-tourism and tourism marketing.

Ana de Almeida is a Computer Science Assistant Professor at the Department of Information Science and Technologies, ISCTE-IUL, Lisbon, Portugal. Before, she was an Computer Science Professor at the Department of Mathematics of the University of Coimbra, Portugal. She received her Ph.D. in Applied Mathematics-Computer Science in 2004 from the University of Coimbra, Portugal. Presently, she also is a Senior Researcher at CISUC Center for Informatics and Systems of the University of Coimbra, an IEEE Senior Member and cofounder of the IEEE Portuguese Chapter in Instrumentation and Measurements. She participated and coordinated national and international funded R&D projects and organized and chaired diverse international conferences. Her research interests include algorithms and data structures, combinatorial optimization, graph theory, complexity theory, machine learning, feature extraction and pattern recognition towards knowledge emergence.

Nuno António holds a Computer Science Engineering degree from Instituto Manuel Teixeira Gomes, Portimão, Portugal (Lusófona University) and a Masters in Hotel Administration and Management from the School of Management, Hospitality and Tourism of the University of the Algarve, Portugal. He is currently pursuing a PhD degree in Computer Science at ISCTE Instituto Universitario de Lisboa, Portugal. He is actually Chief Technology Officer at Itbase/WareGuest, a software development company specialized in the hospitality and retail industries. Itbase/WareGuest has offices in Lisbon and Portimão (Portugal) – where Nuno is based, Barcelona (Spain) and Luanda (Angola). He is also an invited lecturer at the School of Management, Hospitality and Tourism of the University of the Algarve, Portugal. Mr. Nuno is certified in Business Intelligence, specialization of Business Analytics by TDWI – The Data-Warehousing Institute. He is certified as ScrumMaster and member of the Scrum Alliance and, he is also certified as Project Management Associate by IPMA - International Project Management Association and a member of APOJEP – Portuguese project managers' association.

Ramya Rajajagadeesan Aroul, Ph.D., is an Assistant Professor of Real Estate Finance at Ecole Hôtelière de Lausanne. She holds a Ph.D. specializing in Finance and Real estate. Ramya's current research focuses on financial distress and valuation. Her research has been published in journals including the Journal of Real Estate Finance and Economics, Journal of Real Estate Research, and Journal of Sustainable Real Estate. She also has several manuscripts at different stages of review at top journals in finance and real estate.

Vincenzo Asero graduated from the University of Catania in Political Sciences. Economist, expert in regional economics, tourism and local development, he has over 20 years of experience in tourism research. He is Assistant professor in Economics at University of Catania where he currently teaches Economics and Regional Economics. He has been Visiting Lecturer at the University of Malta, FEMA, Tourism Studies Unit, where he taught Regional Tourism Development. His research interests are focused on Tourism, Local Development and Regional Economics. He regularly acts as reviewer for various international journals such as Tourism Management, Current Issues in Tourism and Tourism Geographies. He is actively involved as expert and coordinator in many projects on local tourism and regional development supported by Italian government and EU funds.

Marcello Atzeni is a PhD student at the Departement of Economics and Business, University of Cagliari. His research is related to tourism marketing and authenticity.

Jonnatan F. Avilés-González is a Ph.D. candidate from the Program in Engineering Sciences at the Tecnológico de Monterrey, campus Monterrey. He has a master degree on Quality and Productivity at the Tecnológico de Monterrey. His research interest areas are operations research, optimization, statistical analysis, design of experiments, regressions models, nonparametric analysis, lean, tourism models, and six sigma. He has been part of the group of consultant of operational excellence in the center of Quality and Manufacturing ITESM, and in the department of Systems and Industrial Engineering at University of Knoxville, Tennessee, United States.

Sonia Valeria Avilés-Sacoto holds a Ph.D degree in Engineering Sciences with major in Industrial Engineering at Tecnológico de Monterrey – Mexico, a M.Sc. in Quality and Productivity Systems from Tecnológico de Monterrey, and a B.Sc. in Industrial Engineering from the Universidad de Cuenca – Ecuador. Her research interests include Operations Research, Modeling, Production/Operations Management, Tourism Modeling, Inventory Control and methodology development of Data Envelopment Analysis. Her work has appeared in such journals as Journal of CENTRUM Cathedra, European Journal of Operational Research, OMEGA, Journal of the Operational Society and Annals of Operations Research.

Carlos Azevedo holds a MSc in Economics by the Technical University of Lisbon (ISEG), complemented with several training courses on Financial Management by the Catholic University of Lisbon. He holds the degree of Specialist, awarded by the University of the Algarve in 2011. He is an Assistant Professor at the University of the Algarve since 2010. He held several positions as a manager in GALP and REN, the latest being General Manager of the LNG Terminal at Sines, Portugal.

Paula Serdeira Azevedo holds a PhD in Management Sciences by University of the Algarve, concluded in 2012, with an investigation in the area of Information Systems. She is a Professor at the University of the Algarve since 1997. She published several papers in international congresses and seminars. She is co-author of a book. Her work also includes publishing on international scientific journals as well as participation on scientific committees of international conferences.

Joe Barth joined the faculty of the School of Hospitality, Food and Tourism Management at the University of Guelph in 1989 after sixteen years of management experience in the restaurant industry. He teaches beverage management and operations management courses in the Bachelor of Commerce and MBA programs. Joe has been a visiting Professor in Australia, Austria and Argentina; worked with NGO's in Vietnam and Nicaragua and on a freighter in the Great Lakes. He is a lifetime member of Phi Kappa Phi and holds a PhD from Cornell University.

Marek Běhálek is an Assistant Professor in Department of Computer science in VŠB Technical University Ostrava. He is also a researcher in IT4I National Supercomputing Center. His research interests are the evolution of programming languages and tools especially for programming of parallel/distributed systems. Now, he is focused on visual programming of distributed applications, their analyzing and verification.

Subhajit Bhattacharya received his PhD degree in Business Administration from the University of Burdwan. Dr. Bhattacharya's has more than eight years' experience in academics and teaches different courses in marketing area. His areas of experience and expertise include teaching Marketing Management, Consumer Behaviour, Services Marketing and Managing Retail Business. Dr. Bhattacharya has published articles in quite a few national and international journals and has also presented papers in several national and international conferences. Currently he is working in the areas of Brand Chemistry, Consumer Insight and Consumer Psychology.

Gabrielle Bodenmann is an academic assistant at Ecole hoteliere de Lausanne (EHL) in Switzerland.

Serol Bulkan graduated from Management Engineering Department of Istanbul Technical University in 1986. He got his first MSc degree from the same department in 1989, a second MSc degree in Operations Research from Florida Institute of Technology in 1992 and a PhD degree in Industrial Engineering from Cleveland State University in 1999. Dr. Bulkan is currently at Industrial Engineering Department of Marmara University, Turkey. His teaching research areas are Operations Research, Scheduling Theory, Network Theory, Production Planning.

Leopoldo Eduardo Cárdenas-Barrón is currently a Professor at school of engineering and sciences at Tecnológico de Monterrey, campus Monterrey, México. He is also a faculty member in the department of industrial and systems engineering at Tecnológico de Monterrey. He was the associate director of the Industrial and Systems Engineering programme from 1999 to 2005. Moreover, he was also the associate director of the department of industrial and systems engineering from 2005 to 2009. His research areas include primarily related to inventory planning and control, logistics, and supply chain. He has published papers and technical notes in several international journals. He has co-authored one book in the field of Simulation in Spanish. He is also editorial board member in several international journals.

Prashant Das is an assistant professor of real estate finance at Ecole hoteliere de Lausanne, Switzerland.

Giacomo Del Chiappa, PhD, is an associate professor of marketing at the Department of Economics and Business, University of Sassari (Italy), and Associate Researcher at CRENoS. He is also a senior research fellow, School of Tourism and Hospitality, University of Johannesburg, South Africa. His research is related to destination governance and branding, consumer behavior, and digital marketing. He has published articles in several international journals, among others the International Journal of Hospitality Management, Journal of Services Marketing, Journal of Travel Research, International Journal of Tourism Research, International Journal of Contemporary and Hospitality Management, Current Issues in Tourism, and Information Systems and E-Business Management.

Simona Gozzo received her PhD in Sociology and Methods of Social Sciences from Catania University in 2007. Currently she is Assistant Professor in Sociology at the Department of Social and Political Sciences in Catania University. Her research interests are focused on methods for social sciences, particularly Network Analysis and Multilevel Regression Models. She is also interested in the evaluation of policies and in studies about policies for gender, young and migrants. She is author of many publications on tourism.

Kijpokin Kasemsap received his BEng degree in Mechanical Engineering from King Mongkut's University of Technology, Thonburi, his MBA degree from Ramkhamhaeng University, and his DBA degree in Human Resource Management from Suan Sunandha Rajabhat University. He is a Special Lecturer in the Faculty of Management Sciences, Suan Sunandha Rajabhat University, based in Bangkok, Thailand. He is a Member of the International Association of Engineers (IAENG), the International Association of Engineers and Scientists (IAEST), the International Economics Development and Research Center (IEDRC), the International Association of Computer Science and Information Technology (IACSIT), the International Foundation for Research and Development (IFRD), and the International Innovative Scientific and Research Organization (IISRO). He also serves on the International Advisory Committee (IAC) for International Association of Academicians and Researchers (INAAR). He has had numerous original research articles in top international journals, conference proceedings, and books on the topics of business management, human resource management, and knowledge management, published internationally.

Maximiliano Korstanje is an associate professor at University of Palermo, Argentina. Editor in Chief of the International Journal of Safety and Security in Tourism. With more than 650 published papers in peer review journals, Korstanje is concerned in the study of risk, capitalism and mobility. He co-edits 10 journals, and takes part of editorial board list of other 25 specialized journals. From 2010 his biography is a point of entry in Marquis Who´s Who in the World. He was awarded as Outstading reviewer 2012/2013 for Emerald Group Publishing, UK. Because his contribution to the sociology of tourism, Korstanje has been nominated to three honorary doctorates.

Martin Kruliš received a doctoral degree from software systems at the Department of Software Engineering, Faculty of Mathematics and Physics, Charles University in Prague, Czech Republic and he is currently working as an assistant professor at the same institution. He is a member of similarity retrieval research group and parallel architectures/applications/algorithms research group and his research interests include parallel programming, high performance computing, and concurrency in database systems.

Rohit Vishal Kumar did his schooling from Ram Krishna Mission Ashram Vidyalaya, Narendrapur (WB) and his plus two (+2) from Ram Krishna Mission Residential College, Narendrapur (WB) in the stream of Humanities. Subsequently he did his graduation from St. Xavier's College Kolkata and then he went on to do his MA in Economics from Centre for Economic Studies and Planning (CESP) Jawaharlal Nehru University, New Delhi. Subsequently he shifted focus and joined Indian Institute of Social Welfare and Business Management (IISWBM) Kolkata to do his MBA in the area of Marketing. He was picked up by Indica Research Private Limited from campus and then for the next three years he worked in the Marketing Research Sector with various clients and organisations. He worked in ORG-MARG Research Limited, New Delhi and TNS-MODE, Kolkata and has handled clients like ITC Limited, TATA Steel, Bengal Ambuja Realty, ABP Group, Eveready Industries, Smithkline Beecham and Maruti Udyog Limited. Given his Economics, Marketing and Research combination and his love for reading, he jumped ship to join academics when his alma mater came calling. In 2001 he joined IISWBM Kolkata as Research Associate and within one year he was promoted to Lecturer. In 2003 he registered for his Doctorate with the University of Calcutta and was awarded the degree in 2007. His topic of doctoral research work was "Lifestyle and Psychographic Segmentation & Classification of Metropolitan Consumers in India for Effective Marketing". In 2006 he joined Xavier Institute of Social Service (XISS) Ranchi as Reader. Currently he is Associate Professor in the Department of Marketing at XISS, Ranchi. As on date he has a total of 16 journal publications out of which 08 are in international journals of repute. In his free time Dr. Kumar loves to play with computers and read novels. He is well versed in statistical softwares like R, SPSS, Systat, & Statistica; Data Mining Softwares like WEKA and typesetting languages like LaTeX. His technical document "Making Friends with LaTeX (MFwL)"is available at CTAN Repositories worldwide.

Giuseppe Melis is Associate Professor in Marketing at the Department of Economics and Business, University of Cagliari. His research is related to tourism marketing and co-creation.

Jean Ngnotchouye is an active researcher in theoretical and numerical aspects of system of conservation laws. He also has interest in optimization and optimal control. He has delivered many award winning talks and he has published a number of papers in international journal and conference proceedings. He is a regular participant at the Mathematics in Industry Study Group held at the University of Witwatersrand. Dr. Ngnotchouye has taught analytic method for partial differential equations and applied optimal control at postgraduate level for the past three years.

Natnael Nigussie (MSc., BEd. in Mathematics) is currently a Lecturer of Mathematics in Addis Ababa Science and Technology University (AASTU), Addis Ababa, Ethiopia. He has been teaching in different Universities of Ethiopia since the last eight years. Apart from the rich Teaching experience he shared so far, the author also actively participates in research activities. The journal paper he published entitled "Fractional Programming" (http://ijesat.org/Volumes/2014_Vol_04_Iss_04/IJESAT 2014_04_04_04) witnesses the strength of his contribution to the Science.

Luís Miguel Martins Nunes received a PhD in Computer Engineering from Faculdade de Engenharia da Universidade do Porto (FEUP), M.Sc in Electronics Engeneering and Computers, from Instituto Superior Técnico (IST) at Universidade Técnica de Lisboa (UTL), and graduated in Computer Science from Faculdade de Ciências da Universidade de Lisboa (FCUL). In 1997 entered Instituto Universitário de Lisboa (ISCTE-IUL) where he coordinated and lectured several courses, mainly in the areas of Programming and Machine Learning (ML). In the latest, he created in 2007 the ML course for the MSc in Computer Engineering, still under his coordination to this day. He is researcher at Instituto de Telecomunicações and ISTAR, having performed the same functions at LIACC (FEUP), ADETTI (ISCTE) and INESC (IST) in the past. Has coordinated more several successful M.Sc. thesis, mostly in ML related topics. Has participated in several national and international projects related to ML and Learning in Multiagent Systems. Current research interests include Machine Learning applications, in particular to problems related to Intelligent Transport Systems, Intelligent Home Automation, Swarm controllers, and Decision Support Systems.

Sebastiano Patti has earned a PhD in Public Economics at the University of Catania (Italy). He studied Economics at the School of Economic Science - University of Manchester (UK), where he got his MSc in Economics. He is teaching Finance at the University of Catania. He thought Environmental and Cultural Tourism and Environmental Economics within University's courses and Academic Masters' Degrees. His research interests have led him to deepen tourism economic policies, with respect to the new tourism market trends, local development and the transition from tourism district policies to local production systems, the application of efficiency analysis tools in local public utilities and in industrial organization. He worked in the European research teams like researcher as well as research's manager within Tourism European Research Framework. He is the author of several scientific papers on tourism and environment. Sebastiano Patti is an Alumni of the Manchester University (UK).

Anna Pawlikowska-Piechotka, born in 1952, Warsaw; architect and urban planner, PhD Eng. Associate Prof. Faculty of Tourism and Recreation AWF University Warsaw and Faculty of Architecture Warsaw Technical University, in 2008 - 2014 a visiting professor (short stay) at: University College Birmingham (United Kingdom), Charles University of Prague (Czech Republic), University of Lahti (Finland), University of Saragossa (Spain), University of Evora (Portugal), University of Viseau (Portugal), Holar University (Iceland). In her researches she focused on the use of cultural heritage for the sustainable tourism and responsible planning for tourism, sport and recreation. She participated in the several international research projects (Germany, Sweden, Holland, UK, Poland, Turkey); is the author of over two hundred scientific publications, including several monographs that were published in English, German, Polish and Serbian; co- author of several architectural and urban plans of recreation and tourist space (commissioned in Poland, Germany and Nigeria) and utility models of recreation facilities for disabled - formally registered in the Polish Patent Office.

Maciej Piechotka, born 1952, Warsaw; MSc, Eng. Architect, ECO CLASSIC ARCHITECTS GROUP, Warsaw, Charted Eng. architect by the State of Poland, member of the Chamber of Architects in Poland, member of the Association of Polish Architects in Poland. Author and co-author of numerous architectural designs focused on: housing, public buildings (banks, offices and universities), sustainability of public space, sport and recreation areas – commissioned mainly in Poland and Nigeria; utility solutions of recreation facilities for disabled - formally registered in the Polish Patent Office.

Mariella Pinna received her PhD in Marketing and Moral Philosophy in 2011. She is a Post-Doc Research Fellow in Marketing at the Department of Econmics and Business, University of Sassari, where she currently teaches Business Ethics. Her research interests are related to consumer behavior in ethics and tourism.

Jaroslav Pokorny received his Ph.D. degree in theoretical cybernetics from Charles University, Prague in 1984. He is a full professor in its Faculty of Mathematics and Physics. He has published more than 300 papers and books on data modelling, relational databases, query languages, XML technologies, and data organization. His current research interests include semi-structured data, Web technologies, database architectures, indexing methods, and social networks. Jaroslav Pokorny served as a chair or co-chair in the organization of conferences ADBIS-DASFAA (2000), EDBT (2002), Int. Workshop ClustWeb in conj. with EDBT (2004), Web X.0 a Web Mining Workshop in conj. with IEEE ICDIM (2007), ISD (2010), IDEAS (2012), ADBIS (2013, 2016), etc. He is a member of ACM and IEEE. He works also as the representative of Czech Republic in IFIP.

Mário J. B. Romão is Professor of Information Systems at Technical University of Lisbon (ISEG). He holds a PhD in Management Sciences by ISCTE-IUL and the Computer Integrated Manufacturing (Cranfield University). He accomplished an international certification in Project Management - PMP Best Practitioner Program by PMI.

Karolina Sawicka, born in 1978. Specialist in the subject of tourism and recreation, her scientific research work is focused on the adaptation of green municipal areas (especially historical municipal parks) for sport and recreation. PhD of Warsaw University of Physical Education (specialization - Tourism and Recreation), Master of Geography and Regional Studies (specialization: Tourism) at Warsaw University, Master of Cultural Studies (Specialization: Latin America) - Centre of Latin American Studies of Warsaw University, 2006 – 2007 - Erasmus student of Geography and History Faculty (specialization: Tourism) at Complutense University (Madrid, Spain).

Stefania Skonieczny is the founder and director of Le Mude Ltd, a business strategic consultancy company based in London, UK. While consulting several public institutions and private companies in their process of expansion in new and foreign markets Stefania created the first formal business network contract for the food sector in Sicily, Italy. Stefania holds a JD in International Law from the University of Bologna and a Master's Degree in International Economy and Development Policies from the International Mediterranean Management Institute and the University of Catania. She also obtained a Master's Degree in Public Administration from the same Institution and one in Diplomatic Studies from the Italian Society for International Organization and the Diplomatic Institute of the Italian Ministry of Foreign Affairs, Rome. Stefania is an Alumni of the Global Village for Future Leaders of Business and Industry, Lehigh University and Iacocca Institute, College of Business and Economics, Pennsylvania, US, accredited by the AACSB.

Surafel Tilahun completed his undergraduate study in Mathematics, MSc in Mathematics and MSc in Computational Science from Addis Ababa University in 2004, 2007 and 2009, respectively. His PhD study was on Applied and Computational Optimization which was completed in 2013 from Universiti Sains Malaysia, Malaysia. Currently, he is doing a postdoc research in School of Mathematics, Statistics and Computer Science, University of KwaZulu-Natal, South Africa.

Venera Tomaselli received PhD in Sociology and Methods of Social Sciences. From 1993 to 2002 she was Assistant Professor in Statistics and since 2002 Associate Professor in Social Statistics at the Department of Political and Social Sciences, University of Catania. She was research fellow at Essex University (UK) in 1992 and 1993, member of the European Research Group at Munich University (1996) and in 2015 visiting professor at University of Salford (UK). She was a member of the editorial boards of specialised editions on tourism. She is reviewer for scientific journals specialised in tourism research and Statistics such as Tourism Economics, Tourism Management, Current Issues in Tourism, Economics, Italian Journal of Applied Statistics (IJAS), Rivista Italiana di Economia, Demografia e Statistica (RIEDS). Her research interests are focused on Multivariate Statistical Models, Network Analysis, Tourism and Market Segmentation Analysis, Ecological Fallacy, Accuracy of Surveys and Opinion Polls, Meta-Analysis, Clinical Research in Genetics. She has a rich profile of articles in peer-reviewed journals, book chapters and volumes.

Anna Tryzno is the Head of the Department of Methodology in Tourism (2012-16), 2003-2005 visiting assistant in University of Pedagogic in Warsaw, 2001- 2010 assistant in the Faculty of Tourism and Recreation at the Academy of Physical Education in Warsaw, 2002 Graduate MA in International Relationships at the Warsaw University – Department of Journalism, Political Science and International Relationships; 2001 Graduate MA in Physical Education Sciences at the Academy of Physical Education – Faculty Tourism and Recreation. In her researches she focuses on sustainable tourism, tourism in garden cities, business tourism, healthy lifestyle and promoting physical activity. She is involved in the local activities in the garden city Podkowa Leśna, she cooperates with the NGOs in the field of sport, tourism and recreation. She participated in international research cooperation. She takes part in the organization of various events (sport, tourist, charity), camps for students, tours for different social groups, scientific conferences and meetings. She is tour leader, local tourist animator, instructor of nordic walking, instructor of skiing, instructor of survival, yacht coastal skipper.

Junzo Watada received the B.Sc.(1970) and M.Sc. degrees (1972)in electrical engineering from Osaka City University, Osaka, Japan, and the Ph.D. degree (1983) from Osaka Prefecture University, Osaka, respectively. Currently, he is a professor of Universiti Teknologi PETRONAS, and a professor emeritus, Waseda University from April 2016, as well as the executive chair of WCICME after being Professor of Management Engineering, Knowledge Engineering and Soft Computing at the Graduate School of Information, Production Systems, Waseda University, Fukuoka, Japan. His professional interests include soft computing, tracking system, knowledge engineering, and management engineering. Prof. Watada is a recipient of the Henri Coanda Medal Award from Invention in Romania in 2002 and a Life Fellow of Japan Society for Fuzzy Theory and Intelligent Informatics (SOFT). He is a pricipal ediotor, associate editor and editor of international journals including IDT, ICICEL IJICIC as well as Informaiton Sciences and Fuzzy Optimization and Decision Making.

Koki Yoshimura received the B.Sc. from Polytechnic College and the M.Sc. from the Graduate School of Information, Production and Systems, Waseda Uiniveristy. He is currently a staff of University. His research interests include management, game thoery and real option model as well as hotel management.

Ivan Zelinka (born in 1965, ivanzelinka.eu) is currently associated with the Technical University of Ostrava (VSB-TU), Faculty of Electrical Engineering and Computer Science. He graduated consequently at the Technical University in Brno (1995 - MSc.), UTB in Zlin (2001 - Ph.D.) and again at Technical University in Brno (2004 - Assoc. Prof.) and VSB-TU (2010 - Professor). Prof. Zelinka is responsible supervisor of several grant researches of Czech grant agency GAČR as for example Unconventional Control of Complex Systems, Security of Mobile Devices and Communication (bilateral project between Czech and Vietnam) and co-supervisor of grant FRVŠ - Laboratory of parallel computing amongst the others. He was also working on numerous grants and two EU projects as member of team (FP5 - RESTORM) and supervisor (FP7 - PROMOEVO) of the Czech team. He is also head of research team NAVY http://navy.cs.vsb.cz/. Prof. Zelinka was awarded by Siemens Award for his Ph.D. thesis, as well as by journal Software news for his book about artificial intelligence. He is a member of the British Computer Society, Machine Intelligence Research Labs (MIR Labs - http://www.mirlabs.org/czech.php), IEEE (committee of Czech section of Computational Intelligence), a few international program committees of various conferences, and three international journals. He is also the founder and editor-in-chief of a new book series entitled Emergence, Complexity and Computation (Springer series 10624, see also www.ecc-book.eu).

Index

E

F

G

H

I

K

L

M

T

U

V

W

Y

www.ingramcontent.com/pod-product-compliance
Lightning Source LLC
LaVergne TN
LVHW060349300926
849162LV00006B/1228
9781522510543